ICE Handbook of Urban Drainage Practice

Other titles in the ICE Handbook series:

ICE Handbook of Concrete Durability, Second edition (2023)
Marios Nicou Soutsos. ISBN 978-0-7277-6375-4

ICE Handbook of Geosynthetic Engineering (2021)
Sanjay Kumar Shukla. ISBN 978-0-7277-6500-0

ICE Handbook of Urban Drainage Practice

Edited by Richard Ashley, Brian Smith, Paul Shaffer and Issy Caffoor

Published by Emerald Publishing Limited, Floor 5, Northspring, 21–23 Wellington Street, Leeds LS1 4DL.

ICE Publishing is an imprint of Emerald Publishing Limited

Other ICE Publishing titles:

ICE Manual of Blue-Green Infrastructure
Carla-Leanne Washbourne and Claire Wansbury. ISBN 978-0-7277-6542-0
Sustainable Water
Charles Ainger and Richard Fenner. ISBN 978-0-7277-5773-9
Land Drainage and Flood Defence Responsibilities, Fifth edition
Institution of Civil Engineers. ISBN 978-0-7277-6063-0

A catalogue record for this book is available from the British Library

ISBN 978-0-7277-4178-3

Cover photo: Susdrain

Commissioning Editor: Viktoria Hartl-Vida
Assistant Editor: Cathy Sellars
Production Editor: Rajachitra Suresh

Typeset by Manila Typesetting Company
Index created by Madelon Nanninga-Fransen

All editor royalties of this book are donated to Water Aid in support of the work they do. To learn more, visit wateraid.org.

Dedication

In memory of John Blanksby and Jamie Margetts, who both gave so much to the practice of urban drainage in the UK.

Contents

Foreword

I undertook my first design of a drainage system in the early 1970s. It was not an ambitious project, just modest lengths of surface water and foul sewers to drain a new housing development. The housing layout had already been determined and the roads designed. Any open space between building plots had been allocated and it only remained for me to set out the geometry and size of the drainage conduits for connection to the individual properties and highway gullies. I completed the work promptly with the aid of a slide rule and some simple design standards and formulae. The drainage was all below ground, out of sight and out of mind. As a colleague remarked at the time, there are no votes in drainage. Yet, as I write this, there are certainly many votes in drainage, with the public alarmed by the poor state of our rivers and the more frequent flooding of our major urban areas. The significant impact this has on the environment, the economy and quality of life has reached a stage in many countries where it is no longer acceptable.

In this book you will find a detailed description of how our drainage systems have developed over the years, the changes that are currently taking place, and what the future might look like. This development has not been a progressive evolution, but a series of step changes arising from crises (Bazalgette's development of the London sewerage system in response to the 'Great Stink' in the river Thames) or opportunities (the development of computers, simulation software and low-cost sensors). And as this has progressed, the skills and competences of drainage engineers and others connected with drainage procurement and operation have also had to progress. This has not been an easy task, as drainage systems become ever more complex, the bodies responsible for their development ever more diverse, and society ever more demanding.

Good drainage, together with the plentiful supply of clean water and effective wastewater treatment, are essential to the health and prosperity of society. The benefits of the original aims of drainage systems – to convey wastewater safely to a suitable point for treatment and disposal, to drain paved areas effectually and to safeguard receiving waters from pollution – have been immeasurable. And in recent years we have realised that we can do even more. A systematic and holistic approach to drainage shows us that managing surface water at source, improving the resilience of communities to flooding and harnessing the benefits of nature-based solutions can improve amenity and biodiversity, provide a much-needed alternative source of water and enhance the quality of life. Imagine a future where we all can benefit from surface water that is drained on the surface, creating natural water landscapes within our urban communities. Such landscapes bring with them all the benefits of natural watercourses by mimicking natural processes, supporting flora and fauna and compensating for the urban heat island effect against a background or rising temperatures. In this future, drainage drives new development rather than appearing as an afterthought.

To realise these benefits, we have to move beyond our technical portfolios. Good data and sound science alone will not achieve the very necessary future objectives. Future infrastructure needs the right technologies, but at the same time must align with the culture of the communities it serves and have a secure financial model and effective governance. All four elements must run together for success to be secured, and drainage professionals will need to be comfortable with working in all these areas. Such change will not always be welcome. Innovation brings with it uncertainty and may be resisted by those wedded to more familiar 'tried and tested' methods. This book is designed to challenge that reticence. Here you will find a detailed summary of all the essential elements of the different aspects of future drainage, written by leading experts in their field. It can be digested in the conventional way, from start to finish, or by dipping directly into individual chapters, which mostly are self-contained. It should also be of interest to those concerned with more general infrastructure provision, and its governance and finance. In each chapter, the evidence base is clearly referenced, and this provides a useful source of additional reading for those who wish to delve into specific topics in more detail. I recommend this book to you, not to sit on your bookshelf, but to be kept open on your desk as a day-to-day aid to your work.

David Balmforth, Visiting Professor, Imperial College London, and Past President of the Institution of Civil Engineers

About the editors

Richard Ashley MPhil, CEng, MICE, MCIWEM

Richard is a professional civil, water and environmental chartered UK civil engineer at the University of Sheffield. He is former managing director of EcoFutures Ltd; researcher in flood resilience at IHE in Delft, the Netherlands; and adjunct professor at Lulea Technical University, Sweden. He is recipient of the International Water Association/International Association for Hydro-Environment Engineering and Research (IWA/IAHR) Joint Committee on Urban Drainage 2014 triennial career achievement award, and co-recipient of the 2008 award for the biennial IWA prize for research excellence in support of sustainable urban water management. Richard originally worked in the sewers in London, but during a practice and research career spanning 50 years he has worked on urban drainage systems in many parts of the world. Most recently he has worked in cross-disciplinary activities covering governance, sustainability and resilience of urban water systems. With more than 600 publications, he is co-author of the *IWA Manual on Performance Indicators for Wastewater Services; main editor of Sustainable Water Services: A Procedural Guide* for IWA; and primary editor and contributor of the *IWA Scientific and Technical Report on Sewer Solids*. He has written reports for OECD on the global need for water infrastructure; flood risk management in the Loire Valley; Paris flood risks; and the future need and costs for water and wastewater assets across the EU. Richard has been a visiting professor for the Cooperative Research Centre for Water Sensitive Cities in Australia (led by Monash University). He was formerly a UK government designated UK representative for the EU COST action on urban flood management and a researcher in the EU PREPARED (for climate change) project, and was scientific adviser to the House of Lords inquiry into water management in England and Wales in 2006. Richard was also an adviser to Ofwat and Thames Water on the Thames Tideway Tunnel; Water UK's 21st Century Drainage Programme on economics of combined sewer overflows and retrofitting SuDS; and numerous UK Water Industry Research Ltd projects. Richard is a co-author of several Construction Industry Research and Information Association (CIRIA) guidance documents, including *Retrofit Surface Water Management; Exceedance Flood Management*; and *The SuDS Manual*. He was adviser to Welsh Water and the Welsh government for stormwater in Wales and for the implementation of Schedule 3 to the Flood and Water Management Act 2010. He is a former chairman of the Yorkshire region of the Institution of Civil Engineers (ICE) and former editor of the ICE journal *Engineering Sustainability* and is a trustee of the ICE Benevolent Fund.

Brian Smith I Eng, MICE

Before retiring in 2019, Brian was the drainage strategy manager for Yorkshire Water, one of the English water and wastewater utility companies. With a background in mechanical, electrical and civil engineering and a career spanning more than 47 years, Brian has a broad, diverse skill base and technical expertise across engineering disciplines relevant to the water industry, gained both in public and private sectors, in the UK, the Netherlands and Nigeria, the last 22 years of which were with Yorkshire Water. In 2011 he organised and led a joint UK water and sewerage company visit to Portland, Oregon and Seattle, Washington USA to learn about their approach, the costs and effectiveness of integrated and retrofit sustainable drainage measures and the benefits achieved from their implementation, and to understand the engineering, social engagement and socio-economic challenges to successful delivery and community acceptance. A consummate professional, committed to providing consistent high standards, Brian has an ability to maintain focus and identify issues and risks that are important to the sector. He provided strategic intelligence on matters relating to sewerage, the environment, and strategic planning and development, and established a clear direction and strategy for stormwater management and innovative, intelligence-driven solutions to help ensure Yorkshire Water was prepared to overcome future challenges in this area. He also developed a wider wastewater networks strategy to enable the company to achieve its vision and strategic objectives. Brian represented Yorkshire Water at a national level on technical, regulatory and legislative matters and was the national and international interface on policy and standards development, representing the UK on the development of an international standard for stormwater management. He formed the Water UK Surface Water Management Network and was chair for almost 8 years from its inception in 2012. He was a member of Water UK's 21st Century Drainage Programme board, chairing the workstream on 'enablers for progress'. Although retired, Brian maintains a working interest in the sector, keeps up to date with current thinking, particularly on integrated water management, and is an active member of the All-Party Parliamentary Water Group.

Paul Shaffer BSc, FCIWEM, C.WEM

Paul has worked in the water sector for over 25 years with a focus on delivering better approaches to the use and management of water. He has coordinated, fundraised for and developed dozens of guidance documents and initiatives to support SuDS, blue-green infrastructure (BGI), flood resilience and integrated water management. He is keen to promote the opportunities around water management and has developed numerous capacity-building programmes, fundraised and directed the delivery of two editions of *The SuDS Manual*, driven the development of tools to quantify the benefits of BGI and facilitated a variety of forums to share good practice. Paul is passionate about the opportunities to collaborate on urban water management and keen to create better places for people and wildlife. He strongly believes that water infrastructure should be multi-beneficial, and this can only be achieved with multiple disciplines working in concert to integrate the water cycle within our towns and cities.

Issy Caffoor BSc, PgDip Biom, PhD

Issy has worked within and alongside the water sector for over 49 years, 27 of which were at Yorkshire Water, where he worked as an R&D manager for over 17 years until he retired in 2006. While at Yorkshire Water, he was actively engaged in the management and peer review of national and international research programmes, including DTi LINK, EPSRC, NERC, BBSRC, European framework and COST actions. He has also been director and chairman of SWIG, a special interest group for sensors in the water industry. More recently Issy led the production of the environmental Knowledge Transfer Network (KTN) business cases: Energy Efficient Water and Wastewater Treatment; Towards Chemical-free Water and Wastewater Treatment; and Integrated Urban Water Management, and reports to NERC on mapping the water industry needs against NERC and Living with Environmental Change (LWEC) science priorities, and the business report to the Technology Strategy Board on a vision for a low-carbon water industry. Issy was formerly engaged as a knowledge transfer chair in water sustainability, based at the University of Sheffield Department of Civil and Structural Engineering. Issy has also held honorary visiting professorships at Imperial College London, Bradford University and Sheffield University. Most recently he was a member of the strategy and leadership boards of the high-profile and successful EPSRC funded TWENTY65 programme on water, which was coordinated by Sheffield University.

Acknowledgements

My career in urban drainage has spanned half a century, with a minor divergence into offshore structures. Starting in the sewers in London, under Norman Crane, I developed a passion for being underground and recently renewed my membership of Sewers Synonymous. Despite this, this book has been rather a long time in gestation. I had hoped that semi-retirement would expedite its production, but such forms of retirement incur full-time working and exciting new projects, with the book always in the background. Having inveigled the fourth editor, Issy Caffoor, into early involvement, his enormous early efforts were unrequited as the draft, in various forms, sat untouched for lengthy periods of time, even during the COVID-19 lockdowns. Nonetheless, Issy provided some excellent insights along the way, until the third editor, Paul Shaffer, was looking for something new and was also inducted into the editorial team, having been a chapter author. Almost as soon as Paul started to provide his long experience, the second editor, Brian Smith, having retired from Yorkshire Water, was recruited to provide the needed impetus in the form of kicks in the right places, to get the book writing process really moving. Brian brought valuable 'real-world' inputs into the book, offsetting the, at times, 'academic' perspective that perhaps is too pervasive in places, even in the final document.

Working with the many authors who have contributed chapters and contributed to chapters has been very rewarding, although many of the early drafts became superseded due to the length of time this book has taken to get to publication and some of the early sections no longer feature in this final version. Of the many contributors, we have lost both Jamie Margetts and John Blanksby, both of whom were significant players in the urban drainage world and who continue to leave their mark in the form of legacies that are affecting how urban drainage practice is carried out. Early contributors included Tony Poole, Mike Faram and Berry Gersonius.

Looking wider, the global urban drainage community has matured, especially since the millennium, with the domain becoming a discipline in its own right, albeit only to now be faced with absorption into the broader world of integrated water management and integrated infrastructure management. Personally, I have learnt so much from the members of this community over the years, including: the early pioneers in the UK, Bryan Ellis and others in academia, together with the real engineers Andy Eadon and John Tyson; Ian Clifforde at the Water Research centre, many of the research managers at CIRIA and former HR Wallingford staff Richard May and Martin Osborne; and those who are no longer with us, including Peter Ackers, Bob Crabtree and Chandra Nalluri.

Thanks also to the international leaders who defined and built the domain of urban drainage, including Bernard Chocat, Wolfgang Schilling, Jiri Marsalek, Geoff O'Loughlin, Thorkild Hvitved-Jacobsen, Larry Roesner and Rich Field; Tony Wong, who first created the water sensitivity ideas; and Wayne Huber, together with Roland Price and others, who did so much to create the computer models we now use routinely.

Richard Ashley

The editors of the book also owe much to the ICE editors Viktoria Hartl-Vida and Cathy Sellars, and the former editor Inês Pinheiro, who kept us on the straight and narrow. Also to our wives, Cate, Jo, Clare and Andrea, God bless them.

All editor royalties from this book are donated to Water Aid in support of the work they do. To learn more, visit wateraid.org.

Contributors

Andrew Bailey, HR Wallingford, UK

Peter Bide CGeol, CEng, FGS, MIMMM, CaBA Urban Water Group, London, UK

John Blanksby BSc, Dip CEng, MICE (†)

Jo Bradley, Stormwater Shepherds, Bamber Bridge, UK

Hannah Casswell, Sensors for Water Interest Group, UK

Dr Jacqueline Diaz-Nieto CWEM, MCIWEM, GMICE, Environmental Protection Group, Warrington, UK

Dr Glyn Everett, University of the West of England, Bristol, UK

Richard Kellagher, MSc, CEng, MICE, MCIWEM, HR Wallingford Ltd, UK

Professor Jeroen Langeveld, Delft University of Technology, the Netherlands, and Partners4UrbanWater, Nijmegen, the Netherlands

Jamie Margetts (†)

Dr Andy Nichols, University of Sheffield, UK

Nick Orman, Wastewater Networks, Water Research Centre, Swindon, UK

Adrian Saul, University of Sheffield, UK

Dr Alma Schellart, University of Sheffield, UK

Dr Will Shepherd, University of Sheffield, UK

Professor Simon Tait, University of Sheffield, UK

Dr Kieran Williams, Environmental Monitoring Solutions Limited, Sheffield, UK

Steve Wilson, Environmental Protection Group Ltd, Warrington, UK

Bridget Woods-Ballard MA, MICE, CEng, HR Wallingford, UK

Yiqi Wu, University of Sheffield, UK

Introduction

Purpose of the book

Who is this book for? It is for anyone with an interest in urban drainage or who thinks they might like to know what urban drainage is and how it is managed today. Although typically delivered by the civil engineering profession, 'urban drainage' is not exclusively 'engineering'. In fact, the term itself can be misleading in today's context, as a 'drain' is meant to transport water from surfaces and 'drainage' is the means of draining away. Urban drainage thus means the management of water or wastewater to 'flow away' from something or somewhere, specifically in built-up areas. Current thinking now sees water as a potential resource to be utilised where it falls as rainfall, even in towns and cities, placing this key part of the water cycle in a similar frame to the circular economy.

Water crises have consistently featured among the top-ranked global risks in the World Economic Forum's annual *Global Risks Report* (WEF, 2023). These risks are both highly likely and highly impactful and are the biggest threats facing life on earth over the coming decades. As the water cycle is intrinsically linked with climate change, ensuring there is sufficient water to support the needs of the world's increasing population will become ever more important.

There is a growing awareness and understanding among decision-makers that water is as important for cities as climate, energy or food, and there are risks and opportunities related to water that will affect economic development and prosperity, as well as life itself. 'Drainage' is still a useful term, but increasingly the value of water what was once automatically regarded as 'wastewater' to be drained away is being recognised. Even drainage from toilets, kitchens and other water use areas can provide opportunities for utilisation and resource recovery, as 'wastewater constitutes a valuable resource that, if sustainably managed, is set to become a central pillar of the circular economy' (UN, 2017).

This book provides an overview of the current state of the art that is the practice of urban drainage. It is a practice that has developed significantly over the past half a century, acknowledging the need to move away from a problem-centred perspective, where drainage is seen as a necessary way to remove water quickly and safely from areas where it may cause harm. The new perspective is adopting an approach that seeks to utilise rainfall near to where it lands and to apply the principles and practices whereby this resource can best be utilised.

Structure and contents

This book comprises fourteen chapters describing the development of drainage systems over the years, current challenges and what the future might look like. The contents of the chapters are as follows:

- Chapter 1 sets out the nature and context for urban drainage systems, briefly reviewing their development and limitations and introducing emerging ideas for managing water more appropriately in urban areas.
- Chapter 2 introduces the types of urban drainage systems in use and the most important aspects of the drivers and challenges faced today, focusing on sanitary systems.
- Chapter 3 deals with stormwater, including surface water, covering the origins and nature of stormwater and control to maximise value and minimise harm, and includes combined sewerage systems.
- Chapter 4 outlines the importance of stakeholder engagement in water management and how this is being carried out, including building capacity.
- Chapter 5 considers the complexities of the urban planning, political and regulatory landscapes in relation to urban drainage and their importance for integrated water management.
- Chapter 6 sets out the different ways in which rainfall can be measured and used to estimate the spatial and temporal patterns of rainfall. It explains the different concepts of rainfall estimation, prediction and design, and how these might be affected by climate change.
- Chapter 7 considers the key elements of urban hydrology used for urban drainage analysis and design.
- Chapter 8 outlines the current and increasing use of nature-based systems for drainage, the value of their utilisation and how they contrast with traditional buried piped systems.
- Chapter 9 explores the technologies used to sense and monitor flow properties and infrastructure conditions in urban drainage systems, including emerging technologies and sensor-driven control for water management in the future.
- Chapter 10 introduces asset management for urban drainage systems, with a focus on operation and maintenance, using an example of the UK approach to this for the water industry.
- Chapter 11 deals with the analysis and computational modelling of urban drainage systems, including the development history of these analysis techniques and a summary of their applications.
- Chapter 12 considers the frames being used for contextualising urban drainage analysis, planning and design, including sustainability, security, integrated water management and water sensitivity.

- Chapter 13 looks at how new thinking and alternative, more innovative ways to deliver a sustainable, affordable operating model for drainage services are required. It examines skills and capacity planning, and the need to futureproof the workforce of the sector. Advances in digitalisation and technological innovation to unlock the potential of data-driven decision-making are considered, as is the potential for artificial intelligence to transform sector efficiency.
- Chapter 14 provides a selection of case studies from across the UK and the USA detailing an extensive variety of approaches, practical applications and solutions to the many, and sometimes complex, issues faced and how urban drainage, and its practitioners, has and continues to evolve to provide a broader range of societal, environmental and economic benefits.

REFERENCES

WEF (World Economic Forum) (2023) *The Global Risks Report 2023.* WEF, Geneva, Switzerland.

UN (United Nations) (2017). *Wastewater: The untapped resource.* United Nations World Water development report. United Nations Educational, Scientific and Cultural Organization (UNESCO), Paris, France.

Ashley R, Smith B, Shaffer P and Caffoor I
ISBN 978-0-7277-4178-3
https://doi.org/10.1680/icehudp.41783.001

Chapter 1
Urban drainage

Richard Ashley and Brian Smith

1.1. Background and current perspectives

Just what is urban drainage? Like all definitions, the term should be seen as having a degree of fluidity. Although urban drainage has been provided for human settlements since time immemorial, urban drainage as a service and separate discipline seems to have emerged sometime in the nineteenth century, when engineered water supply and sewer systems were being designed and built using scientific knowledge. While it is arguable that water supplies came first, it can be seen in the left end of Figure 1.1 that drainage systems implicitly included the means to manage both the runoff from water falling onto towns and cities, now known as stormwater or surface water, and the emissions of human, animal, commercial and industrial wastes, which are termed sewage.

The scope of traditional urban drainage is also shown in Figure 1.1 and illustrated in Figure 1.2. In some urban areas, this scope extends also to point and diffuse water pollution management, but even in the first industrialised nations, diffuse pollution from runoff from urban surfaces is largely ignored, despite this

Figure 1.1 Evolution towards the water-sensitive city (Ashley *et al.*, 2013; adapted from Brown *et al.*, 2009)

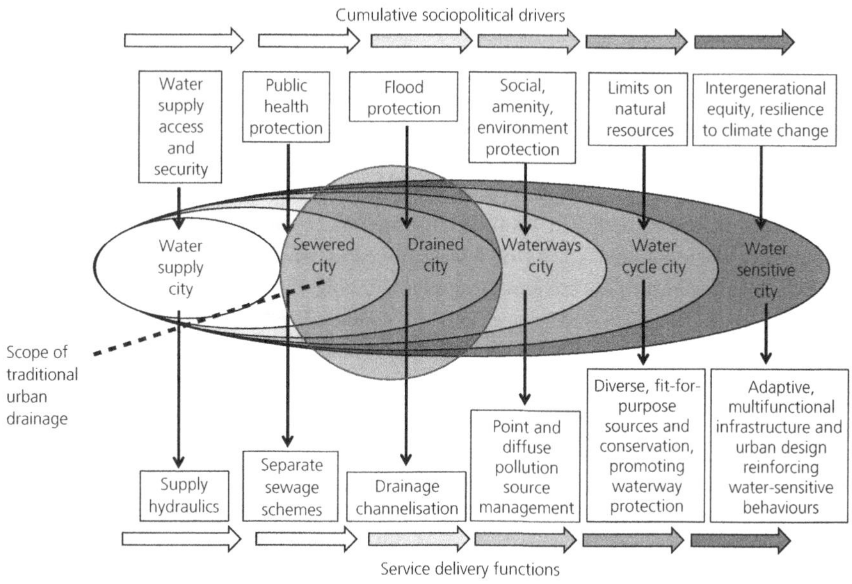

Figure 1.2 Urban drainage in Kolkata (Basu *et al.*, 2013)

being a major factor in watercourse impacts (e.g. OECD, 2017). Figure 1.1 also illustrates that it is unwise to manage urban drainage systems separately from the other components of water cycle service and wider infrastructure provision.

1.1.1 The development of urban drainage

The history of how waterborne wastes have been managed in human settlements stretches back to several millennia BC (e.g. Angelakis *et al.*, 2005; Burian and Edwards, 2002; De Feo *et al.*, 2014; Lofrano and Brown, 2010). Effective wastewater management had emerged by at least the second millennium BC (as illustrated in Figure 1.3), or even before then, in Mesopotamia. The prehistoric settlement of Akrotiri on the island of Santorini (Thíra) had an elaborate drainage system and was made up of sophisticated multistorey buildings. The buildings had water closets with pipework between the walls and drains under the footpaths. In 4000 BC, the area had a better drainage system than much of modern-day Greece. The inhabitants of the town were forced to leave towards the end of the seventeenth century BC following a series of severe earthquakes and a volcanic eruption. Many of these lost early technologies have been ‘rediscovered’, especially in industrialised countries, where what are now considered to be normal technologies have been further developed (see Juuti and Katko, 2005). These modern developments in hygiene were initiated in Paris and London by engineers and others who travelled back and forth between France and England (Bertrand-Krajewski, 2005).

As well as protecting human settlements from flooding, much of the focus of industrial-era sanitation reforms was on improving standards of behaviour of the working population; originally even more so than for maintaining the health of the population (Allen, 2008). Formalised attempts at protecting the environment, and in particular the aquatic environment, can be traced in England to the reign of Richard II. In 1388, an ACT was passed to prevent the pollution of ‘ditches, rivers and other waters’ in

Figure 1.3 Drain in the Temple of Knossos, Crete, built during the Minoan period in the second millennium BC (Richard Ashley)

cities and towns by the dumping of animal remains and dung. Until the cessation of horse-drawn transport systems, animal dung had arguably been more of a problem in towns and cities than human waste. For example, in 1894, New York City experienced the 'great horse manure crisis' as more than 1000 t of manure were being deposited each day in the city. This caused chaos on the streets, where mountains of dung built up, disrupting transport, and resulted in acute and chronic pollution of the city's watercourses, into which the dung was conveyed when it rained.

1.1.1.1 Drainage is one part of how land is used, and towns planned

The integral relationship between drainage and the layout of streets and buildings and the way land is used in urban areas is an important aspect of urban drainage. For example, Figure 1.4 shows a street in Pompeii, illustrating the depth of the kerbs and the stepping-stones for crossing the road. The street is designed to convey surface water during rainfall, which would include various contaminants, while allowing pedestrians to cross safely. As contemporary city planning has evolved, the appropriate incorporation of drainage measures has also changed, much of this reflecting the prevailing modes of transport. Originally, drainage systems were on the surface, an obvious part of the urban landscape, with the polluted nature of wastewater being visible to everyone. Gradually, as urban streams and rivers became open sewers, these were covered over and hidden to enclose the evil-smelling and insanitary wastes they conveyed, many eventually becoming sewers that are still in use now in industrial-era towns and cities. Being 'out of sight, out of mind' has been a mixed blessing to humanity (Chocat *et al.*, 2004). Modern developments invariably do not consider the 'drainage' aspects of the site layout until everything else has been decided, expecting that the drainage can be fitted in below ground, whatever the

Figure 1.4 A road as a drain in the city of Pompeii (Chris Digman)

layout required. Breaking with this perspective and returning surface water drainage to the surface to make the most of what this water can provide is one of today's big challenges for engineers and practitioners. It is equally a challenge for politicians and the public, who will have to be comfortable with water being more visible and managed on the surface.

The other big challenge in changing mindsets, as in the seventeenth to nineteenth centuries, concerns what is conveyed in foul flows and that it can and should be utilised as a resource, using appropriate technologies (WWAP, 2017). Switching from a 'problem-centric' perspective, in which urban drainage deals with the threats to towns and cities from surface water and foul sewage, to an 'opportunity-centric' perspective, in which surface water and foul flows present opportunities to get the most from the rain and from the human and other wastes generated every day, is a theme of this book. Figure 1.5 shows the water supply needs of the City of Melbourne in Australia, together with volume of runoff from rainfall. The unutilised runoff matches the water demand. In view of this understanding, Australia commissioned a focused research and development programme known as 'Cities as Water Supply Catchments', which became a nationwide programme under the Cooperative Research Centre for Water Sensitive Cities (CRCWSC, 2023). This ongoing programme for water sensitivity aims to exploit all opportunities for water in cities and to do this by developing and integrating water management ideas with city planning processes.

How has urban drainage technology advanced? It is important for practitioners to understand that not all progress in the way in which drainage system technologies have evolved into what is now considered to be standard design and operational practice has been straightforward. For example, the original sewers in France were broader, flatter and, in many places, include cunettes or side walkway banquettes, and were of a different cross-sectional shape from the then commonly used egg-shaped sewers in London. These are legacies that both Paris and London have now, owing to the continuing longevity of the mainly brick sewers built in the sanitation revolution of the 1800s.

Figure 1.5 The water supply for Melbourne balances the rainfall on the city (Tony Wong, Personal communication)

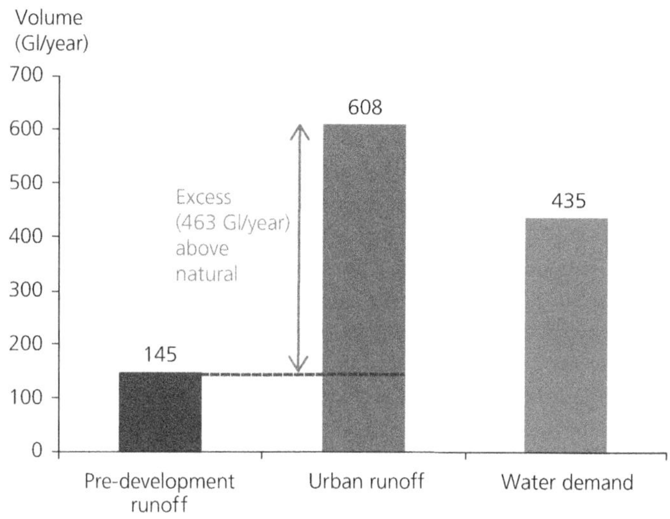

1.1.1.2 The great sanitation reform

Engineers, or even scientists, did not lead many of the developments in sanitation. Much of the credit for the great sanitary reform in the UK that instituted the urban drains and sewers that are still used now is given to a London-based lawyer, Edwin Chadwick, who conducted an investigation between 1839 and 1841. The Sanitary Report of 1842 formed the basis of the Public Health Act 1848 – legislation that would improve the quality of public health in Britain's towns and cities by ensuring more effective provision for improving sanitary conditions. Many engineers in the eighteenth to nineteenth centuries were concerned with their business opportunities and not upsetting the status quo, rather than necessarily with the best way of arranging drainage.

> For Chadwick, mid [19th] century British civil engineers were part of the problems, not the solutions. He saw them as both loyal to a primitive laissez-faire and in cahoots with the most corrupt and irrational institutions of local government: [their works were] hyperexpensive, uninformed by science, even dangerous[...] they clung to obsolete doctrines and rejected truths from outsiders. (Hamlin, 1992)

Even Joseph Bazalgette, the great builder of London's interceptor sewerage system, among other prominent members of the Institution of Civil Engineers (ICE), opposed the reforms to bring in small-bored piped sewers into much of the sewer network on the grounds that these would not work.

Of course, controversies among engineers still prevail, although maybe not on such a significant topic as the sanitary conditions in England's capital city. Nevertheless, even in such an apparently straight-forward field of knowledge and practice as urban drainage, there are ongoing disagreements over innovations, and the risks and lack of certainty and security in not always doing what has always been done before and is 'known to work' (e.g. Palmer, 2000). Some of these controversies are highlighted in this book and what was formerly understood to be 'best practice' in urban drainage may no longer be so (e.g. Chocat *et al.*, 2007). In a world of increasingly scarce resources, population growth and expectations of certain living standards, where the climate is changing at a rate that is noticeable in a single lifetime, the profligate use of materials and the waste of water implicit in the 'get the water away from the urban area as quickly as possible' approach are no longer tenable. A new model is required and needs to be implemented as quickly as possible, not only to provide for the millions of people without adequate water supplies and sanitation, but also to ensure that those who have these services can maintain them in an affordable way (Libey *et al.*, 2020).

No longer can engineers focus on 'the engineering' as this was understood in the past. It is the interaction of different systems in urban areas, the need to engage with and work across disciplines and to understand as never before that 'the client' is society and it is the natural environment that is depended on, not the funder of a particular project (e.g. Dresp-Langley, 2008). Although, it is arguable that engineers have always done this; at least to some extent, and that the 'art' of engineering does in itself bring a many-disciplined perspective into engineers' work (e.g. Gaterell, 2012).

Policies, regulations and enforcement have always been key to ensuring the delivery of safe and effective drainage systems, forming the institutional arrangements for water services and sanitation (see Chapter 5). Service failures are usually due to policy, enforcement and fiscal issues rather than technical problems. For example, in 2021, Southern Water in England was fined a record £90 million for 6971 unauthorised sewer and wastewater treatment plant overflows over a prolonged period (Environment Agency, 2021a). This was not a technical failure, but a deliberate decision by the private company not to invest adequately in the assets needed. It was cheaper to break the law and pay any fines incurred. However, the company's decision should not be seen as anything but a logical product of the policy, regulatory and enforcement regime prevailing in England. From 2010 onwards, such companies were relieved of most of the performance oversight by the Environment Agency, whose budget was reduced, presuming self-reporting was sufficient (e.g. Hammond *et al.*, 2021). The current regulatory regime might encourage some in the English private water companies to focus too much on shareholder dividends, rather than service, as illustrated in this example, in which the consequences of inadequate regulation and poor oversight should have been obvious. The failures in this case are clearly in governance and regulation, encouraging defective institutional arrangements and ineffective enforcement. Yet, most of the English media considered Southern Water as the criminal organisation, rather than the inadequate regulatory regime that allowed such behaviour (e.g. Monbiot, 2021).

The challenges that current needs and perspectives for societal water and sanitation bring now, whether it is the best response to climate change or balancing the many ways of 'seeing' water as an opportunity or a threat, may help to promote a return to engineering's roots, which are to serve society.

1.2. Contemporary urban drainage

Stormwater and wastewater are defined in BS ISO 24513:2019 (BSI, 2019): stormwater is 'water arising from precipitation and snowmelt[;] stormwater is associated with the entire range of rainfall events' and wastewater is 'water arising from any combination of domestic, institutional, commercial or industrial activities, surface runoff and any accidental sewer inflow/infiltration water and which can include collected stormwater, discharged to the environment or sewer'.

These 'waters' potentially impact the environment where they enter or are discharged into surface and groundwater bodies. High flows, erosion of riverbanks and polluted inflows can all damage natural habitats. However, where the assimilative capacity of the water body is high – that is, the drainage discharges are small (in volume or contaminants) compared with the size or flow of the water body, impacts can be small and may even in some cases be beneficial, provided the flow contains only organic materials. The nutrients conveyed in drainage flows can provide food for natural organisms, and provided these are not too concentrated, and the levels of toxic and bioaccumulative substances are low, drainage discharges can even help support industries such as fisheries. However, such clean discharges are rare, and in general pollutant discharges from urban runoff and even treated wastewater are likely to be damaging to water bodies and aquatic ecosystems and impact human health. Currently almost all wastewaters will convey persistent and 'forever' chemicals, including the ever-increasing number of poly- and perfluoroalkyl substances (PFASs) (e.g. Blum *et al.*, 2015), as these are now ubiquitous in human and, increasingly, natural environments. Hence, without treatment, even the 'cleanest' wastewater is likely to be polluting.

Rainfall not only directly supports our water resource and supply systems, but it also delivers valuable environmental and amenity benefits, enhancing our urban environments by way of blue spaces in the form of water bodies. It also supports green spaces by providing essential water for plants, agriculture, irrigation and ecosystems as a whole. Thus, there is a wide range of potentially interested parties who should and do take an interest in drainage and the subsequent management of rainfall and runoff. Although public water supplies, health, sanitation and protection from the extremities of too much and too little water have traditionally been the province of engineers, contemporary understanding is that a far wider group of interested parties needs to be involved in water management (Chapter 4). It is not only relevant to professionals, but individual citizens, property owners and dwellers, private and commercial; all need to be much more engaged with how water is managed, especially in urban areas.

Why should many different people be interested in drainage? Well, the world is now more populous than ever before, resources are more stretched, and the pace of change is greater than ever before. Urbanisation has dramatically changed our environment, especially the built environment. In the developing world, citizens have always had to be intimately engaged with how they get their water, how wastes are dealt with and how best to cope when there is too much rainfall and flooding occurs, or indeed when there is too little water and there is water stress or drought. Hence the need for the worldwide initiative of the Sustainable Development Goals (SDGs), aiming to bring water supply and sanitation (WSS) to those without (Ashley and Horton, 2022), although it applies equally to those countries that already have functioning WSS, but possibly in need of improvement.

In the industrialised world, where provision has been a fact of life for at least a century, delegating responsibilities for secure water services, health protection and public safety to water service providers has become the norm. These services are provided by a variety of organisations, from local community groups through to multinational and multiutility corporations (Juuti and Katko, 2005). Many of these organisations make money from providing these services; other services are provided by the state as part of the community. There is no single management or operational model that appears to serve for all types of community or society. However, such services are provided within defined boundaries, and often those dealing with water supplies are not the same as those managing surface water or wastewater. Even in a single organisation there are many examples of different departments not communicating effectively together, a phenomenon termed 'silo thinking'. In addition, water and drainage system boundaries are typically defined by hydrological catchments, which may be natural or engineered by urbanisation, although boundaries may also be set within political or institutional frames. Disparities between boundaries for different types of service, such as catchment boundaries not matching management, political or policy boundaries, invariably lead to complexities in the provision of services and sometimes service provision gaps, where an expected service is 'someone else's job, not mine'.

This picture is complicated even further by the knowledge that urban drainage systems do not exist independently of the myriad of systems, infrastructure and services in urban areas. There are crucial interactions with land use, natural systems, transport systems, power supplies and telecommunications. Such interactions and the need to take them into account are changing the way in which urban systems and services are being planned, designed and operated, as 'systems of systems'. These systems also need to include the people using and interacting with them. Importantly, the understanding that there is a need to get more from the systems by creating and operating them in a harmonised way for a range of functions – including how land is used – is growing. Urban drainage is, of course, a vital service, even if the traditional ideas as to how best to provide it are being challenged by such new understandings.

From this, it is possible to see that 'drainage' is a topic that concerns everyone. It can be problematic that it only concerns some of the key players for part of the time – for example, when flooding is prolonged, persistent and impacting vulnerable communities. It was only after months of flooding in 2014 in regions of the south and west of England that Prime Minister David Cameron began to take any real notice. Unfortunately, more flooding occurred in 2015. Despite decades of world-leading research, reports and recommendations, before and after the previous major flooding in 2007, there had been very little commitment by policymakers to address the growing flood risk in England, as is apparent from the UK's recurrent Climate Change Committee reports (CCC, 2023). In England, more and more new houses are being built on known flood plains, despite these risks. It also appears that commitments to improve the quality of natural water bodies in England, formerly under the EU's Water Framework Directive (WFD) have slipped, and the quality of waters is continuing to decline in England and Wales (Environment Agency, 2021b), with only 14% of UK rivers meeting good ecological status and not one meeting good chemical status in 2020. Systems thinking, a prerequisite to effective WFD implementation, is something in which the UK falls short. The exit of the UK from the EU has now seen the replacement of environmental directives with the Environment Act 2021, which sets out ambitious plans that should improve our natural water bodies, but only time will tell if this comes to fruition.

The situation is different in many countries around the world where water and drainage are seen as vitally important. This is usually because there are shortages of supply or too much water, with flooding being a regular occurrence. Globally the SDGs, through which water is a common thread, are providing the baseline for planning services and infrastructure within communities and include WSS (Hutton and Varughese, 2016). Urban drainage is frequently overlooked in the application of the SDGs. However, it is the ambition of the European Commission to realise the human right to water, and the implementation of SDG 6 – clean water and sanitation. Water and sanitation are human rights, recognised by the United Nations (EurEau, 2016) to be essential for living a life in dignity and the realisation of all human rights. Clearly, WSS services play a fundamental role in realising these basic human rights (EurEau, 2022). Often, though, the focus is on clean water supplies and basic sanitation, and some aspects of flooding. Surface water drainage is often subordinated, especially in urban areas. This may also happen to the important interactions between drainage and other services, as illustrated in Figure 1.6, where inadequate municipal solid waste systems impair effective drainage provision.

Although clearly relevant to developing countries where there is a lack of WSS, the SDGs are also being utilised in countries such as the UK. For example, the city of Bristol has a vision for 2050, developed in 2020, that is framed around the SDGs, including SDG 6 (Bristol, 2020).

Figure 1.6 Storm drains in Mumbai are compromised by inadequate solid waste management (Richard Ashley)

One of the most advanced countries in effective management of urban water, if not the most advanced, is Australia. Here, a prolonged drought period in the 2000s compelled politicians to engage in water efficiency drives, resulting in world-leading ideas for local water use, wastewater reuse and the provision of different use classes based on quality of water supplies, supported by campaigns to influence public behaviour. This followed decades of attention to managing surface water discharges to clean up receiving waters. The world-leading initiative to engage communities and manage water demand was interrupted by the decision to introduce desalination technology in the mid-2000s, when politicians in various states in Australia saw this as their 'silver bullet' solution to water shortages. Following expensive construction under private finance schemes, many of these Australian desalination plants became underused as the drought came to an end towards 2010. In Melbourne, for example, the plant was first used in 2017 and at a cost of more than £1 billion per year (2017 £/AUD exchange rate). Whether in use or not, it is not a cheap option when compared with promoting water efficiency and it may also detract attention from the need for users to manage their demand (e.g. Lam *et al.*, 2017). Desalination plants have also been built in London. The first new plant was opened in 2010 and hailed as 'a victory for common sense' by then Thames Water's chief executive officer David Owens, as reported in the *New Civil Engineer* magazine on 9 October 2008 (NCE, 2009). Now there are plans for London's sewage to be reused (Thames Water, 2023), and the rainwater falling on London provides a similar opportunity to the direct use of rainfall on Melbourne. Unfortunately, the lack of integration of urban water cycles means that water company jurisdictions (like that of Thames Water) do not include rainwater harvesting. However, there may be a growing need to join up these approaches and consider the role of stormwater harvesting.

The political and policy landscape in which water services, including urban drainage provision, operate is fundamental to how this provision is made and operated. The previous examples of how many policymakers and politicians choose to focus their attention on matters seen as more important than water, drainage and the wider interactions across society also show how they prefer to utilise traditional thinking and ideas that typically rely on 'building their way out of problems' rather than take a careful, systematic look at why things might need to be done differently. This demonstrates why it is so important for everyone to try to move water up the political agenda. To guarantee the intergenerational sustainability of water services, accessibility and affordability should be considered intertwined. The traditional idea that individual items of infrastructure serve only single-use purposes still prevails, despite worldwide initiatives for creating multiuse, multipurpose and multifunctional infrastructure. A good example of what can be achieved when highway engineers and drainage engineers speak to each other can be seen in Kuala Lumpur, Malaysia. A world's first, the dual stormwater and road tunnel provides a unique solution to the Malaysian capital's long-term traffic and stormwater management problems. The tunnel diverts floodwaters from confluence of two major rivers running through the city centre. The central section doubles as a two-deck motorway to relieve congestion. Originally conceived as a flood-relief tunnel to divert one-in-100-year floodwater away from the city centre, municipal officials then realised that it could be utilised in periods of low rainfall as a road tunnel to alleviate Kuala Lumpur's congested highway infrastructure. The 9.7 km structure is the longest multipurpose tunnel in the world and the longest stormwater drainage tunnel in South East Asia. In 2011, the SMART tunnel won the UN Habitat Scroll of Honour award for its innovative and unique management of stormwater and peak hour traffic.

Despite a few examples, such as the Kuala Lumpur multifunctional tunnel, there are signs that engineers are in the early stages of knowing how best to provide multifunctional infrastructure. In contrast, there are numerous reasons why London has elected to build a new 'common-sense' single-function 'super sewer' rather than integrating multifunctional sustainable drainage systems (SuDS) that provide a range of other benefits (Woods-Ballard *et al.*, 2015), as in Philadelphia in their Green City, Clean Waters programme. The super sewer is a financial risk-free business model for the service provider, Thames Water (Loftus and March, 2019), and the way in which this option was selected demonstrates the intrinsic relationship between how schemes are financed and the type of scheme built. Given the aspirations to 'green London' (Mayor of London, 2020), it would seem a sensible option to use SuDS to utilise the rainfall and also help manage combined sewer overflow spills into the River Thames, using the stormwater as a resource in the severely water-stressed area of the south east of England. As in 'the great sanitary revolution', when the new sewers transformed the entire urban area of London (Allen, 2008), were SuDS to be used for this now, this would substantially impact on land use in London for decades into the future. Given the evidential value and security of a sewer, such as that under construction in London, how should and could engineers shift the common-sense ground from large environmentally impacting heavy single-function infrastructure to multibeneficial, less-impacting systems?

1.2.1 Urban drainage in the UK context

In the past 30 years urban drainage practice in the UK has seemingly come a long way; or has it really changed very much? When Larry Roesner gave an award-winning presentation on Best Management Practices (BMPs) to the Urban Drainage Group Conference in Blackpool in 1994 (Roesner and Bristol, 1994), it was hailed as inspirational (it won the best paper prize), and around this time there was a general growing awareness of the possibilities to change how surface water was being managed in the UK, with SuDS in Scotland, CIRIA's first SuDS guidance in 2000, The first edition of *The SuDS Manual* (Woods-Ballard *et al.*, 2007) and other initiatives coming about, including the Flood and Water Management Act 2010 in England and Wales (CIRIA, 2023a).

Yet have engineers properly grasped the opportunity and intent to, as far as practicable, manage water on or near the surface in the best way possible, in so doing mimicking nature and restoring natural hydrology (Walsh *et al.*, 2012) and made it business as usual? In England, according to Defra, a consistent approach to sustainable drainage is to be introduced: 'The new approach to drainage will ensure sustainable drainage systems are designed to reduce the impact of rainfall on new developments by using features such as soakaways, grassed areas, permeable surfaces and wetlands' (Defra and Pow, 2023).

Drainage in the UK is linked to the uptake of SuDS use. In England this is progressing despite the limitations of the regulatory system. Many developers, through the planning system, are required to use

SuDS by local authorities. The London Sustainable Drainage Action Plan (Mayor of London, 2016) sets out a vision for implementing the widespread use of SuDS, shown to be effective and deliverable in the London Strategic SuDS Pilot Study programme (LoTAG, 2021). In Scotland, where there are much more clearly defined responsibilities for long-term maintenance, SuDS are becoming more mainstream (CIRIA, 2023b; D'Arcy, 2013) and in Wales the devolved administration has implemented Schedule 3 of the Flood and Water Management Act 2010 to make the use of SuDS mandatory for all new developments of more than one house or where the construction area is 100 m^2 or more. For example, the Retro Fit Green Infrastructure SuDS Project was Civil Engineering Project of the Year 2014 in the Building Construction Industry Awards.

In Europe, the EU's biodiversity strategy places significant importance on incorporating green infrastructure (GI) in urban spatial planning. The provision of GI has contributed to the EU's sustainable growth objective, and the European Commission wishes to promote the deployment of GI in urban areas, by way of incentives for up-front investment, with the initial aim of restoring at least 15% of degraded (urban) ecosystems by 2020 (EC, 2011). This strategy is impacting urban water systems as the promotion of GI solutions with their multiple benefits offers a wider range of innovative solutions in which water systems are managed, as GI systems potentially provide a wide range of benefits (EEA, 2015; Lundy and Wade, 2011). There are indications that there is a widening of perspective in the UK to consider 'nature-based systems' (NbSs) with their multiple benefits, in part owing to the challenges that the COVID-19 pandemic has brought to urban lifestyles and the desire by many citizens to interact more with nature (e.g. Ashley *et al.*, 2023; Dobson, 2020). Green and blue infrastructure is a key element of on-the-surface drainage systems, including SuDS.

The drivers for a better approach to the management of surface water and delivery of non-piped systems such as SuDS are growing. Engineers are facing climate and ecological crises that require a paradigm shift in how the entire infrastructure is managed. The water management infrastructure (whether grey, blue or green; Thorne, 2020) will increasingly come under scrutiny to see how it can assist in managing and adapting to climate change and reducing the impacts it has on the human, water and terrestrial ecosystems. Within the UK, as in other parts of the world, there is growing commitment to net zero carbon targets and decarbonisation. The drive for net zero carbon, a key policy of the Institution of Civil Engineers, can only help to promote the use of natural infrastructure. Guidance produced in the UK includes PAS 2080:2023, a specification for managing whole life carbon in infrastructure (BSI, 2023). It is incumbent on all professionals to understand the carbon budget for any development and minimise any excess, even at the planning and design stage. Drainage is part of a water cycle, and any such understandings need to take an integrated water cycle approach (BS ISO 24513:2019; BSI, 2019), ideally including the interactions with other services and infrastructure systems.

Mimicking nature, using NbSs such as SuDS, and managing water at or close to the surface, rather than burying the water, will help reduce carbon emissions and also enable drainage infrastructure and towns and cities to adapt better to the impacts of climate change. SuDS can help manage climate change impacts such as droughts and deluges or higher temperatures in urban areas. NbSs such as SuDS are vital approaches to help overcome the ecological and climate crises engineers are facing.

Recognising these challenges, combined with urbanisation and living within environmental constraints, contemporary good practice proposes the need to integrate better the management of the water cycle, not only across the cycle itself, but also within and in partnership with other urban systems and services (Eusgeld *et al.*, 2011). Urban drainage must sit within the drive for and progress towards 'sustainable water services' as integrated systems (Ainger and Fenner, 2016) (Chapter 12).

So-called 'common-sense thinking' is a theme of this book and, as will be shown, such thinking often has literally been costing the earth. Traditional, locked-in ideas and locked-into technologies and governance are often unfit for the challenges ahead, especially for the changing climate (e.g. Ainger and Fenner, 2016).

> The climate is a common good, belonging to all and meant for all. At the global level, it is a complex system linked to many of the essential conditions for human life. [...] we are presently witnessing a disturbing warming of the climatic system. In recent decades this warming has been accompanied by a constant rise in the sea level and, it would appear, by an increase of extreme weather events [...]. Humanity is called to recognize the need for changes of lifestyle, production and consumption, in order to combat this warming or at least the human causes which produce or aggravate it. (Pope Francis, 2015)

Despite the many attempts to define how to set about working towards sustainable urban water management, little real progress has been made in transforming the way in which urban water systems are being managed. Locked-in (i.e., this is how engineers have always done it) or 'entrapped' institutional and regime 'common sense', in which engineers need to add to what infrastructure they already have with similar technologies, still prevails, largely resulting in the reproduction of old solutions to new and complex urban water-related problems (Palmer, 2000). The reasons for this relate to all aspects of society, from the public expecting services to be provided at little or no cost, to politicians who see water services as a minor part of their portfolios. An example of this is in Ireland, where this attitude has led to major and recurrent outbreaks of *Cryptosporidium* since domestic water charges were abolished in 1997 (Ashley, 2007). Other reasons include professionals who are unwilling to challenge clients to do the best for society, and politicians who prefer the low cost, 'private knows best' safe and sure – engineers know it works – solutions.

There are signs that the all-pervasive neoliberal econometrics in which nothing is valued unless it can be monetised is being challenged, with new initiatives to understand and define 'value' based on community and individual perceptions and definitions. For example, under the Enabling Water Smart Communities initiative, a project partnership led by Anglian Water in England is reframing water cycle stewardship with the aim of moving closer to integrated water management (EWSC, 2023) and developing alternative ways of valuing water in the process.

In this book, various authors contribute to explaining why a distorted 'common-sense' approach to urban water management has been prevalent. They identify problems with

- locked-in perspectives in frames of reference within the professional and political sectors relating to fundamental mis-assumptions about designing for slowly changing conditions and outmoded operational conditions
- locked-in regulatory forces supporting complex, fragmented, opaque and inert governance structures that impede institutional learning, the taking of the essential system perspectives and cross-disciplinary and cross-institutional approaches required to cope with uncertain futures, especially for the community at large
- transitioning from a 'problem-centric' perspective for urban drainage to an 'opportunity-centric' mindset in which the appropriate management of surface water and foul flows provides opportunities for better and more integrated water management, and
- a despondence in parts of the scientific, professional and practitioner community, especially in England – that is, a normative lock-in, relating to a loss of hope and belief that the essential step change or complete transition to a more effective and sustainable water management regime will ever be possible.

Countering this somewhat negative view of urban drainage practices in the UK, and elsewhere, there are signs of possible improvements, as described throughout this book and as outlined next.

1.2.2 Advances in urban drainage practice

In Europe, the USA and various parts of the world, the form and nature of what has traditionally been urban drainage has been shifting. The way urban drainage has needed to change has been in response to increasing knowledge about the service requirements over time, reflecting the ongoing societal changes. These have been most significant since the Industrial Revolution, as illustrated in Figure 1.7. There were two major developments in the latter part of the twentieth century that have had the greatest influence on this shift. The development of computational models in the 1970s that can now provide real-time and predictive estimates of water flow rates, volumes and water depths, together with various estimates of conveyed dissolved and suspended substances, introduced a means of properly examining the performance of urban drainage systems. Simultaneously, solid-state memory launched a new wave of instruments that could be used to observe, monitor, record and transmit, in real time, drainage-related phenomena. Advances in analytical chemistry also facilitated a better understanding of the nature of the growing range of substances that can make their way into the natural world (Lofrano and Brown, 2010). At the same time there were initiatives to make drainage greener – that is, mimicking nature's way. Growing awareness of the threats of climate change, population growth and urbanisation has necessitated a major reflection on how drainage can best be provided within the context of sustainable cities, unlocking opportunities to support the delivery of cities that are more ecologically rich, resilient, liveable and connected with the natural environment.

Termed Best Management Practices (BMPs) in the USA, and source control and sustainable drainage systems (SuDS) in the UK, various pioneers were able to utilise these nature-based, on-the-surface drainage systems comprising various green spaces and 'blue' water bodies in preference to piped drainage (Fletcher *et al.*, 2014). At present, the appropriate use of computational modelling of urban drainage systems, instruments monitoring rain, flow and a range of substances, and the utilisation of nature-based 'blue and green' infrastructure systems in preference to piped drainage are becoming more and more the

Figure 1.7 Timeline of some important milestones of significance to urban drainage

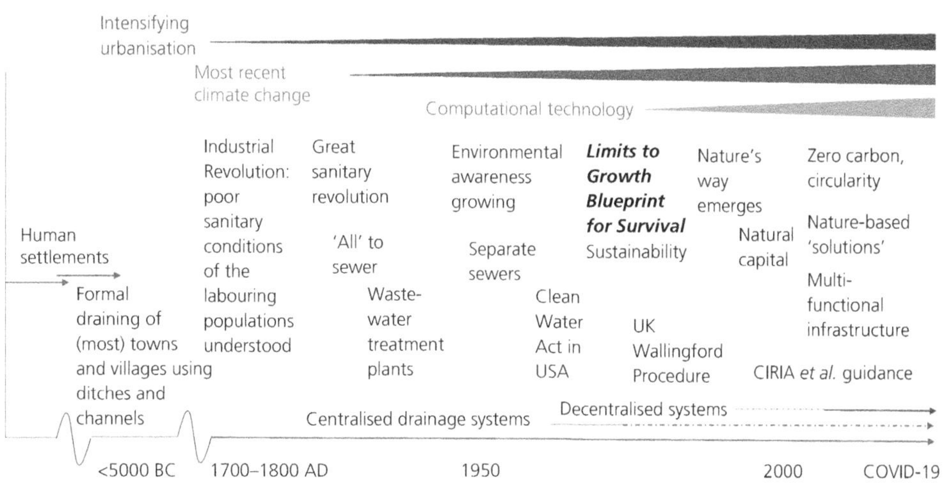

business-as-usual approach to urban drainage design and management. This 'nature-based' approach fits into the drive to utilise nature even in urban areas and to provide support to the ecosystems that deliver so many benefits to humanity (Natural England, 2021). Now engineers are faced with the potential effects of further changes in society and the way people in urban areas choose to live after the COVID-19 pandemic; a shock that should lead to a reappraisal of how best to provide and sustain all of humanity's systems and services (Ashley *et al.*, 2023). This further challenge to urban drainage practice is taken up later in this book.

1.2.3 The water-sensitive city

Figure 1.8 illustrates the concept of 'water sensitivity', in which water is considered as one of the primary needs and also impacts on human and other life (CIRIA, 2013). Figure 1.8 needs to be seen in the context of Figure 1.1, showing the landscape of water service provision. The increased scope of each of the aspects of water sensitivity in Figure 1.1 is shown from provision of a rudimentary set of water services to what is now considered to be a fully water-sensitive city. The concept of water sensitivity provides a vision to identify the most viable options for managing water availability (either too little or

Figure 1.8 Key elements of water sensitivity (Louise Walker, based on Nickerson, 2023)

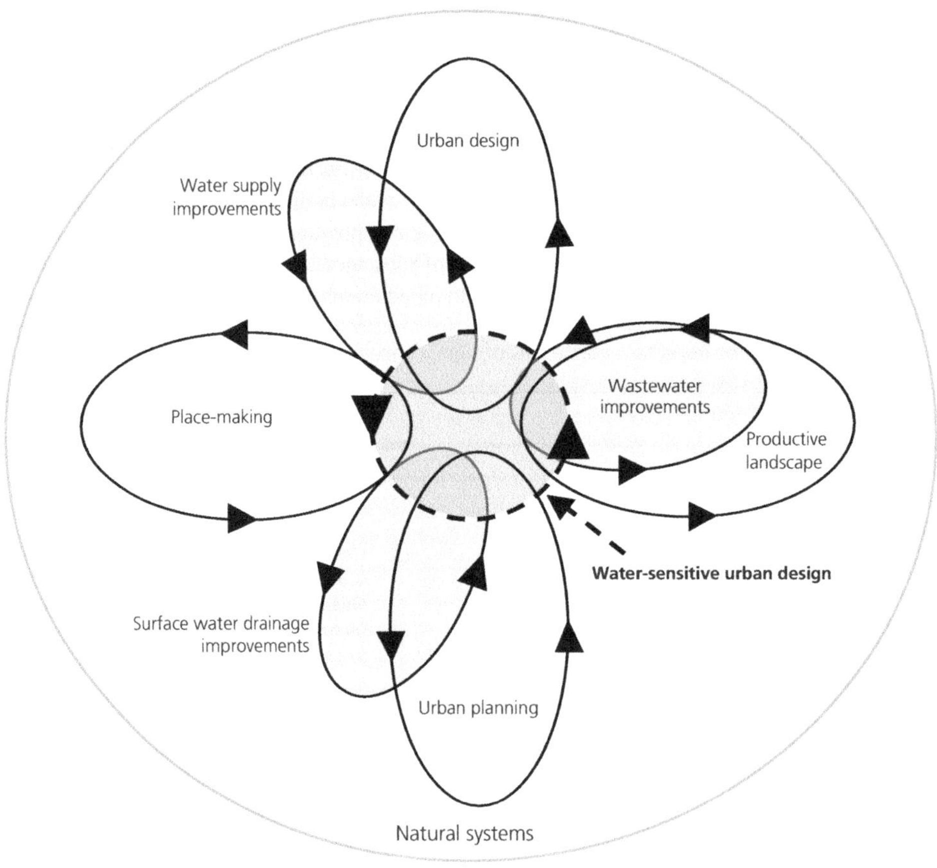

too much) and water quality within the context of other urban systems and services. However tempting as it may be to view the rightward advancing steps in Figure 1.1 as sequential stages in the development of services that are sustainable, this does not necessarily represent a historical or developmental timeline. In the history of humanity there have been many reversals or jumps from positions to the right side of the diagram back to the left side, which is arguably less developed. Examples include the Dark Ages in Europe, during which much of the Roman-built sanitation systems were neglected, and more recently the 'silver bullet' of desalination in Australia (Section 1.2), for which the ultimate goal of water sensitivity appeared to be becoming embedded, only to be subverted to somewhere around the 'drained city' in Figure 1.1.

The Cooperative Research Centre for Water Sensitive Cities in Australia (CRCWSC, 2023) has undertaken a well-established research and development programme to understand and build the key elements needed to transition from a fragmented water urban area, where all the various components of the water cycle are dealt with separately, to one where water is central to urban planning and development (CRCWSC, 2022). Urban drainage is but one part of this. Further consideration of water sensitivity is given in Chapter 12.

Considering the current practices in the UK, Ashley *et al.* (2013) reviewed the potential for water-sensitive urban design, setting out potential benefits from taking such an approach, and CIRIA (2013) set out ideas for water-sensitive design in the UK. Conversely, Green and Anton (2012) show how far England is behind Germany in taking an integrated or more sustainable approach to water management; suggesting a lag of some 30 years due to a lack of incentivisation caused mainly by the regulatory boundaries set in England as to who is responsible for the various aspects of the water systems. This question of the importance of boundaries in moving towards effective or sustainable water management and, within this, drainage is reviewed in Chapter 12, where the concept of sustainability is considered.

1.3. Professional practice

Internationally, a number of groups are advancing urban drainage practice and knowledge, each with a slightly different interpretation of what urban drainage comprises. The Urban Storm Drainage Specialist Group was set up in the 1960s by the International Association for Hydro-Environment Engineering and Hydraulic Research (IAHR) and International Water Association (IWA, 2023) to address the emerging domain of both sanitary and stormwater management. At the end of the twentieth century, the group reviewed the definition of urban storm drainage as understood at that time, how this might need to be redefined and what therefore the role of experts in knowledge development might mean for the future. Box 1.1 shows the current definition used by the group.

The review of urban storm drainage as a discipline by the joint IAHR/IWA specialist group highlighted the need to drop the word 'storm', as the group's activities included aspects of wastewater, often known as sanitary or foul flows, as well as storm- or surface water. The group's is now part of the urban water umbrella domain of IWA. Although not specifically dealing with clean or wastewater treatment processes, the integrative nature of the vision for urban drainage shown in Box 1.1 means that the entire water cycle is potentially within the perspective of 'urban drainage', with wide-ranging interactions with water supply, wastewater treatment, receiving waters and natural and human environments.

In the UK, the Chartered Institution of Water and Environmental Management (CIWEM) includes an Urban Drainage Group (UDG) (CIWEM, 2023). The group was formerly independent and known as the Wastewater Planners User Group (WaPUG) when established in 1984. Box 1.1 shows the definition used

Box 1.1

Aims of the Urban Drainage Specialist Group of IAHR/IWA

To foster and conduct fundamental and applied research on urban drainage and promote innovative approaches to urban drainage worldwide. Core issues include urban rainfall in current and future climates; quantity and quality of urban stormwater and combined sewer overflows; advanced management of stormwater in urban areas, with emphasis on source controls, rainwater harvesting, low-impact development and water-sensitive urban design; operation and control of urban drainage, including sewer systems and receiving waters, in real time; modelling of whole systems or their elements; and applicability of these approaches in regions with various climates, governance and various levels of economic development.

Aims and objectives of the Urban Drainage Group of CIWEM

Aligned with CIWEM's Royal Charter, these are as follows.

1. Advance the science and practice of urban drainage management for the public benefit.
2. Promote education, training, study and research in urban drainage management and publish the results of such research.
3. Establish and maintain appropriate standards of competence for urban drainage management professionals.

by the UDG of its interests. This definition, although more focused than that for the IAHR/IWA specialist group, does take a broad view, although the UDG's activities are noticeably less cross-disciplinary, with scant social science or land use planning interactions or knowledge development. Figure 1.9 illustrates the vision of the UDG.

Unlike the IAHR/IWA specialist group, the UDG's incorporation within CIWEM, which is a professionally accrediting institution, means it has an auditable legitimacy to its functioning.

Elsewhere in the world, urban drainage as a discipline is typically recognised in terms of the association with the activities and perspectives of the IAHR/IWA, or with professional institutions in the USA, especially the American Society of Civil Engineers (ASCE) and various other organisations, such as the Water Environment Federation, whose mission is to protect public and environmental health and who include stormwater management in their activities. An institute within ASCE, the Environmental and Water Resources Institute (EWRI) comprises professionals engaged in multidisciplinary water resources and environmental engineering and science aiming to advance EWRI's vision: 'Advancing water resources and environmental solutions to achieve a sustainable future'. (ASCE, 2023).

Many other countries have their own institutions and established arrangements for urban drainage that provide baselines for how the domain is operated locally. Most of these relate back to the international organisations highlighted previously for scope definition.

Throughout the book consideration is given to ideas on what regime would be required to support a transition to the contemporary and effective management of urban water systems. After all, water is essential to all life and the use of, and interaction with, water by other natural systems must be safe, secure and as low risk as possible.

Figure 1.9 Chartered Institution of Water and Environmental Management (CIWEM) Urban Drainage Group interests overlap with the Rivers and Coastal Group

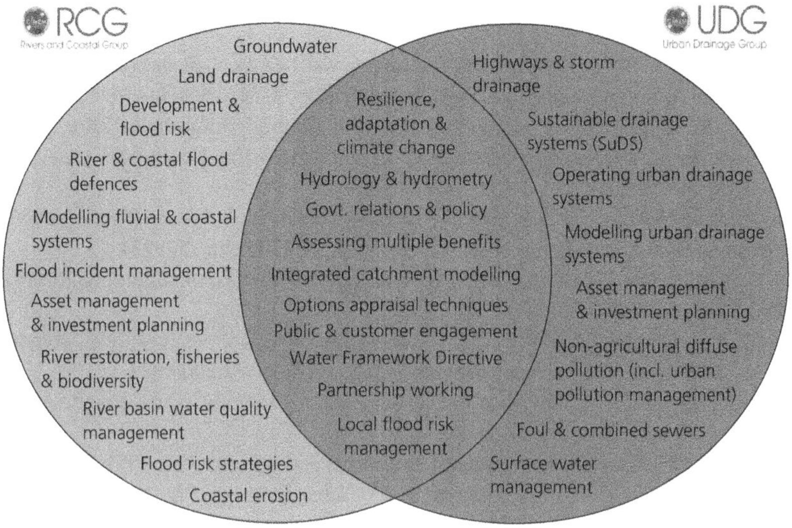

The message here for practitioners is as follows.

- As professionals, remember that your client is society, no matter who is paying for the scheme.
- Begin by thinking as broadly as possible – even for small drainage schemes – both spatially and in time: from the carbon impacts to people and how the scheme will be used and sustained over time. Only then concentrate on the detail.
- Challenge received 'common sense' and traditional 'this is how we have always done it' thinking and innovate to create schemes that are as sustainable as possible.
- Wherever and whenever possible, utilise natural systems in preference to manufactured resources and the need for complex construction.

REFERENCES

Ainger C and Fenner R (eds) (2016) *Sustainable Water*. ICE Publishing, London, UK.

Allen M (2008) *Cleansing the City*. Ohio University Press, Athens, OH, USA.

Angelakis AN, Koutsoyiannis D and Tchobanoglours G (2005) Urban wastewater and stormwater technologies in ancient Greece. *Water Research* **39**: 210–220, 10.1016/j.watres.2004.08.033.

ASCE (American Society of Civil Engineers) (2023) EWRI Environmental & Water Resources Institute. https://www.asce.org/communities/institutes-and-technical-groups/environmental-and-water-resources-institute (accessed 19/07/2023).

Ashley RM (2007) From technology entrapment to adaptability. Editorial. *Proceedings of the Institution of Civil Engineers – Engineering Sustainability* **160(ES3)**: 109–110, 10.1680/ensu.2007.160.3.109.

Ashley RM and Horton B (2022) Characterizing financing needs and financing capacities in different regions: a global perspective on water-related financing flows and drivers for investment needs. In *Financing Investment in Water Security – Recent Developments and Perspectives*. (Leflaive X, Dominique K and Alaerts GJ (eds)). Elsevier, Amsterdam, the Netherlands.

Ashley RM, Lundy L, Ward S *et al.* (2013) Water-sensitive urban design: opportunities for the UK. *Proceedings of the Institution of Civil Engineers – Municipal Engineer* **166(ME2)**: 65–76, 10.1680/muen.12.00046.

Ashley RM, Walker L, Digman C *et al.* (2023) The benefits of nature-based systems in a changing and uncertain world. *Proceedings of the Institution of Civil Engineers – Engineering Sustainability* **176(2)**: 61–71, 10.1680/jensu.21.00102.

Basu NB, Dey A and Ghosh G (2013) Kolkata's brick sewer renewal: history, challenges and benefits. *Proceedings of the Institution of Civil Engineers – Civil Engineering* **166(2)**: 74–81, 10.1680/cien.12.00016.

Bertrand-Krajewski J-L (2005) Sewer systems in the 19th century: diffusion of ideas and techniques, circulation of engineers. *Proceedings of the 4th International Water History Association Conference*, Paris, France. December 2005.

Blum A, Balan SA, Scheringer M *et al.* (2015) The Madrid Statement on Poly- and Perfluoroalkyl Substances (PFASs). *Environmental Health Perspectives* **123(5)**: CID, 10.1289/ehp.1509934.

Bristol (2020) *One City Plan*. See https://www.bristolonecity.com/wp-content/uploads/2020/01/One-City-Plan_2020.pdf (accessed 24/07/2023).

Brown RR, Keath N and Wong T (2009) Urban water management in cities: historical, current and future regimes. *Water, Science and Technology* **59(5)**: 847–855, 10.2166/wst.2009.029.

BSI (2019) BS ISO 24513:2019. Service activities relating to drinking water supply, wastewater and stormwater systems. Vocabulary. BSI, London, UK.

BSI (2023) PAS 2080:2023. Carbon management in buildings and infrastructure. BSI, London, UK.

Burian SJ and Edwards FG (2002) Historical perspectives of urban drainage. *Proceedings of 9th International Conference on Urban Drainage*, Portland, Oregon. Available via the ASCE library http://ascelibrary.org/doi/abs/10.1061/40644%282002%29284?src=recsys (accessed 19/07/2023).

CCC (Climate Change Committee) (2023) Better transparency is no substitute for real delivery. See https://www.theccc.org.uk/ (accessed 19/07/2023).

Chocat B, Ashley R, Marsalek J *et al.* (2004) Urban drainage – out-of-sight-out-of-mind? Position paper. *Proceedingss of the Novatech Conference*, Lyon, France, pp. 1659–1690.

Chocat B, Ashley R, Marsalek J *et al.* (2007) Toward the sustainable management of urban storm-water. *Indoor Built Environment* **16(3)**: 273–285, 10.1177/1420326X07078854.

CIRIA (Construction Industry Research and Information Association) (2013) *Water Sensitive Urban Design in the UK – Ideas for the Built Environment Practitioners*. CIRIA C723F. CIRIA, London, UK.

CIRIA (2023a) England and Wales legislation & regulation, https://www.susdrain.org/delivering-suds/using-suds/legislation-and-regulation/england-and-wales.html (accessed 19/07/2023).

CIRIA (2023b) Scotland legislation & regulation, https://www.susdrain.org/delivering-suds/using-suds/legislation-and-regulation/scotland.html (accessed 19/07/2023).

CIWEM (Chartered Institution of Water and Environmental Management) (2023) Urban Drainage Group, https://www.ciwem.org/special-interest-groups/urban-drainage-group (accessed 19/07/2023).

CRCWSC (Cooperative Research Centre for Water Sensitive Cities) (2022) *Urban Water Guide*. Australian Water Partnership, Canberra, Australia.

CRCWSC (2023) Realising the Vision of a Water Sensitive City, https://watersensitivecities.org.au/ *(accessed 19/07/2023).*

D'Arcy BJ (2013) *Managing Stormwater: From aspirations to routine business*. Gwf-Wasser. Abwassser. International Issue. 2–8. See https://gwf-wasser.de/produkt/managing-stormwater-from-aspirations-to-routine-business/ (accessed 14/12/2023).

De Feo G, Antoniou G, Fardin HF *et al.* (2014) The historical development of sewers worldwide. *Sustainability* **2014(6)**: 3936–3974, 10.3390/su6063936.

Defra (Department for Environment, Food and Rural Affairs) and Pow R (2023) New approach to sustainable drainage set to reduce flood risk and clean up rivers, https://www.gov.uk/government/news/new-approach-to-sustainable-drainage-set-to-reduce-flood-risk-and-clean-up-rivers (accessed 05/12/2023).

Dobson J (2020) Wellbeing and blue-green space in post-pandemic cities: drivers, debates and departures. *Geography Compass* **15(10)**: e12593, 10.1111/gec3.12593.

Dresp-Langley B (2008) Dimensions of environmental engineering. *Open Environmental Engineering Journal* **2008(1)**: 1–8, 10.2174/1874829500801010001.

EC (European Commission) (2011) Communication from the Commission to the European Parliament, the Council, the Economic and Social Committee and the Committee of the Regions. Our life insurance, our natural capital: an EU biodiversity strategy to 2020. (SEC(2011) 540 final). See https://eur-lex.europa.eu/LexUriServ/LexUriServ.do?uri=COM:2011:0244:FIN:EN:PDF (accessed 24/07/2023).

EEA (European Environment Agency) (2015) Green infrastructure: better living through nature-based solutions, https://www.eea.europa.eu/articles/green-infrastructure-better-living-through (accessed 24/07/2023).

Environment Agency (2021a) Record £90m fine for Southern Water following EA prosecution, https://www.gov.uk/government/news/record-90m-fine-for-southern-water-following-ea-prosecution (accessed 19/07/2023).

Environment Agency (2021b) Catchment Data, https://environment.data.gov.uk/catchment-planning (accessed 24/07/2023).

EurEau (2016) *Making the Human Right to Water and Sanitation a Reality in Europe*. EurEau, Brussels, Belgium. See https://www.eureau.org/resources/position-papers/99-making-the-right-to-water-and-sanitation-reality-europe-september2016/file (accessed 19/07/2023).

EurEau (2022) *The Value of Wastewater Services*. EurEau, Brussels, Belgium. See https://www.eureau.org/resources/publications/6720-the-value-of-sanitation-services/file (accessed 19/07/2023).

Eusgeld I, Nan C and Dietz S (2011) 'System-of-systems' approach for interdependent critical infrastructures. *Reliabiliry Engineering & Systems Safety* **96**: 679–686, 10.1016/j.ress.2010.12.010.

EWS (2023) Enabling Water Smart Communities, https://ewsc.org.uk/ (accessed 24/07/2023).

Fletcher TD, Shuster W, Hunt WF *et al.* (2014) SUDS, LID, BMPs, WSUD and more – the evolution and application of terminology surrounding urban drainage. *Urban Water Journal* **12(7)**: 525–542, 10.1080/1573062X.2014.916314.

Gaterell M (2012) The art of John Martin's London sewerage system. *Proceedings of the Institution of Civil Engineers – Engineering History and Heritage* **165(EH1)**: 3–8, 10.1680/ehah.2012.165.1.3. Paper 800004.

Green C and Anton B (2012) Why is Germany 30 years ahead of England? *International Journal of Water* **6(3/4)**: 195–214, 10.1504/IJW.2012.049496.

Hamlin C (1992) Edwin Chadwick and the Engineers, 1842-1854: systems and antisystems in the pipe-and-brick sewers war. *Technology and Culture* **33(4)**: 680–709, 10.2307/3106586.

Hammond P, Suttie M, Lewis VT, Smith AP and Singer AC (2021) Detection of untreated sewage discharges to watercourses using machine learning. *Clean Water* **4**: 18, 10.1038/s41545-021-00108-3.

HMG (Her Majesty's Government) (1848) Public Health Act 1848. The Stationery Office, London, UK.

HMG (2010) Flood and Water Management Act 2010. The Stationery Office, London, UK.

HMG (2021) Environment Act 2021. The Stationery Office, London, UK.

Hutton G and Varughese M (2016) *The Costs of Meeting the 2030 Sustainable Development Goal Targets on Drinking Water, Sanitation, and Hygiene*. The World Bank, Washington, DC, USA, https://doi.org/10.1596/K8543.

IWA (International Water Association) (2023) Urban Drainage Specialist Group, http://www.jcud.org.

Juuti PS and Katko TS (eds) (2005) *Water, Time and European Cities: History matters for the Futures*. WaterTime. See http://www.watertime.net/docs/WP3/WTEC.pdf (accessed 19/07/2023).

Lam KL, Kenway SJ and Lant PA (2017) Energy use for water provision in cities. *Journal of Cleaner Production* **143**: 699–709, 10.1016/j.jclepro.2016.12.056.

Libey A, Adank M and Thomas E (2020) Who pays for water? Comparing life cycle costs of water services among several low, medium and high-income utilities. *World Development* **136**: 105155, 10.1016/j.worlddev.2020.105155.

Lofrano G and Brown J (2010) Wastewater management through the ages: a history of mankind. *Science of the Total Environment* **408(22)**: 5254–5264, 10.1016/j.scitotenv.2010.07.062.

Loftus A and March H (2019) Integration for whom? Financialisation and the Thames Tideway Tunnel. *Urban Studies* **56**: 2280–2296, 10.1177/0042098017736713.

LoTAG (London Technical Advisers Group) (2021) *London Strategic SuDS Pilot Study*. LoTAG, London, UK. See https://www.lotag.co.uk/london-strategic-suds-pilot-study (accessed 19/07/2023).

Lundy L and Wade R (2011) Integrating sciences to sustain urban ecosystem services. *Progress in Physical Geography: Earth and Environment* **35(5)**: 653–669, 10.1177/0309133311422464.

Mayor of London (2016) *London Sustainable Drainage Action Plan*. Greater London Authority, London, UK. See https://www.london.gov.uk/sites/default/files/lsdap_december_2016.pdf (accessed 19/07/2023).

Mayor of London (2020) National Park City, https://www.london.gov.uk/what-we-do/environment/parks-green-spaces-and-biodiversity/london-national-park-city (accessed 19/07/2023).

Monbiot G (2021) Why are England's water companies pumping out a tide of sewage? Because they can. *The Guardian*, 13 Jul. See https://www.theguardian.com/commentisfree/2021/jul/13/water-companies-britain-seas-sewage-fines-environment-agency?CMP=Share_iOSApp_Other&fbclid=IwAR1usepiu9NzHh_zd90AuhUfBOHcP53DkH2djR4OhpYd_x_cYzqv8ms3Prg (accessed 19/07/2023).

Natural England (2021) *Natural Capital Evidence Handbook: To Support Place-based Planning and Decision-making*. Natural England Research Report 092, 2nd edn. Natural England, London, UK.

NCE (*New Civil Engineer*) (2008) Pure solution. *New Civil Engineer*. 9 Oct.

Nickerson C (2023) Latour's Actor Network Theory. *Simply Sociology*. 12 May. See https://simplysociology.com/actor-network-theory.html (accessed 19/07/2023).

OECD (Organisation for Economic Co-operation and Development) (2017) *Diffuse Pollution, Degraded Waters: Emerging Policy Solutions*. OECD Publishing, Paris, France. See https://read.oecd-ilibrary.org/environment/diffuse-pollution-degraded-waters_9789264269064-en (accessed 14/12/2023).

Palmer R (2000) From the inside out. *Environmental Values* **9(4)**: 411–418, 10.3197/096327100129342128.

Pope Francis (2015) Encyclical Letter Laudato Si' of the Holy Father Francis on Care for our Common Home, https://www.vatican.va/content/francesco/en/encyclicals/documents/papa-francesco_20150524_enciclica-laudato-si.html (accessed 20/09/2023).

Roesner L and Bristol C (1994) Joint management of urban runoff quality and combined sewer overflows. *Proceedings of the WaPUG Conference*, Blackpool, UK. November. See http://ftp2.ciwem.org/1994/1994-11-23%20to%2025/AU1994.PAPER10.pdf (accessed 24/07/2023).

Thames Water (2023) Water recycling (reuse) schemes in London, https://www.thameswater.co.uk/about-us/regulation/strategic-water-resource-solutions/water-recycling-reuse-schemes-in-london (accessed 19/07/2023).

Thorne C (ed.) (2020) *Blue-Green Cities*. ICE Publishing, London, UK.

Walsh CJ, Fletcher TD and Burns MJ (2012) Urban stormwater runoff: a new class of environmental flow problem. *PLoS ONE* **7(9)**: e45814, 10.1371/journal.pone.0045814.

Wong THF and Brown RR (2009) The water sensitive city: principles for practice. *Water, Science and Technology* **60(3)**: 673–682, 10.2166/wst.2009.436.

Woods-Ballard B, Kellagher RBB, Martin P *et al.* (2007) *The SUDS Manual*, 1st edn. CIRIA C697. Construction Industry Research and Information Association, London, UK.

Woods-Ballard B, Wilson B, Udale-Clark H *et al.* (2015) *SUDS Manual*, 2nd edn. CIRIA C753. Construction Industry Research and Information Association, London, UK.

WWAP (United Nations World Water Assessment Programme) (2017) *United Nations World Water Development Report 2017. Wastewater: The Untapped Resource*. UNESCO, Paris, France. See https://unesdoc.unesco.org/ark:/48223/pf0000247153.

Ashley R, Smith B, Shaffer P and Caffoor I
ISBN 978-0-7277-4178-3
https://doi.org/10.1680/icehudp.41783.021

Chapter 2
Urban drainage systems

Richard Ashley and Brian Smith

2.1. What urban drainage does

An essential service is provided to society by managing sewerage systems to deal safely with wastewater. This function is critical to human health, flood risk management and environmental management. This is a high priority because increasing pressure on these systems is making it difficult to maintain and enhance this service affordably, and many countries have ageing infructure with limited flexibility to cope with development and climate change (OECD, 2020). Many sewerage systems combine surface water and sanitary drainage and have historically served us well. The impact of stormwater on the drainage infrastructure critically affects society and the environment through flooding and pollution. It also affects the quality of biosolids produced through treatment, reducing their effective use as a fuel source rather than a waste. There is general agreement that improving the management of stormwater is an increasing imperative in the face of the challenges relating to population growth, economic growth, urbanisation and climate change. In many countries, the effective control and management of stormwater is crucial to maintaining legal compliance and regulatory obligations to meet the social, financial and environmental challenges of the present and into the future.

Urban drainage comprises the management of sanitary or foul flows, together with storm or surface water management. As outlined in Chapter 1, any approach needs to take an integrated overview, from catchment to local scale, and consider the future as well as the present conditions. The primary contemporary drivers are fulfilling service needs – that is, effectually draining areas according to present needs – and coping with the likely changes in drivers in the lifetime of the drainage assets in use. This requires an understanding of the demand (control of excess runoff, management of wastewater generation rates and quality) and potential changes due to society's expectations and behaviour as well as environmental changes, including rainfall, runoff, regulation and habits. Increasingly, urban drainage is moving from a 'problem to be solved' (i.e. 'wastewater') to 'an opportunity' in an integrated approach to managing sources of water, resource recovery and local use as part of water cycle management as an interconnected process. However, 'wastewater' is still commonly perceived as a problem.

2.2. Terminology for urban drainage

Terminology related to urban drainage and related activities is varied and standardisation may differ between countries, jurisdictions and professions. International standards and their national implementation provide standards for practice, definitions and a vocabulary. The International Organization for Standardization (ISO) provides an online browsing platform (ISO, 2023), where keywords can be used to access standards for aspects of urban drainage practice and internationally agreed definitions of terms. The most relevant here are

- BS EN 16323:2014: Glossary of wastewater engineering terms (BSI, 2014)

- BS ISO 24513:2019: Service activities relating to drinking water supply, wastewater and stormwater systems. Vocabulary (BSI, 2019a)
- BS ISO 20670:2023: Water reuse (vocabulary) (BSI, 2023).

BS EN 16323:2014 defines urban drainage as a 'group of techniques and actions to achieve satisfactory hygiene and public heath relating to wastewater, including both foul wastewater and surface water'. Types of 'wastewater' are defined in Table 2.1. Separately listed, terms for 'water derived from rainwater' and pertaining to 'flow' are adapted from the appropriate standard above.

Table 2.1 Wastewater engineering terms, adapted from the standards listed in the text

Wastewater component	Definition
Greywater	Domestic wastewater excluding that from toilets or urinals
Blackwater	Domestic wastewater contaminated only with faeces, urine or toilet paper
Domestic wastewater	Water polluted by human living, including greywater and blackwater
Non-domestic wastewater	Water polluted by industrial, craft or commercial activities
Foul wastewater	Water comprising domestic wastewater or industrial wastewater
Sanitary sewer system	Intended to collect only water arising from any combination of domestic, institutional, commercial or industrial activities. In some countries (e.g. European practice), a sanitary sewer is referred to as a foul wastewater sewer
Industrial wastewater	Trade wastewater, trade effluent, wastewater from any industrial or commercial activity
Combined wastewater	Wastewater conveyed in a combined (sewerage) system
Septic wastewater	Wastewater subjected to anaerobic degradation
Water derived from rainwater	**Definition**
Rainwater	Water arising from atmospheric precipitation (which has not yet collected matter from the surface)
Runoff	Water from precipitation that flows off a surface to reach a drain, sewer or receiving water
Surface water	Water from precipitation that has not seeped into the ground and is discharged to the drain or into the sewer system directly from the ground or building surfaces
Groundwater	Water present in the saturated zone of subsurface strata
Stormwater	Water arising from precipitation and snowmelt

(*continued on next page*)

Table 2.1 *(Continued)*

Flow	Definition
Drainage	Operation consisting of the removal of a fluid volume
Stormwater system	For the collection, storage and conveyance of stormwater, including sustainable drainage systems (A stormwater system can include combined sewer systems or separate sewer systems, which can incorporate green infrastructure or source controls)
Sanitary sewer system	Intended to collect only water arising from any combination of domestic, institutional, commercial or industrial activities. In some countries (e.g. European practice), a sanitary sewer is referred to as a foul wastewater sewer
Infiltration (into the ground)	Movement of water into the ground
Infiltration (into the drainage system)	(Unwanted) flow into drainage or sewerage systems from groundwater
Exfiltration	Escape of wastewater from a drain or sewerage system into the ground
Flooding	Condition where wastewater or stormwater escapes from or cannot enter a drainage system
Dry weather conditions	Period during which the influence of rainfall precipitation on flow is negligible

2.3. The nature of urban drainage

The UK, in common with countries that industrialised in past centuries, has an extensive network of sewerage systems of various types. Drains and sewers are both parts of the same system and designed in accordance with the same general principles. The word sewer was in common use centuries before sewers, as they are now known, came into existence, and was applied to watercourses, both natural and artificial. Historically, many watercourses and land drainage channels were culverted or piped and then concealed, as areas became urbanised. The terms watercourse, drain and sewer occur in numerous regulatory processes and are applied to more than one type of asset. Over the years there have been many changes to the way the provision and operation of drainage systems are governed, with, for example, in the UK, different asset types covered by an assortment of legislation that affords specific powers and responsibilities to different agencies (Water UK, 2018). There are also numerous direct interfaces between sewerage systems and other societal systems, such as the drainage of highways and rural land, that interact with and impact sewerage systems.

2.3.1 Urban drainage systems in the UK

Since the latter part of the twentieth century in the UK, drainage systems for buildings, businesses, highways and paved areas have been built to drain the foul and surface water runoff separately using two sets of pipes, termed 'separate' sewer systems, as shown in Figure 2.1. There are some 350 000 km of sewers in the UK, conveying wastewater to treatment from 96% of households (Defra, 2002). A significant proportion of these sewers are combined, comprising some 36% of the sewers managed by the

Figure 2.1 Combined and separate sewer systems (Brombach *et al.*, 2005)

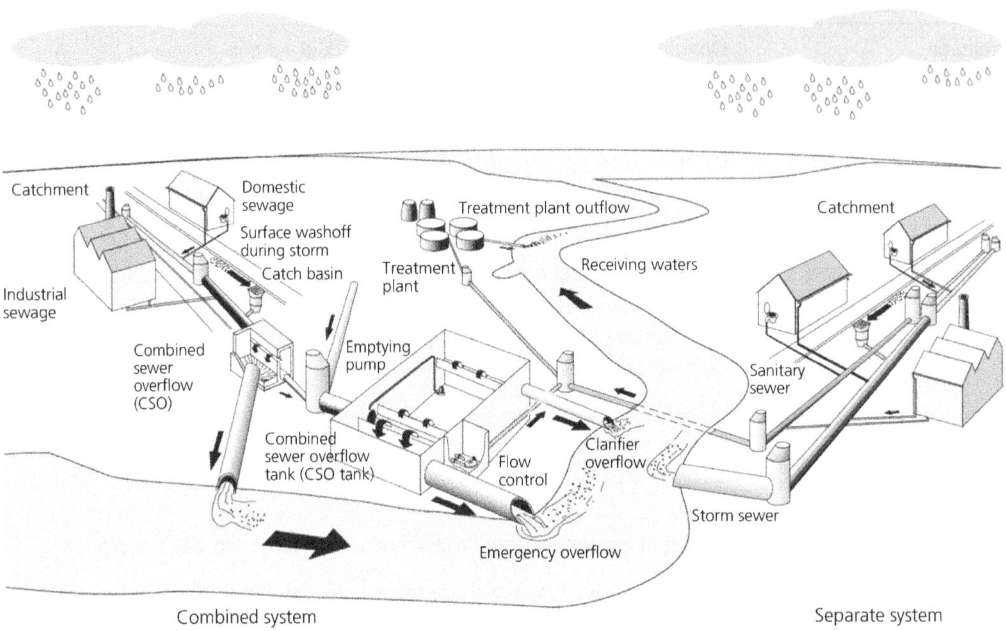

English sewerage undertaker private companies, with a further 36% being foul sewers and 28% surface water only. The term 'combined', in connection with sewerage or drainage, has always been understood to imply the conveyance of foul (or sanitary) flows together with surface water runoff in the same pipes. There are, however, drawbacks to using separate sewer systems. Table 2.2 illustrates the potential benefits and drawbacks of using separate or combined sewer systems. Brombach *et al.* (2005) provide measured data to illustrate that separate sewers do not necessarily lead to less pollution.

Despite the move exclusively to use separate sewer systems, the extensive network of existing combined sewers in the UK means that many of these are separated only locally at the top of a catchment – for example, to service a new housing estate – and subsequently connect into an existing combined system further downstream. Increasingly, surface water had been managed by direct connections into local watercourses, until it was realised that drainage runoff from roads and even houses was not 'clean' and could cause significant pollution in the watercourses due to dissolved and solid associated pollutants (e.g. SOENECS, 2021).

Wrongly or misconnected foul drainage from properties into surface water sewers also causes significant effects on receiving water bodies. Both of these problems persist, and many argue that combined sewer systems are better than separate systems, as they may not cause as great an impact because much of the flow is treated. In addition, the need for only one single pipe for the combined system can mean that construction costs are lower than when using two separate pipes. Many existing systems comprise partial separation. In this, the surface water runoff from certain discrete areas, usually paved, is conveyed into a foul sewer.

Table 2.2 Some of the benefits and drawbacks of separate and combined sewer systems (see Mannina and Viviani, 2009)

Aspect to be considered in design and operation	Separate sewers	Combined sewers
Collection of sanitary flows	Individual pipes sized for peak dry weather flow, laid to appropriate gradient	Sanitary flow is a small component in sizing pipes and selecting gradients, hence solids may deposit in dry weather
Collection of stormwater flows	Individual pipes sized for design storms, laid to appropriate gradient	Pipes sized for design storms as stormwater is the biggest component of flows collected
Flood risk management	Simpler to utilise nature-based systems where appropriate, in conjunction with stormwater pipes. Exceedance flows can be managed more easily	If pipe conveyance capacity is exceeded, flows on the surface are contaminated with sanitary flows and could be harmful
Pollution of receiving water bodies	If the separated stormwater is discharged into water bodies, this can cause significant polluting impacts. Wrong connections in which sanitary flows are erroneously connected to stormwater systems can also cause point and diffuse pollution	Combined systems require combined sewer overflows to alleviate high flows. The combination of storm and sanitary flows discharged into water bodies can cause pollution
Downstream treatment effectiveness	Wastewater treatment plant processes can be managed for optimum performance	Incoming wet weather flows dilute the sanitary flows and need to be temporarily held in stormwater tanks at inlet works. Treatment plant performance is more difficult to optimise
Construction and relative cost	Although there are two sets of pipes/nature-based systems, these can be laid to clearly calculated gradients typically shallower than for combined systems	Sizes and gradients are defined by the greater stormwater flows conveyed. Larger sized pipes need to be designed to convey solids during lower flow periods or to achieve self-cleansing conditions at daily peak dry weather flows

Associated with combined sewers in England are more than 15 000 combined sewer overflows (CSOs), many with pumping stations. In 2019 there were more than 200 000 discharges from these overflows into water bodies (rivers and the coast) (CIWEM, 2021), and in 2020 the overflows in England operated 342 346 times (Stantec, 2021). Overloaded sewers are not included in these figures, and these can also contribute to the ‘overflow crisis’ in many countries, such as the UK, as illustrated in Figure 2.2.

Figure 2.2 An informal overflow from a combined sewer draining an urban area into a tributary of the River Holme in 2021 (courtesy of Richard Ashley)

CSOs are used extensively throughout the industrialised world as a way of alleviating the need to construct sewers large enough to take both foul and stormwater flows. An approximation as to the relative magnitude of foul to stormwater flows is 1:30. Many CSOs were designed to act as a throttle to restrict combined flows downstream to only around six times the normal foul flow during dry weather, termed dry weather flow (DWF). This implies that the discharges to receiving water bodies were, and in many instances still are, at least 24 times the local DWF. The DWF is the average daily flow to a wastewater treatment works during a period without rain, comprising the collected wastewater from domestic sources plus any commercial or industrial inputs and infiltration into the system from the ground – that is, flow without the influence of rainfall.

With the development of computational models (Chapter 11) and sophisticated instrumentation (Chapter 9), the hydraulic and water quality performance of existing and newly designed sewer systems can now be better predicted than in the past and therefore managed with greater confidence. In England at least, the only constraints on managing CSO discharges are costs, institutional regulation and governance frameworks, and not technical knowledge. Stantec (2021) estimated that it would cost between £350 billion and £600 billion to eliminate all overflows from sewers in England and Wales. Much of the remediation would be to separate the combined sewer systems. Alternative less costly approaches would seek to reduce the frequency and volume of discharges, tailoring these to the capacity of the receiving water bodies. Analysis of such scenarios shows costs of between £5 billion (40 spills, on average) and £280 billion (no spills, on average), resulting in equivalent benefits of £2 billion and £39 billion. As private operators manage the sewerage services in England, these costs would be borne by householders and businesses, imposing additional charges of between £9 and £495 per annum, respectively. A further alternative approach could be to construct a new foul system, picking up all property foul flows and utilising the existing combined system as a surface water sewer. Considering the rainfall as a resource, rather than a waste, makes this utilisation attractive. In many parts of the world, the direct use of rainwater reduces the volumes needing to be drained.

The growing use of non-piped drainage systems from the latter part of the twentieth century, retaining surface water on the surface as far as practicable, has changed the perspective on the limitations of separate sewerage systems, as wrong connections become obvious, with their effects readily being seen in the stored or slowly moving surface water. However, the pace of change is different between the UK's constituent countries, with sustainable drainage systems (SuDS) and nature-based systems (NbSs) being used preferentially in Scotland and now Wales, unlike England where the provisions of the Flood and Water Management Act 2010 (HMG, 2010) were not initially fully implemented. This means that there can be greater confidence in the use of separate systems. There are other reasons for utilising separate systems, especially where the surface water is used as a resource (Chapter 8) and better facilitates the use of the foul flows, not diluted with surface water, as a resource. Of course, given the extensive existing legacy network of sewers – combined, separate and partially separate – in practice, combinations of types and mixtures of system will continue to be needed for the foreseeable future.

So far, the types of main drainage systems in use have been outlined and how this may change into the future introduced. Between the rainfall and the main drainage system are properties, which are served by property drainage, a key component in the origin and conveyance of wastewater. In this context, property drainage refers to drainage from buildings on the premises and from land occupied with and appurtenant to the buildings (termed a curtilage) (section 117 of the Water Industry Act 1991; HMG, 1991). Domestic sewerage purposes are defined as the removal of water used for washing, cooking and sanitary purposes. In urban areas, the highways that serve the premises may also drain into the same sewers that drain the premises, where this has been agreed with the relevant sewerage undertaker (Water UK, 2018).

At least in the UK, property drainage is often overlooked in government policy and planning and by the main institutional organisations. A review of responsibilities for surface water (Defra, 2020) points out the need for improved engagement between all stakeholders, but in particular notes that leading local flood authorities in England should improve their engagement with property owners, businesses and communities to help them better understand the risks and to play their part in the management of surface water. However, often property and surrounding curtilage drainage issues are largely the 'Cinderella' of urban drainage, owing to the institutional arrangements, which are outlined in Chapter 5 (focused on the UK).

Over the past few decades there has been a gradual uptake of 'more natural' drainage systems, including the growing use of green and blue systems (Fletcher *et al.*, 2014) in recognition that these provide more than just disposal of wastewater or stormwater. The use of on-the-surface nature-based drainage systems can provide numerous benefits other than just disposal of wastewater (Digman *et al.*, 2015); see also Chapters 3 and 8. Seeking and utilising these benefits should be the starting point for thinking about any urban drainage scheme, particularly in relation to their context and how they will be funded. For example, BS ISO 24536:2019 (BSI, 2019b) sets out the principles for effective stormwater management, as follows.

- Hydrology: minimise the impact of urbanisation and land management practices on the hydrology of a catchment, including base and peak flows.
- Water quality: minimise pollution entering into and discharged from the stormwater system.
- Vegetation: maximise the value of riparian, floodplain and bank vegetation for flood attenuation, erosion control and water quality improvements.
- Aquatic habitat: minimise the negative impacts of stormwater discharges on the integrity of aquatic habitats within the stormwater system.
- Stormwater use: promote opportunities to identify and use collected stormwater as an alternative water source.

BS ISO 24536:2019 promotes the use of surface water runoff as a resource, water conservation and reuse to maintain the total water cycle balance, and maximisation of opportunities to use stormwater through rainwater harvesting to supplement household, commercial or industrial, streetscape and parkland potable water needs, and also to contribute to recharging groundwater aquifers. Even these principles are limited in their recognition of the wider opportunities for stormwater management synergies. Getting the most out of stormwater is a rapidly evolving field of endeavour, spanning multiple disciplines and gaining traction worldwide. Contemporary guidance is provided by the Construction Industry Research and Information Association (CIRIA), British Water and others in the UK. Elsewhere, the US Environmental Protection Agency (USEPA) provides numerous sources of guidance and information that demonstrate current thinking in stormwater management (see guidance relating to Best Management Practices, low-impact developments and green infrastructure) (Chapter 8).

2.3.2 Design guidance

There are numerous sources of information that provide guidance on the design of urban drainage systems. The starting point is to define the sources – that is, rainfall, sanitary flows, commercial, and industrial – and then devise the most suitable drainage system. This needs to comply with national and local standards and regulations and should follow best practice. Worldwide, there are numerous sources of online guidance applicable to different types and configurations of drainage, needs and practices. Where traditional drainage is used (Figure 1.2), various best practice guides are available, notably by way of the USEPA website (USEPA, 1970, 2023). In the UK, CIRIA produces numerous guidance documents for best practice applicable to urban drainage systems (https://www.ciria.org/). Each of these, and other organisations, are continually updating their supporting information to reflect the most up-to-date practice.

2.3.3 Regulation, standards and directives

There are many national and international standards that deal with urban drainage. Regulations and standards have developed to reflect what is considered best practice. Especially important are the international standards, ISOs; European standards, DIN, which are German but also published in English; European standards, ENs; and standards issued by the British Standards Institution (BSI). Some standards are published jointly as DIN and EN, while for the UK joint BS and EN standards are common.

Directives are aimed at politicians and legislators, while standards are aimed at practitioners. Standards and directives complement each other. Standards cannot overturn legislation but do provide a framework for regulators to determine acceptable performance that does not impose unnecessary burdens. Standards provide a consistency of approach and can support higher-level policies and legislation, for example by codifying

- good practice
- best available techniques for solutions
- checklists to ensure all options are considered.

Table 2.3 shows examples of the most significant legislation, regulations and related standards relevant to drainage systems, mainly for England. Table 2.4 provides examples of sources of design guidance for drainage systems in England.

Table 2.3 Examples of the most significant legislation, regulations and related standards relevant to drainage systems

Legislation	
Water Industry Act 1991 (section 94)	Imposes a duty to 'provide, improve and extend... a system of public sewers... to ensure that the area is and continues to be effectually drained'. This explicitly indicates that the sewerage system should be improved to keep pace with growing pressures
Flood and Water Management Act 2010	Imposes a duty for water and sewerage companies (WaSCs) to cooperate with local authorities and the Environment Agency in the exercise of their functions in relation to surface water management and combined sewers
Water Act 2014	To help address resilience issues, this includes changes to some existing flood and drainage measures to encourage the uptake of sustainable drainage systems (SuDS)
Climate Change Act 2008	Legal framework to cut greenhouse gas emissions and adapt the UK to the changing climate. WaSCs are required to participate in the carbon reduction commitment and report adaptation progress to government
Environment Act 2021	Makes provision about targets, plans and policies for improving the natural environment, waste and resource efficiency; air quality; water; nature and biodiversity
Consolidated version of the Treaty on the Functioning of the European Union	TFEU article 191.2 calls for the application of the 'polluter pays' principle should control-at-source measures fail (Legislation.gov.uk, 2020)
Regulations	
Flood Risk Regulations 2009	Transpose the EU Floods Directive on the assessment and management of flood risks into domestic law to implement its provisions. In addition, they set out the duty of cooperation between the Environment Agency and the leading local flood authority
The Building Regulations 2010 (2015 edition) for use in England. Approved Document H Drainage and Waste Disposal	Provide guidance on how to meet the Building Regulations in relation to drainage and waste disposal

(continued on next page)

Table 2.3 (*Continued*)

European Directives	
EU Water Framework Directive 2000/60/EC	Promotes a new approach to water management through river basin management planning to meet water quality requirements. Stormwater management complements the catchment-based approach and is one of the responses that will help deliver compliance
Revised Bathing Water Directive 2006/7/EC	Protects bathers from faecal pollution in coastal waters
EU Habitats Directive 92/43/EEC	Cornerstone of Europe's nature conservation policy for protected sites and system of species protection
EU Freshwater Fish Directive 78/659/EEC	Aims to protect or improve the quality of fresh waters to support fish
EU Urban Waste Water Treatment Directive 91/271/EEC	The objective is to protect the environment from the adverse effects of urban wastewater discharges
EU Directive on the Assessment and Management of Flood Risks 2007/60/EC (the Floods Directive)	Assess and manage flood risks and identify areas at significant risk, with a focus on prevention, protection and preparedness
Standards	
BS 8533:2017	Assessing and managing flood risk in development
BS 8582:2013	Code of practice for surface water management for development sites
BS EN 752:2017	Drain and sewer systems outside buildings – Sewer system management
BS EN 16933-2:2017	Drain and sewer systems outside buildings – Design – Hydraulic design
BS ISO 24536:2019	Guidelines for stormwater management in urban areas
BS 851188:2019	The British standard for flood resistance products
BS EN ISO 14090:2019	Adaptation to climate change – Principles, requirements and guidelines
ISO/TR 24539:2021	Examples of good practice for stormwater management

Table 2.4 Examples of design guidance for drainage systems in England

Guidance	
Sector Guidance in relation to the adoption of sewerage assets by sewerage companies in England Water UK (2022)	An enforceable obligation on WaSCs whose areas of appointment are wholly or mainly in England to develop, in collaboration with customers, drafts of sector guidance and model adoption agreements that are subject to approval by Ofwat
Design and Construction Guidance for foul and surface water sewers offered for adoption under the Code for adoption agreements for water and sewerage companies operating wholly or mainly in England Water UK (2021)	Guidance for use by developers when planning, designing and constructing foul and surface water drainage systems (including pumping stations and rising mains) intended for adoption under an agreement made in accordance with Section 104 of the Water Industry Act 1991
Drainage and Wastewater Management Plans	As part of collaborative and integrated long-term planning, the UK water industry is producing strategic drainage and wastewater plans to maintain, improve and extend robust and resilient drainage and wastewater systems. A new way to plan for the future of drainage, wastewater and environmental water quality, these long-term plans set out the investments needed to make wastewater and drainage systems fit for today and the future. Their production will be made statutory through the Environment Act
Design Manual for Roads and Bridges National Highways (2021)	The *Design Manual for Roads and Bridges* (DMRB) is a suite of documents that contain requirements and advice relating to works on motorway and all-purpose trunk roads for which one of the overseeing organisations is the highway or road authority. CG 501 deals with the design of highway drainage systems
Designing for Exceedance in Urban Drainage (C635) CIRIA (2006)	This guidance provides good practice advice for the design and management of drainage systems to accommodate extreme rainfall events and manage drainage exceedance
Structural and Geotechnical Design of Modular Geocellular Drainage Systems (C737) CIRIA (2016)	Guidance on testing to determine strength and deformation characteristics, geotechnical and structural design and behaviour of modular geocellular structures
Retrofitting to Manage Surface Water (C713) CIRIA (2012)	This guidance sets out a process to achieve the retrofitting of surface water management measures. It integrates the principles of urban design with surface water management
Guidance on the Construction of SuDS (C768) CIRIA (2017)	Good practice guidance to ensure effective delivery and operation of the full range of SuDS components, from understanding design requirements, through site preparation and management to construction and inspection of SuDS components

This section and the next summarise key elements of the design and operation of sanitary sewerage, building drainage and main drainage. Chapter 3 considers surface water drainage in detail.

2.3.4 Developments in urban drainage

The collection and management of sewage (sanitary, foul flows) from domestic, commercial or industrial properties (trade effluent) requires a different perspective from how surface runoff and property and highway drainage can best be managed. While it is possible to utilise such flows, recovering nutrients, materials, energy and water, the need to minimise risk to maintain public health and safety needs to be paramount. As yet, the utilisation of such flows is mainly restricted to recovering the water conveying dissolved and particulate waste. Examples of this include 'Newater' in Singapore (PUB, 2022) and the different classes of water supplied by Melbourne Water in Australia. In the Netherlands the concept of the 'wastewater factory' is being developed as a means of exploiting wastewater in the circular economy. The Energy and Resources Factory recognises not only the potential to recover water from wastewater, but also the myriad of opportunities to recover nutrients and chemicals (Energy and Resources Factory, 2019). This is seen as a business opportunity in the Netherlands, with the total overall annual revenues from the wastewater recovery innovation programmes estimated at some €14 per head per year by 2030. In contrast Kehrein *et al.* (2020) review the field of resource recovery from wastewater treatment plants (WWTPs) and, although various technologies and clear opportunities are available, they point out the current bottlenecks to progress, related to economics, value chain development, environment, health, society and policy issues.

At a more local level, the utilisation of urine from domestic sources has been a key feature of ecovillages in Sweden and is being promoted as a forgotten resource to offset the global use of chemical fertilisers in agriculture (e.g. Wald, 2021). Urine separation is now seen as one of the main planks of delivering the Sustainable Development Goals (SDGs) worldwide, with claims that it may become much more prevalent even in developed countries (e.g. Larsen *et al.*, 2021).

As with all aspects of service provision and associated infrastructure, urban drainage design and functioning are based on the balance between risk and cost. Risks to public health and welfare cannot be eliminated entirely and designers and operators need to plan and manage systems that maintain the service at acceptable cost and acceptable risk. In the rest of this chapter, the risk-based approach to urban drainage provision is emphasised.

2.4. Sanitary sewer systems

Sanitary sewage comprises the sources of foul wastewater: domestic, commercial or industrial discharges. Property drains convey the wastewater under gravity to the sanitary main sewerage system, normally to centralised treatment. However, where topography or other factors inhibit straightforward gravity conveyance, pumping may be needed or local decentralised treatment (e.g. Risch *et al.*, 2021). In some areas, vacuum systems are used to overcome the gravitational limitations. Other means of dealing with sanitary sewage may include onsite management, using, for example, septic tanks or other local treatment (e.g. Crites and Tchobanoglous, 1998). Various proprietary systems are in use for the latter and guidance can be found in USEPA (2002) and a UK review in May and Woods (2015). Discussion here is concerned with the drainage systems, primarily main drainage, but also with property drainage when this has a bearing on the practices used for main drainage design and management. In industrialised countries, the owner or occupier of any premises shall be entitled to have their drains or sewer communicate with the 'public' sewers and thereby to discharge foul water and surface water from those premises.

2.4.1 The nature and movement of sanitary sewage

Sanitary sewage can comprise a wide range of substances, dissolved and in particulate form, and a range of flows and volumes to be managed. Lifestyles, domestic activity and contemporary commercial and industrial activities will create the wastewater that needs to be managed. Hence, the nature of sanitary sewage will change over time and is location-specific to reflect society. In many countries or jurisdictions there is no specific definition of what amounts to either foul or surface water. However, as an example, in England, within the context of sewerage law, 'domestic waste' is taken to mean the 'ordinary' contents of lavatories, and water that has been used for bathing, washing and cooking purposes. Because of this, textbooks on the nature of sanitary flows, such as Tchobanoglous *et al.* (2002), might not fully define the nature of contemporary wastewater. As human health needs will foreseeably require the use of water closets (WCs), showers, baths and washing facilities, each emitting wastewater, the nature of the flows will depend on more the collection of both blackwater, containing urine and faeces, and greywater.

In practice it is possible to define a baseline – for example, using Ainger *et al.* (1998) and Ellis *et al.* (2004), who provide the characteristics of DWFs in sewers, applicable to UK conditions (Table 2.5). Traditional wastewater treatment facilities have been designed effectively to manage such wastewater and also to cope with sanitary flows diluted with rainwater runoff as combined flows (e.g. Bui *et al.*, 2018). However, the effectiveness of traditional treatment at removing the wide range of potential pollutants and harmful substances is increasingly being tested, owing to the ever-changing nature of inputs.

Typical characteristics of domestic sanitary wastewater contributions are given in Table 2.6. These are indicative only and would be expected to vary for days of the week, seasonally; the flow ranges will vary between catchment areas.

Table 2.5 Typical characteristics of DWFs in sewers (from Ainger *et al.*, 1998; Ellis *et al.*, 2004)

Characteristics	Calculated	Terminology
Flow (volume/time)	$DWF = PG + I + E$	P = population connected
		G = average volume input per head per unit of time
		I = infiltration
		E = industrial or commercial inputs
Pollutant load (mass/time)	$DWL = PH + J$	P = population connected
		H = average pollutant mass per head per unit of time
		J = industrial or commercial input mass per unit of time
Illustrative examples (typical ranges)		
Flow ranges:		
Organic loads	500–1000 mg/l	Chemical oxygen demand (COD)
Solids loads	80–300 mg/l	(Total) suspended solids (TSSs)
Microbiological parameters	10^7–10^8 most probable number/100 ml	Total coliforms

Table 2.6 Typical domestic sanitary wastewater contributions (from Ainger *et al.*, 1998; Ashley *et al.*, 2004)

Flows (litre/head per day)	140–150
Flow ranges per day (multiplier on average)	0.87–1.74
Total suspended solids (TSSs) (g/head per day)	55–80
5-Day biochemical oxygen demand (BOD) (g/head per day)	55–60
Faeces (g/head per day)	100–130
Toilet paper (g/head per day)	6.4
Gross solids (g/head per day)	1
Ammonia (g/head per day)	6–7

Gross solids are the larger solids visible to the naked eye, conveyed in the sewage flow along the surface (floating), in the body of the flow or along the bed. These are faecal or vegetable matter, textiles, paper, leaves, plastics and other litter (Ellis *et al.*, 2004). Separate sanitary sewers would be expected to have very little litter and few leaves, owing to the inputs being only from WCs, bathrooms, sinks and kitchens, although commercial or industrial inputs could introduce a variety of particles. Evidence suggests that many gross solids degrade into smaller particles as they are conveyed downstream, provided the retention time is long enough (Ashley *et al.*, 2004). Many gross solids can be deposited on the bed of sewers, to be moved intermittently, depending on the flow regimes (see next).

In general, the characteristics of sanitary flows can be determined by reference to standards and guidance (such as Table 2.3) but need to be tempered with up-to-date knowledge about the origins of such flows and behaviours associated with wastewater generation and disposal. Box 2.1 shows an example of the relative distributions of pollutants conveyed in a former combined system compared with partial separation.

Sewerage systems are vulnerable to discharges that are, at times, difficult to control. As there is a constantly changing societal need for and use of materials and consumer products, what is put into the wastewater system, even from domestic sources, can vary from day to day. It is therefore important that sewer systems are allowed to function for their primary purpose, as a means to maintain public health and welfare, especially reducing the incidence of potentially fatal waterborne diseases. Unfortunately, sewer systems have come under increasing pressure in the past few decades, owing to the inappropriate disposal of a variety of materials. This is commonly referred to as 'sewer abuse'. It is, therefore, important to keep up to date with what is being discharged.

The main areas of concern regarding substances being disposed of and conveyed by sewers are perfluoroalkyl and polyfluoroalkyl substances (PFASs, also known as 'forever chemicals'), poly-aromatic hydrocarbons (PAHs), personal care products (PCPs), fats, oils and greases (FOGs), microplastics, 'flushable' products and healthcare waste. Collectively, these are a major cause of operational problems that impact the economic, social and physical environment through increased operational costs and poor regulatory performance. In the following are some examples of current substances of concern being introduced into sanitary sewage and sewers. All of these are also found in separated surface water drainage systems (Section 3.2.1.2).

Box 2.1

Illustration of the comparative nature of sanitary, stormwater and combined sewage flows and the consequences for sewer separation

The No Rainwater In Sewers (NORIS) EU Interreg North Sea Region project looked at sewer separation for a number of catchments (Keep.eu, 2023). Linden is a catchment in Hanover with residential and commercial areas as well as open spaces. In the 1970s, the city started to separate the existing combined system in the area, to reduce CSO spills to the rivers Ihme and Leine. The original total impervious area drained was 2.4 km^2, composed of 0.7 km^2 roads, 0.8 km^2 roofs and 0.9 km^2 other surfaces, serving a population of 52 305 population equivalent (pe). Some of the system was separated, leaving 0.6 km^2 of road surfaces, 0.2 km^2 of roofs and 0.3 km^2 of other surfaces connected. The population served by the sanitary system was assumed to remain the same in the study.

Model studies were undertaken to assess the benefits of the sewer separation in terms of pollutants discharged to the receiving watercourses. These used bespoke models, STORM for flows and SEWSYS for substance flow analysis, both developed at Chalmers University of Technology in Sweden (Eliasson and Ahlman, 2006). As part of the project, filters were fitted into road gullies to trap pollutants (Figure 2.3). The INNOLET system shown achieved reduction efficiencies of 20–50% for different pollutants (Sommer and Sieker, 2005). Model analyses were carried out to assess

(*a*) the pollutant loads with the original combined sewer system in place
(*b*) the pollutant loads from the separated system, for both the modified combined system, with reduced surface water, and for the separated stormwater flows.

Figure 2.3 The INNOLET filter for highway gullies

Measurements of flows and loads were made and the historical data were used to develop catchment pollutant models. Table 2.7 shows the modelled data for one year for the inputs to the treatment plant, based on 9 years of records for the combined system, as it was before separation. Figure 2.4 shows the distribution of pollutant loads from the treated effluent and CSO spills for the original system.

The modelled results for the distribution of pollutant discharges for the separate system are shown in Figure 2.5, with Figure 2.6 showing the comparative pollutant loads discharged.

Table 2.7

Characteristic	Into treatment plant		
	Total	Stormwater	Sanitary
Flow (m^3)	4.33 million	0.9 million	3.5 million
Total P (kg)	38 200	84.6	38 100
Total N (kg)	264 000	2620	261 000
Cu (kg)	208	45.5	162
Zn (kg)	488	126	362
Pb (kg)	25.7	9.3	16.4
Cd (kg)	1.32	0.45	0.86
Polyaromatic hydrocarbons (PAHs) (kg)	10.1	0.57	0.45

Figure 2.4 Distribution of the pollutants discharged to the watercourses for the original combined system

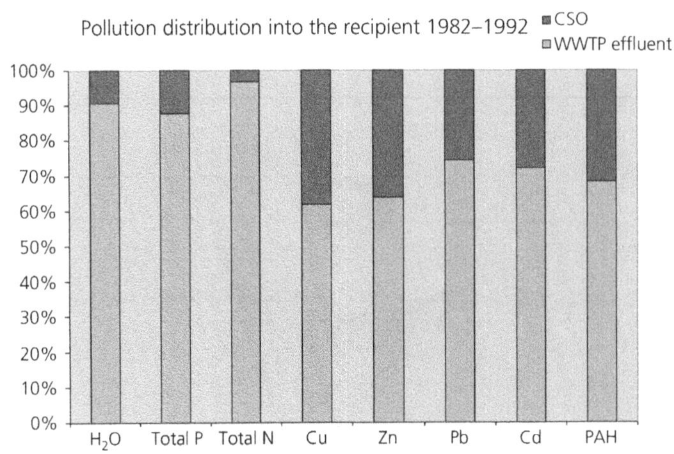

Sewer separation reduces the nutrient loads discharged, owing to a significant reduction in CSO spills and better treatment of the sanitary flows. However, loads of heavy metals and polyaromatic hydrocarbons (PAHs) increase for the separated compared with the combined system. Heavy metal loads increase by some 17–44%, and PAHs by 22% as the stormwater is no longer treated. The installation of INNOLET filters would reduce the heavy metal loads to the same as before separation, but not the PAH load. The overall message is that sewer separation can be a mixed blessing. The conclusion from the NORIS project was that separated stormwater systems need treatment, and not only filtration, but also using other processes, including oxidation (see Chapter 8).

Figure 2.5 Distribution of the pollutants discharged into the watercourses for the new separated system

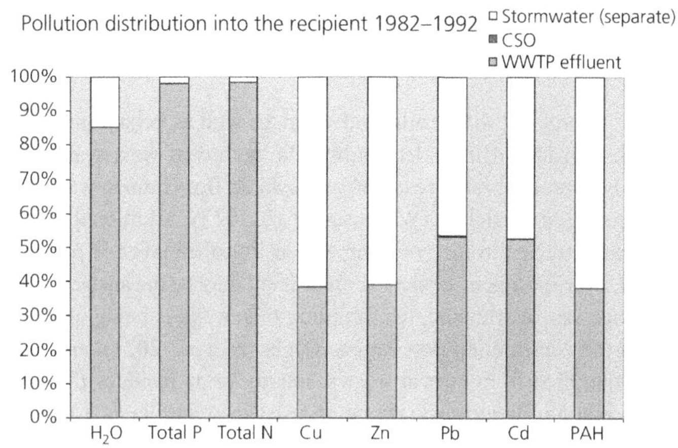

Figure 2.6 Comparison of the distribution of the pollutant loads being discharged into the watercourses for the original combined and new separated systems

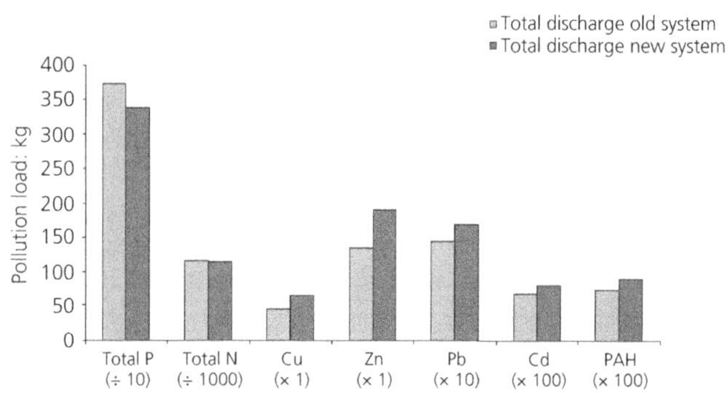

2.4.1.1 PFASs, PCPs, endocrine disrupters, pharmaceuticals and recreational drugs and contaminants of emerging concern

There are at present almost 5000 PFA substances, which are now ubiquitous in human beings and other living organisms and ever present in the natural environment (Salvidge and Hosea, 2023). They are bioaccumulative, and persist in the environment as they break down only slowly. They are known to damage human health (NIH, 2019) and are difficult to remove, not being amenable to conventional wastewater treatment (Speth, 2020).

Endocrine disrupters and pharmaceuticals are some of the other biggest risks. Endocrine disrupters may be present in river water, potentially disrupting hormone systems in both human beings and animals. While wastewater treatment processes are currently effective in preventing such chemicals from reaching drinking water supplies, it is important for the water industry and environmental sectors to work and act together to assess the impact from these chemical substances getting into rivers and the environment in general. If wastewater treatment plants are to embrace circular economy principles and transition into 'resource' factories, as seen in the Netherlands, hazardous substances must be removed at source. Source control is the most effective way to manage such substances, as many cannot be removed during the treatment process and may be released into the aquatic environment.

The current use of PCPs, detergents, soaps, oils, and so on, as well as behaviours in toilet use, washing and cleaning, need to be considered in understanding the nature of wastewater from these sources. Pharmaceuticals and medical products are present in wastewater, from human waste inputs, and also due to the habit of flushing unwanted medicines (Wilkinson *et al.*, 2022). Additional measures to control the levels and types of pharmaceuticals in the environment, as a further layer of protection to people and wildlife, may be required. More stringent wastewater treatment may be the answer, but this would clearly add extra costs to the processes. Worldwide, the sampling of sewage is being used to find evidence for COVID-19 infections in the contributing populations (Dobson *et al.*, 2021), and for recreational drug taking, as these can be identified from human wastes, mainly urine residues (Sulej-Suchomska *et al.*, 2020). Wastewater epidemiology is increasingly being used in Australia as a means of looking more broadly at community health, as part of an early warning public health surveillance programme. A good example of this is a programme to look at the health of the whole community as part of a drugs surveillance programme, which involved developing maps showing levels of cocaine, heroin, alcohol, drugs or metabolites, pharmaceuticals and anti-depressants. This is a new and different way to measure community health. This process is also being used to provide early warnings of polio outbreaks in countries trying to eradicate the disease.

There are many contaminants of emerging concern (e.g. Water Research Foundation, 2022) and as the evolution of chemicals in use in the manufacture of various substances continues at pace, it is difficult to keep up with the challenges of their disposal (e.g. Biel-Maeso *et al.*, 2019). The time lag in identifying emerging pollutants is not expected to change in the future (EU, 2016):

> As technologies for monitoring new pollutants take time to develop and might be expensive during some period, it is assumed that monitoring technology will not drastically change over time. This implies that pollution by emerging pollutants in areas with intensive economic activity (agriculture, industry) and population density will take time to be identified, posing potential health risk to local population.

It is clear that even when municipal wastewater is treated using traditional methods, a variety of chemically and biologically active substances are still being discharged into receiving waters worldwide (Wilkinson *et al.*, 2022).

2.4.1.2 Fats, oils and greases and wet wipes

There are approximately 200 000 sewer blockages throughout the UK every year; around 75% are caused by fats, oils and greases (FOGs) (Arthur and Blanc, 2013), originating mainly from commercial premises, but also locally from domestic disposal by way of the kitchen sink. Behaviour is mainly due to lack of awareness among household customers of the way in which FOGs should be disposed of, the difficulties and effort required in recycling FOGs (compared with disposing of them down sinks and drains), and lack of understanding of what the impact could be of disposal to the sewer system. Changes in waste legislation in the UK, making it illegal to recycle animal foods, are thought to have led to an increase in disposal of FOGs to sewers.

The problems often appear greatest where sewer systems serve a proportionally high number of catering establishments, but some proportion also comes from domestic customers. Ethnic, regional and socioeconomic factors play a part, such as cooking styles, health and dietary attitudes and housekeeping methods, as well as the cumulative volumes of FOGs produced by millions of wastewater emitters each day.

Many countries have legal instruments designed to control such inputs. In England, restaurants could be regulated under trade effluent legislation (Water Industry Act 1991) but this would require all discharges from catering establishments to be treated as trade effluent as the legislation is based on process and not the impact on the sewerage system. The Water Industry Act 1991 only requires that the discharge is likely to interfere with the free flow of its contents, not that it actually has caused a blockage, which is not easy to prove in a court of law. Foden *et al.* (2017), in reviewing the issue and potential responses to change public habits, saw the problem as one needing to be tackled at source, despite the legal requirements in England not to dispose of FOGs into the main drainage system.

Alongside FOGs is the ubiquitous use of wet wipes in the twenty-first century and the erroneous belief that these can be flushed into the WC (promoted by the manufacturers) as they are promoted as 'biodegradable' (e.g. Durukan and Karadagli, 2019). This has contributed to blockage problems in sanitary sewer systems in the industrialised world (e.g. Okwori *et al.*, 2020). The disposal of products labelled as 'flushable' is a growing issue, with the number of sewer blockages caused by wet wipes increasing. The issue is not solely related to wet wipes. Similar issues arise through disposal of products such as feminine panty liners, cotton buds and other plastic-based materials. In the UK, a study on behalf of Water UK (2017) showed that wet wipes were the major cause of blockages in sewers (both sanitary and combined sewers) and at wastewater treatment plant inlets. Disposal of wet wipes into WCs is technically illegal, in respect of section 111 of the Water Industry Act 1991, as the disposal of clothes/woven materials should be by way of the solid waste route. There is also a related issue with regard to product labelling. People are confused by the mixed messages, with many products being labelled as flushable and industry advice that these same products should not be disposed of to sewers. There are even reputable industrial sources in the UK that promote flushing of certain 'flushables' such as wet wipes (e.g. Water UK, 2022). This only leads to confusion in the public domain and is unhelpful in controlling the problem. Numerous studies recommend strengthening public awareness campaigns to stop flushing and also that manufacturers should be stopped from using non-biodegradable materials.

2.4.1.3 Microplastics

Globally, some 22 Mt of plastic materials leaked into the environment in 2019. Microplastics, polymeric particles with a diameter smaller than 5 mm, accounted for 12% of the total, from, for example, tyre abrasion, brake wear or textile washing. Some 109 Mt of plastics have accumulated in rivers (OECD, 2022). These micro- and nano-plastic particles (smaller than 5 mm) are ubiquitous and particularly prevalent in sanitary wastewater, although significant amounts also come from stormwater runoff. A Slovenian study found that 15.2 mg/person per day of microbeads were being input to the wastewater system, with concentrations of up to 21 particles/m^3 in receiving waters downstream (Kalcíkova *et al.*, 2017). Only limited studies have looked at the origins and behaviour of these microplastics in wastewater networks (e.g. Prata, 2018), preferring to concentrate on understanding the make-up of the PCPs and other products being used domestically and the microplastic content, together with the impacts and effectiveness of microplastic removal in traditional wastewater treatment plants, although there is evidence linking wet wipes and sanitary towel flushing to microfibre plastics (e.g. Ó Briain *et al.*, 2020).

Prata (2018) suggests that some 16–86 t/year of microbeads are 'released' by the UK population from their use of hygiene products, especially from washing machines. This is presumed to enter the wastewater network. In a Finnish study of the removal in an 800 000 pe conventional WWTP, only 0.1% of the microlitter (smaller than 5 mm) entering the plant was released in the effluent and the rest was removed with the sludge. Twenty per cent of the microlitter in the removed sludge returned to the process with reject water, to undergo further separation (Talvitie *et al.*, 2017).

The behaviour and conveyance of these particles in the sanitary sewer network has not been extensively studied; it is presumed that these are conservative estimates and what enters the system arrives at downstream treatment plants. These may contaminate sewage sludge/biosolids, preventing their use in agriculture and the circular economy. There may be an increase in abrasion of pump impellers where the concentrations of plastic particles are high. However, there are no reported studies of this.

2.4.1.4 Food waste

Waste management is an increasingly important issue. The European Landfill Directive (EC, 1999) encourages practices that divert waste streams away from landfill and, while there can be little disagreement that this is a prudent long-term approach, the implications of alternative disposal routes has become more evident to wastewater managers.

In terms of 'waste' there is a renewed emphasis on resource utilisation – that is, there are no wastes, rather an opportunity for material, energy and other resource exploitation. This is the principle of the so-called circular economy and is applicable to food wastes (KPMG, 2020). However, understanding the place of and the best way to manage food wastes in this context is weak and there are ongoing projects to seek to develop the required knowledge about the best way to manage food wastes in a given context. Much of the current discussion is emotive and based on tradition, practice and sectoral interests. In many jurisdictions food waste routes are mandated and do not necessarily reflect the circular economy perspective.

One of the alternative food waste management routes of interest here is disposal to sewers through an increased use of domestic macerators (food waste disposers; FWDs). This proliferation and greater use is of much concern to the water industry in the UK and Europe. Food waste, and particularly macerated food waste, would not appear to be specifically covered by the definition of sewage or its component parts. Householders have no express rights to use food waste disposers and it is considered by the sewerage undertakers in many countries (and several governments) that the waste produced would not legally constitute domestic waste and should not be disposed of to the sewers. Through national and local legislation control, many EurEau member states prohibit or discourage discharge of macerated kitchen waste to sewers (EurEau, 2020). This is because traditional sewerage systems were not designed to be used in this way and there is a risk of blockages, gas build up and accretions of fats, oils and greases. This trend to use FWDs has been encouraged not only by commercial marketing but also by municipality promotion, although in the USA FWD usage in homes has been the norm for decades. Municipalities in industrialised countries see FWDs disposing to sewers as a means of reducing the amounts of municipal solid wastes they need to deal with. This is a very serious problem in developing countries, as illustrated in Figure 1.6. Where FWDs are used, it is argued by opponents that, in addition to in-sewer problems, this can increase energy-intensive wastewater treatment needed at works to ensure compliance with discharge consents.

However, there is evidence that food wastes conveyed in sewerage systems also provide a potential means of supplying nutrients to enhance downstream treatment and energy recovery (e.g. Legge, 2022). For example, in Los Angeles there is a concerted programme to reduce food waste going to landfill that includes retrofitting domestic FWDs across a test sewer catchment as a means of recovering energy at treatment plants downstream (LACitySan, 2022). In Helsingborg, Sweden, the 'new city' has a three-pipe drainage system, with one system specifically to convey macerated food wastes to the treatment plant and a separate sanitary sewer for other domestic wastes (Run4Life, 2022).

In England, the Environment Agency does not support the large-scale use of FWDs for kitchen wastes, except where sewage infrastructure is specifically designed to manage and recover value from such waste. Having considered the matter at policy level, the European Commission indicated that solid biodegradable waste should not be shredded and discharged to the sewer because this would bring about an unjustified increase in sewage sludge.

Anaerobic digestion is seen as one of the best ways to treat organic waste. Ideally, a joined-up approach between municipalities, who can collect separated domestic food waste, and the wastewater industry, which already has the basic infrastructure to treat the waste, is an opportunity to contribute to the circular economy. By moving to segregation of waste at source and using the putrescent fraction for energy/fertiliser production, this would also satisfy the aims of the European Landfill Directive.

In the Netherlands, for example, the challenge of managing food wastes in dense housing areas, especially from high-rise apartments, is encouraging the installation of domestic FWDs, but rather than dispose to sewers, basement tanks are being fitted to collect the arisings. Inevitably, this requires the wastes to be trucked to anaerobic digestion plants, where co-digestion is seen as the most effective means of managing such wastes (STOWA, 2015).

As with the other wastes highlighted in Section 2.3.1, food waste disposal is a sector that can potentially influence the performance of sewerage systems. Whether or not sewers have a role to play in food wastes as a part of the circular economy will depend on context and improving our future understanding of the potential impacts from so doing.

2.4.1.5 Healthcare waste

Globally, the healthcare industry impacts the lives of many people and the environment. Healthcare organisations undertake many activities that do, or have the potential to, result in the discharge of various items and substances to sewerage systems. Waste materials are a by-product of medical activities, with establishments generating a vast array of both non-hazardous and hazardous waste streams. The disposal of wastes into water systems is known as 'wastewater discharge' and is subject to conditions and controls set by the sewerage undertakers in whichever jurisdiction these occur – for example, see Water UK (2014).

In the 1990s, the UK National Health Service (NHS) management introduced a policy for the installation and use of macerators to dispose of incontinence pads and disposable bedpans, etc. The policy was introduced as part of infection control to combat the record-breaking increase of antibiotic-resistant infections, which was proving difficult to control, and the increased cases of methicillin-resistant *Staphylococcus aureus* (MRSA) and the highly virulent new strain of *Clostridium difficile*. A key focus was to reduce the volume of clinical waste due to substantial disposal costs.

Manufacturers of healthcare pulp products state that their systems are in use in 94% of UK hospitals and more than 50 countries worldwide. Such systems currently serve six continents and over 5000 sites across the globe, with the manufacturers promoting macerators that provide for the disposal of single-use pulp products through the existing sewer system, including disposal of bedpans and urine bottles. Various manufacturers produce and promote macerators as an alternative for waste discharges, with pad and incontinence product disposal systems that can reduce the municipal wastes going to, for example, landfill. These units disintegrate pads and nappies and convert polyethylene, non-woven products and waste into fine slurry for disposal to the sewerage system.

As well as there being no legal right of discharge to sewers for this type of waste stream in the UK, there is also the potential of introducing to the environment harmful pathogens. While single use, disposable pulp products for use in healthcare may be manufactured to PAS 29:1999 (BSI, 1999), the water industry does not deem them suitable for discharge to the sewerage system. The sewerage undertaker view was that cardboard bed pans and urine bottles might well constitute clinical waste and be subject to a regime of special controls. Clinical waste is defined in The Controlled Waste (England and Wales) Regulation 2012 (HMG, 2012) as follows

> 'clinical waste' means waste from a healthcare activity (including veterinary healthcare) that —
>
> (*a*) contains viable micro-organisms or their toxins which are known or reliably believed to cause disease in humans or other living organisms,
> (*b*) contains or is contaminated with a medicine that contains a biologically active pharmaceutical agent, or
> (*c*) is a sharp, or a body fluid or other biological material (including human and animal tissue) containing or contaminated with a dangerous substance within the meaning of Council Directive 67/548/EEC on the approximation of laws, regulations and administrative provisions relating to the classification, packaging and labelling of dangerous substances, and waste of a similar nature from a non-healthcare activity.

As such it would be 'controlled industrial waste' and also susceptible to control under the Environmental Protection Act 1990 (HMG, 1990).

A related concern is that of care homes installing macerators as a one-off capital cost to dispose of incontinence pads to sewers and reduce operational expenditure for clinical waste disposal and incineration.

2.4.1.6 Sanitary sewer overflows

Combined sewer overflows are considered in Chapter 3. In many places worldwide there are sanitary sewer overflows (SSOs), not to be confused with direct wastewater discharges into the environment, which are also common. Widely prevalent in the USA, SSOs were built as 'relief valves' to reduce the costs of sanitary sewer construction in the nineteenth and twentieth centuries and are still widely used there (USEPA, 2022). Some 23 000–75 000 SSO spills were occurring annually across the USA in 2004 (USEPA, 2004). There are clear links between the prevalence of SSO discharges and gastroenteritis infections in the population in parts of the USA (Jagai *et al.*, 2017).

Concentrating on the problems of discharges of engineered nanomaterials, Loosli *et al.* (2019) highlight the difficulties faced by those managing sewer networks where there is a legacy of infrastructure that presumed receiving waters could assimilate whatever was spilled, from both SSOs and CSOs, and relating to all uncontrolled sewage discharges.

SSOs are no longer seen as acceptable, albeit they will persist even in industrialised countries such as the USA due to the costs of their closure. The move to utilise green or NbS options to manage stormwater on the surface can, by reducing infiltration and inflow (see Section 2.5.2.4) into sewers, help to reduce the need for SSOs, but such retrofits cannot deal with the entire problem (e.g. Shamsi, 2018).

2.4.2 Responding to the ever-changing nature of sanitary sewage

Recognition of the threats from the variety and ever-growing range of substances being input to sanitary sewer systems is focused on the treatment processes for centralised plants, rather than the conveyance in sanitary sewers. For example, in the UK the Chemicals Investigation Programme has been investigating these threats for more than a decade, initially in response to European legislation requirements and aimed at protecting receiving water bodies (*Water Industry Journal*, 2022). The extensive investigations covering some 600 treatment plants across the UK have so far revealed that sewage flows are dominated by domestic wastewater but also contain some specific trade effluents (FWR, 2018). The performance of existing treatment plants is broadly effective for most substances, but in order to control risks now and in the future and avoid costly upgrades to treatment, more needs to be done in influencing manufacturing and treatment at-source. The overall challenge is one of 'problem substances', rather than problem locations or problem treatment plants. More information as to the most problematic substances can be found in Comber *et al.* (2018), who especially highlight that ethinyloestradiol, diclofenac, propranolol, the macrolide antibiotics, fluoxetine, tamoxifen and carbamazepine are not being effectively removed by conventional wastewater treatment processes.

Conceivably, illegal or accidental disposal of corrosive substances into sewers from domestic or industrial sources could contribute to the decay of the sewerage system and cause damage to pumping stations and other ancillaries.

There is very limited evidence that domestic or other users of wastewater systems have any awareness of the implications of what they flush. Hence trying to influence behaviour is difficult. In a recent study, Karadagli *et al.* (2021) found that awareness of potential harm as to what was being flushed by individuals in Turkey depended on the individuals' socio-economic status and their level of education. In the UK, years of 'bag-it-and-bin-it' campaigns have had only limited success in changing behaviours to stop flushing anything other than the three 'Ps': poo, pee and paper. The alternative 'think-before-you-flush' (e.g. Clean Coasts, 2022) campaign, adopted by Irish Water, has the same aims but is founded on detailed research that revealed that the latter slogan was much more effective than bag-it-and-bin-it as a behaviour change vehicle (e.g. Ashley *et al.*, 2005; and the case study in Chapter 14, Section 14.11).

The innovative management of sanitary sewerage systems as part of a circular economy perspective is very dynamic. Sotelo *et al.* (2021) consider how the sewer network itself can be better utilised as a part of the treatment process, concluding that this may be suitable for relatively low-density suburban communities (100 inhabitants/ha), provided the sewer network is long enough. Some 50% reduction in conveyed organic load can be achieved by relatively modest alterations to ensure that the sewage is in contact with the biofilm layers on the pipe walls, helping with pre-treatment. Nagpal *et al.* (2021) review the potential for the recovery of heat from sanitary sewage using heat exchangers and heat pumps. It is concluded that wastewater heat recovery is feasible at all stages of the generation and conveyance of wastewater, within the domestic situation from greywater and further downstream in the sewer network, where ambient temperatures typically range from 10°C to 25°C throughout the year.

In summary, drainage practitioners may need to be prepared to manage a wider range of substances, large, small and in solution, being input to sanitary sewers, along with variable flow rates and volumes. However, this is likely to require a change in legislation and charging regimes. The inputs reflect societal cultures, norms and behaviours, as well as the state of economic development and innovations. Consumerist cultures would be expected to utilise the greatest range and most up-to-date of PCPs, pharmaceuticals and other potentially flushable materials, and more self-reliant communities to emit the least blackwater or greywater. Changes in perspective are now coming about that see the value of managing inputs by way of, for example, novel WC technologies, using the sewer itself as a potential for resource recovery, like heat, and also as a means of enhancing the overall treatment process. This complies with the current thinking that is promoting a circular economy perspective for all aspects of human living.

Although this section has considered in some detail the nature of sanitary sewage, the prevalence of combined sewers means that whatever may be conveyed in sanitary sewers will also be found in combined systems, as discussed further in Section 3.5.

2.5. Design and operation of sanitary sewer systems

2.5.1 Building drainage

Here a brief introduction to the fundamentals of building drainage design is given, focused on the flows.

Domestic, commercial and industrial sanitary inputs are used to define the form and layout of building drainage. In the UK, BS EN 12056-2:2000 (BSI, 2000) provides standards for calculating the relevant flows from various water uses in buildings and dwellings, including appliances, into a gravity drainage system. This uses the concept of discharge units, in which the wastewater flow rate is determined from the sum of the various appliances and other domestic units scaled to account for the frequency of use of the individual units, such as a washing machine, dependent on the nature of the dwelling or building. For example, in a hospital, the use frequency is 0.7, while in a domestic property it is 0.5. The discharge unit can vary up to 2.0 for the WCs dependent on the system installed. There may be other flows, some pumped. Design should accommodate the greatest flow rate, or the highest flow rate from the appliance with the individual biggest flow rate. The Colebrook–White equation (Science Direct, 2022) is recommended for estimating pipe capacity. This method is recommended for systems with connected populations of up to 1000.

Building drainage will experience rapidly varied flows, pipe full or pressurised flow conditions and air entrapment and movement in the branched networks. These conditions are especially likely to occur in high-rise building drainage. There is an added complication of gross solids and their transport along the drainage network. Transport is intermittent and 'packs' of toilet paper and other textiles can accumulate. Design should follow the rules for connections and line and level in BS EN 12056-2:2000 in order to ensure effective solids transport. Consideration needs to be given to the possible changes in water use and domestic behaviours over the lifetime of the drainage system.

2.5.2 Main drainage

2.5.2.1 Gravity sewers

Where practicable, sewers and drains need to be designed to convey flows by gravity. Only where the gradients are too shallow to ensure effective conveyance of flows and solids should other systems be considered. However, where gradients are steep, pumping stations may be needed to avoid flow velocities that are too high.

Outside of buildings, for UK applications, BS EN 752:2017 (BSI, 2017a) sets out how to design the connecting drains and main sanitary sewer system. When these are to be transferred into ownership of the local sewerage undertaker from a private developer, as in the UK, the designs formerly needed to follow the principles and details in *Sewers for Adoption* (SfA) for England or Sewers for Scotland (SfS) – for example Scottish Water (2018). Each of the UK constituent countries has unique, albeit overlapping, standards for sewer adoption, and in England these have been replaced by *Design and Construction Guidance for foul and surface water sewers offered for adoption* (Water UK, 2019, 2021). Irrespective of adoption, the property drainage up to the curtilage of the property will remain the responsibility of the property owner. Although in England, where a sewer is shared by more than one connected property, the private sewer and any associated assets, including pumping stations, are now the responsibility of the sewerage undertaker. The connection between the property drainage and the main sewer, that is, between the private and public networks, is termed the 'lateral drain'.

The design and construction guidance for England states that the flow rates should be designed in accordance with BS EN 12056-2:2000, as set out previously, or 4000 l/dwelling per day (0.05 l/s per dwelling) as a peak flow rate. Even reducing water consumption will not necessarily diminish the peak flow rate, although the average flow will change. Industrial or commercial units should be designed to accommodate the domestic flows as well as the trade effluents, and should be based on a metered discharge from premises or assumed as 0.5 l/s per hectare for normal industry, 1 l/s per hectare for wet industry. When the proportion of wet industry is unknown, an average flow of 0.7 l/s per hectare should be used.

Drains should be designed to operate at 75% of hydraulic capacity, although minimally sized and commercially sized pipes should be used that may provide greater than this capacity. Peaking factors are suggested as four to six times dry weather average flow, based on a discharge of 220 l/head per day (Kellagher, 2003). The latter provides extensive supporting information for sanitary sewer design for new developments.

BS EN 752:2017 references BS EN 16933-2:2017 (BSI, 2017b) for the hydraulic design of drains and sewers outside of buildings. Where connected populations are such that the effects of individual appliances on the variability of flow rates are minor, presumed to be where more than 500 are connected, local per-capita water supply data can be used to estimate the average and peak flow rates. In the design, a minimum flow velocity of 0.7 m/s should occur once per day in order to ensure movement of solids through the sewer system. Alternatively, a minimum pipe gradient of 1/(the pipe diameter in mm) should be used.

Where sanitary sewers are large or need to be laid to steep gradients to achieve these self-cleansing conditions, a more detailed analysis of solids transport could potentially reduce the gradients needed and hence save on excavation costs. This is further considered in Section 3.5.

2.5.2.2 Alternatives to gravity sewers

There are alternatives to gravity sewer systems, which typically require minimum gradients of at least 2%. These alternatives include vacuum sewerage and pumped (rising) mains. As indicated previously, gravity sewers may also require pumping stations, especially where the topography is shallow or steep, to avoid excessive excavation. Although some gravity systems may require pumping stations, especially where gradients are shallow or very steep, vacuum and low-pressure systems depend on pumping. Typically, the latter will have smaller bore pipes than for the equivalent gravity sewers.

Vacuum sewer systems are not new. They were first installed in Europe in 1882. Vacuum sewerage has gained popularity in the past decades, particularly in regions with flat topography, high water tables and poor ground conditions, such as East Anglia in England. They may also be used to minimise excavation or where the ground is rock and costly to excavate (Islam, 2017). Low-pressure systems may be installed in domestic properties by using grinder pumps, and in domestic or commercial properties by using small-bore pipe connections to the main vacuum network. In isolated areas, vacuum systems may be connected to on-site septic tanks.

Figure 2.7 (Mohr *et al.*, 2018) shows how vacuum sewers utilise air pressure differential and convey a mixture of air and water (wastewater). Vacuum pumps at downstream locations produce the negative internal pressure to draw wastewater along the sewer from chambers or pits where it has collected flows enter the system. As they are airtight, vacuum sewers do not suffer from the infiltration or exfiltration problems endemic in gravity sewers. Any leaks are rapidly identified from the changes in vacuum pressure that result. However, vacuum systems experience greater problems than gravity sewers, owing to failures of the control valves and problems at the collection pits. Offsetting this is the rapidity with which failures are identified and the ease of rectification in vacuum systems (Mohr *et al.*, 2018). Comparing gravity and vacuum systems, Islam (2017) points out that the latter are useful where separate conveyance of blackwater is required. For example, in Helsingborg, Sweden, a major city expansion (Oceanhamnen) is utilising vacuum systems to collect blackwater separately, while, in a separate system, kitchen wastewater is connected to a low-pressure pipe, each in order to recover the nutrients and energy

Figure 2.7 Example of a vacuum sewerage system (adapted from Mohr *et al.*, 2018)

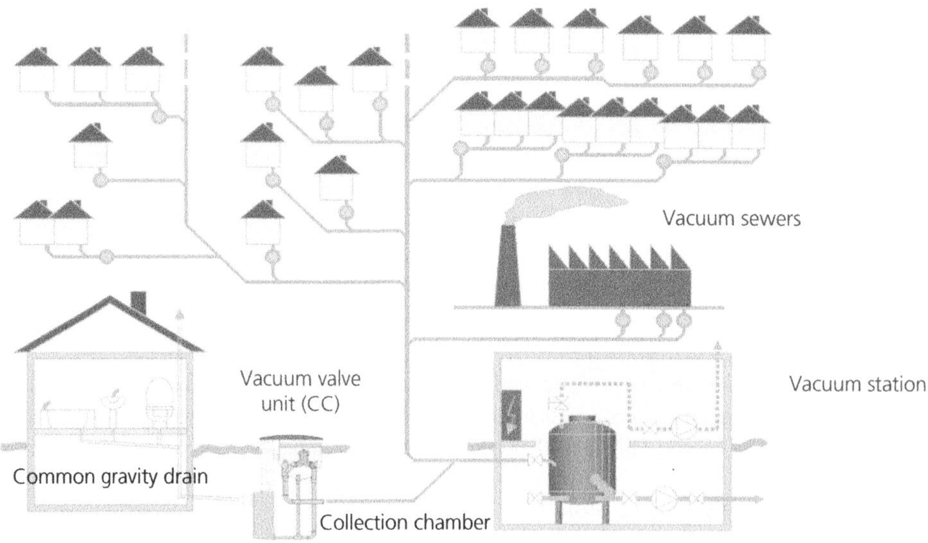

Overview of a vacuum sewer system involving the vacuum valve unit at the valve pits, vacuum mains and the central vacuum station

Table 2.8 Comparisons between aspects of vacuum and gravity sewerage (adapted from Islam, 2017)

Characteristic	Vacuum sewerage	Gravity sewerage
Pipe size	Comparatively small	Comparatively large
Construction	Shallow and narrow trench, easy to excavate	Large trench may be at significant depth
Obstacle crossing	Straightforward	More difficult
Cleaning	Minor	May need flushing of sediment
Costs	Capital lower	Capital higher
Treatment of conveyed wastewater	Concentrated and optimal	Typically diluted with infiltration – less efficient
Performance monitoring	In-built to system by way of valve operation observation	Usually fit and forget
Control of flows	In-built	None

conveyed (Run4Life, 2022). The vacuum system is connected to low water use (1 l) vacuum toilets. Table 2.8 provides some comparative information when considering gravity or vacuum sewerage alternatives; further details are given in Islam (2017).

Looking at Table 2.8, it could be concluded that vacuum sewerage is preferable to gravity. However, as pointed out, these systems are more reliant on active control systems, necessitating numerous vacuum pumps to maintain operation and continual use of energy, and hence incur carbon impacts. They also depend on the reliability of the electronic control valves throughout the network. Mohr *et al.* (2018) suggest that where new sewerage is required in rural areas, vacuum systems may be more cost-effective and efficient. This may be why vacuum systems are selected where a major expansion of a settlement, like Oceanhamnen above, is planned.

BS EN 16932-3:2018 (BSI, 2018) includes guidance on vacuum and pumping systems and specifically deals with vacuum systems and their design. Design depends on the relative changes in height of the system (highest to lowest point in a pipeline that rises and falls with the ground level), and on the air/water mix, which depends on connected population density and pipeline length. Islam (2017) provides further design and construction recommendations, including for pipe bedding and any associated gravity sewerage.

Pumping stations are relatively common, even in gravity sewer networks, whereas vacuum sewers are less common. Often, pumping stations are designed to provide added head to gravity sewers to avoid excessive excavation depths – that is, the sewer line follows a saw-tooth pattern, falling with gravity to a low point, where the flow is pumped up to a higher level to fall again by gravity to the next low point or treatment works inlet.

The design and operation of pumping stations and rising mains intended to be adopted by the sewerage undertaker in England is set out in Part D of the Sewerage Sector Guidance Appendix C (Water UK, 2021).

These systems are recommended for use only where their whole-life costs can be shown to be lower than those of gravity sewer systems over a period of 40 years. Where pumped flows are discharged at or near the inlet of a wastewater treatment plant, it is likely that additional storage will be required at the inlet in order to help balance the inflows to the plant and avoid overloading.

The general basis for the design of sewage pumping stations is that pumps are normally required to be capable of pumping peak flows as they occur within the sewerage network. Diurnal patterns and daily variations may range from low flows to perhaps flows approximating to 6 DWFs. Current design is based on 6 DWFs for combined systems; 4 DWFs may be used for foul-only systems, but this should be used with extreme caution, owing to the potential for infiltration or increased surface water from houses, yards, patios and other impermeable areas.

Consideration should be given to design, performance and application. The final choice on pump selection is likely to be dependent on duty, in terms of discharge rate and head conditions. The majority of sewage pumping stations utilise submersible pumps, requiring a wet well and valve chamber arrangement, as illustrated in Figure 2.8 (Water UK, 2021). The wet well needs to be carefully designed with benching such that flow paths are efficient and sedimentation and the evolution of hydrogen sulphide (H_2S) minimised. If the flow entering the wet well is from a pumped rising main, this may risk the release of hydrogen sulphide or methane into the chamber, where conditions in the rising main have allowed formation (e.g. Shypanski *et al.*, 2018).

Figure 2.8 Typical wet well submersible pumping station (refer to original publication for references to clauses and other specifications) (Water UK, 2021)

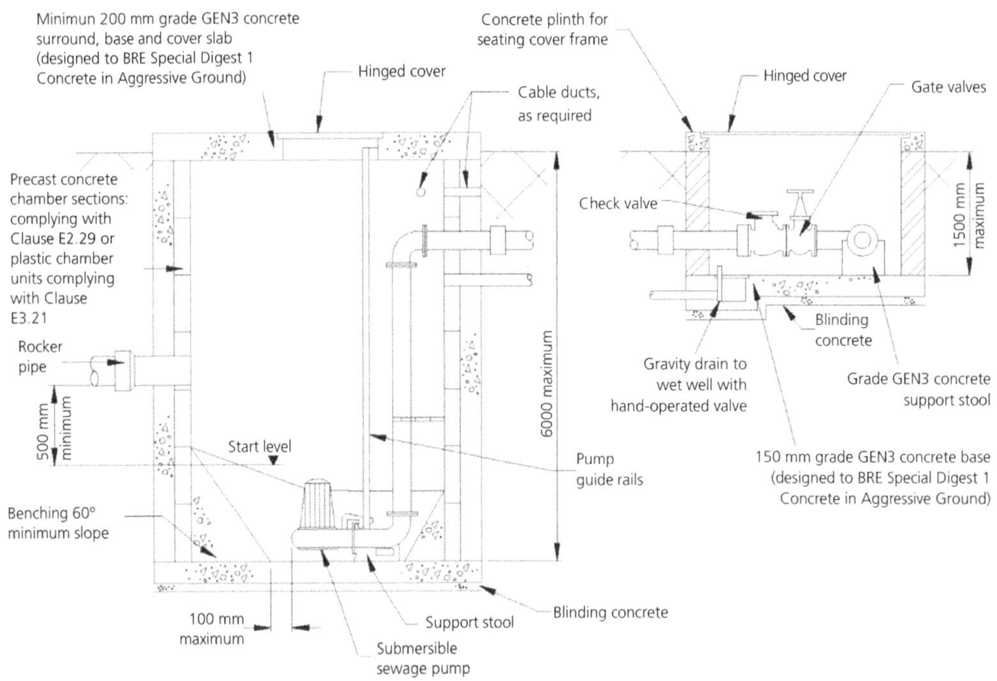

The mechanical design of pumps should not be considered alone, but should be considered alongside the civil engineering aspects of pumping station design, means of access, and source and availability of power supply. It is important to consider the three engineering disciplines of civil, mechanical and electrical as a complete entity in terms of pumping station design. Pumping station design may differ, depending on the population served. The station may be a package unit or a bespoke design for the situation. The Sewerage Sector Guidance Appendix C specifies that the design flow rate for a pumping station serving 500 dwellings or fewer should be half the incoming design flow rate and sufficient to ensure self-cleansing – that is, in the range 0.75–1.8 m/s when the pump unit is operating. Rising mains should be at least 80 mm in diameter and ideally laid in as straight lines as practicable.

Owing to the general low-lying location of pumping stations there is an increased need to consider location in terms of potential flood levels, types of motors and electrical equipment. Sewage pumps operate 'on demand'; therefore, reliability and standby capability are important to prevent flooding, pollution and environmental impact.

Pressure decrease can result in the formation of vapour-filled bubbles and cavitation. This results in pressure waves in the liquid, which are likely to cause excess wear and mechanical damage to the pump and pipework. Net positive suction head (NPSH) is one of the most important design criteria. This is the elevated pressure at the inlet to the pump. It is recommended that the required NPSH from the hydraulic design for the pump/pipeline system should be maintained but with an additional 2 m of head in order to avoid cavitation as the pressure falls though the pumps (Water UK, 2021).

In any pumping system, the change in velocity arising from pump starts or stops and operation of valves can lead to pressure surges, depending on the speed of valve closure. Pressure surges are particularly important for uPVC rising mains as they can result in fatigue and shortened design lifetimes. Pump control regimes are of paramount importance and if set correctly can reduce or eliminate surge. Where necessary, to prevent damage occurring to the whole pumping system, surge controls may be needed, for example, by active flow control measures or providing expansion vessels, although these are normally common in water mains. Surge is a complex issue. It is therefore important to consider and address potential surge problems at the design stage.

Septicity occurs under anaerobic conditions for prolonged periods of time. This may be accelerated at high temperatures, such as during summer or warm weather conditions. The generation of hydrogen sulphide results in the production of sulphuric acid, which can cause erosion of concrete sewers and corrosion of metal parts such as valves. Hydrogen sulphide production can also result in odours and dangerous working conditions. Reducing the retention time in the pumping main can prevent septicity. Aeration and chemical dosing can help avoid septicity in long-retention pumping mains, reducing the production of hydrogen sulphide and controlling odour. Wet wells can, and should, be designed to reduce retention time to a minimum, and flow velocities within rising mains should be self-cleansing. If problems cannot be designed out, then consideration should be given to a septicity control regime.

One of the most important factors in pumping station design is to consider the system as a whole – that is, pumping station, rising main and any ancillary air valves and washouts, and the receiving gravity sewer network.

2.5.2.3 Operation of sewer systems

Dynamic wastewater systems, such as pumping stations and rising mains, are especially amenable to smart control, whereas gravity sewer flows are normally only controllable where there are pumps and at inlets to wastewater treatment plants. The modern generation of flow and water quality sensors is also providing opportunities to manage flows more dynamically, using valves, pump rates and storage within sewer networks. For example, real-time control can help to optimise the operation of an entire wastewater system, utilising excess capacity and often delaying the need to construct or augment additional capacity.

Growing interest in managing the different types of sanitary waste stream individually, including blackwater and greywater, as in Helsingborg, described previously, together with the rapidly evolving development of flow sensors and controllers, is raising interest in smart system operation. For example, McDermott *et al.* (2020) show how, under current design methods, there is a potential problem relating to building drainage layouts and the transport of solids into main sewers by WC flushes. It was shown by on-site studies of a group of 156 rural properties that smart sewer layouts and operation could enhance the efficiency of the transport of the sewer solids emitted from the properties into the main sewer network. This finding is especially important for future wastewater management, as domestic water use and flow rates are expected to decrease, but with solids production remaining similar to the present. Hence, where there are conventional sewer layouts, faecal solids' transport capacity from WC flushes may be compromised. By ensuring greywater emissions are located in the same sewer line and operated smartly, the deficiency in transport capacity could be remedied. McDermott *et al.* (2020) set out a designer checklist that helps ensure the maximum solids transport capacity can be utilised in the property drainage.

Sewer operation is further considered in Section 3.5, where combined sewers are described, and in Chapter 11 on modelling.

2.5.2.4 Infiltration and inflow

Sewers are constructed to convey flows and, in the main, remain watertight. However, ageing, poor construction or damage from other services can lead to gaps, cracks, misalignments and opportunities for leakage out of the sewer (exfiltration) or inflows from surrounding ground. Inflow and infiltration (I & I) – where inflow is water entering the sewer directly and infiltration is water entering from the ground – is a complex problem. Inflow is surface water discharged to, or entering, the sewer network by either a direct or indirect route – for example, through manhole cover lids, land drainage connections and cross-connections, or backflow from rivers by way of surface water outfalls that either are without flap valves or have defective flap valves.

Infiltration can be from groundwater, rainfall induced or tidal. Groundwater infiltration occurs when the water table is higher than the sewer, and may be seasonally influenced. Rainfall-induced infiltration is caused by the seepage of rainfall through the subsoil or permeable areas, and tidal infiltration occurs as a direct result of sea water from tidally saturated ground and will coincide with tidal cycles. Other inflows can occur from, for example, adjacent leaking water mains or elevated groundwater levels in times of flooding or wrongly connected stormwater drainage. The impact from inflow, infiltration and exfiltration affects the following three key areas: sewer flooding, environmental compliance, and mitigating the carbon footprint and adapting to climate change. White *et al.* (1997) reported for CIRIA that, at that time, using data for 1646 UK sewer catchments, in some 28% of these catchments more than 25% of the flow arriving at treatment plants were attributed to infiltration. Nine per cent of the catchments reported that 50% of flows were infiltration. The primary sources were normal and high groundwater levels and adjacency of tidal waters and watercourses. Leaking water mains were less often found to be the cause.

Detecting I & I is very difficult; differentiating and tracing the sources to remediate has historically proved elusive (e.g. Sowby and Jones, 2022). Traditionally it has been presumed that infiltration can best be assessed by observations of flows during times of minimum domestic and trade discharges – that is, during night times, which also presumes that I & I remains constant over a daily cycle. With current flow monitors, such measurement is feasible (Chapter 9), although given the changes to lifestyles and commercial activity it cannot be an assured means of detecting or measuring I & I rates. Other recent technological developments now provide the means to measure in-sewer temperatures along lengthy stretches, using fibre-optic distributed temperature sensing. With this, even small changes in water temperature can be detected, with temperature reductions indicating I & I inputs (e.g. Beheshti and Saegov, 2018).

Design or operation of sewers to include infiltration in the UK is considered in Ainger *et al.* (1998). With the I & I presumed to be the 'minimum night-time flow', estimated infiltration in the study ranged up to 89% of the total measured flows. It was recommended that ammonia concentrations could be measured as a means of identifying infiltration, due to the presence or lack of ammonia in DWFs and I & I, respectively.

USEPA (2014) guidance explains how to estimate I & I using inflow data from wastewater treatment plants. This guidance is framed so as to assist with compliance with the US National Pollutant Discharge Elimination System (NPDES) and thus qualify for funding support. One check of the relative magnitude of I & I is to use contributing population data together with average domestic wastewater discharge rates to estimate expected treatment plant inflows. If these estimates are found not to match the recorded inflows – that is, inflows are much higher than expected – then there is most likely significant I & I, although other sources, such as unrecorded trade flows, should also be considered.

2.6. Challenges for sewer design and operation

As introduced in Section 2.5.1, with moves to manage water demand, especially from domestic users, wastewater discharge rates could be reduced. Domestic appliance water use is being reduced, with, for example, washing machines and dishwashers using less water, and the widespread use of low flush toilets (Savic and Banyard, 2011). This may mean flows in the sewer network being below what is presumed in standards or design codes, including those shown in Table 2.3. Because of this, there has been much speculation about the implications of water savings domestically and in commercial and industrial premises on the functioning of existing sewers and the need for additional sanitary sewers. Basupi (2020) and Mattson *et al.* (2015) consider how reductions in water use would affect the economics and efficiency of existing and newly designed sanitary sewer systems. Lower flows and velocities would potentially cause more deposition in the heads of branched networks. Systems recently designed to accommodate future additional inputs, as for staged developments, would also potentially have increased solids deposition and blockage risks. Longer retention times in existing systems, following significant water saving measures being introduced, would also risk in situ solids biodegradation and dissolved substance problems in more concentrated flows, especially the production of hydrogen sulphide and methane, with attendant corrosion as well as safety risks from gas build-up and emissions (e.g. Ren *et al.*, 2022).

Jiang *et al.* (2015) explain how methane, sulphides and volatile organic compounds are formed in sewer systems. Matias *et al.* (2018) consider the use of computational models for sulphide production for both gravity sewers and rising mains (see also Chapter 11). One model, Wastewater Aerobic/Anaerobic Transformations in Sewers (WATS), has been used in conjunction with a treatment plant simulator model to evaluate the effects of prolonged residence times in sewer networks on the treatment efficiency at downstream plants, including when using storage tanks in combined systems (Ashley *et al.*, 2001). The implications are further discussed in Section 3.5. These models can illustrate the risks of anaerobic conditions when operating sanitary sewer systems, where there are low oxygen concentrations and excessive sulphide production. As an example, Kraakman *et al.* (2018) used the WATS model to evaluate the implications of introducing domestic food waste slurry into a main sewer in Los Angeles in order to reduce the need to collect food waste using trucks in the city. The slurry introduced into the sanitary sewer is to be conveyed some 29 km, along with sanitary sewage, to the downstream plant. The nutrient-rich wastewater is to be treated to recover energy. Modelling showed that hydrogen sulphide levels in the main sewer would rise, but they would not be problematic, especially as the sewer has already been appropriately lined to resist corrosion. Despite not adding significantly to the flows, the addition of the food waste would mean an additional 59 800 kg of chemical oxygen demand at the treatment plant every day, helping to increase significantly the recovery of energy by way of anaerobic processes.

These models can also be used to analyse the hydraulic behaviour and sulphide risks in pressure sewers (rising mains), pumping stations and inverted siphons. Each of these has significant opportunities for oxygen depletion by biological processes, and being closed to the atmosphere, this deficiency cannot be rectified as it can in gravity sewers with manholes and vents. Inverted siphons may also have a greater risk of blockages if poorly designed or operated, as these are gravity flow, such as reductions in pipe diameter downstream in order to pass beneath obstacles, waterways or other infrastructure (e.g. May, 2003; USEPA, 2015). Worldwide there have been experiences with operating and maintaining siphon and pressurised sewers. For example, Vasilyev and Vasilyev (2015) show how the new 12 km long siphon sewer in St Petersburg has been designed to manage both blockages and potential corrosion problems. The formation of sulphides in rising mains has also been extensively investigated. A detailed discussion and guidance for management is given in Hvitved-Jacobsen *et al.* (2013).

Where there is an intermittent water supply, such as in low-income areas, alternative low-cost, settled or simplified sewerage systems can provide an alternative to the traditional sanitary systems so far discussed. While it is clear that these can be much cheaper to install than traditional systems, as they are smaller bore, laid to shallower gradients and require more community engagement in terms of what is input to the systems and how they are maintained, so far their longer term effectiveness, reliability and whole life costs are unclear (e.g. Daudey, 2018). Nevertheless, they may be required to deliver the sanitation requirements of the SDGs (Hutton and Varughese, 2016).

REFERENCES

Ainger CM, Armstrong AJ and Butler D (1998) *Dry Weather Flow in Sewers*. CIRIA R177. Construction Industry Research and Information Association, London, UK.

Arthur S and Blanc J (2013) *Management and Recovery of FOG (fats, oils and greases)*. CREW project CD2013/6. See https://www.crew.ac.uk/sites/www.crew.ac.uk/files/sites/default/files/publication/CREW_FOG.pdf.

Ashley RM, Dudley J, Vollertsen J *et al.* (2001) Interaction between extended in-sewer storage and wastewater treatment plant performance – a protocol to avoid problems. *WaPUG Conference*, Blackpool, UK. November 2001.

Ashley RM, Bertrand-Krajewski J-L, Hvitved-Jacobsen T and Verbanck M (eds) (2004) *Solids in Sewers*. IWA Scientific and Technical Report No. 14. IWA Publishing, London, UK.

Ashley RM, Blackwood D, Souter N *et al.* (2005) Sustainable disposal of domestic sanitary waste. *ASCE Journal of Environmental Engineering* **131(2)**: 206–215, 10.1061/(ASCE)0733-9372(2005)131:2(206).

Basupi I (2020) Performance analyses of existing sanitary sewers with the selection of household water-saving schemes. *Water Science and Technology* **82(11)**: 2400–2414, 10.2166/wst.2020.492.

Beheshti M and Saegov S (2018) Quantification assessment of extraneous water infiltration and inflow by analysis of the thermal behavior of the sewer network. *Water* **10**: 1070, 10.3390/w10081070.

Biel-Maeso M, Corada-Fernandez C and Lara-Martín PA (2019) Removal of personal care products (PCPs) in wastewater and sludge treatment and their occurrence in receiving soils. *Water Research* **150(2019)**: 129–139, https://doi.org/10.1016/j.watres.2018.11.045.

Brombach H, Weiss G and Fuchs S (2005) A new database on urban runoff pollution: comparison of separate and combined sewer systems. *Water Science Technology* **51(2)**: 119–128, doi.org/10.2166/wst.2005.0039.

BSI (British Standards Institution) (1999) PAS 29:1999. Disposable pulp products for use in healthcare. BSI, London, UK.

BSI (2000) BS EN 12056-2:2000. Gravity drainage systems inside buildings. Sanitary pipework, layout and calculation. BSI, London, UK.

BSI (2014) BS EN 16323:2014. A glossary of wastewater engineering terms. BSI, London, UK.

BSI (2017a) BS EN 752:2017. Drain and sewer systems outside buildings. Sewer system management. Incorporating corrigenda October 2019. BSI, London, UK.

BSI (2017b) BS EN 16933-2:2017. Drain and sewer systems outside buildings. Design. Hydraulic design. Incorporating corrigendum 2018. BSI, London, UK.

BSI (2018) BS EN 16932-3:2018. Drain and sewer systems outside buildings. Pumping systems. Vacuum systems. Incorporating corrigendum July 2018. BSI, London, UK.

BSI (2019a) BS ISO 24513:2019. Service activities relating to drinking water supply, wastewater and stormwater systems. Vocabulary. BSI, London, UK.

BSI (2019b) BS ISO 24536:2019. Service activities relating to drinking water supply, wastewater and stormwater systems. Stormwater management. Guidelines for stormwater management in urban areas. BSI, London, UK.

BSI (2023) BS ISO 20670:2023: Water reuse. Vocabulary. BSI, London, UK.

Bui X-T, Chiemchaisi C, Fujioka T and Varjani S (eds) (2018) *Water and wastewater treatment technologies*. Springer Nature Singapore Ptd Ltd., Singapore, 10.1007/978-981-13-3259-3.

CIWEM (Chartered Institution of Water and Environmental Management) (2021) *Combined sewer overflows – it's time we cut the crap*. See https://www.ciwem.org/news/combined-sewer-overflows-its-time-we-cut-the-crap (accessed 18/07/2023).

Clean Coasts (2022) *About Think Before You Flush*. See https://thinkbeforeyouflush.org/about/ (accessed 18/07/2023).

Comber S, Gardner M, Sörme P, Leverett D and Ellor B. (2018) Active pharmaceutical ingredients entering the aquatic environment from wastewater treatment works: a cause for concern? *Science of the Total Environment* **613–614**: 538–547, 10.1016/j.scitotenv.2017.09.101.

Crites R and Tchobanoglous G (1998) *Small and Decentralised Wastewater Management Systems*. McGraw Hill, Boston, USA.

Daudey L (2018) The cost of urban sanitation solutions: a literature review. *Journal of Water, Sanitation and Hygiene for Development* **8(2)**: 176–195, 10.2166/washdev.2017.058.

Defra (Department for Environment, Food and Rural Affairs) (2002) *Sewage Treatment in the UK*. PB 6655. Defra, London, UK.

Defra (2020) *Report of a Review of the Arrangements for Determining Responsibility for Surface Water and Drainage Assets*. PB 14629. Defra, London, UK. See https://www.gov.uk/government/publications/surface-water-and-drainage-review-of-responsibilities (accessed 18/07/2023).

Digman CJ, Horton B and Ashley RM (2015) *Benefits of SuDS Tool (BeST)*. CIRIA W045. Construction Industry Research and Information Association, London UK. See https://www.susdrain.org/resources/best.html (accessed 10/11/2023).

Dobson B, Jovanovic T, Chen Y *et al.* (2021) Integrated modelling to support analysis of COVID-19 impacts on London's water system and in-river water quality. *Frontiers in Water* **3**: 641462, 10.3389/frwa.2021.641462.

Durukan S and Karadagli F (2019) Physical characteristics, fibre compositions, and tensile properties of nonwoven wipes and toilet papers in relevance to what is flushable. *Science of the Total Environment* **697**: 134135, 10.1016/j.chemosphere.2019.03.097.

EC (European Community) (1999) Directive 1999/31/EC of the European Parliament and of the Council of 26 April 1999 on the landfill of waste. *Official Journal of the European Communities* **L182**.

Eliasson M and Ahlman S (2006) *SEWSYS-modelling at the Linden pilot site, Hannover.* NORIS: No Rainwater In Sewers.

Ellis JB, Chocat B, Fujita S, Marsalek J and Rauch W (eds) (2004) *Urban Drainage – A Multilingual Glossary.* IWA Publishing, London, UK.

Energy and Resources Factory (2019) *Water Factory – The New Source* [in Dutch]. See https://www.efgf.nl/producten/water/ (accessed 18/07/2023).

EU (European Union) (2016). *Study Supporting the Revision of the EU Drinking Water Directive.* Final Report. EU, Brussels, Belgium, 10.2779/650074.

EurEau (2020) *EurEau Position Paper on Overflows from Collecting Systems.* See https://www.eureau.org/resources/position-papers/4955-position-paper-on-overflows-from-collecting-systems-1/file (accessed 18/07/2023).

Fletcher TD, Shuster W, Hunt WF *et al.* (2014) SUDS, LID, BMPs, WSUD and more – the evolution and application of terminology surrounding urban drainage. *Urban Water Journal* **12(7)**: 525–542, 10.1080/1573062X.2014.916314.

Foden M, Browne A, Evans D, Sharp L and Watson M (2017) *Fats, Oils, Grease and Kitchen Practices: Implications for Policy and Intervention.* University of Sheffield, UK, 10.13140/RG.2.2.13146.24005. See https://changepoints.uk/2017/10/06/report-fats-oils-grease-and-kitchen-practices-implications-for-policy-and-intervention/24005 (accessed 10/11/2023).

FWR (Foundation for Water Research) (2018) *Water Industry Chemical Investigations – CIP1.* Foundation for Water Research. See http://www.fwr.org/wastewat/WWF%20Bellamy%20June%202018.pdf (accessed 18/07/2023).

HMG (Her Majesty's Government) Environmental Protection Act 1990. The Stationery Office, London, UK.

HMG (1991) Water Industry Act 1991. The Stationery Office, London, UK.

HMG (2012) The Controlled Waste (England and Wales) Regulations 2012. The Stationery Office, London, UK.

HMG (2010) Flood and Water Management Act 2010. The Stationery Office, London, UK.

Hutton G and Varughese M (2016) *The Costs of Meeting the 2030 Sustainable Development Goal Targets on Drinking Water, Sanitation, and Hygiene.* The World Bank, Washington DC, USA.

Hvitved-Jacobsen T, Vollertsen J and Nielsen AH (2013) *Sewer Processes: Microbial and Chemical Process Engineering of Sewer Networks.* CRC Press, London, UK.

Islam MS (2017) Comparative evaluation of vacuum sewer and gravity sewer systems. *International Journal of System Assurance Engineering and Management* **8**: 37–53, doi.org/10.1007/s13198-016-0518-z.

ISO (International Organization for Standardization) (2023) Online Browsing Platfrom (OBP), https://www.iso.org/obp/ui.

Jagai JS, DeFlorio-Barker S, Lin CJ, Hilborn ED and Wade TJ (2017) Sanitary sewer overflows and emergency room visits for gastrointestinal illness: analysis of Massachusetts data, 2006–2007. *Environmental Health Perspectives* **125(11)**: 117007-1–117007-7, 10.1289/EHP2048.

Jiang G, Sun J, Sharma KR and Yuan Z (2015) Corrosion and odor management in sewer systems. *Current Opinion in Biotechnology* **33**: 192–197.

Kalcíkova G, Alič B, Skalar T, Bundschuh M and Gotvajn AZ (2017) Wastewater treatment plant effluents as source of cosmetic polyethylene microbeads to freshwater. *Chemosphere* **188(2017)**: 25–31.

Karadagli F, Theofanidis F and Eren B (2021) Consumers' evaluation of flushable products with respect to post-disposal effects in wastewater infrastructures. *Journal of Cleaner Production* **278(2021)**: 123680, 10.1016/j.jclepro.2020.123680.

Keep.eu (2023) *NORIS: No Rainwater In Sewers*. EU Interreg North Sea Region. See https://keep.eu/projects/320/NORIS-No-Rainwater-in-Sewers-EN/ (accessed 19/07/2023).

Kehrein P, van Loosdrecht M, Osseweijer P *et al.* (2020) A critical review of resource recovery from municipal wastewater treatment plants – market supply potentials, technologies and bottlenecks. *Environmental Science: Water Research Technology* **6**: 877–910, 10.1039/c9ew00905a.

Kellagher RR (2003) *Drainage of Development Sites – A Guide*. CIRIA X108. Construction Industry Research and Information Association, London, UK.

KPMG (2020) *Fighting Food Waste using the Circular Economy*. KPMG, Australia.

Kraakman NJR, Johnson B, Randsom C, Bassett D and Kedan G (2018) Food waste impact study: sewer process modelling for future options assessment. *Proceedings of the Water Environment Federation Conference*. Water Environment Federation, March Issue 2018 pp. 62–91, 10.2175/193864718822846742.

LACitySan (2022) Residential food waste diversion program – Being green is easy. See https://www.lacitysan.org/san/sandocview?docname=cnt020080 (accessed 19/07/2023).

Larsen T, Larsen TA, Gruendl H and Binz C (2021) The potential contribution of urine source separation to the SDG agenda – a review of the progress so far and future development options. *Environmental Science: Water Research Technology* **7**: 1161, 10.1039/d0ew01064b.

Legge A (2022) *Food Waste Disposers, the Sewers and the Circular Economy*. PhD thesis, Department of Civil and Structural Engineering, University of Sheffeild, UK.

Legislation.gov.UK (2020) TFEU article 191.2. *Consolidated Version of the Treaty on the Functioning of the European Union*. See https://www.legislation.gov.uk/eut/teec/article/191 (accessed 19/07/2023).

Loosli F, Wang J, Rothenberg S *et al.* (2019) Sewage spills are a major source of titanium dioxide engineered (nano)-particles into the environment. *Environmental Science Nano* **6(3)**: 763–777, 10.1039/C8EN01376D.

Mannina G and Viviani G (2009) Separate and combined sewer systems: a long-term modelling approach. *Water Science Technology* **60(3)**: 555–565, 10.2166/wst.2009.376.

Matias N, Matos R, Ferreira F, Vollertsen J and Matos JS (2018) Release of hydrogen sulfide under intermittent flow conditions – the potential of simulation models. *Water Science and Technology* **77(3–4)**: 777–787, 10.2166/wst.2017.595.

Mattson J, Hedström A, Ashley RM and Viklander M (2015) Impacts and managerial implications for sewer systems due to recent changes to inputs in domestic wastewater – a review. *Journal of Environmental Management* **161(2015)**: 188–197, 10.1016/j.jenvman.2015.06.043.

May R (2003) Preventing sediment deposition in inverted sewer siphons. *Journal of Hydraulic Engineering* **129(4)**: 283–290, 10.1061/(ASCE)._0733-9429(2003)129:4(283).

May L and Woods H (2015) *A Review of the Effectiveness of Different On-site Wastewater Treatment Systems and their Management to Reduce Phosphorus Pollution*. Natural England Commissioned Reports, Number 179. See http://publications.naturalengland.org.uk/publication/6636267020222464 (accessed 19/07/2023).

McDermott R, Strong WA, Griffiths P, Littlewood K and Doherty MG (2020) Smart sewers 1: sewer configuration at rural houses. *Journal of Water Resource and Protection* **12**: 825–834, doi.org/10.4236/jwarp.2020.129048.

Mohr M, Beckett M, Schliessmann U, Erlbeck R and Trosse R (2018) Vacuum sewerage systems – a solution for fast growing cities in developing countries? *Water Practice and Technology* **13(1)**: 157, 10.2166/wpt.2018.028.

Nagpal H, Spriet J, Murali MK and McNabola A (2021) Heat recovery from wastewater – a review of available resources. *Water* **13**: 1274, 10.3390/w13091274.

NIH (National Institutes of Health) (2019) *Perfluoroalkyl and Polyfluoroalkyl Substances (PFAS)*. NIH/US Department of Health and Human Services. See https://www.niehs.nih.gov/ (accessed 19/07/2023).

Ó Briain O, Mendes ARM, McCarron S, Healy MG and Morrison L (2020) The role of wet wipes and sanitary towels as a source of white microplastic fibres in the marine environment. *Water Research* **182**: 116021, 10.1016/j.watres.2020.116021.

OECD (Organisation for Economic Co-operation and Development) (2020) *Financing Water Supply, Sanitation and Flood Protection: Challenges in EU Member States and Policy Options*. OECD Studies on Water. OECD Publishing, Paris, France, 10.1787/6893cdac-en.

OECD (2022) *Global Plastics Outlook: Economic Drivers, Environmental Impacts and Policy Options*. OECD Publishing, Paris, France, 10.1787/de747aef-en.

Okwori E, Viklander M and Hedstrom A (2020) Performance assessment of Swedish sewer pipe networks using pipe blockage and other associated performance indicators. *H_2Open Journal* **3(1)**: 46, 10.2166/h2oj.2020.027.

Prata JC (2018) Microplastics in wastewater: state of the knowledge on sources, fate and solutions. *Marine Pollution Bulletin* **129**: 262–265, 10.1016/j.marpolbul.2018.02.046.

PUB (Public Utilities Board) (2022) *NEWater*. Singapore National Water Agency. See https://www.pub.gov.sg/watersupply/fournationaltaps/newater (accessed 19/07/2023).

Ren D, Zuo Z, Xing Y *et al.* (2022) Simultaneous control of sulfide and methane in sewers achieved by a physical approach targeting dominant active zone in sediments. *Water Research* **211**: 118010, 10.1016/j.watres.2021.118010.

Risch E, Boutin C and Roux P (2021) Applying life cycle assessment to assess the environmental performance of decentralised versus centralised wastewater systems. *Water Research* **196**: 116991, 10.1016/j.watres.2021.116991.

Run4Life (2022) https://run4life-project.eu/demosites/helsingborg-se/ (accessed 19/07/2023).

Salvidge R and Hosea L (2023) Revealed: scale of 'forever chemical' pollution across UK and Europe. *The Guardian*, 23 Feb.. See https://www.theguardian.com/environment/2023/feb/23/revealed-scale-of-forever-chemical-pollution-across-uk-and-europe (accessed 18/07/2023).

Savic D and Banyard JK (eds) (2011) *Water Distribution Systems*. ICE Publishing, London, UK.

Science Direct (2022) *Colebrook–White Equation*. See https://www.sciencedirect.com/topics/engineering/colebrook-white-equation#:~:text=The%20Colebrook%E2%80%93White%20equation%2C%20sometimes,and%20inside%20diameter%20of%20pipe (accessed 19/07/2023).

Scottish Water (2018) *Sewers for Scotland – A technical specification for the design and construction of sewerage infrastructure*. Version 4.0 October. See https://www.scottishwater.co.uk/-/media/ScottishWater/Document-Hub/Business-and-Developers/Connecting-to-our-network/All-connections-information/SewersForScotlandv4.pdf (accessed 19/07/2023).

Shamsi UM (2018) Green first approach for wet weather programs. *Journal of Water Management Modeling* **26**: C439, 10.14796/JWMM.C439.

Shypanski AH, Yuanab Z and Sharma K (2018) Influence of pressure main pumping frequency on sulfide formation rates in sanitary sewers. *Environmental Science: Water Research Technology* **4**: 403. Doi: 10.1039/c7ew00428a

SOENECS (Social, Environmental & Economic Solutions) (2021) *Insight Report: 2030. Futures Insight Report no.1: Invisible Ocean Pollutants from our Roads*. SOENECS) Ltd, Burgess Hill, UK. Copyright SUEZ Ltd. April.

Sommer H and Sieker H (2005) Reduction of pollution load from street-runoff by an inlet filtration-system filled with adsorptive material. *Proceedings of the 10th International Conference on Urban Drainage*, Eriksson E *et al.* (eds), 21–26 August 2005, Copenhagen, Denmark.

Sotelo TJ, Sloen GB and Satoh H (2021) Circling the drain: A systems analysis of opportunities for enhanced sewer self-purification technologies in wastewater management. *Journal of Environmental Management* **288(2021)**: 112451, 10.1016/j.jenvman.2021.112451.

Sowby RB and Jones DR (2022) A practical statistical method to differentiate inflow and infiltration in sanitary sewer systems. *Journal of Environmental Engineering* **148(1)**: 06021006, 10.1061/(ASCE)EE.1943-7870.0001962.

Speth TS (2020) *PFAS Treatment in Drinking Water and Wastewater – State of the Science*. USEPA PFAS Science Webinars for EPA Region 1 and State and Tribal Partners. Session 3. 3 September. See https://www.epa.gov/sites/default/files/2020-09/documents/r1-pfas_webinar_day_1_session_3_speth.pdf (accessed 19/07/2023).

Stantec (2020) *Storm Overflow Evidence Project – Final Report*. Prepared for Water UK, November. See https://assets.publishing.service.gov.uk/government/uploads/system/uploads/attachment_data/file/1030980/storm-overflows-evidence-project.pdf (accessed 19/07/2023).

STOWA (2015) *Principles for Implementing LCA: Food Waste in the Water Chain*. Stichting RIONED/STOWA 2015-W-02. Foundation for Applied Water Research STOWA, Amersfoort, the Netherlands.

Sulej-Suchomska AM, Klupczynska, Dereziński P *et al.* (2020) Urban wastewater analysis as an effective tool for monitoring illegal drugs, including new psychoactive substances, in the Eastern European region. *Nature – Scientific Reports* **10**: 4885, 10.1038/s41598-020-61628-5.

Talvitie J, Mikola A, Setälä O, Heinonen M and Koistinen A (2017) How well is microlitter purified from wastewater? A detailed study on the stepwise removal of microlitter in a tertiary level wastewater treatment plant. *Water Research* **109**: 164–172.

Tchobanoglous G, Burton FL and Stensel HD (2002) *Wastewater Engineering – Treatment and Reuse*, 4th edn. McGraw Hill, New York, USA.

USEPA (United States Environmental Protection Agency) (1970) *Control of infiltration and inflow into sewers*. Water pollution control research series. 11022 EFF 12/70. See https://nepis.epa.gov/Exe/ZyPDF.cgi/9100WGY8.PDF?Dockey=9100WGY8.PDF (accessed 19/07/2023).

USEPA (2002) *Onsite Wastewater Treatment Treatment Systems Manual*. USEPA, Washington, DC, USA. EPA/625/R-00/008.

USEPA (2004) *Report to Congress on Impacts and Control of Combined Sewer Overflows and Sanitary Sewer Overflows*. USEPA, Washington, DC, USA. EPA 833-R-04-001. See https://www.epa.gov/sites/default/files/2015-10/documents/csossortc2004_full.pdf (accessed 19/07/2023).

USEPA (2014) *Guide for Estimating Infiltration and Inflow*. USEPA, Washington, DC, USA. See https://www3.epa.gov/region1/sso/pdfs/Guide4EstimatingInfiltrationInflow.pdf (accessed 19/07/2023).

USEPA (2015) *Wastewater Siphon Elimination*. USEPA, Washington, DC, USA. See https://www3.epa.gov/region1/sso/pdfs/SiphonElimination.pdf (accessed 19/07/2023).

USEPA (2022) *Sanitary Sewer Overflow (SSO): Frequent Questions*. USEPA, Washington, DC, USA. See https://www.epa.gov/npdes/sanitary-sewer-overflow-sso-frequent-questions (accessed 19/07/2023).

USEPA (2023) *National Pollutant Discharge Elimination System (NPDES)*. USEPA, Washington, DC, USA. See https://www.epa.gov/npdes (accessed 19/07/2023).

Vasilyev V and Vasilyev F (2015) Innovations in Saint Petersburg's 12 km siphon sewer, Russia. Paper 1400022. *Proceedings of the Institution of Civil Engineers – Waste and Resource Management* **169(WR1)**: 42–51, 10.1680/jwarm.14.00022.

Wald C (2021) *Pipe Dreams*. Avid Reader Press, New York, USA.

Water Industry Journal (2022). Lessons learnt from chemical investigation programmes. See https://www.waterindustryjournal.co.uk/lessons-learnt-from-chemical-investigation-programmes (accessed 19/07/2023).

Water Research Foundation (2022) *Constituents of Emerging Concern*. See https://www.waterrf.org/sites/default/files/file/2022-04/Constituents-Emerging-Concern.pdf (accessed 19/07/2023).

Water UK (2014) *National Guidance for Healthcare Waste Water Discharges*. See https://www.dropbox.com/s/mzywcrz3iza9pnv/Water%20UK%20National%20Guidance%20August%202014.pdf?dl=0 (accessed 19/07/2023).

Water UK (2017) *Wipes in Sewer Blockage Study – Final Report*. Report 21CDP.WS. See https://www.water.org.uk/publication/wipes-in-sewer-blockage-study-final-report/ (accessed 10/11/2023).

Water UK (2018) *Sewers for Adoption – A Design and Construction Guide for Developers*, 8th edn. Water UK/WRc plc.

Water UK (2019) *Protocol for Correctly Classifying Culverted Watercourses and Sewers*. See https://www.water.org.uk/guidance/protocol-for-correctly-classifying-culverted-watercourses-and-sewers/ (accessed 19/07/2023).

Water UK (2021) *Design and Construction Guidance for foul and surface water sewers offered for adoption under the Code for adoption agreements for water and sewerage companies operating wholly or mainly in England ('the Code') Approved Version 2.1*. 25 May 2021. See https://www.water.org.uk/wp-content/uploads/2021/07/SSG-App-C-Des-Con-Guide.pdf (accessed 19/07/2023).

Water UK (2022) *Fine to Flush*. See https://www.water.org.uk/wp-content/uploads/2022/10/Sewerage-Sector-Guidance-v2-update-2.pdf (accessed 19/07/2023).

White M, Johnson H, Anderson G and Misstear B (1997) *Control of Infiltration to Sewers*. CIRIA R175. Construction Industry Research and Information Association, London, UK.

Wilkinson JL, Boxall ABA, Kolpin DW and Teta C (2022) Pharmaceutical pollution of the world's rivers. *PNAS* **119(8)**: e2113947119, 10.1073/pnas.2113947119.

Ashley R, Smith B, Shaffer P and Caffoor I
ISBN 978-0-7277-4178-3
https://doi.org/10.1680/icehudp.41783.059

Chapter 3
Stormwater control

Richard Ashley, Brian Smith and Jo Bradley

3.1. Surface water systems

There are many issues that add to the complexity of the drainage landscape, of which stormwater is one aspect. In Chapter 2, the types of sewer were introduced (Figure 2.1). Chapter 2 focused on sanitary sewerage and related aspects for urban drainage. Here, both separate stormwater sewers and combined sewers are considered as these both convey stormwater or surface water.

Stormwater is rainwater or snowmelt that occurs on surfaces and can enter the sewer or drainage system from all types of surfaces, such as roofs, roads or paved and unpaved areas. It may also be referred to as surface water, especially if it ponds or puddles. Stormwater, being less predictable than sanitary sewer flows, can exert an uncontrolled demand on drainage and sewer systems and be the dominant cause of flooding[1] and hydraulic overloading causing excessive discharges into rivers from outfalls and overflows. Stormwater impacts will, if unmanaged, increase in the future as the effects of climate change and development growth, accompanied by sealing of surfaces, lead to more flooding and pollution where the drainage or sewer network lacks capacity to handle the increasing peak flows and volumes.

Stormwater, being ubiquitous, is perhaps the most important of all sources of wastewater. It finds its way into formal and informal drainage systems and, despite the best efforts at prevention, will get into supposedly separate drainage networks. As an example, some 80–90% of the flows in sewers in a Norwegian municipality in winter is due to inflow and infiltration (see Chapter 2). Although the aged state of infrastructure bears some of the responsibility for this, much is due to sewer misconnections (Ellis and Butler, 2015; Envirotec, 2023). The on-costs from pumping and unnecessary treatment of stormwater at downstream plants are excessive. This has been detected because of real-time monitoring using the integrated Internet of Things and machine learning technologies, along with pre-existing real-time data from supervisory control and data acquisition (SCADA) systems.

Rainfall leads to runoff from surfaces, the volume of which is dependent on the relative permeability of the surfaces. Some rainfall will infiltrate into soils and add to groundwater, some will be intercepted by vegetation and can evaporate or be evapotranspired from plants, especially trees. Runoff occurs after these 'losses', and typically in temperate climates only following some 1–2 mm of rainfall. The speed and amount of runoff depends on surface type, including flow pathways. A more detailed explanation of rainfall estimation is provided in Chapter 6, and Chapter 7 details runoff and infiltration.

1 Flooding can occur from many sources other than rainfall, including the sea and water main bursts.

There is some debate as to the origins of the understanding of how rainfall translates into runoff. The so-called 'rational method' for calculating runoff was demonstrated in France in the 1600s, Ireland in the 1800s, used in the USA in 1800s for sewer design, and known in England as the Lloyd–Davies formula. More recently in the UK, the Wallingford Procedure (DoE/HR Wallingford, 1983) proposed the use of a modified rational method, which refined the original formula to account better for flow pathways and catchment size (see Chapters 7 and 11).

According to Water UK (2021) 'The purpose of surface water drainage is to carry water away from buildings and yards belonging to buildings in a manner that manages flood risk and water quality'. While a key component of satisfying the need for safety, this should be considered as only the last resort for surface water drainage. As storm or surface water may be used directly at the location to flush toilets or for other purposes, such as irrigation of nature-based systems, it is a valuable potential resource (e.g. San Francisco Water Power Sewer, 2022). Surface water management is thus about managing the risk while maximising the potential for use, rather than focusing solely on safe disposal.

It is helpful to think about stormwater in terms of the linked 'source–pathway–receptor' (SPR) concept. Although initially aimed at ecosystem impact control, it has since become used as a frame in flood risk management (Chapter 12). SPR can help in the assessment of sources of high flows and pollution, where water goes and what it does when it flows along pathways, and where it ends up – in a receptor. In well designed and managed systems the source will be where the runoff is generated by rainfall landing on surfaces. The pathway is a managed conveyance conduit – either above or below ground. The receptor is a safe and ideally welcome destination, such as a watercourse.

Managing stormwater flows and volumes is important and can now be done at each stage of the SPR system, as part of strategic flood risk management (see Chapter 12). If the stormwater is directed to surface water drains and sewers, it will help minimise the risks of flooding and pollution of receiving waters resource use; if to combined sewers, it can minimise sewer flooding and pollution, reduce pumping and energy costs and improve treated sludge quality. It can further help maintain sewer and wastewater treatment capacity to accommodate future growth without detrimentally impacting assets, service or the environment.

At the same time as controlling unwanted impacts, better stormwater management will provide many other benefits for society, such as enhancements to aesthetic value, biodiversity, health, wellbeing and recreation, and potentially create a new water resource.

Box 3.1 sets out a broad vision for managing stormwater and combined sewer flows, taken from BS ISO 24536:2019 (BSI, 2019).

Box 3.1

The objectives of stormwater management systems as set out in BS ISO 24536:2019

The objectives of stormwater management systems include effective control and management of flows; protection of water quality; preservation of water quantity; protection of the built, public and natural environments; water conservation and reuse; protection or enhancement of ecosystem health; protection or enhancement of public health, safety and welfare; protection or enhancement of social values; and facilitation of sustainable development and climate adaptation.

Although all types of drainage system should be designed, constructed and managed to meet particular standards of performance, this should be risk based in order give maximum performance for minimum cost. Traditionally, stormwater systems have been planned and operated to manage a certain frequency of storm event safely, depending upon location, such as a once in 100 years rainfall event (Chapter 6).

Contemporary approaches to stormwater management look at the probability of occurrence of a set of rainfall and consequential flood events in what is termed a 'risk-based' approach (Figure 3.1), as part of the strategic approach to flood risk management (Sayers *et al.*, 2022). In this way, the management of stormwater can be linked better to the cost of management measures and the value of the outcomes, as illustrated in Figure 3.2.

BS ISO 24536:2019 sets out five principles for managing stormwater, based on holistic catchment management, as shown in Table 3.1.

Table 3.1 does not include the potential impacts of overflows from combined drainage systems, for which alternative measures are likely to be required.

Considering the principles in Table 3.1, surface or stormwater systems should be designed and operated to encompass a wide range of rainfall events, as illustrated in Figure 3.2 from Digman *et al.* (2014) and Ashley *et al.* (2015). This illustrates the four-points approach to surface water management, managing surface flooding and maximising the use of surface water where practicable. Each of the four points (domains or zones, as for any given application the design requirements may be specific) as defined may also be used to maximise the value from the stormwater managed. These are set out in Table 3.2.

Table 3.3 provides examples of some of the design standards for below-ground piped sewers based on the rainfall return period. In the UK, BS EN 752:2017 (BSI, 2017) provides details for design of all types of drainage system. The standard overlaps with various other standards, as listed in Table 2.3.

Local authorities' Surface Water Management Plans and water and sewerage companies' Drainage and Wastewater Management Plans (DWMPs) (Atkins, 2021) in England now aim to manage surface water

Figure 3.1 Risk curves related to loss due to flood (BSI, 2019)

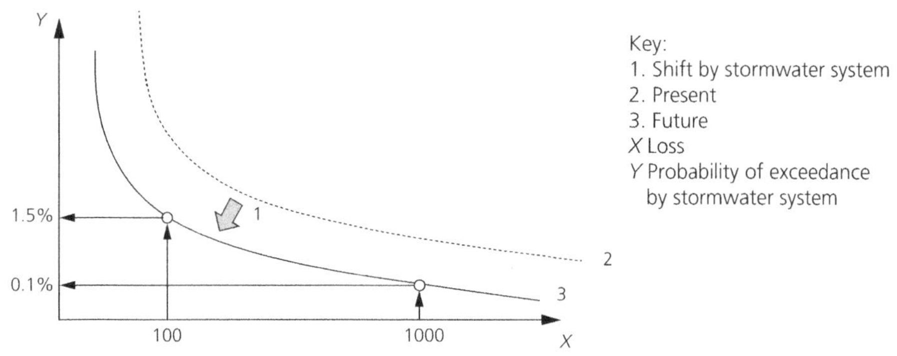

Figure 3.2 The four-points approach to storm and surface water management (modified from Ashley *et al.*, 2015)

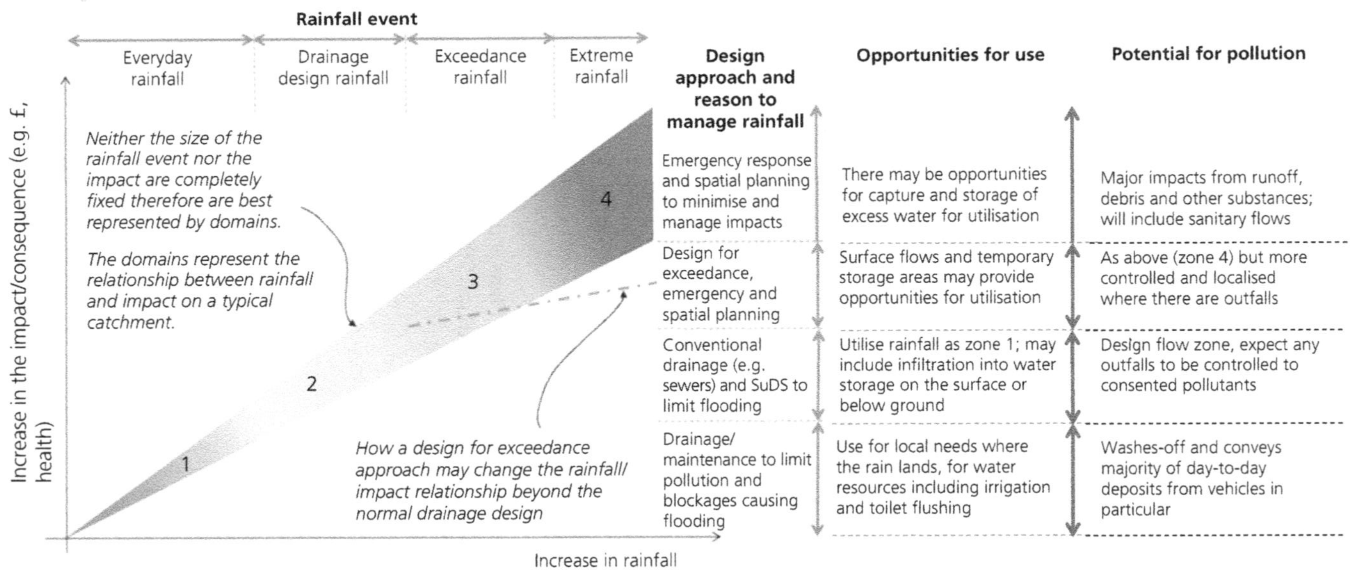

Table 3.1 Principles for managing stormwater, adapted from BS ISO 24536:2019

Principles	Measures related to urban areas
1. Hydrology – minimise the impact of urbanisation and land management practices on the catchment hydrology	Reduce and avoid sealed surfaces, maintaining permeable surfaces by using, for example, SuDS wherever practicable (Fallast *et al.*, 2021) Retain/reduce the runoff from upland catchments using, for example, natural flood management (Wren *et al.*, 2022)
2. Water quality – minimise pollution entering and discharged from the stormwater system	Install interceptors and ensure highway gullies are effective (National Highways, 2022) Education to avoid polluting inputs from domestic, industrial and other sources, including avoiding wrong connections of sanitary into stormwater systems and infiltration of groundwater (Chapter 2) Use SuDS where practicable to control pollution at source and prior to discharge into water bodies (Chapter 8)
3. Vegetation – maximise the value of riparian, floodplain and bankside vegetation for flood attenuation, erosion control and water quality improvements	Maximise the use of nature-based systems both upstream and in urban areas (e.g. Green Building Council UK, 2022). Retain/reduce the runoff from upland catchments using, for example, natural flood management (Wren *et al.*, 2022)
4. Aquatic habitat – minimise the negative impacts of stormwater discharges on the integrity of aquatic habitats	Use SuDS and other treatment systems prior to any surface water discharges into water bodies to reduce pollutant loads and flushes (Chapter 8)
5. Stormwater use – promote opportunities to identify and use collected stormwater as an alternative water source	Make the most of the opportunities to utilise surface water where it originates (e.g. Gould and Nissen-Petersen, 2008)

to minimise property flooding to a once in 50 years standard, and in Scotland (Scottish Government, 2022) to up to once in 200 years.

3.2. A risk-based approach to stormwater management

Stormwater management is all about providing a standard of infrastructure performance (level of service) that is acceptable in urban areas. Typically, this may be ensuring the absence of water on surfaces where it is not usually found – a classic definition of flooding. Unfortunately providing such a service may be unaffordable and, in some cases, not feasible. Hence, there is a need to define both what level of service is expected and what level is realistically achievable. Society may expect a completely risk-free urban environment but, in general, is not willing to pay for this expected level of service, which in any case is impossible to attain.

Table 3.2 The four points approach to managing stormwater (Digman *et al.*, 2014)

Domain/zone in Figure 3.2	Design	Utilisation potential
Zone 1 Everyday or frequent lower intensity rainfall events	Designed drainage to zone 2 service levels will provide effective management provided system is maintained and blockages avoided	Aim to use these events as water resources for local use where the rain lands, including irrigation and toilet flushing
Zone 2 Design standard rainfall events	As shown in Table 3.3 this is what is aimed for (service standard) to minimise impacts and provide safe and secure drainage	Utilise rainfall as above; may include infiltration into water storage on the surface or below ground
Zone 3 Exceedance drainage rainfall events	Purposeful design to store and route flows in excess of the drainage capacity to minimise impacts as far as practicable. This will rely on surface water on streets and other areas, hence spatial planning and emergency contingencies need to take this into account	Where the rainfall exceeds the design performance in zone 2, surface flows and temporary storage areas may provide opportunities for utilisation
Zone 4 Extreme and flooding rainfall events	Design for emergencies, with safe and reliable evacuation routes in spatial planning processes. The largest rainfall events will not be accommodated by the designed drainage system, nor will the exceedance arrangements be able to handle all flows	Where water resources are a potential problem, there may still be opportunities for capture and storage of excess water for utilisation

A risk-based approach is one in which identification and assessment is made of as wide a range of risks as possible in relation to the service being provided. This is recommended when considering a problem and the response options, such as water on a highway that vehicles could drive into. Assessing such risks can be based on a common framework, such as the BS ISO 31000:2018 risk management principles and guidelines, interpreted for the domain of interest, including stormwater management (BSI, 2018a). Ideally, risks need to be linked and considered in the light of interdependencies between infrastructures. A structured approach enables the risks from the differing perspectives and infrastructure domains to be properly recorded, considered and, where appropriate, mitigated – for example, by using signs to warn motorists that the 'road is liable to flooding'. Society needs to be more understanding and accepting of 'exceedance' of drainage systems, where water will occur more often and more severely on surfaces in urban areas (Section 3.4.1). Professionals have a duty to balance the risks with the service provided, taking fully into account the four zones in Figure 3.2.

Urban drainage design has only relatively recently begun to utilise a risk-based approach equivalent to that used to manage other sources offlooding, such as from rivers, instead using a rainfall 'return period' standard to assess the problem and the performance of systems (Table 3.3). A risk-based approach typically considers the impacts for various rainfall events of different frequency and the consequences,

Table 3.3 Examples of underground storm drainage pipe design capacity

Standard	Applicability/performance requirement	Criteria – pipe full unless otherwise stated
Water UK (2021)	Sites with average ground slopes greater than 1%	1 year return period event
	Sites with average ground slopes 1% or less	2 year return period event
	Sites where consequences of flooding are severe (e.g. existing basement properties adjacent to new development)	5 year return period event
National Highways (2022)	No surcharge of the drainage system	1 year return period event
	No flooding from the drainage system	5 year return period event
BS EN 752:2017 (BSI, 2017)	Rural area locations	1 year return period event
	Residential areas	2 year return period event
	City centres/industrial/commercial	5 year return period event
	Underground railway underpasses etc.	10 year return period event

such as to inventories, human lives, or property damage and loss of business – that is, risk is considered as a combination of how often the drainage system 'fails' due to excessive load (here the rainfall) and the consequential impacts from that failure. Defining performance requirements and thus what is understood to be failure is critical in this context. Monetising and calculating the annual average damages from flooding can help in prioritising areas to investigate and target for flood alleviation. This type of approach is vital if we are to make our urban areas more resilient, even in areas where flooding has not been recorded. The risk-based approach is also routinely applied to structural aspects of sewerage assets, as described in Chapter 10. Taken together, the potential for hydraulic failure and/or structural failure, here meaning a failure to provide the expected service levels, can be considered jointly within a risk-based perspective.

Urban drainage risk management is often based on minimising costs or maximising net present value (NPV). However, an alternative perspective is one which aims to maximise societal utility, which might not necessarily be the lowest cost or highest NPV solution. This perspective is essential when considering how best to manage flood risks from all sources, not urban drainage systems alone (see Section 3.2.1.1). In reviewing Semple (1991), Ashley (2009) shows how the application of a risk-based approach to urban storm drainage devised as part of the Wallingford Procedure (Department of the Environment/HR Wallingford, 1983) is still relevant. Figure 3.3 illustrates Semple's recommended approach to balancing utility (i.e. beneficial and equitable outcomes) against consequential environmental impacts due to the remedial works required. This is an illustration of the complexity of managing urban stormwater systems, which becomes even more complex when considering multiple sources of flooding, other impacts (see next) or multiple infrastructure interactions, including green-grey (e.g. EcoShape, 2021). Hence the 'simple' concept of selecting a design storm return period, as illustrated in Table 3.3, is not as simple as it appears, albeit for small schemes a fully risk-based approach may not be worthwhile due to the costs of the analysis required.

Figure 3.3 Semple's risk-based approach to urban storm drainage – the optimum design condition (Semple, 1991)

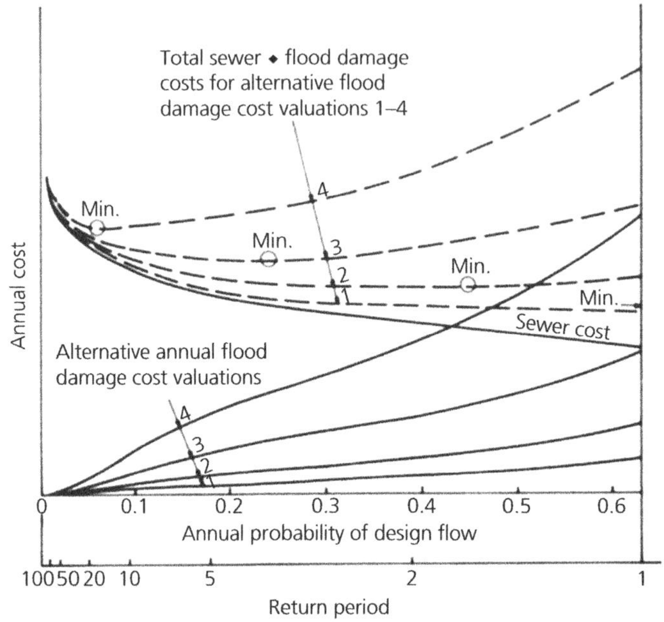

Other risks include potential polluting incidents (see Section 3.2.1.2). Here risk assessment is based on integrated analysis of both the drainage system and the receiving waters. While the procedures for doing this are well developed for buried piped systems, they are less so for surface-based systems (Chapter 8). Flows, contaminant loads and the assimilative capacity of the receiving watercourse are all important. For UK applications, the *Urban Pollution Management Manual* (UPM Manual) was developed in the 1990s (Foundation for Water Research, 1998) to set out a risk-based process for managing urban drainage systems in conjunction with controlling the impacts on receiving waters, primarily from combined sewer systems. With the principal aim of meeting in-sewer storage requirements or flow retention needs in a rational way, the UPM Manual provides a means of arriving at integrated sewerage catchment management. This has been the primary means for deciding on cost-effective combined sewer overflow (CSO) spill control in the UK ever since.

The introduction of mandatory DWMPs in England (Atkins, 2021) has brought together a range of aspects of stormwater management processes. The aim is to ensure better inter-stakeholder working to deliver more concerted, coordinated and effective stormwater plans into the future. Although primarily for and by the private sewerage undertakers, the DWMP framework aims to deal with the complexity of the stakeholder groups, institutions and even individual property owners and dwellers and their roles and responsibilities in England. A comprehensive framework like this has been desperately needed, albeit the framework so far covers only sanitary and combined sewers.

The aim of the DWMP framework is to ensure robustness and resilience in performance of drainage and wastewater assets. Atkins (2021) cites the resilience definition from the water industry regulator in England (Ofwat) as 'Resilience is the ability to cope with, and recover from, disruption, and anticipate trends and variability in order to maintain services for people and protect the natural environment now and in the future'. Although this has been developed specifically for the water industry in England and Wales, it is a general definition that can be applied widely. Here, it is not intended to provide a treatise on the various definitions of resilience in use, or misuse (see Chapter 12). However, the definition could be extended to include the more generally used definition, which includes the restoration of the drainage system to functioning after a significant impact. This may mean that the system has changed – that is, the assets have been damaged but are still able to provide the service intended. In Figure 3.2 this would correspond to zone 4 and possibly zone 3. Robustness is an important attribute of any system, and zones 1 and 2 in Figure 3.2 correspond to the drainage system performing robustly – that is, coping with the events as designed. A major risk factor in the drawing up of DWMPs is the potential for future changes in external drivers, as well as in the assets themselves, and the need to consider possible changes explicitly.

In the UK, guidance on securing and maintaining value from water and wastewater assets into the uncertain future provides various supporting frameworks and tools for this purpose (e.g. UKWIR, 2022). Focusing on asset health, their aim is to support asset planning by water and wastewater agencies. The frameworks include one for scenario planning, to standardise way in which states of systems that could exist in the future, and hence the services needed, under a range of conditions are envisioned.

Traditional drainage systems have been served over the past few decades by a well-developed knowledge base, particularly regarding their design and management. This has emphasised the need to take a risk-based approach, primarily to maximise the benefits in use, maintain functionality and invest wisely. Now that nature-based systems (NbSs) are gaining much more ground and are the preferred types of measure in many countries, the knowledge available from selecting and using grey infrastructure needs to be further developed and applied to systems that are in themselves, self-sustaining, evolutionary and dynamic. This is considered in more detail in Chapter 8 and Chapter 10.

As NbSs bring a range of benefits beyond simply controlling water quantity and quality (Chapter 8), this opens a new dimension of risk, one that needs to include all the various benefits that are considered important in the urban domain. In each case the way in which the drainage or surface water system performs over time must be considered. An example of how the value of a SuDS scheme may change over time, based on scenario planning, is given in Box 3.2. In many domains, catastrophic failures or even 'black-swan' events are considered in analysis and to inform planning for longer-term operation.

Box 3.2

Illustration of the potential changes over time of the value of the wider benefits of a drainage system, assessed using the CIRIA B£ST valuation tool (from Ashley *et al.*, 2018)

The figure shows how a proposed SuDS flood risk management scheme may have varying added benefits over three time periods under four different future scenarios. The benefits are primarily from reducing flood risks, which stay consistent over the 100 years, whereas other constituent benefits, such as water quality and recreational value, change to reflect societal changes.

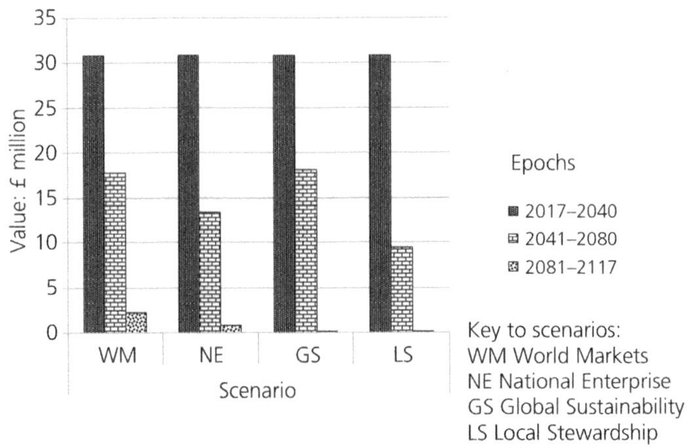

3.2.1 Impacts from stormwater

3.2.1.1 Flooding

With the growing changes in climate, there are clear signs that flooding is increasing in frequency and impacts, especially from surface water (for England, see NIC, 2022). The increases in population in countries such as England will also add to the scale of impacts. Table 3.4 shows the numbers of properties in England at risk from two levels of rainfall event (NIC, 2022). These impacts have been determined using national standardised databases for population and property distributions (Gov.UK, 2022), together with high-level computational modelling (e.g. Sayers *et al.*, 2020).

Table 3.4 Number of residential properties exposed to surface water flooding in England by settlement type (NIC, 2022)

Settlement type	Properties exposed to flooding – present day	
	High event (more frequently than 1 in 30 years, on average)	Moderate event (more frequently than 1 in 100 years, on average)
Rural town and fringe	24 700	58 806
Rural town and fringe in a sparse setting	1486	3219
Rural village and dispersed	18 731	37 632
Rural village and dispersed in a sparse setting	1664	3241
Urban city and town	139 317	349 100
Urban city and town in a sparse setting	671	1511
Urban major conurbation	130 834	348 084
Urban minor conurbation	8284	22811
Total	325 587	824 404

Figure 3.4 Adaptation measures – surface water flood hazard management (modified from Sayers *et al.*, 2022)

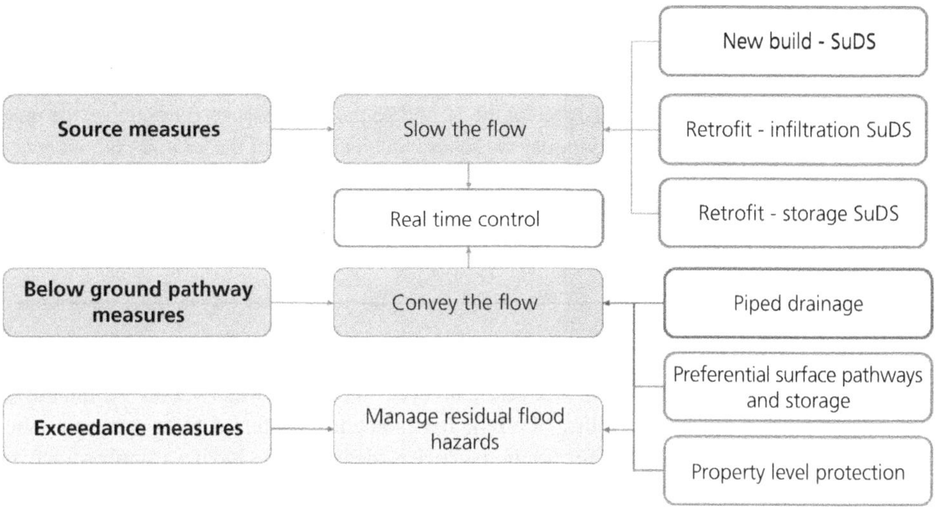

If not responded to, changes underway already will add to these property numbers and potentially to the severity of impacts – property damage, disruption, human health and ecological impacts (Environment Agency, 2019a). Ideally, responses need to address the physical flood management system: the regulatory, institutional and planning processes that frame the responsibilities and also define where and how property can be built (Chapter 5). Responses need to include social networks to help support people impacted or likely to be impacted. Non-structural measures include early warnings and, if necessary, evacuations. Responses require a range of measures, including infrastructure adaptation (or adaptability), which is to be maintained into the unknown future. Hence futures planning is required using, for example, scenario analysis. Figure 3.4 shows the potential responses to deal with surface water flooding hazards, categorised as source, pathway and receptor measures (NIC, 2022). Ideally, a balanced portfolio of measures needs to be selected, which collectively provide the protection required at a cost that is acceptable. Collectively, the various measures can provide a range of benefits other than flood control, depending on the measure. SuDS potentially bring many more benefits than below-ground systems (Figure 3.2 and Chapter 8), as recognised by NIC (2022), which calls for their greater usage in England.

In this chapter we consider only static interventions. Dynamic control of the piped drainage systems (real-time control) will increasingly have a place in the management of drainage systems (Chapter 9).

Adapting piped drainage to increase capacity requires augmenting existing assets and installing new assets – for example, increased pipe sizes and new storage tanks. New developments require below-ground assets to be large enough to accommodate the storm events that might occur in the future. Urban flood events are more likely to be brought about by shorter duration storm, but it is only relatively recently that predictions of these have been developed (e.g. for the UK, Chan *et al.*, 2021). It is incumbent on those dealing with infrastructure to make sure that the most up to date projections for climate change phenomena are accessed for planning, design and operation. Box 3.3 provides an example from NIC (2022).

Box 3.3
Example of short-duration rainfall uplifts in intensity used by NIC (2022)

Projections of uplifts in short-duration rainfall intensity were based on a 5 km national grid for 2°C and 4°C global temperature rise for the 2050s and 2070s. This was for a range of return periods and critical storm durations (1, 3, 6, 12 and 24 hours) and for the 50 and 95 percentile values. For example, the uplift in the 1 in 50 year rainfall from the present day was based on an average of the <3 and <6 hour critical storm duration assuming a 4°C temperature rise by 2100 (from pre-industrial times). Across England, this gave an uplift in short-duration rainfall intensity from 10% to more than 22.5% by 2100. The increases occur roughly in a south-east to north-west gradient across the country, with the greatest uplifts in the Lake District and Cumbria.

3.2.1.2 Diffuse pollution

Diffuse pollution occurs from dispersed sources, conveyed in surface water runoff continuously or intermittently into receiving water bodies (RWBs). It is sometimes referred to as non-point source pollution, to distinguish from point source pollution, which is identifiable as having a single source, such as a pipe outflow. Diffuse pollution can occur everywhere and anywhere that runoff occurs from surfaces, including fields, roofs, paved and unpaved areas, roads, etc., depending on how the runoff is managed. Discharges from piped drainage systems can readily be identified and hence managed, whereas often the source of diffuse pollution is difficult to track down. Sources could include a sanitary sewer wrongly connected into a surface water drain, a combined sewer overflow that discharges intermittently, or infiltrating farm runoff that emerges from shallow groundwater drainage. There are numerous sources that individually might not cause major problems to RWBs, but which collectively can lead to significant impacts. D'Arcy *et al.* (2000) provide a definition, shown in Box 3.4.

The principal sources and pathways for pollutants in the urban area are shown in Figures 3.5 and 3.6. Diffuse pollution can arise from all the sources shown, depending on how effectively the sources and pathways are managed. In the UK, detailed studies in the 1990s revealed the extent of the impacts from this pollution, matching these in other countries (see House (1993) for a scoping review). Details of its origins and nature in urban areas are given in Muller *et al.* (2020) and the scale and extent of the diffuse pollution problem are addressed by the Organisation for Economic Co-operation and Development (OECD, 2017), which points out that solutions are as much about policies as technical responses.

The amounts of diffuse pollution that arise from the different urban surfaces depend on the various activities locally. For example, grassed areas, pedestrian areas, cycleways and school playgrounds create very little pollution as there are no vehicles or very few chemicals on site. Car parks, haulage yards and some industrial and commercial sites cause more pollution, as vehicles and lorries move around and foodstuffs and chemicals may leak or be spilled. Heavy pollution loads may be created on busy roads,

Box 3.4
Diffuse pollution as described in D'Arcy *et al.* (2000)

Pollution arising from land-use activities (urban and rural) that are dispersed across a catchment, or sub-catchment, and do not arise as a process effluent, municipal sewage effluent, or an effluent discharge from farm buildings.

Figure 3.5 Principal sources, pathways and receptors for pollutants in urban areas

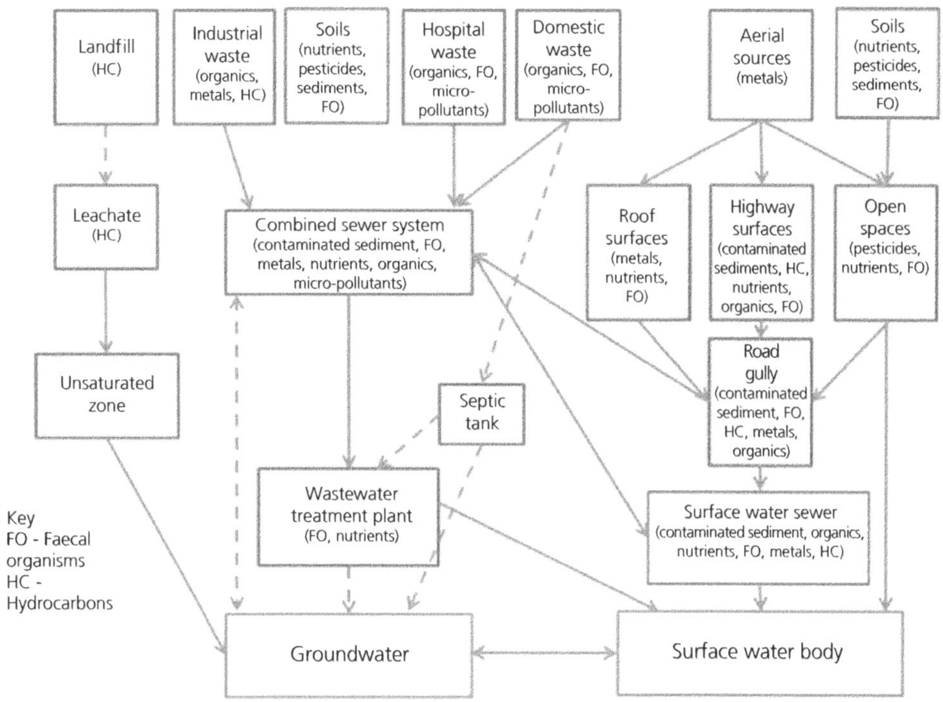

congested roads, waste transfer sites, waste storage areas and other areas where vehicles, fuel oils, chemicals, wastes and spills are present. The runoff from these surfaces will ordinarily be classified as surface runoff or stormwater and will be directed into the surface water sewer or the combined sewer.

There is now a much greater understanding of the extent to which the environment has been fundamentally damaged as a result of human activities, and those associated with urbanisation in particular. The Water Framework Directive (WFD), which came into force on 22 December 2000, established a new, integrated approach to the protection, improvement and more sustainable use of Europe's rivers, lakes, estuaries, coastal waters and groundwater. Previous European water legislation set objectives to protect users of the water environment from the effects of pollution and to protect the water environment itself from especially dangerous chemical substances. These types of objectives are taken forward in the directive's provisions for protected areas and priority substances, respectively. However, the directive also introduced new, broader ecological objectives, designed to protect and, when necessary, restore the structure and function of aquatic ecosystems themselves, and thereby safeguard the sustainable use of water resources. Future success in managing Europe's water environment will be judged principally by the achievement of these ecological goals, and reducing diffuse pollution will be a key contributor.

As for sanitary systems, there is an ever-growing range of substances that find their way into drainage flows. These 'cocktails' vary depending on land use, location, behaviours and a wide variety of other factors. While the cocktails in sanitary drainage systems are likely to be more easily definable as they are

Figure 3.6 Grouping of urban stormwater pollution sources and pollution transport pathways (adapted from Muller *et al.*, 2020)

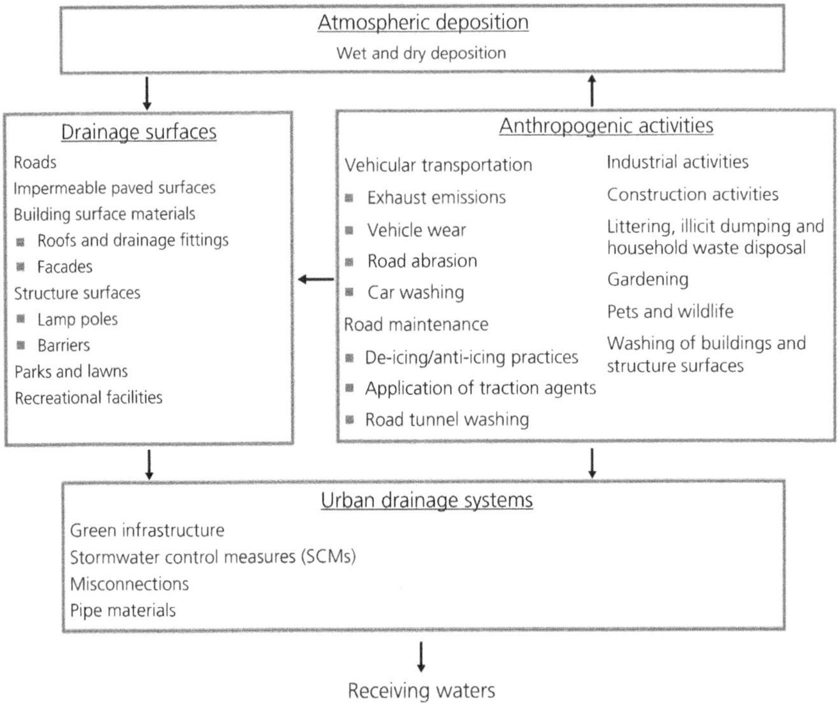

mainly from point source inputs (Chapter 2, Section 2.4.1), this is not the case for surface water systems. Sources include atmospheric fallout and the wide range of natural processes and human activities, and pollutants include the entire range conveyed in sanitary systems.

In 2012, Defra accepted that, in the heavily urbanised Thames river basin, road runoff accounted for the majority of water body pollution failures and more than 1000 were caused by diffuse pollution (Woods-Ballard and Cherrier, 2019). In England in 2015, about 11% of water bodies were identified as being damaged by pollution from towns, cities and transport. By March 2019 this figure had risen to 18% of water bodies, some 879 out of a total of 4950 assessed water bodies (Environment Agency, 2019b).

Ugarelli *et al.* (2011) suggest that diffuse pollution is likely to become more of a problem as the climate changes in the future in the western hemisphere. This is largely due to increased rainfall intensities and volumes (sources), but may also be as a result of changes in urban surfaces (pathways), ecology and even dominant microorganisms. This will be manifest in higher flow volumes, more frequent discharges from intermittent sources and possible changes in the nature and occurrence of the pollutants conveyed. In hotter drier periods, pollutants may accumulate on surfaces, ultimately being washed off in highly concentrated flushes into RWBs. Coping with such pollution may be more difficult for the RWBs when flows and dissolved oxygen are lower in hotter drier periods. Impacts may be offset if more nature-based systems are used instead of pipes as these are more effective at pollutant removal/retention (Chapter 8).

Whatever drainage treatment scheme is selected, it must effectively capture a proportion of the suspended solids, the particulate metals, the polyaromatic hydrocarbons and the microplastics particles from, for example, tyre wear. These pollutants are toxic, carcinogenic, bioaccumulative and affect the reproductive success of aquatic organisms (e.g. Honda and Suzuki, 2020). If they are allowed to discharge into rivers, streams and groundwater unabated, the pollution affects the entire ecosystem, causing immeasurable, insidious and enduring harm. A well-designed drainage and treatment system can control this pollution and help protect the environment.

3.2.1.3 Water quality standards

Untreated stormwater runoff containing high levels of sediment, oils, hydrocarbons and other pollutants flows through surface water sewers and often into inland surface waters and coastal bathing waters. Stormwater management controls provide an opportunity to reduce the levels of contaminants entering the environment through biological breakdown or settlement. Several recent approaches to protecting water quality have been presented within various EU Directives and are defined in the UPM Manual (Foundation for Water Research, 1998) for UK applications. These can be divided into two main groups

- emission standards
- receiving water standards.

The emission standards define the pollutant concentrations, loads and discharge frequencies that represent thresholds that need to be observed in order to maintain the environmental status of the RWB. In contrast, receiving water standards define the pollutant concentrations and persistence times that can compromise the environmental status of the RWBs if their frequency is higher than a specified threshold. The evident difference between the two approaches is that the former does not consider the effective characteristics of the RWBs or its self-cleansing capacity.

Gaining information about the status of an RWB from emission-based indicators – that is, establishing a direct connection between emissions and the RWB quality (immission) — is highly desirable. Such a connection would potentially reduce monitoring costs, as only pollutants that have a discernible impact on the RWBs would have to be monitored. In the past some attempts have been made to develop such links, but this work remains only theoretical in practice (Engelhard *et al.*, 2008; Freni *et al.*, 2010). The formulation of a connection between the RWB quality indicators determined by environmental regulations (i.e. WDF or UPM Manual) and emission-based indicators that consider both continuous and intermittent discharges of pollution are undoubtedly a key issue that must still be addressed and represents an outstanding crucial challenge for the effective application of the WFD across Europe.

3.3. Main surface water drainage

At the beginning of the twenty-first century some 62% of all main sewers operated by the English sewerage undertakers were receiving surface water runoff, both to separate storm sewers (26%) and to combined systems (36%). The extent of combined systems varied between the operators, from 9% up to 65% of the total proportion of public sewers. Around the world there are numerous countries where stormwater systems comprise surface ditches or open channels, both natural and engineered. The large open channels in the USA, usually trapezoidal in shape, are particularly famous because these often appear in popular films. A great number of such channels are contaminated by sanitary connections and refuse dumping (Figure 1.6). Combined sewers, with attendant CSOs, are found mostly in northern industrialised countries. Discharges from combined sewers into the environment have been monitored and, as an example, a French and German study found them to comprise some 69% to 95% stormwater (Petrie, 2021).

This section considers the principal components of stormwater drainage systems. First, property drainage is introduced, then highway and land drainage systems. These are considered in terms of separate stormwater systems. Combined sewer systems, including the associated assets (pumping stations and overflows) and potential impacts on receiving waters, are considered in Section 3.5.

3.3.1 Building drainage

In many countries and jurisdictions, property drainage is the responsibility of the property owner or dweller (where rented). However, in many industrialised countries, owners and dwellers are often neglectful of these responsibilities due to presumed reliance on institutions or municipalities. This is especially true of the management of flood risks, not only on the property, but also downstream, where any excess runoff can lead to problems for others. As traditional drainage techniques have relied on buried pipes and culverts, with only limited surface evidence of stormwater assets, an 'out of sight, out of mind' attitude has been understandable (NAO, 2004). With the growing use of surface-based drainage systems, this attitude must change. Clear visual evidence of how stormwater is collected and conveyed on surfaces will ensure that owners and dwellers will pay greater attention to the effectiveness of measures for this. However, this will take some time to come about, due to the networks of existing underground drainage systems which will remain in use for decades to come.

In developing countries there is an opportunity to avoid the 'everything underground' approach, employing, as far as practicable, the surface-based drainage systems described in Chapter 8, which can provide many added benefits compared with piped systems. In a process termed 'leap-frogging', developing countries could avoid the use of the less-sustainable systems prevalent in industrialised countries, instead utilising drainage systems that are more sustainable (e.g. Larsen *et al.*, 2021) (see also Chapter 12).

Building and property drainage has long been something of a Cinderella regarding opportunities, flood risks and flood control. In many jurisdictions the institutions responsible for planning and managing public space drainage have neglected to consider the associated drainage in the adjacent private spaces, particularly in terms of the potential impacts of excess rainfall, flow pathways and destinations. A notable example is England, where the arrangements for managing surface water are complex (Chapter 4). The myriad actors, stakeholders and often antiquated regulations have frequently ignored or left house- and property holders uncertain as to responsibilities. For example, in Bradford (Yorkshire) in the 2000s, some highway flooding was determined to be caused by a blocked land drainage culvert that passed beneath properties and gardens. The owners had no idea that the culvert existed and that they were responsible for maintaining the length on their land and keeping it clear. This and several other examples related to local drainage issues are outlined in Ashley *et al.* (2007).

Figure 3.7 shows the sources and pathways taken by property drainage from roofs and surfaces in the area within a property boundary (termed the curtilage). Drainage should be safely managed on the property and, ideally, the surface water utilised for toilet flushing or irrigation. New property drainage in the UK should comply with BS EN 12056-3:2000 Gravity drainage systems inside buildings – Roof drainage, layout and calculation (BSI, 2000) and Part H of the Building Regulations in England (HMG, 2010). Systems should be designed with due regard to the nature and use of the building, ideally using

Figure 3.7 Simple illustration of building drainage sources and pathways

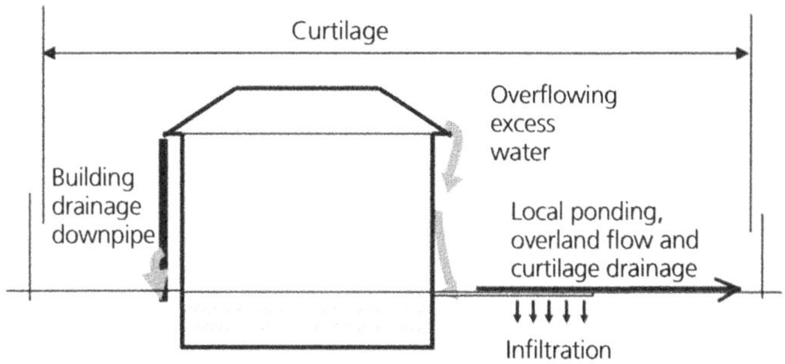

rainfall statistics for the location; otherwise, the standard provides rainfall intensity rates (litres per second per square metre). There are two main types of roof drainage system: siphonic systems and gutters and downpipes. The former provides a more rapid movement of water from roof surfaces and therefore can utilise smaller pipes. As yet there are no standards specifically for siphonic systems, although it is recognised that these may be affected by downstream backing up (as the outlet must be sealed), and in turn can cause downstream problems by concentrating flows very rapidly.

Recently there has been growing interest in blue roofs. These contain deliberately constructed roof spaces that retain the stormwater, which is then slowly released into the main drainage system. Managing stormwater at roof level can alleviate the need to enhance ground-level drainage, but it has the disadvantage of adding significant loading to the building structure and increasing the risk of waterproofing membrane failure. Blue roofs, alongside green roofs (Chapter 8), provide multiple benefits, especially by aiding cooling and protecting against solar impacts, which are important owing to the growing problems of urban heat islands. Usually, such roofs also include green or SuDS spaces, which create many added benefits.

The age of an existing property will also help define the likely standard of performance of its drainage systems. For most UK properties this should be for a once in 30 years storm and/or a rainfall intensity of 75 mm/hour as this has been the design standard. When the property has been well maintained and there is some confidence that construction has been to prevailing standards, these rainfall events should be adequately accommodated by the building and surrounding area drainage. Further analysis can then proceed on this assumption. However, given that climate changes have proceeded over the past decades, older properties will not necessarily be able to accommodate what is now the changed once in 30 years event – that is, the return periods may now be less than this depending on location.

There is a growing use offlood-protection measures at properties to address increasing flood risks, which are mainly due to climate change, but also because properties are being built in high flood risk areas. Flood resistance and resilience measures may be used to stop the flood water entering areas where it causes damage, or to help a propery to recover quickly when it floods (e.g. Dhonau, 2022; Kelly *et al.*, 2021).

For most purposes the analysis of property drainage systems, which are typically small scale, can be done using specified rainfall (e.g. Table 3.3 and Chapter 5) and simple rational method runoff estimates (Chapter 7). For larger properties and complex roof and paved areas, including mixed green-grey surfaces, a more detailed approach may be appropriate. With the growing use of green roofs, and in some cases blue roofs, together with pervious surfaces (Chapter 8) and water conservation efforts, such modelling is becoming more important for the understanding of drainage flows in each part of the building drainage system shown in Figure 3.7. There are computational tools available for this, such as the US-based Storm Water Management Model (SWMM; USEPA, 2022), although such models are aimed mainly at main drainage systems (see Chapter 11). Most of the available models do not explicitly represent the complex flow conditions in a building, including branched downpipes or siphonic systems. Specialist computational models are needed where these are to be analysed, for example for multistorey buildings (e.g. DRAINET, e.g. Mushin *et al.*, 2016). These models are also being used to understand air flows and ventilation in closed surface and sanitary sewer draining buildings.

Everywhere water occurs, even wastewater, there are opportunities to use it as a resource. Direct rainfall on to rural land provides irrigation; it is usually where spaces are sealed, where people live, work and use public and private spaces, that rainfall can be a problem. However, runoff can occur from pervious surfaces, even from agricultural land, leading to what is termed ‘muddy flooding’. Figure 3.8 illustrates the potential for utilisation of stormwater. This can be for direct use, termed harvesting; for irrigating vegetated areas, natural or planted; and even for aesthetic purposes in blue and green spaces. Rainwater can be used directly in buildings for a variety of purposes, including toilet flushing, clothes washing and external purposes. In many countries the unit price of supplied mains water is so low that the economics of direct rainwater use are not conducive to the use of this source. In places where water stress is significant, including many parts of Australia, the direct use of rainwater, even for potable purposes, has

Figure 3.8 Potential for all types of water to be opportunities for use (Ashley *et al.*, 2013)

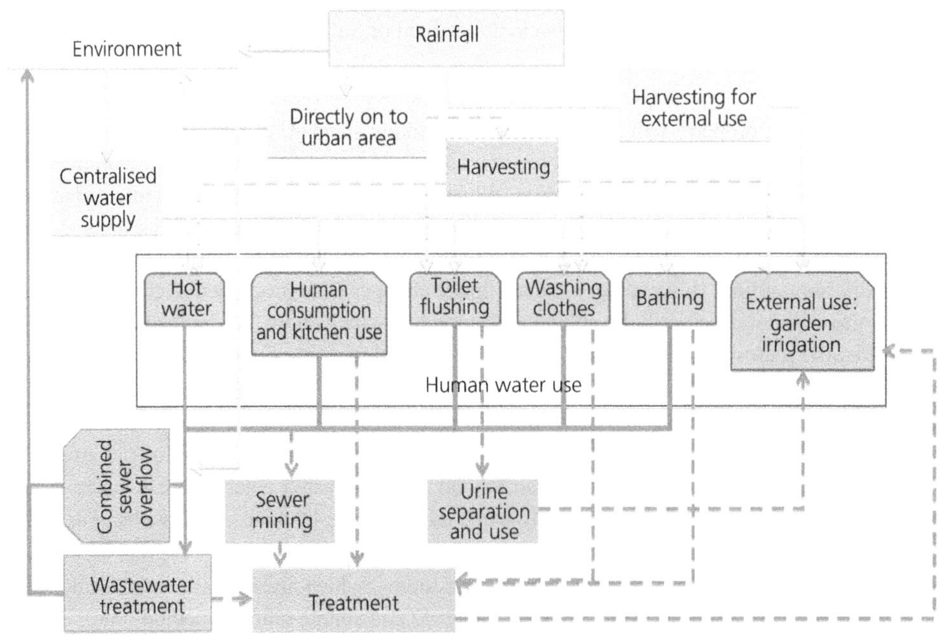

become common, despite the obvious risks from contamination by bird droppings, atmospheric fallout, leaching of roof materials and other substances (e.g. Chubaka *et al.*, 2018). Where rainwater is sourced from ground level the potential contaminants are even more diverse (Leggett *et al.*, 2001). Rainwater harvesting requires tanks for storage, which need to be large where this is a major source of supply. Figure 3.9 shows a storage tank in Melbourne, Australia, which is large in relation to the size of the property served.

Public buildings and larger premises, especially where significant amounts of water are used, may benefit from the use of harvested rainwater and/or recycled wastewater. Analysis requires inventories of usage, to determine the relative economics of so doing (e.g. Roebuck *et al.*, 2012). Recently, there has been growing interest in the role of rainwater harvesting systems in downstream flood mitigation (e.g. Cristiano *et al.*, 2021). For this to be effective, there needs to be assurance that the storage can be as available as possible prior to a storm event, which often requires intervention to ensure that it is. Novel dynamic control systems have been developed to empty storage tanks and rainwater barrels automatically when heavy rain is forecast (e.g. Oberascher *et al.*, 2021).

There are no specific regulations regarding the direct use of rainwater in the UK, although recommendations are provided in BS EN 16941-1:2018 (BSI, 2018b). However, where there is any risk of contamination of water supplies, by backflows from the reclaimed water, regulations need to be followed as explained by Water Regs UK (2022). More information is provided by the UK Rainwater Management Association (UKRMA, 2022) about harvesting and other aspects of direct rainwater use.

Figure 3.9 Rainwater storage tank in Melbourne, Australia (courtesy of Darren Bos)

3.3.2 Highway drainage

Highway drainage systems are designed to capture surface water runoff to alleviate flooding, both to protect the fabric of the road and to ensure that it is safe for vehicles to travel. In the first industrialised countries, the responsibilities for and approach to drainage of highways may differ depending on the highway. Major roads, such as motorways and turnpikes, may be considered differently from minor roads. Higher speeds and greater volumes of traffic on major roads may require different considerations. There are also climatic considerations, with different approaches taken in more northerly climes, with frequent freezing conditions, compared with warmer southerly climates, where ice rarely, if ever, occurs on surfaces.

Currently some 30–70% of storm flows in combined sewers in England are from roads and footpaths. These flows dominate the capacity of drainage networks, can lead to overflows and pollution (Section 3.8), compromise downstream treatment capacity, cause flooding, increase pumping needs, add to wastewater treatment costs and require increased energy use, with attendant carbon emissions.

In the UK, highway drainage design should comply with National Highways (2022). For a one in 1 year storm event there should be no surcharge of the drainage system; and for the one in 5 years event, no flooding from the drainage system. Also, highway surface water flooding should not extend beyond the highway boundary up to the one in 100 years rainfall event, including an allowance for climate change. Climate change should be accounted for by using a 'minimum design life of 60 years' and '[f]or the design of carriageway drainage, calculation of a 20 % uplift in peak rainfall intensity together with a sensitivity test to 40 % uplift in peak rainfall intensity'.

National Highways (2022) provides tables that list the advantages and disadvantages of various drainage options for highways, including pollution control.

Highway authorities will, in general, aim for surface water to be from road surfaces as soon as practicable. Traditionally, stormwater would be conveyed rapidly into natural watercourses, by way of limited lengths of pipework, or into a sewer network. As the potential damaging impacts from both high flows and large volumes of highway runoff and the pollutants conveyed became better understood in the twentieth century, highway drainage practices became more cognisant of the need to control both flows and diffuse pollutant impacts (D'Arcy *et al.*, 2000). A complex cocktail of physical, chemical and biochemical substances is conveyed in runoff from highway surfaces, which varies seasonally (e.g. Mengistu *et al.*, 2021). In many countries some control of substances washed from highway surfaces has been in place for decades, in the form of gully pots. These are small settlement chambers used to collect solids in sumps as the stormwater runoff flows through. They also collect some of the associated pollutants, including hydrocarbons that adhere to the solids settled (e.g. Rietveld *et al.*, 2020). Typically, these can only convey runoff from storms of 1–2 years return period.

Although gully pots can effectively capture some of the pollution in the road runoff, the maintenance operations that they require (emptying out the captured sediment every 6 or 12 months) may be perceived as too onerous. Such interceptors are not used everywhere. In southern France, inlets from streets lead directly by way of drops into sewers, with attendant inputs of large solids, such as plastic bottles and cans, which accumulate at the bottoms of these shafts (Ashley *et al.*, 2005). In countries such as the USA, street inlets can be very large, more than a metre in width, due to the intensity of the rainfall needed to be coped with.

In some regions, there is a shift away from using gully pots in new road building schemes for trunk roads and motorways (Highways England, 2020). Larger, more centralised treatment devices are often deployed, which can be located away from the road and emptied more easily and more quickly. These devices can be vortex grit separators or oil/water separators. Where they discharge into a particularly sensitive environment, or there is a high risk of spillage, they can include a remotely operated automatic closure device on the outlet, which can be activated if there is a spill of hazardous materials on the highway. The requirement for these closure devices will be even more important as we move to electric vehicles, and the control of firewater runoff should be considered.

On new road schemes, the default destination for the highway drainage is often the existing sewer network, either separate or combined. In countries such as the UK this effectively transfers the burden of managing the runoff from the highway agency to the sewerage undertaker and is perceived as the 'low or zero' cost option. In new developments, property layouts and highways are typically designed with scant regard for drainage, as it is perceived that the drainage can be fitted in as required. There is often a disconnect between those responsible for roads and highways and the drainage engineers, with the former presuming that 'drainage is easy once the road has been designed'.

The drainage system should, where possible, include vegetative treatment devices as part of a nature-based scheme. Such systems can provide attenuation of flow and partial treatment of the pollution contained in the road runoff. Schemes now include ponds, reedbeds and bioretention zones alongside road schemes. However, the pollutants in road runoff include toxic organic compounds, toxic metals and microplastic tyre wear particles, and the level of pollution can be too high for the receiving watercourse or groundwater. It is essential to recognise that road runoff from trunk roads, motorways, congested roads and busy junctions should be treated and a proportion of the pollution captured before it enters any vegetative treatment device. This treatment can be provided by filter drains, vortex grit separators, oil/water separators or in situ built sediment traps (Chapter 8). These devices must be maintained in accordance with the manufacturer's or designer's instructions to ensure that they continue to work as intended. They will capture between 50% and 80% of the gross pollutants and allow downstream vegetative devices to 'polish' the runoff and capture some of the residual pollution.

Linear treatment devices, which are installed along the length of the road, such as filter drains and swales, should be considered and be selected if possible. They capture and treat the pollutants as close to the point of discharge as possible so that each linear metre of the device only has to treat the runoff from the adjacent linear metre of road surface. This provides the maximum capacity for stormwater attenuation and treatment, and is preferable to conveying the stormwater downstream in a piped drainage system for end-of-pipe treatment.

Scheme selection must also take account of the nature of the receiving environment, whether it is a small, slow-flowing stream, a vulnerable aquifer, an estuary or another aquatic environment. The sensitivity of the receiving water should dictate the level of treatment that the road runoff receives. Careful consideration must also be given to the waste disposal route and destination of the hazardous sediment that is captured in treatment devices.

The most important design criterion to define is the treatment flow rate for the treatment devices. Whereas there is extensive guidance for the management of runoff volume during design storms, there is little guidance on the rate or volume of road runoff that must be treated by the devices, so, to design a good scheme and prevent environmental harm, the designer must ascertain how much treatment is required.

There are currently no incentives for highway authorities in countries such as the UK to minimise the costs of dealing with runoff from the roads or seek alternative receptors, although it is accepted that there may be limited opportunities within urban areas. In most parts of the world, highway runoff is either conveyed directly to a watercourse, with scant thought for the potential impacts, or put into main sewers, on the presumption that they can cope with it. Due to increasing demands on sewer networks in denser populated areas, a more comprehensive and robust procedure for managing discharges from the highway to the sewer is required than is evident in most countries.

3.3.3 Land drainage

Land drainage refers to the management of water on or in the land, to improve the land for purposes such as agriculture, development and navigation. In this context, 'drainage' includes defence against: water (including sea water); irrigation; warping (deliberately flooding land to improve the soil); and any other practice that involves management of the level of water in a watercourse. Land drainage functions can be performed by landowners and/or a range of different organisations. In the UK, while there is no legal right of connection for land drainage to sewers, it is not uncommon for these systems to be deliberately, or erroneously, connected to sewer systems. This type of inflow can have a significant and detrimental effect on the capacity and condition of the sewer.

3.4. Management of stormwater

Increased peak flows often result in hydraulic overloading of the sewer system and are a major contributor to flooding. Highway drainage flows and surface runoff are often the source of large quantities of sediment, which make up a large proportion of the sediment entering sewers. This sediment can cause reduced sewer capacity, blockages and increased emissions/premature discharge from combined sewer overflows and consequential environmental impact (Ashley *et al.*, 2005). In addition, the abrasive particulate matter can cause excessive wear on the mechanical apparatus at pumping stations, reducing performance and increasing the risk of failure.

Climate change and urban creep are placing increasing demands on all drainage systems. The effective management and control of stormwater underpins the functioning and performance of sewer systems for the benefit of public health and safety. To support expected social, economic and environmental outcomes, new thinking and alternative ways of managing stormwater are critical. To help prepare more effectively for current and future needs, inputs to the sewer need to be managed better by reducing and controlling stormwater flows, where appropriate to do so, providing flexibility and increased capacity for adaptive management. This, however, is a complex problem with highly dynamic hydraulic conditions. Fortunately, the use of contemporary computational models and monitoring systems can provide the information needed to optimise the performance of above- and below-ground drainage systems, as outlined in Chapters 9 and 11.

The approach needs to reflect the various occurrences of surface water as illustrated in Figure 3.2 and outlined in Table 3.2. This should not be considered as a once-and-for-all 'solution' to stormwater management problems. Given the dynamic nature of the systems involved, both natural and engineered, regular monitoring and assessment of the performance of the drainage system is required, with remedial measures being implemented as necessary. The shift in practice from 'below-ground grey infrastructure is best' to managing more water on the surface using natural systems has come about due to the need to respond to these changes. Human needs have remained much the same: safe, secure and pleasant environments, but the drivers of change have compelled a response that is changing practice far more rapidly than in the past. This includes the need to consider surface water as a resource rather than just a problem.

3.4.1 Designing for exceedance

The place of exceedance drainage is illustrated in Figure 3.2, zone 3. Design for exceedance is different from 'normal' design of urban drainage systems, which traditionally sought to manage stormwater below ground. In the context of surface water in an urban environment, exceedance can be defined as (Digman *et al.*, 2014): 'flow that is conveyed or stored on the surface because the capacity of a drainage system carrying stormwater has been exceeded'. This may or may not be perceived as 'flooding'. The public traditionally consider all forms of water on the surface as 'puddles', and large puddles on roads as flooded areas. This view is also taken by many engineers responsible for highways, who try to avoid puddles at all costs. The safety of drivers of motor vehicles does depend on safe driving conditions, but, in towns at least, it will be necessary to store more water in puddles on road surfaces if surface water is to be managed affordably. Designing for exceedance helps to manage the excess flow that may occasionally occur on the surface within the existing urban environment to reduce the impact on homes and other vulnerable assets (zone 3 in Figure 3.2). This normally takes place on the surface, making the most of the available space and creating or utilising multifunctional infrastructure, as illustrated in Figure 3.10 for the Manor housing development in Sheffield, which holds water only for storms of one in 100 years or greater and is otherwise used as a community amenity.

Urban redevelopment can be exploited to build in capacity to manage pluvial flood risk on the surface by alteration of storage volume and location, flow paths and surface water generation (Diaz-Nieto *et al.*, 2012). By moving on from traditional design, the landscape can be better used as a means of managing surface water runoff, integrating this within place-making (Chapter 5). It is possible to use roads as flood

Figure 3.10 Exceedance basin in Manor housing area, Sheffield (courtesy of R. Nowell)

pathways to channel and convey exceedance flows by strategically increasing the height of kerbs or profiling the carriageway to direct flow or ponding away from vulnerable locations (NIC, 2022). Similarly, footpaths and central reservations to be designed to be lower than the road, allowing water to drain naturally into vegetated systems to be conveyed, stored and treated. Often, small changes to the surface, such as changing the slope of a footpath or removing a low point in a kerb, can be the most effective way to manage surface water. By using a least-cost path analysis, which in addition to topography incorporates land use constraints, an optimum path can be identified and engineered into the busy urban landscape to manage 'nuisance' overland flow paths (Diaz-Nieto *et al*, 2016).

Effective measures to avoid surface water flooding require a holistic approach to water management at both local and landscape scales, with actions needed across a range of sectors and including both built infrastructure and nature-based solutions (HM Treasury, 2021) as well as engagement (Chapter 4). The use of stormwater management measures and blue-green infrastructure or NbSs options will necessitate changes in land use and planning policies, together with modified design standards. The operation and maintenance of such multifunctional spaces as shown in Figure 3.10 requires closer cooperation between stakeholder organisations, communities and individuals.

Designing for exceedance involves combining a variety of measures, depending upon the nature of the catchment, ambient drainage features and the opportunities (type and layout of space) available (Balmforth *et al.*, 2006). The right combination of green, blue and grey infrastructure can provide a natural and more sustainable approach to the management of severe rainfall and exceedance events. BS ISO 24536:2019 (BSI, 2019) includes the following (see Figure 3.11).

> During rainfall events of small to moderate size, stormwater runoff is either captured by pervious surfaces and underlying permeable media (from where it may infiltrate or evapotranspire) or contained within the underground stormwater pipes and/or open-air transport channels (low-flow). During larger, less frequent storms, runoff will exceed transport capacity of the stormwater system leading to controlled or uncontrolled flooding (high-flow).

This approach recognises the value of making the most of multifunctional infrastructure and sharing space. Highways offer one of the largest shared spaces in urban areas, as do public open spaces, such as playing fields or parks. Each of these can be utilised in appropriate circumstances (Digman *et al.*, 2014).

Figure 3.11 Surface water flow and management regions (BSI, 2019). (Permission to reproduce extracts from British Standards is granted by BSI Limited. No other use of this material is permitted. British Standards can be obtained in PDF or hard copy formats from the BSI online shop: www.shop.bsigroup.com)

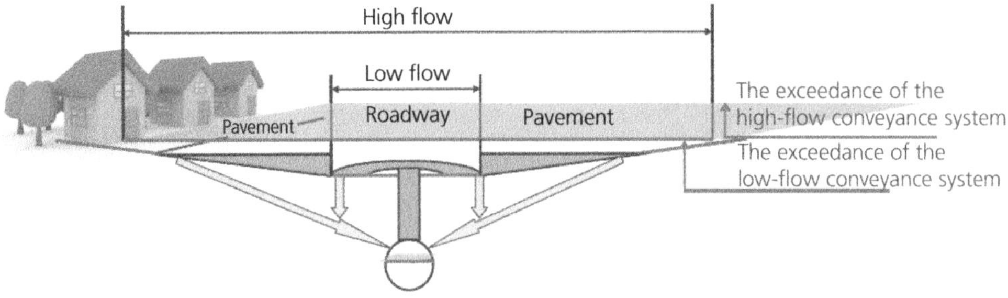

As designing for exceedance makes use of available space and often creates (or formalises) multifunctional assets, professional disciplines need to work together, especially if multiple-benefit opportunities are to be realised (Chapter 3). Emergency responders also need to recognise that exceedance events need to be dealt with in different ways to flooding events and to engage with other professionals, thus ensuring exceedance remains an inconvenience rather than a flooding problem. Disciplines working together will help to avoid imposing inappropriate solutions and to mitigate the perceived or real risks of a scheme.

Despite many jurisdictions having a planning framework that implies designing for exceedance should be an intrinsic part of all design, including for development planning in England, it is still not standard practice for new developments. The National Planning Policy Framework (NPPF) explicitly stated that exceedance flows needed to be considered in design, although this is now no longer mentioned (DLUHC, 2023). To strengthen these requirements for drainage of new developments, municipalities can and should produce clearer planning documents to ensure that designing for exceedance is appropriately accounted for. (See also Chapter 4.)

3.5. Combined sewer systems

The various types of sewer are introduced in Chapter 2. Across Europe there are estimated to be more than 3 million km of sewers and 650 000 combined sewer overflows (CSOs) (EurEau, 2020). Most of these sewers are combined, conveying both sanitary and stormwater flows (see Section 2.3.1). Combined sewers are the most common type in developed countries. In developing countries where sewerage has been introduced over the past century or so, most of the infrastructure is notionally separated. Despite aspirations to separate sanitary and storm flows, wrong connections are ubiquitous, and even a few misconnections of sanitary flows into separate storm sewers can invalidate the purpose of separation (i.e. to reduce receiving water pollution) and compromise the designed sewer capacity. Figure 3.12 shows a notionally separate stormwater drain in the capital of Colombia, Bogota. Significant volumes of incorrectly connected sanitary flows in this drain adversely impact the local park through which it flows and lower the quality of the receiving water, the Arzobispo River. It can be concluded that, despite the best of intentions, the use of separate sewer systems is highly risky in terms of impact on both water pollution and the capacity of the network. Recently, with the increased recreational use of riverine and coastal waters, the risks of infections and toxic impacts on humans from overflows and wrong connections have become more apparent. In addition, any flooding from combined sewers is known to increase human health risks, albeit separate stormwater system flooding also conveys a variety of health-impacting pathogens (e.g. Han and He, 2021).

Table 3.5 summarises the relative merits and drawbacks of combined sewers and separate sewers (see also Chapter 2). Partially separate sewers – which convey both sanitary flows and limited surface runoff from discrete area, such as properties – are not included.

The use of sewer overflows is a form of designed pollution (Section 2.3.1), based on the presumption that when overflows occur from combined systems during rainfall, receiving waters are also more able to cope with the added pollutants due to higher streamflow and dilution. The nature, range, concentrations and frequencies of the discharges have been routinely monitored only relatively recently. There are special concerns regarding present emerging pollutants (e.g. Petrie, 2021), although historically in industrialised countries uncontrolled emissions from industries into water bodies were obvious to the population, when 'rivers ran red, green or orange' with toxic and carcinogenic substances. This is still occurring in many parts of the world (e.g. Hooton, 2014) where industrial and commercial practices are

Figure 3.12 Notionally separate stormwater channel in Bogota, Colombia (courtesy of Richard Ashley)

poorly controlled, or where people are misbehaving (e.g. *Mail Online*, 2012). It is easy to pour a polluting substance such as engine oil into a stormwater gully or drain; indeed, in the last century it was common practice for city dwellers in countries such as the UK to dispose of used car engine oil in this way.

National programmes to separate combined sewer flows were implemented in many countries, including the Netherlands and Germany, in the twentieth century and into the start of the twenty-first century. No-dig technology provided the means to install a new set of pipes to convey either the sanitary or stormwater runoff flows, separated from the original combined sewers. Unfortunately, these programmes soon found that separation was not a panacea and did not necessarily yield the expected beneficial outcomes, as outlined in Box 3.5.

Since 2006, when NORIS was completed, the emerging perspective on management at source (Chapter 2) has changed from asking 'What wastewater goes into what sewer?' to 'Does the surface water need to be included in the wastewater at all?'. The Dutch case example in Box 3.5 needs now to be considered in terms of the shift in professional perspective that puts control at source, resource utilisation and nature-based infrastructure at the forefront of wastewater management (STOWA, 2012). Unfortunately, the reluctance or incapability of policymakers to take this shift seriously and effect the change in policy, regulation and funding from 'all to sewer' to one of resource utilisation is apparent worldwide, with a few notable exceptions. As Ampe *et al.* (2020) state in relation to the situation in the Netherlands: 'If a water or sanitary engineer from over 100 years ago was to return to the present time [...] they would

Table 3.5 Benefits and problems of the main types of sewerage system

Characteristic	Combined system	Separate system
1. Flow regimes	Predictability uncertain. Typically, wet weather flows some 30x DWF System cannot be 'self-cleansing'	Sanitary system flow is quantifiable based on contributing population Wet weather system flows statistically predictable Wrong connections may make flows less predictable
2. Pollutants	Mixed flow quality may contain complex and varied mix of particulate and dissolved substances. Wet weather dilutes dry weather flow strength but erodes deposits. CSO spills cause pollution System likely to have most sediment deposits due to variability of flow regimes	Surface runoff may be polluted. Without treatment, the storm drainage may cause significant receiving water impacts Wrong connections, infiltration and inflow may cause significant pollution Sanitary DWF may convey mix of physical and other substances, but provided trade effluent is controlled, this should be within expected ranges
3. Effects at downstream treatment plant	Shock loads at downstream wastewater treatment plants during wet weather. Prolonged rainfall can cause loss of treatment biomass from plant (washout)	Quality and flow regimes most consistent in sanitary system, most likely to optimise treatment efficiency

Box 3.5

The NORIS project on separation of combined sewer flows

No Rainwater in Sewers (NORIS, 2022) was a European project that looked at sewer separation in several countries using no-dig technology and in-sewer relining. This also included retrofitting strainers into stormwater gullies to reduce solids inputs to combined sewers. A key finding was that separated surface water systems were severely impacting receiving waterbodies due to customer misuse. In a Dutch catchment, a plasterer was routinely washing tools into house drains that drained to a pond and wetland. Bioassays determined that this was depleting the ecological vitality of the natural environment. It was also found that when set in the context of sustainability assessment, a one-solution panacea approach like separation was not applicable everywhere, with solutions best suited to local circumstances and context (Hurley *et al.*, 2008).

probably marvel that we continue to use the very same infrastructure that they and their colleagues installed'. Wastewater continues to be low down the list of important aspects of society to be managed by governments and, despite what professionals believe to be the better approach to wastewater management, it is hard to engage governments in the need to change practice.

Combined sewers will continue to be as common for the foreseeable future as they are now. The vast legacy of existing, mainly combined sewers in developed countries, many more than a century old and still functioning perfectly well (Figure 3.13), would require huge resources to replace or separate.

Figure 3.13 The interceptor sewer in Dundee, Scotland, built in 1888 and still in use today (courtesy of Richard Ashley)

Separation would lead to problems regarding low-flow conveyance of sanitary flows in such a large cross-section sewer, with sediment deposition and toxic gas build-up (see below). Over time, surface water flows may be reduced where managed at source using NbSs or for direct utilisation of the resources needed for supply of water. This will gradually reduce the flows in combined sewers, potentially diminishing the problems caused, including from overflows during rainfall. It is important to understand the functioning of combined sewers, including the nature of the processes involved, if impacts from overflows and the consequential effects on downstream wastewater treatment plants are to be better managed.

3.5.1 Processes in sewers

Sewers may be described as reactors in the wastewater treatment sense (Section 3.5.2). All key processes occur: physical, chemical and biological. These processes have been more closely studied over the past few decades, thanks to improved instrumentation (Chapter 9) and analytical procedures (e.g. Gao *et al.*, 2023). There are some issues that have been affecting sewers ever since they were first used, including sewer solid deposition and sewage gas production, responsible respectively for blockages, loss of capacity, toxic and odorous compounds and corrosion, but it was only in the latter part of the twentieth century that the science was investigated in detail. Through studies undertaken at laboratory scale and in situ, the development of understanding of sewer processes was largely driven by the need to use this knowledge for computational models of hydraulic performance (Chapter 11).

Figure 3.14 shows sediment inside the Dundee interceptor sewer prior to cleaning. Figure 3.15 illustrates diagrammatically the various processes taking place inside a combined sewer such as the interceptor. These processes are different for separate storm and sanitary sewers. In storm sewers there are fewer biochemical process interactions and in sanitary sewers there should be less granular material (Ashley *et al.*, 2005; Hvitved-Jacobsen *et al.*, 2013). Box 3.6 illustrates some of the issues related to sewer solids and biochemical processes.

Figure 3.14 Sediments in the combined sewer in Dundee (courtesy of Richard Ashley)

Figure 3.15 Diagrammatic illustration of sewer processes (adapted from Ashley *et al.*, 2005)

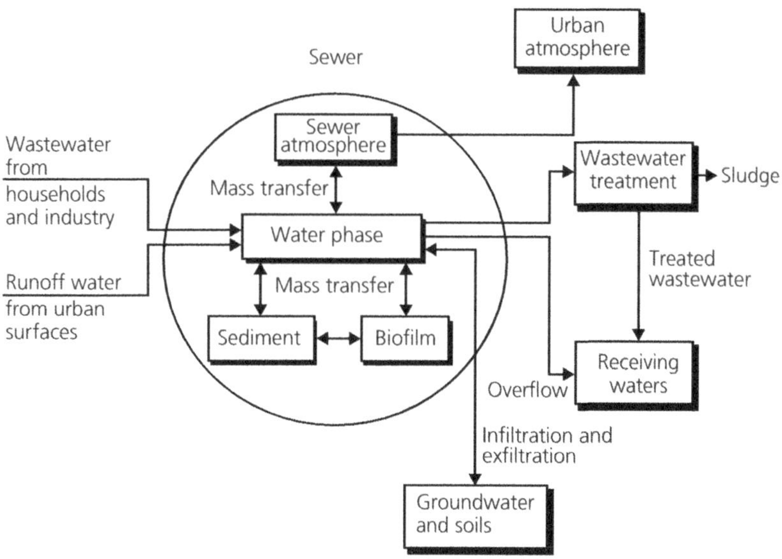

Box 3.6

Sewer solids and biochemical processes

In a UK study CIRIA (1986) concluded that the problems of sewer sediments were widespread in the UK and that up to 25 000 km of sewers and drains were likely to be affected, at a cost of some £60 million per annum. There was a lack of knowledge about the nature, properties and occurrence of sewer sediments, and criteria for the design of sewers had traditionally been based on empirical and experiential information derived largely from local knowledge and engineering judgement. Sixty-seven per cent of the sample of sewerage authorities questioned about the problems caused by sewer sediment failed to attribute water quality problems directly to sediment deposits. At that time the new generation of computational models of sewer flows was being developed, and the occurrence, behaviour and effects of sediments needed to be better understood for modelling.

The deposition of solids, formation of biofilms and complex biochemical and biophysical reactions occur both transversely across the sewer section and longitudinally (Figure 3.15). In general, the transformation of carbon along a sewer length will depend on retention time, changing from being readily available for bacterial metabolism to less readily available. With weak sewage, or long retention times, the wastewater arriving at a downstream plant may lack the key substrates, primarily carbon and nitrogen, for effective biological treatment. The transport capacity during dry weather is usually low, and organic and granular solids will deposit. Processes in these deposits and the attendant biofilms will potentially act to bind the deposits, and conditions in the deeper deposits may become anaerobic. During prolonged periods of dry weather this cohesion can result in permanent deposits (Figure 3.14) and septicity. With wet weather, the less solidified deposits will be eroded, with the organic material more susceptible to transport, both in the suspended phase and near the bed.

Control of sewer solids at source should be the starting point for managing potential problems (Ashley *et al.*, 2005). This requires engagement with users, to prevent wet wipes and other textiles being flushed into toilets. The growing use of wet wipes in the twenty-first century and misleading marketing information that these are 'flushable' illustrate how the nature of what is introduced into the wastewater system can change over time (e.g. Harter *et al.*, 2021; Beamont, 2022). Wet wipes, together with disposal of fats, oils and greases to sewers, have been identified as causing the build-up of 'fatbergs', which block sewers in lengthy plugs, requiring difficult and costly removal. Wet wipes and other textiles and plastics can also cause blockages of screens and pumps when disposed of to the sewer. Solids can also originate from commercial and industrial activities, including building construction, where they are poorly controlled (Murnane *et al.*, 2006). The growing problem of plastic particles, especially of micro and smaller sizes, has emerged in the twenty-first century as a major issue in the environment and in surface water runoff and sanitary and commercial effluents (Chapter 2). It is recognised that these are ubiquitous and found in sewers; in deposits on the bed and on walls, in suspension and transported into the environment by way of CSOs and through wastewater treatment plants. Control at source through changes to manufacture and disposal habits is needed to prevent major problems, although in some countries, costly removal facilities are being installed at treatment plants.

The other main source of solids deposited in sewers is the conveyance of granular particles in runoff, encompassing the entire range of grit sizes (Ashley *et al.*, 2005). Gully pots, inlet traps and other devices are intended to restrict the entry of these solids from highway surfaces (Section 3.3.2), but require regular cleaning to be effective (Butler and Clark, 1993). This surface washoff also brings organic material, especially seasonal leaf fall, which can contribute to blockages (Rietveld *et al.*, 2020).

Towards the latter part of the twentieth century, a lot of efforts were made to control the larger textile and plastic solids conveyed in sewer flows from being discharged by way of CSOs. Screens have long been used at treatment plants, but the problems of screen blockages at CSOs were recognised, as these required costly operational intervention to clear. Development programmes mainly in the USA and UK led to novel technologies for collecting and removing these solids that are now used routinely, comprising various automatic cleaning screen designs and a range of vortex and other devices that separate out these and the heavier granular particles (Ashley *et al.*, 2005).

Traditional sewer design has utilised sediment transport equations to minimise deposition and maintain sewer capacity. Typically, a minimum flow velocity of 0.75 m/s occurring at least once per day was presumed to be effective in sanitary sewers. In combined sewers the storm flows were presumed intermittently to scour sediments that have been deposited. However, given the range and nature of solids in sewers, ranging from low density large organic material to fine granular particles, together with the cohesiveness of fatty and organic materials in mixtures conveyed in different ways, these presumptions are known to be too simplistic. Design and analysis of existing sewers to manage sediment problems should consider the more complex phenomena as recommended in Ackers *et al.* (1996) and Ashley *et al.* (2005).

Computational models available (Chapter 11) for modelling flows in sewer systems now include sediment transport and deposition estimation. These often use default data and, in any application, field studies should be undertaken to characterise the sediments and verify the performance of the models. Guidance is given in CIWEM (2017) and CIWEM (2021).

Biochemical processes in sewers are intrinsically linked with solids in the flows, in the deposits and in biofilms on surfaces (Hvitved-Jacobsen *et al.*, 2013). The most important aspects include sulphide formation, low oxygen and septicity. Computational models may be used to examine processes in sewer networks (Chapter 11). Models dealing with biochemical processes can estimate where and when septicity can occur for the range of sewer operational conditions. For example, Vollertsen *et al.* (2015) use a process modelling tool (Wastewater Aerobic/Anaerobic Transformations in Sewers; WATS) to estimate the sulphide and gas production in the San Francisco Bayside drainage area catchment with the objective of refining the current operational processes of adding various chemicals, including sodium hypochlorite, into the intercepting sewers to control gas, odours and corrosion. This process model also predicts oxygen depletion along sewers and rising mains and can be used for separate sanitary sewers as well as combined sewers. WATS can be accessed by way of https://thewatsguys.com/references where numerous examples of applications are outlined.

3.5.2 Interactions between sewers, treatment plants and receiving water bodies

Advances in biochemistry in the past few decades have provided the means to examine better the conditions and processes in sewer systems and their implications for both overflows and downstream treatment plants (e.g. Gao *et al.*, 2023). Emerging contaminants such as pharmaceuticals, illicit drugs and personal care products continue to be developed and, as outlined in Chapter 2, the potential range of substances entering a combined sewer system is continually growing. Of growing interest in sewer systems in urban areas is the formation of volatile organic compounds, odours, bioaerosols and greenhouse gases, and their potential for release through vented manholes.

Complex mixtures of material eroded from sewer deposits and dissolved substances will be spilled from CSOs and arrive in 'shock waves' at downstream treatment plants and CSOs. In the highest flows, even boulders (larger than 375 mm) can be conveyed. Together with large textiles, these solids can block screens at overflows and at treatment plant inlets. The altering quality of the wastewater from dry weather to wet weather flows requires special measures at treatment plants to ensure efficient treatment (e.g. Bauwens *et al.*, 1996). Most treatment plants have storm flow retention tanks to store the initial flush of mixed wet and dry weather flows. Many CSOs also have chambers with screens which provide temporary storage during wet weather flows prior to spilling into receiving waters. As the rainfall-related flow recedes, these online or offline storage chambers and tanks will allow flow to pass to treatment. However, the concept of a 'first foul flush' observed as a concentrated pollutant wave in combined sewers is contested. It is not always observed and often a 'second flush' appears which is more concentrated than the first. Sewer catchments vary and the natures of the flows, although having certain common characteristics (Gao *et al.*, 2023), will differ depending on catchments. Some are on steep catchments, others on shallow, and they can have differing inputs depending on connected commercial and highway surfaces. Therefore, it is essential to measure both the flows and the nature of the flows in a system before making any design analyses and/or changing the system operation.

3.5.2.1 Integrated catchment management

The concept of integrated catchment management was introduced in the twentieth century as a perspective for controlling the emissions of final effluent from treatment plants and from any accompanying CSO spills, both from CSOs in the catchment and from the storm tanks at the treatment plant (e.g. Mascher *et al.*, 2017). The integrated approach seeks to balance the emissions in a way that satisfies regulatory standards at each point of emission, from the point source at the plant and from the diffuse sources composed of the various CSOs in the contributing catchment. Emission controls may either aim to ensure a 'fixed emission standard' of maximum concentrations of specified characteristics or aim to balance the assimilative capacity of the receiving waters with the discharge quality. Fixed emissions may, for example, specify limits to oxygen demand (e.g. maximum 20 mg/l biochemical oxygen demand; BOD 5 days) and solids in suspension (e.g. 30 mg/l), as had been defined historically in the UK and known as the Royal Commission Standard (HMSO, 1915). Receiving water assimilative capacity varies seasonally, depending on water temperature, flow regimes and other inputs of pollutants upstream of any discharge point. In order to define pollutant concentrations in discharges, long-term data need to be available and computational simulations are needed to examine the performance of the receiving waters and the wastewater system. This is the approach used in the UK in urban pollution management (UPM), as outlined in the following. Regardless of assimilative capacity, there are numerous substances that are so damaging to environmental systems that concentrations need to be zero or nearly so. These specified de minimis substances are often classified as 'hazardous substances' and defined under the EU's Water Framework Directive. In the UK, the UKTAG was established to specify environmental standards for directive compliance, including for hazardous substances (e.g. UKTAG, 2008, and https://www.wfduk.org/resources).

It has been realised that the provision of added storage at CSO chambers can lead to unanticipated problems for treatment plants. The storage, when drained down, can cause a variety of problems (Ashley *et al.*, 2001), most notably nitrogen shock loading, from displaced flows from the primary settlement tank, and biomass washout from the reactor at the plant. The extended in-sewer storage can also lead to hydrogen sulphide and volatile fatty acid production, with attendant septicity, odours, toxicity and gas generation and in-sewer corrosion.

More recent initiatives are using surface-based measures. As well as removing surface water from combined systems by way of NbSs, treatment of CSO flows prior to spillage is gaining ground, for example by using wetlands (e.g. Rizzo *et al.*, 2020; Tao *et al.*, 2014), although there are major problems in providing wetlands for CSO spill treatment. In a report for UKWIR, Georges *et al.* (2023) stated that 'The prevention of rainfall entering the sewer in the first place should be the first option considered in any situation', rather than the use of technological options. Studies of the potential for maximising multiple benefits from surface water reduction to alleviate CSO spills from using NbSs (Ashley *et al.*, 2017) revealed that, in cost–benefit terms, not all CSO problems would be amenable to this approach: 'To create significant benefits, distributed multi-functional options upstream and downstream of overflows are likely to be necessary'.

The planned use of sewers as bioreactors was a theme of numerous studies at the end of the twentieth century (e.g. Hvitved-Jacobsen *et al.*, 2013). From this work, it has been concluded that while some in-sewer pre-conditioning of the wastewater may be appropriate, the operation of combined sewers as linear treatment systems was not efficient, with sewers remaining as conveyors to treatment plants. The growing interest in decentralised water resource management has raised questions about the use of large sewer networks and single treatment plants, with efforts now to understand when, where and how best to utilise systems that are of a more local scale, potentially obviating the need for sewers, as wastewater is being seen more as a resource (e.g. Capodaglio, 2017).

Box 3.7

The UK *Urban Pollution Management Manual* (Foundation for Water Research, 1998)

'Urban pollution management (UPM) is defined as the management of wastewater discharges from sewer and sewage treatment systems under wet weather conditions such that the requirements of the receiving water are met in a cost-effective way'.

The manual provides guidance for the procedure to be followed. This entails scoping the problems in an existing wastewater system, planning an investigation, data collection, and using a suite of computational models to assess the performance of the existing system

- rainfall modelling
- sewer flow and quality modelling
- sewage treatment works modelling
- river impact modelling
- estuary impact modelling
- marine impact modelling, and
- integrated urban pollution modelling.

Guidance is provided on the options for modifications to the system, including sewerage, storage and CSOs.

The required regulatory standards provide the targets for the expected performance of the system. Modifications to reach these targets are introduced into the modelled systems and their performance assessed to evaluate compliance under a range of wet weather conditions. A range of rainfall conditions is used, together with continuous simulations to assess the receiving water impacts and compliance. In this way the most effective modifications can be selected. The manual recommends post-project monitoring to assure performance.

In the UK, the UPM Manual sets out a procedure that deals with the integrated management of the wastewater system, aimed at optimising usage and minimising receiving water impacts in compliance with regulatory standards (Foundation for Water Research, 1998). Box 3.7 outlines the purpose and process.

Across the EU, the Water Framework Directive and associated directives provide the standards required for compliance for wastewater systems (Woods-Ballard and Cherrier, 2019). Following secession, the UK is setting separate standards. However, it is likely that these will be broadly consistent with the EU in that discharges from CSOs and from wastewater treatment plants will be regulated in volume, frequency and quality to control impacts on receiving waters, but within the perspective of affordable and practicable solutions.

3.5.2.2 Real-time control

Dynamic operation of wastewater systems using real-time control (RTC) is increasingly feasible due to the advancement of sensor technologies and reduction in relative costs (Chapter 9). Automation of wastewater treatment plant operations using sensors to assess the influent quality and manage the processes accordingly has been practised since the latter part of the twentieth century (e.g. Olsson *et al.*, 2005). More recently automation is proving useful in managing energy use and production, together with carbon at treatment plants, and especially in detecting and controlling infiltration and inflow (Envirotec, 2023).

There are several examples of the integrated operation of sewer networks together with treatment plants. However, most sewer networks are passive, relying on gravity flows, and dynamic control opportunities are constrained to system assets such as pumping stations and, in some cases, gates. The technology and associated computational models have been developed and are routinely in use in sectors other than drainage. This is an area where continuous development of the technologies in other sectors, such as artificial intelligence and dynamic data, will provide new opportunities in drainage.

Reviews on the use and applications of RTC are provided by Schutze *et al.* (2004) and van der Werf *et al.* (2022). One of the earliest RTC applications in the UK for CSO control was for the Bolton sewerage system (Sharman and Tidswell, 1993; Williams and Tidswell, 1994). This was made possible by the development of computational models and electronic instrumentation, in particular telemetry-facilitated predictive models that could be operated to simulate the links and controls to manage flows in systems. Many sewerage systems do and have had dynamic controls to manage flows, especially into treatment plants and to help to manage CSO spills. These have typically consisted of pumping stations and in some cases controllable gates and moveable weirs, often situated in the sewer network. Their operation was based on simple rules concerning flows and their diversion above certain rates, for example in Aberdeen and Liverpool (Ashley *et al.*, 1995; Brown and Osborne, 1995).

There are examples of successful RTC in operation. Cincinnati has a network of valves and control gates for its smart sewer operation (GE Digital, 2023), which minimises the cost of wastewater management and reduces CSO spills. Vienna was one of the only European cities to have a complete sewer network in the 1700s. In the 2000s, a city-wide RTC system was implemented, which included numerous monitoring devices: 25 for rainfall measurement, 40 measuring flow and 20 measuring water depth (Fuchs and Beeneken, 2005). Around this time, RTC was also being installed in the city sewers of Berlin and Tokyo. PD ISO/TR 24539:2021 (BSI, 2021) provides various case examples of good practice for stormwater management. One of these is the RTC system for Biarritz (France), which, since 2016, has reduced sewer system overflow volumes by 40% while also controlling flooding and loads to the treatment plant. The global controlled system is predictive and operates gates and weirs to maximise the in-system storage.

Implementation costs for retrofitting RTC are often lower than those for retrofitting NbSs to manage stormwater. Operational costs, however, are sometimes perceived to be significant for RTC for wastewater systems, and there can be a suspicion that such systems bring added risks to performance. For these reasons, while there are many proponents of the use of RTC, there are as many opponents, who take the view that the costs and risks are too great, preferring to utilise passive drainage systems. Nevertheless, RTC in the drainage domain is being used successfully in many parts of the world (van der Werf *et al.*, 2022). Most recently, RTC ideas and strategies are being applied in relation to the use of green infrastructure for CSO spill reduction in Montreal, and to green-grey infrastructure in Quebec, where the combination of green and grey infrastructure with RTC control has been found to be the most cost-effective option after taking into account uncertainties (Jean *et al.*, 2021, 2022). Eulogi *et al.* (2022) and Brasil *et al.* (2022) illustrate how local RTC, with attendant flow controls, can be used in an optimisation process to site various controls, together with NbSs, to enable better management of surface water runoff, flows and CSO spills. In the future it would be expected that RTC will play a major role in improving the management of the entire range of types of wastewater systems and also in energy and carbon control (van der Werf *et al.*, 2022).

With the advent of sophisticated computational models for wastewater systems, there has been a tendency in the UK, at least, to rely on modelling and not to invest in the monitoring required to ensure that the models are good representations of reality and that schemes are performing as designed. Effective monitoring was at one time expensive and often difficult in sewer networks, due to access and the complexity of some systems. Post-project monitoring to verify performance is even less implemented than routine monitoring. There has been and still is a presumption that what the 'model says' provides the entire answer to how best to design and modify systems to achieve performance requirements. Suggestions in the 1990s that optimum value from investments could not be achieved without adequate pre- and post-project monitoring (Price and Catterson, 1995) are largely being ignored even now.

3.6. Opportunity for change

The economic costs from increasing flood risks, the legal and regulatory requirements for water quality improvements, such as those set out in the EU WFD, and the recognition of the importance of future adaptability of our urban infrastructure in the face of a potentially dramatically changing climate have highlighted the inadequacies of the traditional approach to drainage of the urban environment using piped and below-ground storage systems. The lack of flexibility and relative difficulties of adaptation present a significant and costly challenge to coping with changes to wastewater in the future using underground infrastructure. This situation is exacerbated by the state of existing infrastructure in the first countries to industrialise, which was constructed decades, or in some cases centuries, ago. Stormwater systems need continual renewal and in many countries this is costly and may be unaffordable. For example, ASCE (2021) estimates that some US $8 billion is needed per annum just to maintain existing stormwater infrastructure sufficiently to comply with regulatory requirements. Tighter regulation and climate change are adding to these costs. However, the trend to use more NbSs has been evident since 2017, with many more projects qualifying for national funding. The recommendations include a call to make more use of the catchment-based approach.

Natural land drainage systems are known to be extremely effective. Small, regular rainfall events seep into the ground wherever there is permeable soil cover, promoting the growth of vegetation, filtering and treating the water through the upper soil horizons, evapotranspiring from surface water and vegetation, and helping with recharge of river base flows and groundwater resources wherever possible. During larger events, excess flows are directed into small streams, which in turn discharge into rivers of increasing size as the water is conveyed downstream. Each element provides a degree of flood attenuation and control, depending on the slope, topographic features and vegetation, and improves the quality where the runoff is contaminated. Land drainage practices in the past few centuries have, in many countries, promoted the rapid drainage of land for agricultural and other purposes. This has led to rapid rises in flows in natural water courses, with attendant downstream flooding, usually in urban areas. The increasing awareness and use of 'natural flood management' (NFM) has changed practice in many countries, such as the UK, with greater emphasis being placed on retaining water in the uplands to prevent problems downstream (e.g. Kvitsjøen *et al.*, 2022), albeit the 'rediscovery' of NFM requires more understanding of the performance of these measures for greater confidence in their design and selection (Ellis *et al.*, 2021).

Our urban development practices to date have involved covering our natural landscapes with impermeable 'skins' – usually asphalt and concrete – which provide a barrier to infiltration and natural drainage processes. Apart from storage within local depressions, rainfall falling on the land surface has tended to be directed towards subsurface, piped systems. In addition to more severe rainfall, there is also a significant and increasing risk of reductions in the permeability of urban surfaces, as residents build extensions, conservatories and parking over garden areas, a process known as 'soil sealing' (European Commission, 2022). The London Assembly has estimated that around two-thirds of front gardens in the London area are already at least partially paved over, primarily to provide off-road parking spaces (e.g. Zhang *et al.*, 2021). Ironically, this soil sealing is likely to increase significantly in the drive towards a net-zero carbon economy. Decarbonisation of transport and the move to electric cars will lead to a commensurate increase in the paving over of front gardens to accommodate off-street parking for the charging of vehicles.

Piped drainage systems have generally been designed with the aim of removing water from the built-up area as quickly as possible to preserve human health and protect urban areas. Without infiltration and natural attenuation, runoff is conveyed downstream in larger volumes and at higher rates than would have occurred from natural surfaces (Chapter 6). Recent realisation that greening surfaces is a key component of managing carbon, especially in urban areas, is a major factor in changing this behaviour (e.g. European Commission, 2022). Greening some 35% of European urban surfaces would transpire about 10 km^3 per year of rainwater, with 17.5% of runoff becoming useful 'green water', helping to reduce pollution and urban flooding.

Human activities in the urban environment give rise to a wide range of pollutants, the majority of which are damaging to the water environment. These contaminants are either adsorbed onto sediments or other particulates, which become entrained in runoff, or dissolve in rainwater, as outlined in Section 3.2.1. The water quality characteristics of urban stormwater runoff therefore become directly correlated with the type and intensity of the activities occurring within the upstream areas.

In drainage systems constructed in the UK prior to the 1950s, foul and surface flows were generally designed to be managed together in a combined sewer (Sections 2.3.1 and 3.1). During flood events, to limit the flows passed to wastewater treatment, discharges are made from such systems directly to the watercourse by way of combined sewer overflows (CSOs). Although sewage will be diluted by rainwater, the receiving watercourse might only be experiencing average flows, and the environmental damage of high overflow rates on ambient river flows can often be significant. These spills frequently breach environmental standards and are increasingly unacceptable (Section 3.5.2). More recent separate sewerage systems reduce the risk of contamination from foul sewage, but can nevertheless discharge urban stormwater runoff, with all its associated untreated contaminants, directly into water bodies such as streams, rivers, lakes, etc. Ineffective stormwater management is a serious problem that has resulted in the modification of water bodies. Effective stormwater management using blue-green infrastructure and nature-based solutions can better help to reduce or mitigate the adverse hydrological and water quality impacts of urban development and land use changes. However, this requires an integrated and holistic approach (Chapter 13).

We are already seeing regular urban flooding around the world resulting from surcharging and incapacity of drainage systems (e.g. Schmit and Scheid, 2019). The 2009 *Flooding in England* report stated that there were over 2.4 million properties at risk of flooding from rivers or the sea in England, one million of which are also vulnerable to surface water flooding, with a further 2.8 million properties susceptible to surface water flooding alone (Environment Agency, 2009), but every property where rainfall occurs has a finite flood risk. The *Foresight Future Flooding* report (Foresight Directorate, 2004) at the start of this millennium estimated 80 000 properties to be at very significant risk from surface water flooding in the UK (10% annual probability or greater), causing on average £270 million of damage each year. In addition, the report estimated that sewerage undertakers could potentially have to spend around £1 billion per year to stop sewer flooding getting worse, given the increased intensity of rainfall events due to climate change. These problems were exemplified in the floods of June 2007, highlighting the vulnerability of the British public to flooding. Extreme rainfall over the Midlands and the north of England led to large-scale urban flooding, with over 55 000 flooded properties. The Environment Agency estimated that one-third of these were flooded from rivers, and two-thirds from other sources, predominantly surface water runoff overloading drainage systems. The combined cost of damage from the June and July 2007 floods was estimated to be around £3 billion. Since then, the National Infrastructure Commission (NIC, 2022) has reinforced the message about both the significant risks from surface water flooding and the likely costs of coping. Yet governments, despite these warnings, continue to delay effective responses, notwithstanding gradual (but inadequate) increases in investments. This is not only in the UK, as across the EU the floods directive effectively ignores pluvial flood risks, instead concentrating on river systems and coastal protection (e.g. Nones, 2015).

Sea level rise and coastal flooding will not only increase the risks of flooding of arable land and salinisation of soils, but will also impact on urban drainage infrastructure. There are conflicting interests between preventing high-quality arable land from flooding, to protect it from saline damage, and allowing it to flood in order to protect the urban environment from flooding, as required by taking an integrated catchment approach. Despite the Dutch polder system, there is a relentless encroachment of salt and brackish water into the polders and agricultural land in the Netherlands, which is accelerating due to climate change (Climate Change Post, 2023). Therefore, should low-lying parts of countries build polders, like the Dutch, to protect their farmland, or, by allowing more saltwater on to the land, might there be an opportunity to transition from agriculture to aquaculture?

The question then is when is water on the surface a flood? In a Dutch survey of publicly reported flooding incidents, it was found that the majority of these were simply water standing on the surface on roadways or pavements in places where water is not normally found – that is, puddles (Ashley *et al.*, 2007). The water had caused only minor inconvenience to drivers and pedestrians and the economic cost was negligible. When flooding does occur, however this is defined, we can choose to manage it (preferably, and most cost-effectively, on the surface) or to accept it and deal with the consequences.

The ongoing development of low-cost sensors, together with RTC, will increasingly offer opportunities to improve the performance of existing assets and reduce maintenance costs (enabling better targeting) (Section 3.5.2.2). As yet, there are few UK examples of larger scale control, although some significant trials are emerging, such as Newcastle Smart City Index (Newcastle City Council, 2023), that include urban water control. Advanced exemplars do exist internationally; in Japan, for example, such active control has been effective at keeping installation costs down in urban areas (Maeda *et al.*, 2005), and RTC is seen as highly cost-effective for CSO control in Canada (Jean *et al.*, 2021). But, as outlined in Section 3.5.2.2, applications of RTC will need to begin to encompass the widest range of drainage infrastructure, including NbSs, and also address the challenges of carbon reduction and energy use.

3.6.1 Governance

The differentiation between pluvial, fluvial and coastal flood risks exhibited by policies, regulatory instruments and institutional responsibilities, together with the attitudes and behaviours of ‘those in charge’, is a reflection of the way in which water, in common with many other natural phenomena, is ‘parcelled up’. Different players, funding sources, deliverers and rationales are found across the entire water domain. Nowhere in the world is there a joined-up approach to water management, even in Australia, the home of water sensitivity. National, state, city and even town governance share responsibilities for different components of the water cycle. Importantly, funding comes from a variety of these sources, as well as from private investors. Citizens, however, do not differentiate between flooding sources nor who is responsible for providing protection from them (e.g. Ashley *et al.*, 2007). Many of those with responsibility deal only with parts of the flooding system, so citizens and local drainage systems may be overlooked. It is for this reason that the Environment Agency in England, responsible for main river flood protection, continually misrepresents surface water flood risk; reporting that only a few million properties are at risk, whereas in fact it rains everywhere so potentially every property is at some finite risk.

The disenfranchisement of citizens is a major issue with regard to flooding and water systems in general. Alongside democratic process norms in any given country, arrangements for the management of flooding and water systems reflect cultures, and this is exhibited in institutions, policymaking and decision processes (Juuti and Katko, 2005). In countries such as France, local democratic processes are at the heart of how water systems are managed and decisions are made, whereas in England, citizens have no direct voice in water management other than to pay charges assigned by the water services economic regulator (Ofwat) (Page and Bakker, 2005). There are many and varied types of water governance around the world.

The current administrative hierarchy of stormwater management, like water management, in many parts of the world is fragmented, disjointed and spread among multiple agencies and organisations (OECD, 2018). With shared but competing interests, there is often no one single body managing the water cycle in a coherent and integrated way. However, there are broad links between stormwater management systems and those of spatial planning and flood risk management (see Chapter 6). Therefore, does it require a single, controlling agency, such as the Ministry of Sustainability and the Environment in Singapore (https://www.mse.gov.sg/)? Having a single, responsible, accountable body for drainage and other aspects of the water cycle, including nature-based surface water assets, would enable a better understanding of the interactions of different networks and mechanisms of failure, allowing a more effective and cost-efficient approach to 'effectually draining'. This would facilitate a catchment approach to investment, thereby creating capacity to facilitate development growth, defer infrastructure investment and smooth the investment cycle (see Chapter 5).

Professionals are at an incredibly exciting and promising time in stormwater management in most parts of the world. However, there are tensions between short-term affordability and long-term funding, with the needs of the environment and society seemingly considerably different, and in certain contexts more acutely than, those which pertained even relatively recently and certainly as long as 30 years ago. Envisaging what is to come is never easy and a number of potential scenarios have been postulated as regards water futures (e.g. Kemira, 2023). Scenario planning – that is, considering what coherent and possibly extreme futures may look like (see Chapter 13) – provides a way of preparing better for uncertainty (e.g. Ashley, 2020). Unfortunately, few water system operators do this, perceiving futures planning, other than for the next 5 to 10 years, only in terms of probable trajectories (Ashley *et al.*, 2022).

The keys to success will be systems thinking, governance and institutional change and an improved way of regulation, with stricter economic and environmental controls that reflect more fully the public policy challenges of the next 50 years, alongside good governance and appropriate legislation (OECD, 2015; Jacobson *et al.*, 2015). Strategic planning is required to balance long-term against short-term investment to ensure intergenerational equity. The regulatory framework should be sufficiently flexible to enable companies to consider a broad range of measures and to use the most efficient means to meet their obligations, including catchment solutions and investment in third party assets. However, any change in regulation for delivering water services that are more sustainable must ensure that we are adequately equipped to meet the future challenges and needs to be a pragmatic, long-term change.

REFERENCES

Ackers JC, Butler D and May R (1996) *Design of Sewers to Control Sediment Problems*. CIRIA 141. Construction Industry Research and Information Association. London, UK.

Ampe K, Paredis E, Asveld L, Osseweijer P and Block T (2020) A transition in the Dutch wastewater system? The struggle between discourses and with lock-ins. *Journal of Environmental Policy and Planning* **22(2)**: 155–169, 10.1080/1523908X.2019.1680275.

ASCE (American Society of Civil Engineers) (2021). Report Card for America's Infrastructure: Stormwater. https://infrastructurereportcard.org/ (accessed 22/07/2023).

Ashley R (2009) Briefing: N.G. Semple's 'Principles of environmental protection'. *Proceedings of the Institution of Civil Engineers – Engineering Sustainability* **162(ES1)**: 3–11, 10.1680/ensu.2009.162.1.3.

Ashley RM (2020) Changes in the way we live and value urban spaces. *Journal of Delta Urbanism* **1(1)**: 2666–7851, 10.7480/jdu.1.2020.5456.

Ashley RM, Budge F and Fleming R (1995) HYDROWORKS RTC modelling for Aberdeen. *Proceedings of the WaPUG Conference*. See http://ftp2.ciwem.org/1995/1995-11-23%20to%2024/AU1995.PAPER15.pdf (accessed 22/07/2023).

Ashley RM, Dudley J, Vollertsen J *et al.* (2001) Interaction between extended in-sewer storage and wastewater treatment plant performance – a protocol to avoid problems. *Proceedings of the WaPUG Conference*, Blackpool, UK. November 2001. http://ftp2.ciwem.org/2001/2001-11/a2001_4.pdf.

Ashley RM, Bertrand-Krajewski J-L, Hvitved Jacobsen T and Verbanck M (eds) (2005) *Solids in Sewers: Characteristics, Effects and Control of Sewer Solids and Associated Pollutants*. IWA Publishing, London, UK, vol. 4.

Ashley RM, Blanksby JR, Cashman A *et al.* (2007) Adaptable Urban Drainage – Addressing Change in Intensity, Occurrence and Uncertainty of Stormwater (AUDACIOUS). *Built Environment* **33(1)**: 70–84, 10.2148/benv.33.1.70.

Ashley RM, Lundy L, Ward S *et al.* (2013) Water-sensitive urban design: opportunities for the UK. *Proceedings of the Institution of Civil Engineers – Municipal Engineer* **166(ME2)**: 65–76, 10.1680/muen.12.00046.

Ashley RM, Walker L, D'Arcy B *et al.* (2015) UK sustainable drainage systems: past, present and future. *Proceedings of the Institution of Civil Engineers – Civil Engineering* **168(3)**: 125–130, 10.1680/cien.15.00011.

Ashley RM, Digman C, Futter M *et al.* (2017) *Valuing the Benefits of Storm Discharge Improvements for Use in Cost-Benefit Analysis.* Report NL5946. Water UK, London, UK. See https://www.water.org.uk/wp-content/uploads/2018/12/Storm-Overflow-Benefits-Valuation-Final-Report.pdf (accessed 22/07/2023).

Ashley RM, Digman C, Horton B *et al.* (2018) Evaluating the longer-term benefits of sustainable drainage. *Proceedings of the Institution of Civil Engineers – Water Management* **171(WM2)**: 57–66, 10.1680/jwama.16.00118.

Ashley RM, Horton B, Walker L *et al.* (2022) The benefits of nature-based systems in a changing and uncertain world. *Proceedings of the Institution of Civil Engineers – Engineering Sustainability* **176(2)**: 61–71, 10.1680/jensu.21.00102.

Atkins (2021) *A Framework for the Production of Drainage and Wastewater Management Plans.* Atkins, Epsom, UK. See https://www.water.org.uk/wp-content/uploads/2021/10/DWMP_Framework_Report_Main_Report_September_2021.pdf (accessed 22/07/2023).

Balmforth DJ, Digmand CJ, Kellagher R and Butler D (2006) *Designing for Exceedance in Urban Drainage – Good Practice.* CIRIA C635. Construction Industry Research and Information Association, London, UK.

Bauwens W, Vanrolleghem P and Smeets M (1996) An evaluation of the efficiency of the combined sewer–wastewater treatment system under transient conditions. *Water, Science and Technology* **33(23)**: 199–208.

Beamont E (2022) Wet wipes still biggest culprit in sewer blockages, Yorkshire Water says. *The Independent*, 12 Jan. See https://www.independent.co.uk/climate-change/news/yorkshire-water-yorkshire-hull-b2260628.html (accessed 21/07/2023).

Brasil J, Macedo M, Lago C *et al.* (2022) Nature-based solutions and real-time control: challenges and opportunities. *Water* **13(5)**: 651, 10.3390/w13050651.

Brown C and Osborne M (1995) The use of RTC to control flows to Sandon Dock WwTW Liverpool. *Proceedings of the WaPUG Conference.* See http://ftp2.ciwem.org/1995/1995-11-23%20to%2024/AU1995.PAPER16.pdf (accessed 22/07/2023).

BSI (British Standards Institution) (2017) BS EN 752:2017. Drain and sewer systems outside buildings. Sewer system management. BSI, London, UK.

BSI (2018a) BS ISO 31000:2018. Risk management. Guidelines. BSI, London, UK.

BSI (2018b) BS EN 16941-1:2018. On-site non-potable water systems. Systems for the use of rainwater. BSI, London, UK.

BSI (2019) BS ISO 24536:2019. Service activities relating to drinking water supply, wastewater and stormwater systems. Stormwater management. Guidelines for stormwater management in urban areas. BSI, London, UK.

BSI (2000) BS EN 12056-3:2000. Gravity drainage systems inside buildings – Roof drainage, layout and calculation. BSI, London, UK.

BSI (2021) PD ISO/TR 24539:2021. Service activities relating to drinking water supply, wastewater and stormwater systems. Examples of good practices for stormwater management. BSI, London, UK.

Butler D and Clark P (1993) *Sediment Management in Urban Drainage Catchments.* CIRIA 134. Construction Industry Research and Information Association, London, UK.

Capodaglio AG (2017) Integrated, decentralized wastewater management for resource recovery in rural and peri-urban areas. *Resources* **6**: 22, 10.3390/resources6020022.

Chan SC, Dale M, Fowler HJ and Kendon EJ (2021) *Extreme precipitation return level changes at 1, 3, 6, 12, 24 hours for 2050 and 2070, derived from UKCP Local Projections on a 5 km grid for the FUTURE-DRAINAGE Project.* NERC EDS Centre for Environmental Data Analysis. See https://catalogue.ceda.ac.uk/uuid/18f83caf9bdf4cb4803484d8dce19eef (accessed 22/07/2023).

Chubaka CE, Whiley H, Edwards JW and Ross KE (2018) A review of roof harvested rainwater in Australia. *Journal of Environmental and Public Health* **2018(1)**: 1–14, 10.1155/2018/6471324.

CIRIA (Construction Industry Research and Information Association) (1986) *Sediment Movement in Combined Sewerage and Stormwater Drainage Systems.* Research project 366. CIRIA, London, UK.

CIWEM (Chartered Institution of Water and Environmental Management) (2017) *Code of Practice for the Hydraulic Modelling of Urban Drainage Systems 2017.* CIWEM, London, UK. See https://www.ciwem.org/assets/pdf/Special%20Interest%20Groups/Urban%20Drainage%20Group/Code%20of%20Practice%20for%20the%20Hydraulic%20Modelling%20of%20Ur.pdf (accessed 22/07/2023).

CIWEM (2021) *Integrated Urban Drainage Modelling Guide.* CIWEM, London, UK. See https://www.ciwem.org/assets/uploads/IUD_1.pdf (accessed 22/07/2023).

Climate Change Post (2023) Salt intrusion The Netherlands. https://www.climatechangepost.com/netherlands/salt-intrusion/ (accessed 22/07/2023).

Cristiano E, Farris S, Deidda R and Viola F (2021) Comparison of blue-green solutions for urban flood mitigation: a multi-city large-scale analysis. *PLoS ONE* **16(1)**: e0246429, 10.1371/journal.pone.0246429.

D'Arcy B, Ellis JB, Ferrier R, Jenkins A and Dils R (eds) (2000) *Diffuse Pollution Impacts.* Terence Dalton Publishers, Lavenham, UK. See https://eprints.mdx.ac.uk/16140/.

Dhonau M (2022) *Property Flood Resilience.* MDA. See https://edition.pagesuite.com/html5/reader/production/default.aspx?pubname=Property+Flood+Resilience&pubid=b3dddc3f-8c89-4711-86f3-4f0f9ce9a713&fbclid=IwAR12PeAbRX4orTTe5YySx-2eqbFvxXOHiWRByZA7jCGsxc2SKhtV8eJo0HI (accessed 22/07/2023).

Diaz-Nieto J, Lerner DN, Saul AJ and Blanksby JR (2012) GIS water balance approach to support surface water flood-risk management. *Journal of Hydrologic Engineering* **17(1)**: 55–67, 10.1061/(ASCE)HE.1943-5584.0000416.

Diaz-Nieto J, Lerner DN and Saul AJ (2016) Least-cost path analysis to identify retrofit surface water conveyance solutions. *Journal of Hydrologic Engineering* **21(3)**: 1–14, 10.1061/(ASCE)HE.1943-5584.0001315.

Digman CJ, Ashley RM, Blanksby JR and Gill E (2014) *Managing Urban Flooding from Heavy Rainfall: Encouraging the Uptake of Design for Exceedance. Recommendations and Summary.* CIRIA C738a. Construction Industry Research and Information Association, London, UK.

DLUHC (Department for Levelling Up, Housing and Communities (2023) *National Planning Policy Framework (NPPF).* See https://www.gov.uk/government/publications/national-planning-policy-framework–2 (accessed 22/07/2023).

DoE (Department of the Environment)/HR Wallingford (1983) *Design and Analysis of Urban Storm Drainage: The Wallingford Procedure.* Standing Technical Committee Report No. 28. Department of the Environment, National Water Council, Hydraulics Research Ltd, Wallingford, UK.

EcoShape (2021) *Integrated System-based Asset Management.* See https://northsearegion.eu/media/17284/s210109-rapport-ecoshape-businesscase-nl-p4-003.pdf (accessed 22/07/2023).

Ellis JB and Butler D (2015) Surface water sewer misconnections in England and Wales: pollution sources and impacts. *Science of the Total Environment* **526(2015)**: 98–109, 10.1016/j.scitotenv.2015.04.042.

Ellis N, Anderson K and Brazier R (2021) Mainstreaming natural flood management: a proposed research framework derived from a critical evaluation of current knowledge. *Progress in Physical Geography: Earth and Environment* **45(6)**: 819–841, 10.1177/0309133321997299.

Engelhard C, De Toffol S and Rauch W (2008) Suitability of CSO performance indicators for compliance with ambient water quality targets. *Urban Water Journal* **5(1)**: 43–49, 10.1080/15730620701736902.

Environment Agency (2009) *Flooding in England: A National Assessment of Flood Risk*. Environment Agency, Bristol, UK. See https://assets.publishing.service.gov.uk/government/uploads/system/uploads/attachment_data/file/292928/geho0609bqds-e-e.pdf (accessed 22/07/2023).

Environment Agency (2019a) *What is the Risk of Flooding from Surface Water Map?* Environment Agency, Bristol, UK. See https://assets.publishing.service.gov.uk/government/uploads/system/uploads/attachment_data/file/842485/What-is-the-Risk-of-Flooding-from-Surface-Water-Map.pdf (accessed 22/07/2023).

Environment Agency (2019b) 2021 River basin management plans. Pollution from towns, cities and transport. https://consult.environment-agency.gov.uk/++preview++/environment-and-business/challenges-and-choices/user_uploads/pollution-from-towns–cities-and-transport-challenge-rbmp-2021.pdf (accessed 22/07/2023).

Envirotec (2023) Norwegian utility claims to calculate true cost of sewer ingress. https://envirotecmagazine.com/2023/01/23/norwegian-utility-claims-to-calculate-true-cost-of-sewer-ingress/ (accessed 22/07/2023).

Eulogi M, Ostojin S, Skipworth P *et al.* (2022) Optimal positioning of RTC actuators and SuDS for sewer overflow mitigation in urban drainage systems. *Water* **14(23)**: 3839, 10.3390/w14233839.

EurEau (2020) *EurEau Position Paper on Overflows from Collecting Systems*. EurEau, Brussels, Belgium. See https://www.eureau.org/resources/position-papers/4955-position-paper-on-overflows-from-collecting-systems-1/file (accessed 22/07/2023).

European Commission (2022) Green Surfacing: An EU-wide Assessment of Costs and Benefits. Science for Environment Policy. Issue 573. European Commission, Brussels, Belgium.

Fallast M-T, Pansinger S, Krebs G, Moser M and Zobl A (2021) Systematically retrofitting city streets: meeting the demands of climate change through multifunctional climate-responsive street gardens. *Urbani Izziv* **32(1)**: 111–122.

Foresight Directorate (2004) *Foresight Future Flooding: Executive Summary*. Government Office for Science, London, UK. See https://assets.publishing.service.gov.uk/government/uploads/system/uploads/attachment_data/file/300332/04-947-flooding-summary.pdf (accessed 22/07/2023).

Foundation for Water Research (1998) *Urban Pollution Management Manual*, 2nd edn. Foundation for Water Research, Marlow, UK. See version 3.1 (2018): http://www.fwr.org/UPM3/.

Fratini CF, Geldof GD, Kluck J and Mikkelsen P-S (2012) Three points approach (3PA) for urban flood risk management: a tool to support climate change adaptation through transdisciplinarity and multifunctionality. *Urban Water Journal* **9(5)**: 317–331, 10.1080/1573062x.2012.668913.

Freni G, Mannina G and Viviani G (2010) Emission standards versus immission standards for assessing the impact of urban drainage on ephemeral receiving water bodies. *Water Science Technology* **61(6)**: 1617–1629, 10.2166/wst.2010.887.

Fuchs L and Beeneken T (2005) Development and implementation of a real-time control strategy for the sewer system of the city of Vienna. *Water Science Technology* **52(5)**: 187–194.

Gao Y, Shi X, Jin X, Wang X C and Jin P (2023) A critical review of wastewater quality variation and in-sewer processes during conveyance in sewer systems. *Water Research* **228(Part B)**: 119398, 10.1016/j.watres.2022.119398.

GE Digital (2023) Cincinnati's "Smart Sewer" reduces overflows and cuts costs from $0.23/gallon to $0.01/gallon. https://www.ge.com/digital/customers/cincinnatis-smart-sewer-reduces-overflows-and-cuts-costs-023gallon-001gallon?utm_medium=Paid-Search&utm_source=Google&utm_campaign=SG-MFG-Water-MOF-GLOB-Search&utm_content=sitelink&keyword=smart%20sewer%20system (accessed 22/07/2023).

Georges K, Piekarniak L and Tan S (2023) *Treatment Options for Storm Overflows*. UK Water Industry Research, London, UK.

Gould J and Nissen-Petersen E (2008) *Rainwater Catchment Systems for Domestic Supply*. Practical Action Publishing, Intermediate Technology Publications, Warwickshire, UK.

Gov.UK (2022) Flood and Coastal Erosion Risk Management Report: 1 April 2020 to 31 March 2021. https://www.gov.uk/government/publications/flood-and-coastal-risk-management-national-report/flood-and-coastal-erosion-risk-management-report-1-april-2020-to-31-march-2021 (accessed 22/07/2023).

Green Building Council UK (2022) Connecting Nature. https://www.ukgbc.org/solutions/connecting-nature-framework/ (accessed 22/07/2023).

Han J and He S (2021) Urban flooding events pose risks of virus spread during the novel coronavirus (COVID-19) pandemic. *Science of the Total Environment* **755(1)**: 142491, 10.1016/j.scitotenv.2020.142491.

Harter T, Bernt I, Winkler S and Hirn U (2021) Reduced dispersibility of flushable wet wipes after wet storage. *Scientific Reports* **11**: 7942, 10.1038/s41598-021-86971-z.

Highways England (2020) *Vortex Separators for Use with Road Drainage Systems*. Design Manual for Roads and Bridges. CD 528. Revision 0. Formerly HD 220/18. Highways England, Guildford, UK.

HMG (Her Majesty's Government) (2010) The Building Regulations 2010. The Stationery Office, London, UK.

HMSO (1915) *Report of the Commissioners Appointed in 1898 to Inquire and Report What Methods of Treating and Disposing of Sewage (Including Any Liquid from Any Factory or Manufacturing Process) May Properly be Adopted*. Volume **1**. Eyre and Spotiswood, London, UK.

HM Treasury (2021) *The Economics of Biodiversity. The Dasgupta Review*. HM Treasury, London, UK. See https://assets.publishing.service.gov.uk/government/uploads/system/uploads/attachment_data/file/962785/The_Economics_of_Biodiversity_The_Dasgupta_Review_Full_Report.pdf (accessed 13/11/2023).

Honda M and Suzuki N (2020) Toxicities of polycyclic aromatic hydrocarbons for aquatic animals. *International Journal of Environmental Research and Public Health* **17(4)**: 1363, 10.3390/ijerph17041363.

Hooton C (2014) River runs orange in Huddersfield. *The Independent*, 18 Feb. See https://www.independent.co.uk/news/uk/river-runs-orange-in-huddersfield-9135854.html (accessed 22/07/2023).

House MA (1993) Urban drainage impacts – on receiving water quality. *Water Science Technology* **27(12)**: 117–158, 10.2166/wst.1993.0293.

Hurley L, Mounce SR, Ashley RM and Blanksby JR (2008) No Rainwater in Sewers (NORIS): assessing the relative sustainability of different retrofit solutions. *Proceedings of the 11th International Conference on Urban Drainage*. Edinburgh. https://www.academia.edu/22805087.

Hvitved-Jacobsen T, Vollertsen J and Nielsen AH (2013) *Sewer Processes: Microbial and Chemical Process Engineering of Sewer Networks*, 2nd edn. CRC Press, Boca Raton, FL, USA.

Jacobson M, Meyer F, Oia I, Reddy P and Tropp H (2015) *User's Guide on Assessing Water Governance*. Oslo Governance Centre, UNDP, Oslo, Norway. See https://www.undp.org/sites/g/files/zskgke326/files/publications/Users%20Guide%20on%20Assessing%20Water%20Governance1.pdf (accessed 22/07/2023).

Jean M-E, Morin C, Duchesne S, Pelletier G and Pleau M (2021) Optimization of real-time control with green and gray infrastructure design for a cost-effective mitigation of combined sewer overflows. *Water Resources Research* **57(12)**: e2021WR030282, 10.1029/2021WR030282.

Jean M-E, Morin C, Duchesne S, Pelletier G and Pleau M (2022) Real-time model predictive and rule-based control with green infrastructures to reduce combined sewer overflows. *Water Research Volume* **221**: 118753, 10.1016/j.watres.2022.118753.

Juuti PS and Katko TS (eds) (2005) *Water, Time and European Cities: History Matters for the Futures*. Water Time. See http://www.watertime.net/docs/WP3/WTEC.pdf (accessed 23/11/2023).

Kelly D, Barker B, Lamond J, McKeown S, Blundell E *et al.* (2021) *Code of Practice for Property Flood Resilience*. CIRIA C790A. Construction Industry Research and Information Association, London, UK.

Kemira (2023) *Water Management 2040. Future Scenarios*. Kemira, Helsinki, Finland. See http://pages.kemira.com/rs/784-UGU-140/images/kemira-water-management-scenarios-2040.pdf?&utm_source=

marketo&utm_medium=email&utm_campaign=iw-scenarios&utm_content=report-future-scenarios-2040 (accessed 22/07/2023).

Kvitsjøen J, Braskerud BC, Borge A, Nilsen V and Zühlke U (2022) Natural flood protection: streamlining the planning of flood detention in natural landscapes for the reduction of urban flooding. *Water Science and Technology* **85(1)**: 367, 10.2166/wst.2021.628.

Larsen T, Gruendlb H and Binz C (2021) The potential contribution of urine source separation to the SDG agenda – a review of the progress so far and future development options. *Environmental Science: Water Research Technology* **7**: 1161, 1039/d0ew01064b.

Leggett D, Brown R, Stanfield G, Brewer D and Holiday D (2001) *Rainwater and Grey Water Use in Buildings*. CIRIA Project Report 80. Construction Industry Research and Information Association. London, UK.

Maeda M, Mizushima H and Ito K (2005) Development of the real-time control (RTC) system for Tokyo sewage system. *Water Science Technology* **51(2)**: 213–220.

Mail Online (2012) The river that DID run red: residents of Chinese city left baffled after Yangtze turns scarlet. 7 Sept. See https://www.dailymail.co.uk/news/article-2199800/The-river-DID-run-red-Residents-Chinese-city-left-baffled-Yangtze-turns-scarlet.html (accessed 21/07/2023).

Mascher F, Mascher W, Pichler-Semmelrock F, Reinthaler FF and Zarfel GE (2017) Impact of combined sewer overflow on wastewater treatment and microbiological quality of rivers for recreation. *Water* **9**: 906, 10.3390/w9110906.

Mengistu D, Heistad A and Coutris C (2021) Tire wear particles concentrations in gully pot sediments. *Science of the Total Environment* **769**: 144785, 10.1016/j.scitotenv.2020.144785.

Muller A, Österlund H, Marsalek J and Viklander M (2020) The pollution conveyed by urban runoff: a review of sources. *Science of the Total Environment* **709**: 136125, 10.1016/j.scitotenv.2019.136125.

Murnane E, Heap A and Swain A (2006) Control of Water Pollution from Linear Construction Projects. Site Guide. CIRIA C649. Construction Industry Research and Information Association, London, UK.

Mushin S, Saunders C and Campbell D (2016) Analysis of building drainage and sewer system performance utilising a tipping tank with water-conserving measures. In *Proceedings of Elfa*, s.r.o, pp. 364–374. *42nd International Symposium of CIB W062 on Water Supply and Drainage for Buildings*, Kosice, Slovakia, 29 August 2016. See https://pure.hw.ac.uk/ws/portalfiles/portal/15430671/CIB_W062_2016_Proceedings.pdf (accessed 22/07/2023).

NAO (National Audit Office) (2004) *Out of Sight – Not Out of Mind: Ofwat and the Public Sewer Network in England and Wales*. The Stationery Office, London, UK. See https://www.nao.org.uk/reports/out-of-sight-not-out-of-mind-ofwat-and-the-public-sewer-network-in-england-and-wales/ (accessed 22/07/2023).

National Highways (2022) *Design of Highway Drainage Systems*. Design Manual for Roads and Bridges. CG501. Version 2.1.0. August. National Highways, Guildford, UK.

Newcastle City Council (2023) Newcastle named UK's smartest city in global Smart City Index. https://www.newcastle.gov.uk/citylife-news/digital/newcastle-named-uks-smartest-city-global-smart-city-index (accessed 22/07/2023).

NIC (National Infastructure Commission) (2022) *Reducing the Risk of Surface Water Flooding*. NIC, London, UK. See https://nic.org.uk/studies-reports/reducing-the-risks-of-surface-water-flooding/surface-water-flooding-final-report/.

Nones M (2015) Implementation of the floods directive in selected EU member states. *Water and Environment Journal* **29**: 412–418, 10.1111/wej.12129.

NORIS (2022) NORIS – No Rainwater in Sewers. See https://keep.eu/projects/320/NORIS-No-Rainwater-in-Sewers-EN/ (accessed 22/07/2023).

Oberascher M, Kinzel C, Kastlunger U *et al.* (2021) Integrated urban water management with micro storages developed as an IoT-based solution – The smart rain barrel. *Environmental Modelling and Software* **139**: 105028, 10.1016/j.envsoft.2021.105028.

OECD (Organisation for Economic Co-operation and Development) (2015) *OECD Principles on Water Governance*. OECD, Paris, France. See https://www.oecd.org/cfe/regionaldevelopment/OECD-Principles-on-Water-Governance-en.pdf (accessed 22/07/2023).

OECD (2017) *Diffuse Pollution, Degraded Waters: Emerging Policy Solutions*. OECD Publishing, Paris, France. See http://www.keepeek.com/Digital-Asset-Management/oecd/environment/diffuse-pollution-degraded-waters_9789264269064-en#.WnGoApOFgWo#page9 (accessed 22/07/2023).

OECD (2018) *IImplementing the OECD principles on water governance*. OECD Publishing, Paris, France. See https://www.perlego.com/book/866602/implementing-the-oecd-principles-on-water-governance-pdf (accessed 22/07/2023).

Olsson G, Nielsen M, Yuan Z, Lynggaard-Jensen A and Steyer J-P (2005) *Instrumentation, Control and Automation in Wastewater Systems*. IWA Publishing, London, UK.

Page B and Bakker K (2005) Water governance and water users in a privatised water industry: participation in policy-making and in water services provision. A case study of England and Wales. *International Journal of Water* **3(1)**: 2005.

Petrie B (2021) A review of combined sewer overflows as a source of wastewater-derived emerging contaminants in the environment and their management. *Environmental Science and Pollution Research* **28**: 32095–32110. https://doi.org/10.1007/s11356-021-14103-1.

Price R and Catterson G (1995) Monitoring and modelling – a question of balance. *Proceedings of the WaPUG Conference*. See http://ftp2.ciwem.org/1995/1995-11-23%20to%2024/AU1995.PAPER06.pdf (accessed 22/07/2023).

Rietveld M, Clemens F and Langeveld J (2020) Solids dynamics in gully pots. *Urban Water Journal* **17(7)**: 669–680, 10.1080/1573062X.2020.1823430.

Rizzo A, Tondera K, Pálfy TG *et al.* (2020) Constructed wetlands for combined sewer overflow treatment: a state-of-the-art review. *Science of the Total Environment* **727**: 138618, 10.1016/j.scitotenv.2020.138618.

Roebuck RM, Oltean-Dumbrava C and Tait SJ (2012) Can simplified design methods for domestic rainwater harvesting systems produce realistic water-saving and financial predictions? *Water and Environment Journal* **26(3)**: 352–360, 10.1111/j.1747-6593.2011.00295.x.

San Francisco Water Power Sewer (2022) *San Francisco's recycled water ordinance*. See https://sfpuc.org/sites/default/files/documents/RecycledWaterBrochure_July_2022.pdf (accessed 22/07/2023).

Sayers PB, Horritt MS, Carr S *et al.* (2020) *Third UK Climate Change Risk Assessment (CCRA3): Future Flood Risk – Main Report* Sayers and Partners and the National Infrastructure Commission, London, UK. See https://www.ukclimaterisk.org/wp-content/uploads/2020/07/Future-Flooding-Main-Report-Sayers-1.pdf.

Sayers PB, Eccleston P, Horton B *et al.* (2022) *Surface Water Future Risk and Investment Needs. Report Prepared for the National Infrastructure Commission*. Sayers and Partners. See https://nic.org.uk/app/uploads/NIC-Sayers-and-Partners-Surface-Water-Future-Risk-and-Investment.pdf (accessed 22/07/2023).

Schutze M, Campisano A, Colas H, Schilling W and Vanrolleghem PA (2004) Real time control of urban wastewater systems – where do we stand today? *Journal of Hydrology* **299(2004)**: 335–348, 10.1016/j.jhydrol.2004.08.010.

Scottish Government (2022) Surface water management planning: guidance (2018). https://www.gov.scot/publications/flood-risk-management-scotland-act-2009-surface-water-management-planning/pages/6/ (accessed 22/07/2023).

Semple NG (1991) Principles of environmental protection. *Proceedings of the Institution of Civil Engineers* Part 1. **90(4)**: 837–851.

Sharman BJ and Tidswell RG (1993) Sprint Project Sp. 226: Real time control of urban drainage and sewage systems – Bolton Town Centre Telemetry Control System. *Wastewater Planners User Group Conference*. See http://ftp2.ciwem.org/1993/1993-11-24%20to%2026/AU1993.PAPER04.pdf (accessed 22/07/2023).

Schmit TG and Scheid C (2019) Evaluation and communication of pluvial flood risks in urban areas. *WIREs Water* **7(1)**: e1401, 10.1002/wat2.1401.

STOWA (2012) *Source-Separation in the Urban Water Infrastructure*. Report W14. STOWA. Stationsplein 89 3818 LE Amersfoort POSTBUS 2180 3800 CD AMERSFOORT. See https://www.stowa.nl/sites/default/files/assets/PUBLICATIES/Publicaties%202012/STOWA%202012-W-14.pdf (accessed 22/07/2023).

Tao W, Bays JS, Meyer D, Smardon RC and Levy ZF (2014) Constructed wetlands for treatment of combined sewer overflow in the US: a review of design challenges and application status. *Water* **6**: 3362–3385, 10.3390/w6113362.

Ugarelli R, Leitão JP, Almeida MC and Bruaset S (2011) *Overview of Climate Change Effects Which May Impact the Urban Water Cycle*. PREPARED Report No. 2011.011.. See http://repositorio.lnec.pt:8080/handle/123456789/1006408 (accessed 22/07/2023).

UKRMA (UK Rainwater Management Association) (2022) https://ukrma.org/. UKRMA, Collingham, UK.

UKTAG (UK Technical Advisory Group)(2008) *UK Environmental Standards and Conditions (Phase 1)*. Final report. WFD UK TAG. See https://southeastaquatics.net/resources/pdfs/UK_enviro_stds_conditns%20Phase%201.pdf (accessed 22/07/2023).

UKWIR (UK Water Industry Research) (2022) *Future Asset Planning – Scenarios, Frameworks and Measures*. 22/RG/05/58. UKWIR Ltd, London, UK. See https://ukwir.org/water-research-reports-publications-viewer/d8116c77-b639-4828-b4ac-6e47a20f054f (accessed 22/07/2023).

USEPA (United States Environmental Protection Agency) (2022) *Storm Water Management Model (SWMM)*. See https://www.epa.gov/water-research/storm-water-management-model-swmm (accessed 22/07/2023).

van der Werf JA, Kapelan Z and Langeveld J (2022) Towards the long-term implementation of real time control of combined sewer systems: a review of performance and influencing factors. *Water Science and Technology* **85(4)**: 1295, 10.2166/wst.2022.038.

Vollertsen J, Revilla N, Hvitved-Jacobsen T and Nielsen AH (2015) Modeling sulfides, pH and hydrogen sulfide gas in the sewers of San Francisco. *Water Environment Research* **87(11)**: 1980–1989, 10.2175/106143015X14362865226752.

Water Regs UK (2022) https://www.waterregsuk.co.uk/ (accessed 22/07/2023).

Water UK (2021) *Design and Construction Guidance for foul and surface water sewers offered for adoption under the Code for adoption agreements for water and sewerage companies operating wholly or mainly in England ("the Code")*. Approved version 2.1. See https://www.water.org.uk/wp-content/uploads/2021/07/SSG-App-C-Des-Con-Guide.pdf (accessed 22/07/2023).

Williams WD and Tidswell RG (1994) Bolton RTC – a glimpse at the future. *Proceedings of the WaPUG Conference*. See http://ftp2.ciwem.org/1994/1994-11-23%20to%2025/AU1994.PAPER09.pdf (accessed 22/07/2023).

Woods-Ballard B and Cherrier V (2019) Review of the UWWTD implementation, and the treatment of CSOs in the EU. *CIWEM UDG Conference*. See http://ftp2.ciwem.org/2019/2019-Autumn/Paper%2010%20Evaluation%20of%20UWWTD%20implementation%20FINAL%20DRAFT-Formatted.pdf (accessed 22/07/2023).

Wren E, Barnes M, Janes M *et al*. (2022) *The Natural Flood Management Manual*. CIRIA C802. Construction Industry Research and Information Association, London, UK.

Zhang Z, Paschalis A and Mijic A (2021) Planning London's green spaces in an integrated water management approach to enhance future resilience in urban stormwater control. *Journal of Hydrology* **597**: 126126, 10.1016/j.jhydrol.2021.126126.

Ashley R, Smith B, Shaffer P and Caffoor I
ISBN 978-0-7277-4178-3
https://doi.org/10.1680/icehudp.41783.105

Chapter 4
Engagement for effective urban drainage

Paul Shaffer, Glyn Everett and Brian Smith

4.1. Introduction

Communities and stakeholders, through their effective engagement, play a vital role in shaping the places we live, work and play. Their contribution influences the vision for the built environment, approaches to delivering and managing infrastructure, interactions with the natural environment and how spaces and places are used. Box 4.1 sets out what is meant by the various terms: engagement, community and stakeholder, as used in this book.

Box 4.1
What are engagement and community and who is a stakeholder

Engagement

Describes an active involvement of stakeholders. Approaches range from basic information giving, gathering and consultation to full deliberation and empowerment.

Community

While the term 'community' is widely used, answering the question 'Who is the community?' is not straightforward (MacQueen *et al.*, 2001). As Giddens and Sutton (2014) put it, community is 'a contested concept, but, simply, a group of people living in a particular locality, or who have a certain shared interest, who engage in systematic interactions with one another'.

Practitioners need to consider if their 'communities' are geographically located or more identity, practice or socially based. Meikle and Jones (2013) suggest we should consider five types of community, as follows.

- Interest (people sharing similar interests or passions)
- Action (people trying to effect change)
- Place (people connected by geographical space)
- Practice (people in similar professions, or undertaking similar activities)
- Circumstance (people connected by external events/situations).

Stakeholder

A stakeholder can be defined as 'anyone affected by the problem or solution or interested in the problem or solution. They can be individuals or organisations and include the public and communities' (Defra, 2010a).

Engagement has an important role in urban design and spatial planning, facilitating local ownership and legitimacy for policies that shape land use, development and people's quality of life. While facilitating local ownership, the policies it helps shape may also be regional and national. Effective engagement is necessary to deliver a more sustainable approach to urban drainage, providing multiple benefits and contributing to ecosystem services and natural capitals accounting. Flooding that occurs where surface water management systems are the main source of the flooding (e.g. that experienced in England in 2007 and London in 2021) highlights the institutional complexity of managing urban drainage (NIC, 2022). This complexity is, in part, due to the variety of public sector, private sector and community organisations, which typically work with different levels of funding, governance and regulations, that have an interest in surface water management and the built environment. The precise responsibilities and funding mechanisms will vary worldwide, as outlined in Chapter 5.

For some time, it has been recognised that collaboration between stakeholders is required for effective local flood risk management (e.g. Daly *et al.*, 2015). Creating the right environment for collaboration, engagement and obtaining participation in local flood risk management and urban drainage presents new opportunities and challenges.

The roles and responsibilities of stakeholders working in and around urban drainage are often unclear, particularly to the public, as outlined for England and Wales by Pitt (2008). Most people do not care who is responsible for flooding or drainage infrastructure; they expect flooding to be prevented and are in many places coming to terms with the reality that flood risk can only be managed and not eliminated. However, urban drainage flooding has long been misunderstood by people, especially in the developed world; from the need to manage their own property drainage to water on the streets, there continues to be an expectation that this is being managed by 'others' (Ashley *et al.*, 2020).

Those working in the built environment and water sectors in developed countries are increasingly collaborating to ensure that potential risks of flooding and pollution from urban drainage are effectively and cost-efficiently managed and potentially shared, and that the opportunities provided from contemporary approaches to urban drainage can be identified and exploited. Collaboration is essential to providing local flood risk management, resolving urban drainage problems and improving river health in ways that can also deliver against diverse drivers and provide multiple benefits – for example, water resources, the quality of places, biodiversity, climate change adaptation and general quality of life.

Collaboration between organisations with flooding responsibilities can reduce each organisation's requirements for resources, both financial and human, and skills. There is also a pressing need to ensure effective engagement with individuals, property owners, dwellers and users as part of this collaborative process. There are a growing number of tools and approaches available to value the multiple benefits from seeing and using water in different ways to the past, and that can underpin partnerships (e.g. in the UK, Hortan *et al.*, 2019). Collaboration and partnership can facilitate a more effective, cost-efficient and multifunctional approach to infrastructure delivery and management.

Collaborating with individuals and communities improves transparency, often enabling trust, 'buy-in' and support for key decisions or activities. Working closely with the public can help empower them, by meaningfully involving them in decision-making and, sometimes, delivery and maintenance of the infrastructure, as illustrated in Box 4.2.

Box 4.2

Community engagement in Bristol, England

Community engagement was embraced and has been studied in Bristol, where a self-initiated group of volunteers started to maintain a blue-green infrastructure (BGI) feature. Initially, the relationship with the local authority was distant. Over time, as the project developed, the group was able to provide more regular on-the-ground targeted basic maintenance and could access sources of funding that the local authority could not. The group now engages regularly with representatives from the local authority and contributes to ongoing maintenance and development conversations. Elsewhere in Bristol, residents of a modern eco-housing development featuring BGI were initially hampered in their desire to undertake site maintenance and improvements by land ownership issues, which had not yet transferred to them when their management company took over the development.

The review by Reed (2008) of literature on stakeholder engagement in environmental management refers to eight best practice features of stakeholder, or community, participation (see Box 4.3). Fundamental factors include: grounding in a trusting, inclusive, equitable and empowering philosophical outlook; early and ongoing attempts at engagement; development of appropriate and case-specific methods; professional and impartial facilitation; and an acceptance and inclusion of both scientific and local lay knowledge. These guiding principles only provide a starting point for any engagement process, as all projects have their specificities, requiring adaptation and a bespoke approach.

Worldwide there have been growing numbers of initiatives to ensure participation and engagement in water-related issues. For example, the European SWITCH project utilised learning alliances in many countries to start up local participatory activities, which in many places have continued since the project was completed (e.g. Butterworth *et al.*, 2011; Hall *et al.*, 2002).

The approach to collaboration and partnerships around urban drainage in England is most recently rooted in the Integrated Urban Drainage pilots managed by Defra in 2007. These explored the role that partnerships and local flood risk assessment could have in the management of surface water flooding.

Box 4.3

Best practice features of stakeholder participation (Based on Reed, 2008)

- Stakeholder participation needs to be underpinned by a philosophy that emphasises empowerment, equity, trust and learning.
- Stakeholder participation should be considered as early as possible and throughout the process.
- Relevant stakeholders need to be analysed and represented systematically.
- Clear objectives for the participatory process need to be agreed among stakeholders at the outset.
- Methods should be selected and tailored to the decision-making context, considering the objectives, types of participants and appropriate level of engagement.
- Highly skilled facilitation is essential.
- Local and scientific knowledge should be integrated.
- Participation needs to be institutionalised.

Fifteen pilot studies explored stakeholder engagement and technical solutions in detail (Gill, 2008), and demonstrated the importance of developing a shared understanding and approach to dealing with flood risk. They also highlighted the importance of being able to use clear language when communicating risk to non-technical stakeholders to obtain a shared understanding. Box 4.4 outlines one of the pilot projects, illustrating community engagement.

This approach has evolved in England with the introduction of Local Flood Risk Management Strategies, Surface Water Management Plans and, more recently, Drainage and Wastewater Management Plans to facilitate a more collaborative approach to urban drainage, making specific linkages to spatial planning and drainage management (Defra, 2010a). Regulators and government suggest that water and sewerage companies can no longer develop effective drainage strategies without partnership (Defra, 2022; Halcrow, 2013).

In England, the approach to funding for improvements in flood risk management and resilience is underpinned by a requirement to attract a proportion of partnership funding and deliver efficiencies. Effective engagement and collaboration are crucial; however, while there are frameworks to provide funding and support partnerships, they often favour the larger fluvial and coastal flood risk management schemes. Those who work in surface water management struggle to obtain funding for local flood risk management (CIRIA, 2017, Ainley, 2021).

Box 4.4

Engagement in West Garforth Integrated Urban Drainage pilot study

West Garforth is a community to the east of the city of Leeds in Yorkshire, UK. Numerous properties, roads and community areas were flooded frequently in the twentieth century. The backbone of the drainage infrastructure is a system of inadequate culverted watercourses, passing through hundreds of private properties, many of which flood annually. Under the extant regulatory system, responsibility belonged the property owners on whose land a buried culvert passed. Although the local authority designed a relief sewer in the 1990s, the revision of drainage responsibilities, privatising water and sewerage, passed some responsibilities to the local sewerage undertaker, who had no statutory duty to relieve the problems. The onus was therefore on the property owners. As a Defra Integrated Urban Drainage pilot, funds were available to investigate the problems and recommend solutions (Gersonius *et al.*, 2012). Crucially, the project worked with the local community, hosting two public meetings (Sharp, 2008). The project steering group, comprising the principal agencies and consultants involved, also co-opted a public representative following the first public meeting. At the first public meeting, details of the issues were presented, together with illustrations of contemporary drainage systems. The professionals involved were careful not to offer 'solutions' at this meeting, leaving possibilities for the community to consider. Despite the concerns from some of the steering group members about this approach, their fears were allayed at the second meeting, when the community was able to suggest that several public areas in the catchment were suitable to direct excess flood waters in order to protect properties and reduce highway flooding. The project team was able to use these ideas in the subsequent computational modelling and hence determine what would be the most effective solution (Gersonius *et al.*, 2012; Leeds City Council *et al.*, 2008). A Garforth Flood Support Group continues to meet due to the ongoing flood risks to the community (https://www.facebook.com/GarforthFlood/).

For many years, organisations championing a more sustainable approach to drainage have been encouraging stakeholders, particularly developers, to undertake early and effective engagement with local authorities and regulators to enable delivery of high-quality BGI schemes. Early and effective engagement can help overcome common barriers to the delivery of BGI, such as minimising land take, securing adoption and overcoming health and safety concerns.

Engagement with communities when retrofitting BGI is particularly important, so that organisations can better understand how areas are used based on current activities and the desired uses of a space. It is also useful to know what impact BGI might have on local communities – for example, providing areas for people to enjoy, or understanding the criticality of car parking spaces. Deciding who will undertake the engagement and when and how it will be delivered become even more important when dealing with local communities.

The challenges within urban drainage are complex, often requiring collaboration to define problems, deliver more beneficial solutions and try to meet the needs of the different organisations and disciplines working within the urban drainage sector. The strong agenda for better local flood risk management arguably focuses on surface water management as being a challenge, rather than as an opportunity to deliver better quality places and spaces. Integrating urban drainage and alternative ways to manage surface water into institutions and natural and built environments can provide opportunities to respond positively and adapt to a changing climate and urban growth. In doing so, engagement with communities becomes increasingly important.

4.2. Stakeholders

Legislation on the drainage of urban areas has evolved over hundreds of years, influenced by a variety of statutes and case law. In many parts of the world these place duties and responsibilities (some statutory, others permissive) on various organisations, stakeholders, landowners and the public at large. This is complicated by regulatory processes, whereby responsible bodies may include public, private, regulated and unregulated organisations (e.g. Juuti *et al.*, 2005; Sharp, 2017).

Reviews on flood risk management and surface water management, particularly following flooding in England (e.g. the Pitt review in 2008 (Pitt, 2008) and the Jenkins review in 2020 (Jenkins, 2020)), suggest that regulatory and institutional complexity is a potential obstacle to the effective delivery of flood risk management both at the local level and the strategic level. Internationally, there is similar institutional fragmentation. In responding to this, the UK government developed the Flood and Water Management Act 2010 for England and Wales and the Flood Risk Management (Scotland) Act 2009 for Scotland. The Scottish approach has embedded sustainability within its framework, and the Flood and Water Management Act 2010 has introduced a framework in which the environmental regulators are responsible for strategic flood risk management, with the Lead Local Flood Authority responsible for delivering and maintaining a Local Flood Risk Management Strategy for its area.

In many countries, through heightened public, media and political awareness, there is now a sharper focus on the quality of waterbodies, with increased scrutiny of how urban drainage is managed, particularly outfalls from highway drainage and combined sewer outflows (CSOs) (Chapter 3). Greater cooperation between organisations and engagement with communities will be required to provide alternative collaborative solutions – that are both acceptable and affordable, some of which may include BGI – to improve river water quality.

A variety of stakeholders are involved at both a strategic level and in more localised, site-specific developments. As an example, the potential range of stakeholders in England is summarised in Table 4.1.

Table 4.1 Key stakeholders involved in surface water management

Stakeholder		Function
National government		Set national policy on water management Provide guidance on, and funding for, flood risk management Manage the spatial planning process
Local government (municipalities)		Lead Local Flood Authority (LLFA) responsible for leading local flood risk management May also act as highway authority maintaining non-trunk roads A Risk Management Authority (covering all natural hazards)
Regulators	Environment	Responsible for maintaining, operating and improving flood defences Provision of flood warning and emergency response Provide a strategic overview and supervisory duties for all flooding. A Risk Management Authority
	Economic	Protect the interests of consumers Ensure that water companies can finance and properly carry out of their statutory functions Ensure water companies secure the long-term resilience of their water supply and wastewater systems
Sewerage undertaker		Duty to provide and maintain a public sewerage system to effectually drain its area A Risk Management Authority
Drainage boards		Autonomous public organisations with permissive powers for the assessment and management of local watercourses and the provision land drainage and maintenance of some flood prevention and defence schemes A Risk Management Authority
Developers		Ultimately responsible for the type of surface water management system used on their developments
Third sector organisations		National and local groups with an interest in the natural and built environment
Regional flood and coastal committees		Approve the annual programme of flood defence work in their region and set the local levy
The public and communities		Vital role in the vibrancy of a development and the acceptance of sustainable drainage

4.3. Local decision-making

In many countries, planning decisions are made locally, with accountability through locally elected members (Chapter 5). This, combined with restrictive financial constraints, particularly within the public sector, a variety of permissive duties and a duty to cooperate, means that wide engagement is essential.

Municipalities, communities and residents can often participate in planning decisions through a local or neighbourhood planning system. In England and Wales the duty to cooperate is inherent in the Flood and Water Management Act 2010.

In England, the National Flood and Coastal Erosion Risk Management Strategy from the Environment Agency (Environment Agency, 2020) requires local approaches to deliver flood resilience. The strategy suggests that (flood) Risk Management Authorities should work in partnership with communities to help them understand the risks and encourage their involvement in both decision-making and the initiatives. Any approach to prioritisation and/or funding should reflect the local circumstances, and accountability rests with Lead Local Flood Authorities. In addition, the strategy focuses on the benefits that delivery of BGI can bring to flood risk management.

4.4. Approaches to engagement

Integrated urban drainage can only effectively be delivered through successful engagement, collaboration and dialogue and is integral to flood risk management.

Urban drainage has historically been influenced by coastal and fluvial flooding initiatives. In England, Defra (2020) stresses the importance of engaging with anyone interested in, or affected by, approaches to managing flood risk. This should enable policymakers and decision-makers to do the following.

- Better understand challenges, establish a common understanding and identify the opportunities to deliver multiple benefits.
- Build trust with those involved with and affected by the project appraisal process supporting partnership working.
- Identify wider approaches to managing flood risk and deliver multiple objectives.
- Assist stakeholders, communities and customers experience open and transparent decision-making.
- Have democratic input and potentially a sense of ownership with the decision-making.

Engagement is required both at strategic level and development or community level when managing local flood risk and deciding which approach to use to deliver urban drainage. This enables communities to provide input into the way that urban drainage is integrated into their natural and built environment, helping to deliver better quality developments and improved local neighbourhoods (e.g. BEGIN, 2023; Willems *et al.*, 2022).

4.4.1 Engagement at a strategic level

Strategic stakeholder engagement and, more importantly, the establishment of partnerships in local flood risk management has primarily been driven by Local Flood Risk Management Strategies and Surface Water Management Plans, in particular when difficult decisions about investment and approaches must be made.

While stakeholders such as municipalities have undertaken engagement on several different environmental and regeneration projects, application to urban drainage is less common, with some employing third sector organisations (such as rivers trusts, specialist consultancies or charities) to undertake engagement on surface water management, particularly retrofit BGI delivery within existing communities. Property and other developers have an important opportunity to so engage, promoting acceptance of 'different types of drainage' among purchasers and the public at large (e.g. Wavin, 2023).

Approaches to improving flood resilience and delivering better surface water management must focus on the role of collaborative leadership, development of partnerships and stakeholder engagement to progress flood risk management and deliver required outcomes. This prevents a fragmented, complex or ad hoc approach to collaboration, and ensures that key roles and responsibilities are understood and risks proactively managed. This is especially important when multiple outcomes are to be achieved, whereby flood risk management is delivered, together with water quality protection and place-making.

4.4.2 Engagement at a development and community level

Effective management is more than just consultation, it requires active listening, involvement and participation, with the intention of sharing information, mutual learning and building trust. Negotiation and strengthened relationships enable compromise, leading to better decisions and the avoidance of conflict. Consultation typically involves the lead organisation engaging in dialogue with stakeholders to obtain information, expertise and opinions before taking a decision or selecting a course of action.

Engagement and participation are more active processes, in which there is dynamic feedback of information combined with active and transparent involvement in collaborative decision-making processes (Bell *et al.*, 2022).

Figure 4.1 presents the full range of engagement activities and demonstrates the interaction between stakeholders and the levels of participation.

Stakeholder engagement should resemble a 'conversation', with two-way feedback and an intention to listen, learn and reflect on new approaches, as well as to maintain trust and fairness. This represents a move away from the traditional decide–announce–defend (DAD) approach towards a more open and inclusive engage–deliberate–decide (EDD) approach.

Figure 4.1 Processes and techniques for engagement

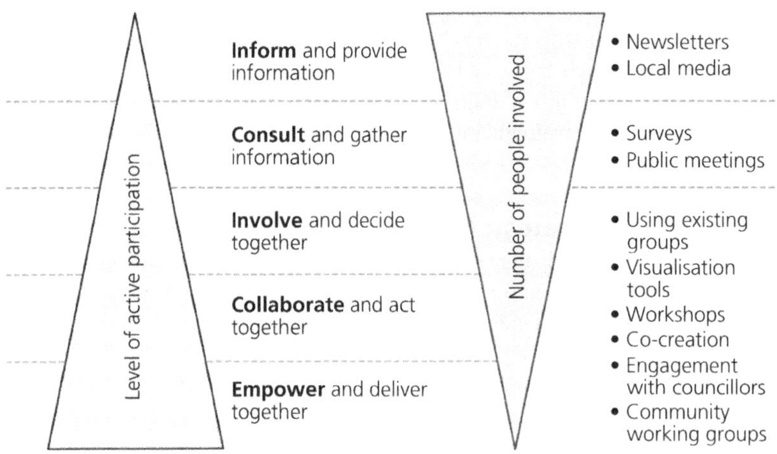

EDD is a more considered approach, which starts with seeking to identify those who might potentially be interested or affected. The next step is to engage participants in scoping, defining and understanding the problem or problems, before identifying and evaluating options and influencing the way forward.

The EDD approach develops a multiway process that should benefit all parties. Mutual trust between all stakeholders and the lead organisation (regulator, municipality, water company, etc.) is critical to the delivery of a more sustainable approach to urban drainage.

Engagement and collaboration should be a well-planned and transparent process, with the core stakeholder group developed from the outset and its objectives, scope and terms of reference agreed among all participants. Both time and space should be allocated to allow for further potential stakeholders to engage with the process after it has started, otherwise some may be missed or unable to participate.

Comprehensive guidance on engaging for local flood risk management is provided by Daly *et al.* (2015) *Communication and Engagement in Local Flood Risk Management.* The authors provide a framework for engagement that first considers the objectives, then the processes and mechanisms that are suitable for the situation. Using the most effective approach comes from having a good understanding of the stakeholders, opportunities and the techniques that might most productively be used. Table 4.2 gives an overview of the framework.

Table 4.2 A framework for engagement (after Daly *et al.*, 2015)

Part A: Define opportunities and challenges	Before engaging with stakeholders, it is important to understand fully the context and the challenges and opportunities. This will help scope the communication and engagement activities and adopt an appropriate approach.
Part B: Identify those to involve	A 'stakeholder mapping' exercise to develop a list of individuals, groups and organisations who may be affected or influence any outcomes. This is essential to identify and evaluate who should be involved in the process and how they can be engaged.
Part C: Prepare the engagement plan	A communication and engagement plan sets out the method and approach for engagement. It details who is going to be engaged with and how. It sets out the overall aim of the project, the specific objectives, the key messages, the action plan for delivery, the skills and resources needed to implement the plan, how success will be measured and how progress will be evaluated and communicated.
Part D: Deliver the engagement plan	Delivery of the plan should constantly refer to the aims and objectives of the communication and engagement plan.
Part E: Monitor and evaluate the engagement plan	Feedback on progress provides reassurance that the process has been worthwhile, that stakeholder involvement adds value to the management of local flood risk.

4.4.3 Delivering sustainable drainage

At a local level, development or community, engagement is essential for delivering a more sustainable approach to drainage; an approach that manages the risks of flooding and pollution and provides additional benefits for the community. These benefits might relate to biodiversity, amenity and place-making (see Chapter 8; Woods-Ballard *et al.*, 2015).

Municipalities and, increasingly, those responsible for other aspects of the water cycle play a crucial role in identifying and responding to the needs of communities. They need to work in partnership with a range of public and private organisations and agencies to coordinate better urban drainage within the context of the other services those organisations and agencies provide – that is, housing, transport, green infrastructure and wastewater management, etc. (e.g. Susdrain, 2012).

During the planning and design stages of urban drainage schemes, many different stakeholders are collaborating. Sustainable drainage that delivers multiple benefits requires a multidisciplinary approach to design, often involving the following disciplines

- drainage engineers
- highway engineers
- landscape architects
- spatial planners
- urban designers
- community engagement specialists (including social scientists)
- ecologists
- geotechnical engineers.

Evidence around BGI contributing to better health and wellbeing is becoming more compelling. Local educators (schools, etc.) and specialists who understand public health, welfare and social cohesion are becoming more involved in the planning and delivery of BGI.

As with strategic stakeholder engagement, it is important to identify key stakeholders early in the project. Their needs should be acknowledged and understood from the outset, with a transparent process in which the aims and objectives of the project are negotiated, designed and agreed.

The overall process for a new development should include the planning and design of the drainage as well as consideration of how the infrastructure will be used and maintained. The scale of the project will dictate the necessary representation from the disciplines mentioned above and whether local businesses, residents and politicians need to be involved, in addition to stakeholders identified at a strategic level.

Approaches to stakeholder engagement are identified in Figure 4.1. However, engagement activities should be tailored to the local situation and those involved in the process – for example, if the public are involved, references to technical detail may be reduced and graphical representations or other visioning exercises could be useful.

The awareness of the benefits of stakeholder engagement for new developments is increasing, yet the quality of stakeholder engagement in decision-making for water management in many parts of the world, including the UK, remains poor. When tackling the challenges and exploiting the opportunities within our existing infrastructure in the built environment, effective engagement, leadership and partnerships are vital.

4.5. Capacity building

The holistic management of water is complex, with many challenges that need to be understood before they can be overcome. There is much that can be done to manage the water cycle and its integration the built environment more efficiently.

Legislation regarding water management (water quality and flood risk management) increasingly requires for the effective engagement and participation of organisations and communities. This is likely to require input from several participants at institutional, organisational and individual levels.

Several reviews have previously identified the lack of skilled practitioners as a limiting factor to innovation, sustainable development and effective surface water management (e.g. for the UK; Jenkins, 2020; ICE, 2005). Consultations on the implementation of the Flood and Water Management Act 2010 and the delivery of surface water management (Defra, 2018) identified the lack of skills and resources (together with funds) to tackle the requirements and responsibilities as being a significant challenge. This situation is more acute within local government, which plays a lead role in local flood risk management. Some local authorities do not have the resources in place or the skills to understand the extent of the challenge, which are needed to deliver an effective way forward (Defra, 2017).

The UK government has been considering approaches to tackle the shortage of skills, and following the introduction of the Flood and Water Management Act 2010 it has primarily focused on technical and engineering disciplines, although the benefits of a broader engagement with other disciplines has been recognised.

Addressing these deficiencies requires a continuous approach to resourcing and upskilling, more frequently called capacity building. Capacity building is a highly dynamic process. It requires programmes flexible enough to provide adaptive capacity and tackle challenges at several different levels and scales, with the aspiration of delivering long-term improvements in skills and capacity. It is often a fundamental part of the transitioning process to an improved approach to manage resources.

In addition to building capacity, efforts can very usefully be made to ensure that existing resources are being deployed effectively – for example, the encouragement of communication and conversations between disciplines, professions and bodies to encourage discussion and knowledge-sharing to support innovation.

Learning and Action Alliances (LAAs) present an opportunity to develop integrated solutions to problems that are not solvable simply through reference to one discipline, professional skillset, group or network. This approach brings together a range of interested professionals (water, flooding, biodiversity, community and city representatives, etc.) to meet routinely and talk openly, within a 'safe space', and to 'think outside the box', thus allowing new and innovative proposals to be considered and developed (Blanksby and Ashley, 2013; O'Donnell *et al.*, 2018).

Ashley *et al.* (2012) provide a guide for the establishment of LAAs for flood risk and water management. Bringing professionals and stakeholders together in informal, non-hierarchical settings can help to initiate open and interdisciplinary knowledge exchange and create transformative thinking. This includes the development of trusting relationships, a deepening of understanding and challenging fundamental assumptions, which in turn leads to the development of new knowledge and shifts in behaviours, attitudes and practice.

An example of this can be found in a Newcastle-based LAA. In this case, several years of collaborative LAA working progressed to a situation where the City Council, alongside five founding partners, were able to make a formal declaration about the proactive position of the city regarding the development and installation of extended blue and green infrastructure in 2016 (see Box 4.5).

4.5.1 Approaches to capacity building

In the UK in 2010, Defra produced a draft strategy for skills and capacity building in local authorities for local flood risk management to assist local authorities deliver their new role under the Flood and Water Management Act 2010 and Flood Risk Regulations 2009 and other actions recommended in the Pitt flooding review (Defra, 2010b). The strategy focuses on short-term actions, but it also looks at which elements of capacity should continue to be developed in the medium to long term.

Three key themes were identified to provide the structure for the Defra strategy, these were as follows.

- Developing knowledge and skills for existing staff: considering the use of partnership approaches, secondments, on-the-job training and mentoring. It also explores the contribution of educational resources – for example, e-learning and training and retraining of existing staff.
- Building capacity through provision of educational courses for new staff: exploring the role of professional qualifications alongside the encouragement of succession planning and knowledge management. Recruitment of already skilled individuals is supportive.
- Providing information and tools: this includes portals to share and record information, expertise, guidance and asset data. This approach also includes initiatives that allow those working in and with local authorities to share experiences and information either through face-to-face or electronic forums.

Box 4.5

Learning and Action Alliances

The Newcastle LAA was established in 2014, to facilitate the development and implementation of a blue-green vision for the city under the EPSRC-funded Blue-Green Cities research programme (O'Donnell *et al.*, 2018). This was in response to the major 2012 pluvial floods which impacted more than 1200 properties and flooded more than 500 homes (Newcastle City Council, 2016). Researchers at the Newcastle LAA meetings observed multiple levels of social learning, through collaboration and negotiation between members who would not normally be brought together in the same room, a gathering of evidence around existing BGI and identification of topics of mutual interest for further work (O'Donnell *et al.*, 2018). Outcomes from the LAA work include the development of a 'Newcastle Declaration on Blue and Green Infrastructure' in 2016, committing six signatories to collaborating and prioritising BGI use for flood and water management issues. The declaration was relaunched in 2019, now with ten organisations from several disciplines including planning, ecology, flood and water management, water resources and estates management (UFR, 2019).

The Defra strategy regards capacity building as being more than training to improve competencies. It includes the process of enabling individuals and organisations to understand challenges, access information and develop the skills needed to help them perform more effectively. It also includes approaches to improve the interaction between individuals and organisations, along with the development and support of management frameworks. This should assist with the management of relationships between different organisations and sectors (public, private and the community), including the development of 'champions' to deliver progress.

Research undertaken in Melbourne, Australia, to identify the potential catalysts for increasing a more sustainable approach to water management, suggests that the transition process to water-sensitive urban design is complex (Brown and Clarke, 2007) (see also Chapter 13). It states that this process requires a range of interconnected activities and initiatives, some planned and others opportunistic and reactive, to respond to the opportunities and challenges encountered. It is suggested that fundamental to successful transitioning is the development of champions. Champions are leaders who have a vision for change and can communicate this across the variety of institutions, organisations, disciplines and stakeholders involved in managing the water cycle. In the UK, it is often these champions who enable real change in the way that improved surface water management and urban drainage is delivered.

Delivering this vision of a more sustainable approach to drainage is dependent on developing social capital (Box 4.6). Here, professionals may respond to the challenges by improving their communication, interaction and relationships. Such behaviour changes may require support from additional resources; financial, guidance or skills, or a combination of all three.

The following sections explore approaches to improve social and institutional capital and the way the challenges in urban drainage can be tackled.

4.5.2 Leadership development

Complex challenges are likely to require composite ways of providing direction, cohesion and commitment. Institutional capital relies on research, building trust and timely service provision. It is fundamentally about the efficiency of institutions and 'institutional governance'. Therefore, the culture of an organisation in terms of communication, beliefs and working together will influence collaboration, capacity for action, speed of the decision-making process and the response to the challenges.

Box 4.6

Social capital

Social capital is defined as follows.

> Relationships among actors (individual, groups and/or organisations) that create a capacity to act for mutual benefit or a common purpose. Social capital is the social resource that is embodied in the relations between people. It resides in and stems from the contact, communication, sharing, co-operation and trust that are inherent in ongoing relationships.
>
> (Spellerberg, 2001)

Leadership development requires an adaptive approach to enable an organisation to become more collective, moving the focus away from the individual and onto a group with a shared awareness and ownership of the problem. It may enable people to understand how to relate and onto others, coordinate efforts, build commitments and develop extended social networks (Day, 2001). Globally, there have been efforts to support the development of leadership skills in relation to water systems (e.g. NWPP, 2017); however, focusing on a single domain such as water might not be what is needed in an increasingly interconnected world, where cross-disciplinarity working will be essential.

4.5.3 Action learning

At its simplest level, action learning is learning by doing – that is, working on 'real world' challenges in a collective and supportive way. It is widely appreciated that traditional approaches to learning through classroom-based lectures have limited effectiveness. Action learning processes can be described as a continuous process of learning and reflection, supported by colleagues and with an emphasis on getting things done.

There are examples of the development of alliances specifically to build capacity for flood risk management and the delivery of BGI in the UK (Maskrey *et al.*, 2020). These are often referred to as learning alliances or Learning and Action Alliances (LAAs). These alliances, underpinned by action learning, assist in the development of the capacity of different stakeholder groups to build mutual trust, accept a different view on risk, interact and find potential solutions. This requires that the traditional 'comfort zones' of the different organisations and disciplines are challenged in a supportive but delivery-focused learning environment (Ashley *et al.*, 2012).

Some key principles for developing LAAs from Ashley *et al.* (2012) include the following.

- LAAs should be formed around real, potentially complex problems (which may be hard to define) with many stakeholders, and an initial group that is committed to change. The LAA members will share a mutual vision, developing common approaches.
- LAAs benefit from being representative, capturing the institutional complexities that define the challenge that is trying to be overcome or the opportunity to be delivered.
- The emphasis is focused on cooperative delivery, changing practices and behaviours rather than research and devising new technologies.
- Innovations and changes relevant to the local context are likely to lead to more appropriate, adaptive and integrated solutions.
- New understanding of knowledge and learning should be promoted. Sharing knowledge is more complex when based on interaction between and within organisations and professionals in which active learning, communication, reflection and cooperation feature strongly.

4.5.4 Networking

Networking is viewed by some as an approach to breaking down functional barriers that occur between organisations and disciplines (Day, 2001; O'Donnell *et al.*, 2018). The main aim is to move leaders from knowing the *what* and *how*, to knowing the *who* in terms of making connections and problem-solving capability.

Networking also provides an opportunity for sharing experiences, thinking and learning with other participants. It enables the subtle challenging of basic assumptions and knowledge within a supportive environment for learning. In some circumstances, it may also enable members to develop relationships with individuals and disciplines they may not usually engage with.

Within urban drainage, such networks are starting to be established and are a fundamental part of capacity building. Participants benefit from an informal network and an opportunity for peer-to-peer learning in which they can engage on relevant topics. In addition, this approach is being mirrored by electronic, virtual networks and forums, providing an opportunity for participants to share knowledge and experiences.

REFERENCES

Ainley S (2021) The Benefits of Distributed SuDS in London. Susdrain Briefing, https://www.susdrain.org/files/resources/briefings/2105_briefing_the_benefits_of_distributed_suds_in_london_final.pdf (accessed 11/07/2021).

Ashley R, Blanksby J, Newman R *et al.* (2012) Learning and Action Alliances to build capacity for flood resilience. *Journal of Flood Risk Management* **5(1)**: 14–22.

Ashley R, Gersonius B and Horton B (2020) Managing flooding: from a problem to an opportunity. *Philosophical Transactions of the Royal Society A* **378(2168)**: 20190214, 10.1098/rsta.2019.0214.

BEGIN (2023) *Blue Green Infrastructures through Social Innovation*. Interreg North Sea Region. See https://climate-adapt.eea.europa.eu/en/metadata/projects/blue-green-infrastructures-through-social-innovation (accessed 17/12/2023).

Bell S, Boyle E, Canton J, Khan Z, Quinn R *et al.* (2022) Establishing a statement of principles for community engagement with civil engineering. *Proceedings of the Institution of Civil Engineers – Civil Engineering.* **175(3)**: 133–140, 10.1680/jcien.22.00007.

Blanksby JR and Ashley R (2013) *Learning and Action Alliances in relation to Urban Water and Flood Risk Management*. Report. INTERREG IVB. University of Sheffield, UK. See https://eprints.whiterose.ac.uk/86610/ (accessed 03/05/2022).

Brown RR and Clarke JM (2007) *Transition to Water Sensitive Urban Design: The story of Melbourne, Australia*. Report No. 07/1. Facility for Advancing Water Biofiltration, Monash University, Melbourne, Australia.

Butterworth J, McIntyre P and da Silva Wells C (eds) (2011) *SWITCH in the City: Putting Urban Water Management to the Test*. IRC International Water and Sanitation Centre, The Hague, the Netherlands, https://www.yumpu.com/en/document/read/26990728/switch-in-the-city-putting-urban-water-management.

CIRIA (2017) *Support for Partnership Funding in Flood and Coastal Erosion Risk Management*. CIRIA C770. Construction Industry Research and Information Association, London, UK.

Daly D, Jodieri R, Mccarthy SS and Pygott K (2015) *Communication and Engagement in Local Flood Risk Management*. CIRIA C751. Construction Industry Research and Information Association, London, UK.

Day D (2001) Leadership development: a review in context. *Leadership Quarterly* **11(4)**: 581–613.

Defra (Department for Environment, Food and Rural Affairs) (2010a) *Surface Water Management Plan Technical Guidance*. Defra, London, UK.

Defra (2010b) *Draft Strategy for Skills and Capacity Building in Local Authorities for Local Flood Risk Management*. Defra, London, UK.

Defra (2017) *Evaluation of the Arrangements for Managing Local Flood Risk in England*. Defra, London, UK.

Defra (2018) *Review of Local Approaches to Surface Water Flood Risk Management*. Defra, London, UK.

Defra (2020) Flood and Coastal Erosion Risk Management Appraisal Guidance, https://www.gov.uk/guidance/flood-and-coastal-erosion-risk-management-appraisal-guidance/1-about-the-fcerm-appraisal-guidance (accessed 06/07/2022).

Defra (2022) *Storm Overflows Discharge Reduction Plan*. HMSO, London, UK. See https://assets.publishing.service.gov.uk/government/uploads/system/uploads/attachment_data/file/1101686/Storm_Overflows_Discharge_Reduction_Plan.pdf (accessed 16/12/2022).

Environment Agency (2020) *National Flood and Coastal Erosion Risk Management Strategy for England*. HMSO, London, UK. See https://www.gov.uk/government/publications/national-flood-and-coastal-erosion-risk-management-strategy-for-england–2 (accessed 11/07/2021).

Gersonius B, Ashley RM, Pathirana A and Zevenbergen C (2012) Adaptation of flood risk infrastructure to climate resilience. *Civil Engineering Special Issue* **165(CE6)**: 40–45.

Giddens A and Sutton PW (2014) *Essential Concepts in Sociology*. Polity Press, Cambridge, UK.

Gill E (2008) *IUD Pilot Summary Report. Making Space for Water: Urban Flood Risk and Integrated Drainage*. Project HA2. Department for Environment, Food and Rural Affairs (Defra), London, UK.

Halcrow (2013) *Drainage Strategy Framework. Good Practice Guidance Commissioned by the Environment Agency and Ofwat*. Halcrow Group, Swindon, UK. See https://www.ofwat.gov.uk/wp-content/uploads/2015/12/rpt_com201305drainagestrategy1.pdf (accessed 21/07/2021).

Hall D, Lobina E, Viero OM and Maltz H (2002) *Water in Porto Alegre, Brazil*. International Research Unit (PSIRU) and Municipal Department of Water and Sanitary Sewage (DMAE). See http://www.psiru.org/sites/default/files/2002-08-W-dmae.pdf (accessed 23/07/2023).

Horton B, Digman CJ, Ashley RM and McMullan J (2019) *B£ST Guidance – Guidance to Assess the Benefits of Blue and Green Infrastructure using B£ST*. CIRIA W047b. Construction Industry Research and Information Association, London, UK. See http://www.susdrain.org/resources/best.html (accessed 11/07/2021).

ICE (Institution of Civil Engineers) (2005) *Action on Engineering Skills Shortages in Flood Risk Management*. ICE, London, UK. See https://assets.publishing.service.gov.uk/government/uploads/system/uploads/attachment_data/file/292914/geho0205bhrw-e-e.pdf (accessed 11/07/2021).

Jenkins D (2020) *Report of a Review of the Arrangements for Determining Responsibility for Surface Water and Drainage Assets*. Defra, London, UK. See https://assets.publishing.service.gov.uk/government/uploads/system/uploads/attachment_data/file/911812/surface-water-drainage-review.pdf (accessed 16/12/2022).

Juuti PS and Katko TS (eds) (2005) *Water, Time and European Cities: History Matters for the Futures*. WaterTime. See http://www.watertime.net/docs/WP3/WTEC.pdf (accessed 23/11/2023).

Leeds City Council *et al.* (2008) *West Garforth Integrated Urban Drainage Pilot Study: Final Report*. See https://democracy.leeds.gov.uk/documents/s20933/West%20Garforth%20DEFRA%20ON%20REQUEST.pdf (accessed 23/07/2023).

MacQueen KM, McLellan E, Metzger DS *et al.* (2001) What is community? An evidence-based definition for participatory public health. *American Journal of Public Health* **91**: 1929–1938.

Maskrey S, Vilcan T, O'Donnell E and Lamond J (2020) Using Learning and Action Alliances to build capacity for local flood risk management. *Environmental Science and Policy* **107**: 198–205.

Meikle H and Jones D (2013) Pedagogy of oppressed community engagement: Socially inclusive visioning of urban change. In *Proceedings of the State of Australian Cities Conference*, Sydney, NSW, Australia. Ruming K, Randolph B and Gurran N (eds), pp. 1–13.

NIC (National Infrastructure Commission) (2022) *Reducing the Risk of Surface Water Flooding*. NIC, London, UK. See https://nic.org.uk/app/uploads/NIC-Reducing-the-Risk-of-Surface-Water-Flooding-Final-28-Nov-2022.pdf (accessed 16/12/2022).

Newcastle City Council (2016) *Local Flood Risk Management Plan*. See https://www.newcastle.gov.uk/sites/default/files/Flooding/2016.03.31%20NCC%20LFRMP%20Complete%20and%20Final%20Report%20accessible.pdf (accessed 22/11/2023).

NWPP (2017) *New Water Policy and Practice Journal* **4(1)**: Special Issue on Water Governance Leadership. See https://www.clearwatervic.com.au/user-data/resource-files/taylor-2017—leadership-traps-paper—nwpp-journal-4-1.pdf (accessed 23/07/2023).

O'Donnell E, Lamond J and Thorne C (2018) Learning and Action Alliance framework to facilitate stakeholder collaboration and social learning in urban flood risk management. *Environmental Science and Policy* **80**: 1–8.

Pitt M (2008) *Learning Lessons from the 2007 Floods*. Cabinet Office, London, UK. See https://webarchive.nationalarchives.gov.uk/20100812084907/http:/archive.cabinetoffice.gov.uk/pittreview/_/media/assets/www.cabinetoffice.gov.uk/flooding_review/pitt_review_full%20pdf.pdf (accessed 11/07/2021).

Reed MS (2008) Stakeholder participation for environmental management: a literature review. *Biological Conservation* **141(10)**: 2417–2431.

Sharp L (2008) User interactions in strategic research: the example of two UK 'Integrated Urban Drainage Pilots'. *Proceedings of the 11th International Conference on Urban Drainage, Edinburgh*, https://www.researchgate.net/profile/Liz-Sharp/publication/27312887_User_interactions_in_strategic_research_The_example_of_two_UK_%27Integrated_Urban_Drainage_Pilots%27/links/53f5bda70cf2fceacc6f5dbd/User-interactions-in-strategic-research-The-example-of-two-UK-Integrated-Urban-Drainage-Pilots.pdf (accessed 23/07/2023).

Sharp L (2017) *Reconnecting People and Water*. Routledge, Abingdon, UK.

Spellerberg A (2001) *Framework for the Measurement of Social Capital in New Zealand*. Research and Analytical Report 2001. Statistics New Zealand, Wellington, New Zealand. See http://citeseerx.ist.psu.edu/viewdoc/download?doi=10.1.1.198.4273&rep=rep1&type=pdf (accessed 11/07/2021).

Susdrain (2012) Community involvement – the key to great SuDS projects, www.susdrain.org. See https://tinyurl.com/l3jhjyf (accessed 11/07/2021).

UFR (Urban Flood Resilience) (2019) Newcastle Blue and Green Declaration, http://www.urbanfloodresilience.ac.uk/newcastle-blue-and-green-declaration/newcastle-blue-green-declaration.aspx (accessed 26/01/2023).

Wavin (2023) *The Value of SuDS to New Home Buyers*. Wavin, Doncaster, UK. See https://blog.wavin.com/en-gb/value-of-sustainable-drainage-systems-suds-to-new-home-buyers (accessed 23/07/2023).

Willems P, Minkman E, Veerbeek W, Ashley R and van Buuren A (2022) Policy transfer in urban water management – svidence from ten BEGIN cities. In *Routledge Handbook of Urban Water Governance* (Bolognesi T, Pinto FS and Farrelly M (eds)). Routledge, London, UK, pp. 257–271, 10.4324/9781003057574.

Woods-Ballard B, Wilson S, Udale-Clarke H *et al.* (2015) *The SuDS Manual*. CIRIA C753. Construction Industry Research and Information Association, London, UK. See http://www.susdrain.org/resources/SuDS_Manual.html (accessed 23/07/2023).

Ashley R, Smith B, Shaffer P and Caffoor I
ISBN 978-0-7277-4178-3
https://doi.org/10.1680/icehudp.41783.123

Chapter 5
Urban drainage and development

Richard Ashley, Peter Bide, Brian Smith and Paul Shaffer

5.1. Introduction

There are many benefits to be gained through linking planning for flooding, water management and biodiversity. Taken at all spatial scales from a catchment to an individual plot, an integrated approach to water management can provide significant benefits for stakeholders and communities. Through integrating internal and external stakeholders, a catchment-based approach can simultaneously support the regeneration of communities, help facilitate essential housing, enhance biodiversity and water availability and quality, reduce flood risk and improve public spaces (Catchment Based Approach, 2023; Defra, 2013).

For local planners, stakeholder relationship management and strategic alignment are essential in developing catchment partnerships. Stakeholder-influenced mapping is a useful tool to aid decision-making. This affords the opportunity to create continuous dialogue with all stakeholders and involve them in the design, implementation and longer-term operation. Planners can also influence what happens in other parts of the catchment, not only in a development area, positively impacting development and growth (e.g. White, 2010).

In areas where development growth is constrained by the availability of water resources or inadequate wastewater management, catchment partnerships can help provide innovative solutions that deliver multiple benefits. Managing river flows and surface water to improve surface water quality and reduce flood risk can improve the local environment, thereby unlocking development potential (Collins *et al.*, 2020).

Integrated water management and nature-based solutions should be developed strategically to link neighbourhood and catchment-wide plans to local actions to manage surface water at source. These should also incorporate integration across the water cycle, with water resources and wastewater simply considered as part of a closed cycle. Working in partnership and with co-funding solutions, this can enable schemes to go ahead that may not be affordable for individual partners on their own, thus improving overall cost-effectiveness. Box 5.1 outlines a case example of such an approach in Australia (CRCWSC, 2022).

5.2. Urban planning

Good urban planning and design is essential for society and the development of beneficial social, economic and environmental outcomes. It can positively influence healthy behaviours and improve public health in general. The planning system and health outcomes are intricately linked and can reflect current and future needs. Well-designed places support communities' health, social and cultural wellbeing and can be supported by effective water management (Bide and Coleman, 2019; DLUHC, 2023).

Box 5.1

Integrated water management in the Elizabeth Street catchment, Melbourne (Australia) – extracts from CRCWSC (2022)

'Elizabeth Street is a 308-ha flood-prone catchment in Melbourne's central business district. The City of Melbourne has adopted a catchment approach to mitigate these flood risks, using a series of stormwater harvesting systems to hold water in the top of the catchment and thereby reduce flooding at the bottom of the catchment. In addition to reducing the social and economic impacts of flooding, these projects will reduce the urban heat island effect that previously cost local businesses A$37 million in lost revenue during a 2013 heat wave.'

'The Royal Exhibition Building and Melbourne Museum stormwater harvesting project is an example of a completed project. Rainwater is harvested from the extensive roofs and surrounding paved areas of these two adjacent buildings and stored in a 1.35 million litre underground tank. The system supplies 6.4 million litres of water for reuse in toilets, fountains, lakes, and garden irrigation.'

The planning system can influence the design of urban developments and, through urban greening, help to reduce our impact on the environment, such as improving local air quality by reducing the level or concentration of harmful pollutants. A review by the UK's Air Quality Expert Group concluded: 'Overall, vegetation and trees in particular are regarded as beneficial for air quality, but they are not a solution to the air quality problems at a city scale' (AQEG, 2018). To continue and accelerate the reduction in outdoor air pollution requires action in many sectors and needs a focus on areas where people live, study, work and have leisure. Urban planning can help support reducing air pollution concentrations locally (CMO, 2022), using blue-green infrastructure (BGI) wherever and whenever feasible. Box 5.2 illustrates some of the issues associated with air pollution and the potential benefits from using nature-based systems (NbSs).

Box 5.2

Air pollution (based on CMO, 2022, and Ashley *et al.*, 2022)

The Bradford Metropolitan local authority district, including the city of Bradford, is the seventh largest populated in England, with over 546 000 residents (ONS, 2022). Pollution is a key contributor to ill health within the city. Routine monitoring of air shows that both nitrogen dioxide (NO_2) and particulate matter (PM) are above World Health Organization guidelines across large parts of the city, particularly near busy roads. NO_2 is of particular concern, as exceedances of the Air Quality (England) Regulations 2000 have been identified in the city by both the government and the council. Pollution is linked to a range of health outcomes. The most recent figures in Bradford suggest that 500 people die from respiratory-related disease each year. There are over 13 000 diagnosed cases of chronic obstructive pulmonary disease (COPD) and more than 41 000 people are diagnosed with asthma (City of Bradford Metropolitan District Council, 2019). It is estimated that a third of asthma cases in the city are attributable to air pollution (Khreis *et al.*, 2019).

Recent plans for a major highway realignment in the Canal Road corridor of the city were conceived with a view to maximising the potential use of NbSs for the multiple benefits they can bring. As well as renaturalisation of the adjacent Bradford Beck, sustainable drainage systems (SuDS) have been incorporated into the highway drainage plans. Application of the Construction Industry Research and Information Association (CIRIA) B£ST tool showed that the present value of benefits for air pollution reduction were some £112 684, benefits that buried piped drainage would not provide (Ashley *et al.*, 2022).

Incorporating green infrastructure into developments as a part of an integrated approach to planning and water management can introduce a variety of benefits. These can include improved air quality, amenity, biodiversity and ecology, carbon reduction/sequestration, cool runoff temperatures, creation of green spaces, educational benefits, groundwater recharge, health and therapeutic improvements, increased infiltration, natural evapotranspiration, precipitation retention, increased property values, rainwater harvesting, recreational benefits, reduced surface runoff, traffic calming, urban agriculture and improved wildlife habitats (see Chapter 8).

BGI can also support resilience to climate change, by mitigating the urban heat island effect or reducing the impact of flood events, and by improving social cohesion, through encouraging safe and accessible places for community interaction (Public Health England, 2020).

5.3. Political and regulatory landscape

The political and regulatory landscape is continually changing, with revisions to policies and changes to practices and responsibilities for water management. Around the world there are numerous examples of recently developed strategies for flood and coastal erosion risk management using an integrated portfolio of approaches that reflect both national and local priorities. This has been to reduce the threat to people and their property; and to deliver the greatest environmental, social and economic benefits, consistent with sustainable development principles. These envision more flood and coastal erosion solutions working with natural processes within the urban landscape (e.g. Bridges *et al.*, 2021, 2022). Coastal flood and erosion risk management is one of the major domains where NbSs are making a major contribution. NbSs are also increasingly being employed in urban drainage, and planning processes are increasingly reflecting this, although there are still many examples of disconnects between the planning process and water and, especially, flood risk management. The attitude of some planners that 'we do the plans, the drainage has to fit them' is not only arrogant but also contributes to increased flood risks and water stress (e.g. Heijs, 2023).

The nature and structure of key institutions and especially, the statutory obligations that they must discharge can help promote collaboration and integrated management of infrastructure systems. However, where an organisation is constrained by statutory and other obligations, and must focus too closely on their own responsibilities, this can militate against delivery of the form of infrastructure that is best for society and that is likely to be as sustainable as possible for society as a whole (Fenner *et al.*, 2022). Collaboration and partnerships are therefore key to delivering better water management, and the planning system needs to reflect this (e.g. Bide and Coleman, 2019). Forms of urbanism and the means to regulate their delivery require appropriate policies, regulations and institutional interactions (Bibri, 2020; Bide and Coleman, 2019).

Primary legislation at the privatisation of the water industry in England and Wales set down a core requirement for water and sewerage companies to provide an essential service for society to 'effectually drain'. Legal duties, regulatory obligations and government priorities, along with the need to improve customer service, underline the compelling need for change. In response to the widespread flooding of 2007 and recognising that the extant approach was unsustainable due to future pressures from a changing climate and population, government and regulators clarified, updated and extended legal and regulatory requirements, for which a new approach is now expected in the management of stormwater. The former approach to managing stormwater, through underground sewers, tanks and shafts, has led to great benefits, such as public health, river water quality and flood control. However, ageing infrastructure requires increasing maintenance and is relatively inflexible to future pressures (see Chapter 10). Regulators, government and service providers now widely recognise that continuing solely with an approach of maintaining and enhancing the traditional wastewater network is not sustainable (Defra, 2011, 2012; Halcrow, 2013).

Defra sets an expectation for stormwater management and SuDS. For example, *Water for Life* (HM Government, 2011) states: 'Given pressures on our infrastructure capacity and the risks from climate change we want to encourage the use of SuDS where they will be effective'. The *Statement of Obligations* (Defra, 2012) which describes the government's understanding of the main environmental statutory obligations that applied in particular to water and sewerage undertakers over the price review period 2015–2020, covering both domestic legislation and requirements of European directives, states: 'PR14 should see strong foundations put in place for longer term sewerage planning' and that 'Government expects these strategic plans to include appropriate sustainable measures for managing flood risks, such as sustainable drainage systems (SuDS).' The National Planning Policy Framework (DLUHC, 2023) includes a presumption in favour of sustainable development, of which stormwater management is an integral part.

The Drainage Strategy Framework (Halcrow, 2013) sets expectations for water and sewerage companies to have strategies for stormwater management and to recognise the need for sustainability to be considered in asset management planning. Ofwat's recent review of international management of surface water drainage (Stantec, 2011) found that the alternative management of stormwater delivers multiple benefits: 'The principle that … sewerage systems should accommodate future flows is unsustainable in the long-term context of climate change.'

Inadequate policies, regulations and duties can lead to problems in particular parts of the drainage network due to ongoing practices. An important background feature to the drainage system in England is that, under extant legislation, developers and house builders have an absolute right to connect their properties to the sewerage system. This has traditionally made it very difficult for water undertakers to control flows, particularly of surface water, into their networks, with the control of the flows being determined as part of the planning process, which is administered by local authorities. The abandonment of a definitive right to control the flow of surface water into the sewer system was and indeed remains a concern.

The report of the Independent Review of Charging for Household Water and Sewerage Services (Defra, 2009) recognises the need for local highway authorities to reduce the volume of highway drainage runoff to sewerage systems and seek alternatives to the use of the sewerage system to help tackle the effects of climate change. Local highway authorities have a duty to cooperate with sewerage companies to minimise the total costs of draining highways. It should be noted that there is no automatic right of communication for highway drainage to the public sewer network. However, section 115 of the Water Industry Act 1991 empowers the the highway authority and the sewerage undertaker to enter into a reciprocal agreement to allow

- highway drainage to be connected to a public sewer provided by the sewerage undertaker
- a public surface water sewer to be connected to a highway drain maintainable by the highway authority.

This is a discretionary agreement, but requests for connections must not be unreasonably refused.

It remains to be seen whether the planning-led approach to this issue will ultimately result in the creation of high-quality NbSs (see Chapter 8) that will materially reduce the volumes of surface water entering the sewer systems in England. While there have been improvements over the past 10 years in England, the increasing use and delivery of high-quality, multibeneficial NbSs is still embryonic. The planning system is sclerotic and does not have as much expertise as it would like or needs in this area (e.g. Washbourne and Wansbury, 2023). Planning legislation and policy setting need to be strengthened as much as possible to encourage different practices. The removal of the automatic right of connection for developers to connect surface water to sewers in England would solve many of the issues faced in getting a greater uptake of green infrastructure. Nevertheless, over recent years, both government and other interested parties, including drainage engineers and environmental non-governmental organisations, have come to the view that traditional sewerage systems, relying largely on pipes and hard engineered structures, are not always the best way to deal with the drainage of surface water and that an integrated approach to water management is seen as the way forward. In Wales and Scotland, there are different approaches, with planning and other regulations designed to promote the preferential use of NbSs compared with buried pipes (Scottish Government, 2001; Welsh Government, 2019). However, despite the best efforts of professional bodies and others in promoting urban greening and alternatives (e.g. TCPA, 2023), development planners often still cling to the easy option of 'connect all to sewer'.

5.4. Urbanisation and the impact on urban drainage systems

5.4.1 The benefits of integrated water management

Urbanisation is dramatically changing the built environment. The largest proportion of the world's population now live in urban environments, and this growth is predicted to continue. This will have both direct and indirect effects on local and regional water resources and environmental impacts; and will put water and wastewater infrastructure under increasing strain. The needs of the environment and society are considerably different and, in certain contexts, more acute than those which existed 100, 50 or even 30 years ago. The status quo as regards urban drainage is not only unsustainable, but is never likely to be sustainable (Ainger and Fenner, 2016).

There are very clear links between urban drainage systems and spatial planning. The effective control and management of stormwater underpins society. There is now a pressing need to move rapidly to integrated water resource management, using an optimal combination of above- and below-ground infrastructure. Integrating wastewater networks, abstraction and discharge consents using dynamic, optimally controlled systems would deliver increased water supply security, river quality and environmental improvements, and significant reductions in flooding, energy use and carbon output. Utilising real-time data and intelligence, it is now possible to be more effective and efficient in the way that the entire water cycle is managed. However, this would require flexibility in how water is regulated – for abstraction and for wastewater management and environmental impacts – and in the way in which development is planned and managed. Surface water systems not only reduce the likelihood of urban drainage being overloaded, but also provide attractive and healthy environments. Cities of the future will need to be far more 'water-sensitive' and utilise as far as practicable the benefits of water at source; where it lands (e.g. Bell, 2018; Howe and Mitchell, 2012; White, 2010). Development planning is at the heart of how this can be achieved.

Integrated water management can provide many positive impacts, such as the following (Bide and Coleman, 2019).

- Reducing flooding through balancing river flows from source to sea.
- Providing alternative supplies of water for non-potable purposes, such as livestock, agriculture, and irrigation.

- Creating and supporting valuable habitats resulting from increased use of nature-based solutions.
- Providing improved amenity and recreation space in the form of parks and bathing waters; and activities such as angling, walking and water sports.
- Supporting local tourism and local community wellbeing.
- Facilitating significant contributions to climate change mitigation and providing a means for adaptation.

To achieve these benefits in a coordinated and sustainable way, all activities within a catchment need to be considered together as part of the water cycle. Through understanding these interactions, planners and developers must ensure that new developments do not compromise existing urban areas, while other stakeholders can ensure that their plans do not compromise wider objectives, such as water quality and the protection and enhancement of the environment. This may necessitate protection of some urban spaces as potential 'buffers' to moderate the future impacts of climate changes, reflecting what is often done in rural areas (e.g. Xu *et al.*, 2022). However, there is a danger that adaptation changes may disadvantage the most vulnerable members of society (e.g. Warner and Weigel, 2021). The management of wastewater systems has, since the sanitation revolution in the 1800s in the industrial nations, been a major element in the way in which urban areas are planned and laid out. In London, for example, the Bazalgette reform of wastewater management resulted in large-scale demolition and reformation of the city (Allen, 2008).

Although here we are concerned with urban water, much of city planning is driven by broader priorities. As an example, the city of Helsingborg in Sweden is aiming to be carbon neutral by 2035. This is necessitating a complete revamp of all aspects of city living and is framing the way in which a large new housing and commercial district is being planned and developed, as outlined in Box 5.3.

Box 5.3

Climate neutral Helsingborg (WWF, 2021)

Aiming to be climate neutral by 2035, Helsingborg in Sweden needs to reform all the existing city services and the way in which people live, work and play. These are linked to the Sustainable Development Goals and quality of life (Helsingborg, 2021). Stormwater has moved from being seen as a threat to an essential component of this reform: 'a resource to increase the attractiveness of urban areas with respect to recreation, natural values, and biodiversity. Open streams and detention and retention ponds in strategic locations are used for flood control and stormwater runoff treatment to improve water quality'. The new development at Oceanhamnen is intended to be a sustainable urban demonstrator in the harbour area. Residents, who are termed 'environmental heroes', enjoy a three-pipe wastewater system that comprises a pipe for ground food waste, a pipe for baths, dishwashing and washing machines, and a vacuum pipe for toilets. Each of these leads to world-leading resource recovery facilities to ensure overall carbon neutrality.

5.4.2 Traditional and sustainable drainage

Figures 5.1 and 5.2 illustrate the different water management characteristics of traditional drainage and sustainable drainage.

Figure 5.1 shows an urban catchment with traditional drainage. In this case, rain landing on the urban area quickly runs off, so has little opportunity to soak into the ground (infiltrate) and recharge the groundwater. Because of the channelised waterways and hard impermeable surfaces, which contribute to runoff, the catchment is very responsive to rainfall, resulting in flooding and potential pollution. Figure 5.2 contrasts, showing how SuDS can be utilised in the same catchment, creating more blue and green spaces and enabling the use of surface water as a resource..

5.4.3 Integrating water management at different spatial scales

The different parts of a river catchment and the various land uses within it are connected in such a way that whatever happens in one area affects other areas, especially those downstream. If not appropriately managed, these interactions can have negative impacts; for example, poor agricultural and river management practices upstream can result in increased flood risk downstream. Bringing together the different activities at the full range of spatial scales across a catchment[1] into a management strategy can help ameliorate any negative impacts.

Good agricutural and land management practices and tree planting can increase water retention in the upper levels of a catchment. This helps to control runoff and reduce peak flows and siltation, while also providing farmers and landowners with a more uniform and reliable water supply.

Both natural habitats upstream and well-designed and managed landscapes can absorb and temporarily store rainwater, thus reducing peak flows. Not only do such areas affect the volume and speed of water arriving in the town, they also deliver other benefits in terms of recreational and public open space. Nature-based features in a town, such as green roofs, permeable surfaces, swales and detention basins, reduce the runoff from rain that falls directly on the town.

Over the past decade there has been a renewed interest in natural flood management (NFM). Ever since humans have occupied settlements, agriculture has been a useful way of managing downstream flood risk. However, the need to provide food has often supplanted the need to restrict flows from catchments and historically there have been repeated switches in emphasis from draining land quickly to retaining water in the uplands (e.g. Newman *et al.*, 2011).

Currently, NFM is considered to be an essential way of managing excess flows to protect downstream areas (Bridges *et al.*, 2021; cbec and Environment Agency, 2017; Wren *et al.*, 2022), albeit much of the detailed science underpinning understanding of the effectiveness and hence design processes is still being collected (Dadson *et al.*, 2017). There is a blurred boundary between NFM, NbSs (which include BGI and SuDS) and other natural process initiatives. Planning needs to encompass all of these in taking advantage of the many benefits they provide. Bide and Coleman (2019) provide a platform for this which focuses on planning in relation to urban water systems in a UK context.

[1] In many countries, the term 'watershed' is used instead of 'catchment'.

Figure 5.1 An urban catchment with traditional drainage

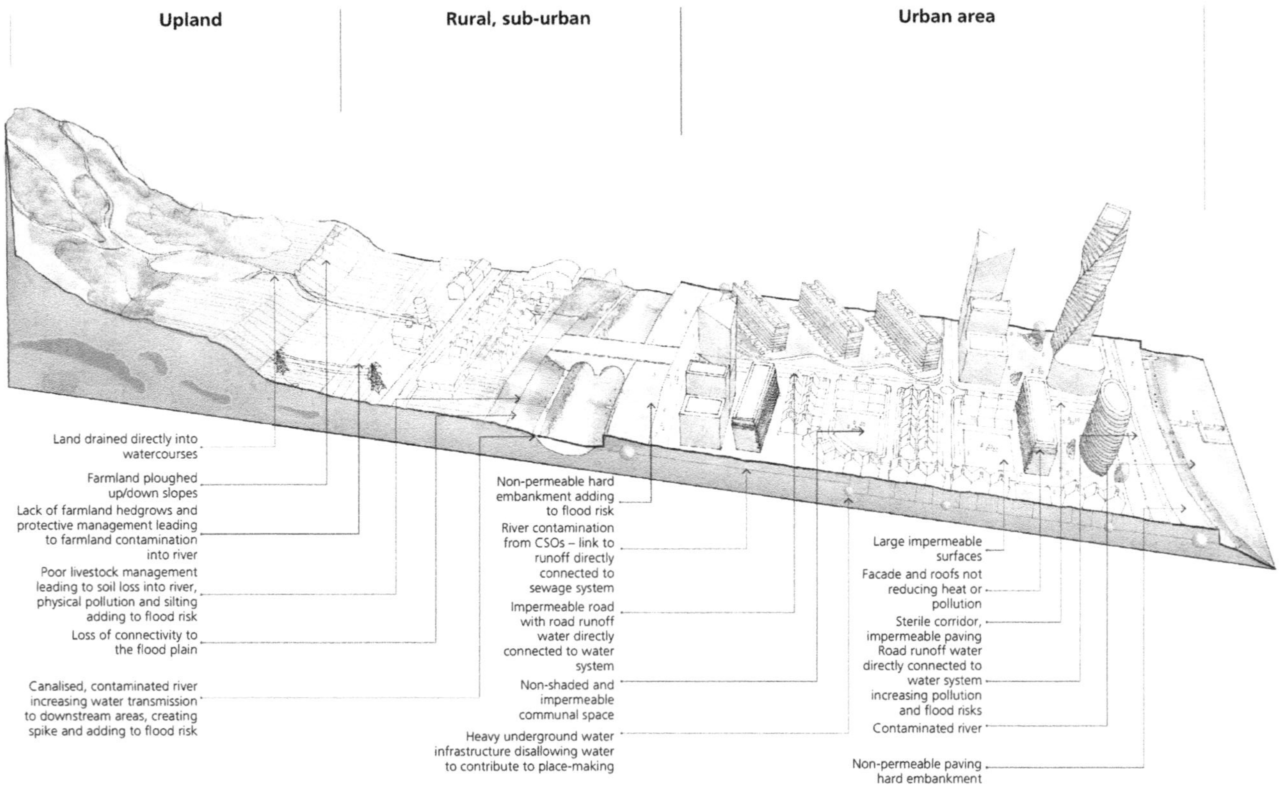

Figure 5.2 An urban catchment using sustainable drainage systems (SuDS)

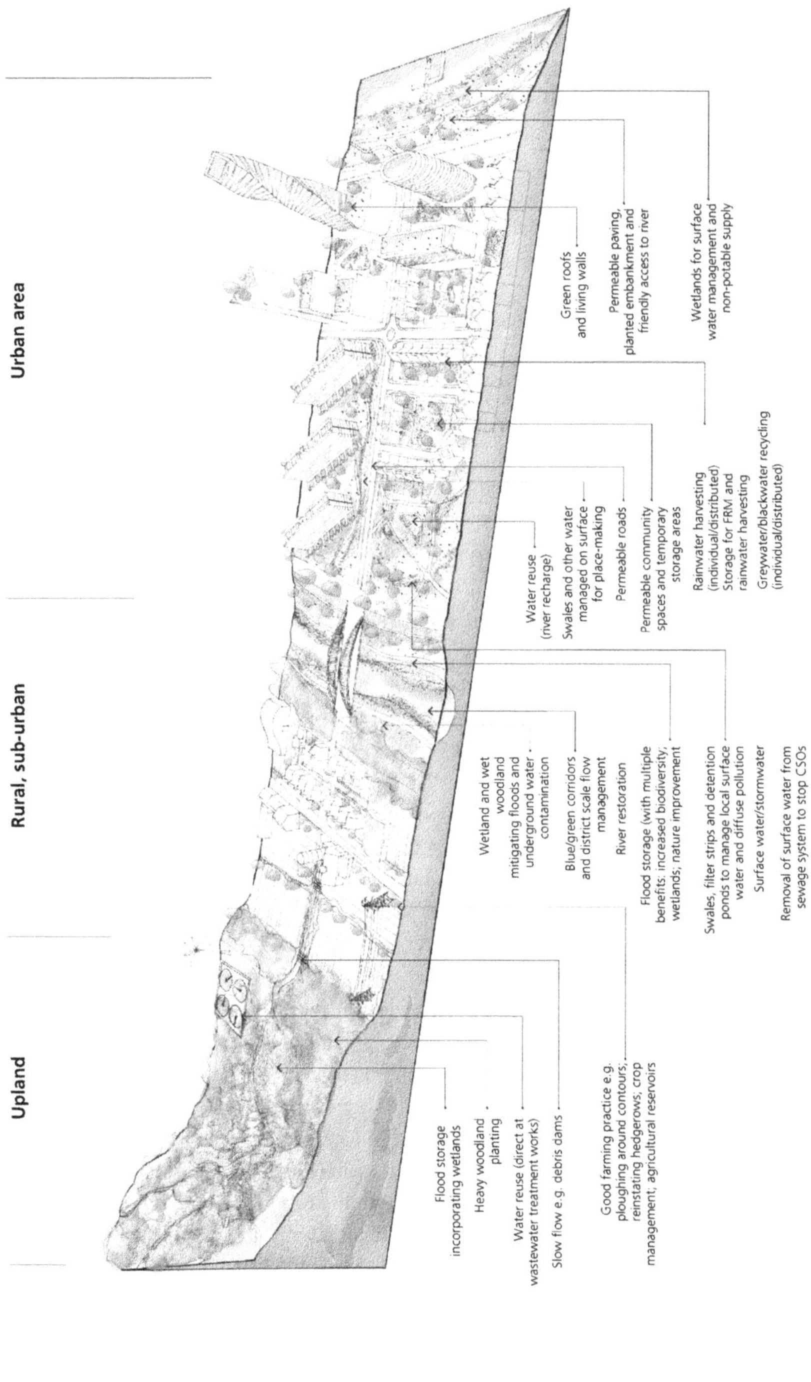

5.5. The catchment-based approach in an urban context

There are risks and opportunities related to water that affect economic development and prosperity. These cut across all societal and industry sectors. Land and water management in river catchments is used in many ways, including for agriculture, public and private water supplies, wastewater management, industry, production, housing and other built development, conservation, recreation and navigation. In the EU, for example, water-dependent sectors generate €3.4 trillion or 26% of the EU's annual gross value added (EC, 2019). This was determined by gathering data for catchments as a whole, in order to capture the important interactions across and within them. As water bodies within a catchment are connected and part of the water cycle, human activities can affect the quality and quantity of water and economic activities in other parts of the catchment. This can have direct impacts, but it can also affect the ability of catchment waters and associated wildlife to provide wider benefits to society, such as clean water supply, flood protection and recreational opportunities, known collectively as 'ecosystem services', as outlined in Section 5.5.1.

Since 2009, it has been mandatory under the Water Framework Directive (WFD) for EU countries to develop River Basin Management Plans (RBMPs) (EC, 2000). The first cycle was completed in 2015 and the second in 2021. Each RBMP is accompanied by a comprehensive Programme of Measures (PoM) for water bodies, which will be required to ensure that the WFD key objectives are achieved. A detailed understanding of the risks associated with all pollution sources in the catchment is required in order to deliver a PoM that can achieve the standards required under the WFD and at the same time ensure adequate water services for human and ecosystem use.

Although based on RBMPs, the WFD requires integrated management of both water and land systems, together with the human interacting water cycle. Under the WFD, the system of interest is a catchment, as defined by the member state, and it is essential to understand the state (health), behaviour and 'performance' (the structure and function) of these various systems of interest in order to define their ecological status and potential in the various cycles of operation of the WFD.

Under the WFD, any deviation of the ecological status of a system of interest from its undisturbed/reference condition is based on indicative assessments for three groups of quality elements: biological, hydromorphological and physiochemical. The quality element that is or could be most impacted by the anthropogenic driver/pressure becomes the target for remediation or control – that is, the element most sensitive to pressure defines the status of the water body.

Many other countries, outside the EU member states, are adopting catchment-based approaches to water and associated human activities (e.g. Kaval, 2019), primarily for economic reasons. Despite this anthropogenic focus, integrated catchment management (ICM) is an essential systems-based process for supporting and maximising the various components that comprise catchments, including non-human environmental and ecological systems. In practice, ICM is not a panacea that will deal with all the water-related issues within a catchment. The complexity of the interlinked web of systems and interests (human and nature, many of which are conflicting), coupled with the need for accurate, continuous and dynamic data, mean that attainment is unlikely and partial understandings and responses to issues are only ever likely to be achieved (e.g. Daly *et al.*, 2019). That said, ICM and RBMPs are important frameworks and process steps to ensuring humans and nature can co-exist to mutual benefit (see also Chapter 12).

ICM addresses water-related issues through good environmental planning and resource management. All the environmental, economic and social issues within, or related to, a river catchment are integrated into an overall management strategy that will deliver the greatest possible mix of benefits for future generations in the communities in the area while protecting the natural resources on which these communities rely. The process links national and regional policy for water, flood risk and the environment, but also requires the coordinated development and management of water, land and related resources by the local communities and businesses that are involved directly or indirectly in implementation. There is an argument that as everyone lives their lives in one or other river basin district, it is incumbent on us all to behave responsibly as regards the environment in general and water systems in particular. While partnerships of key organisations are necessary, local initiatives, such as citizen science projects, can help to ensure effective and prominent engagement of communities and individuals (e.g. Collins *et al.*, 2020; Laville, 2022).

Working with communities towards an agreed vision of sustainable land and water resource management for their catchment will ensure that economic and social benefits are maximised and the environment, public amenity and biodiversity are improved (UK Government, 2022).

In summary, the catchment-based approach recognises the catchment as the appropriate organising unit for understanding and managing

- water resources
- surface water
- flood risk from all sources
- land use
- ecosystems.

Good outcomes are dependent on local and community action and are reliant on collaboration for the development of a shared vision. By acting together, informed decisions can be made for the effective management of the land and natural resources of a catchment. Such decisions would consider the effect on all the systems, resources and people within the catchment (Chapter 4).

There are two key challenges for ICM. The first is linking local communities, businesses and developers, who can make things happen, with the regional and national plans and policies that set out what needs to happen. It is important that those promoting national and regional strategies communicate and engage effectively with local communities about water issues to ensure that people participate actively. Conversely, local communities need to have their aspirations recognised and incorporated into the strategy by policymakers. An example of a current challenge within the RBMP process is that of nutrient neutrality in England. This is an example of one policy (to promote housing development) seemingly conflicting with another (to protect water quality). The issue is illustrated in Box 5.4. Nutrients can be managed by using SuDS for surface water runoff. CIRIA provides guidance for this (Bradley *et al.*, 2022), especially where SuDS management trains are used (see also Chapter 8).

The second challenge is linking businesses and organisations that can fund actions with each other and with the community groups that can do things, in ways that ensure that all will benefit. In many countries, it will be possible to unlock local contributions for partnership funding, particularly from the

Box 5.4

Nutrient neutrality in England seemingly constraining housing development

The issue from a developer's point of view is explained by Jessel (2022) in the UK *Architects' Journal*. Planned housing developments potentially add to the problem of excess nutrients being discharged into receiving water bodies due to the additional wastewater arising from new homes, as well as from the temporary runoff from sites during construction. This pollution raises levels of nutrients and risks of eutrophication. Recent housebuilding has been found to be damaging protected sites, some of which are now classified as in 'unfavourable condition'. In 2018, an EU Court of Justice ruling resulted in Natural England issuing advice that no developments should be permitted near these sites unless it can be proved that they are 'nitrate and phosphate neutral'. This has resulted in delays to the construction of more than 100 000 houses across most of the English municipal areas. Developers must ensure that plans show that developments are nutrient neutral both during construction and over the longer term, when the properties are occupied. Neutrality may be achieved by offsetting additional nitrate and phosphorous loads from a development by reducing these loads elsewhere in a catchment. As with most wastewater challenges, this is not a technical issue, as appropriate treatment processes can be installed within a development, but rather an economic issue. It is estimated that managing nutrients locally will require an investment of some £5000 per property. Developers are also unhappy that any treatment facilities would take up valuable land space.

private sector. However, this is especially challenging for the larger-scale projects, which may deliver the biggest benefits. Nonetheless, there are examples of public–private partnerships delivering significant improvements to catchments. Water finance and investment needs are heterogeneous, especially in countries where basic infrastructure is inadequate and improvements to catchments should be seen as going hand in hand with ensuring safe and effective water supply and sanitation (WSS). For, this reason, the Organisation for Economic Co-operation and Development (OECD, 2018) considers a typology of water security investments and a typology of financiers to help match specific investment needs with the most appropriate financing available. Potential classifiers for water investments include: scale (from catchment to household); function (water supply, wastewater management, flood protection, etc.); and operating environment (ownership, governance and regulation). Not all donors or investors will wish to support all aspects of RBMPs or WSS but will likely wish to focus on one or other areas of most interest.

As an example, the UK Government (2018) undertook a review of partnership funding for flood and coastal erosion risk management in England. Surface water flood risk management schemes were found to attract funding mainly from government or governmental agencies. However, for some schemes, significant funding came from private sources, albeit much of this was from individual households. However, many potential private sources of funding that would benefit from flood risk schemes were of the view that funding was a public sector issue and were therefore unwilling to contribute.

Partnerships, as illustrated in Figure 5.3, at the catchment and sub-catchment scales are key to responding to the challenges of water management. They are important in local planning, for linking local actions to RBMPs. By working together, different stakeholders in a catchment can appreciate each other's drivers and objectives to create a shared understanding of the challenges and opportunities within

Figure 5.3 How integrated catchment management links people, water and services

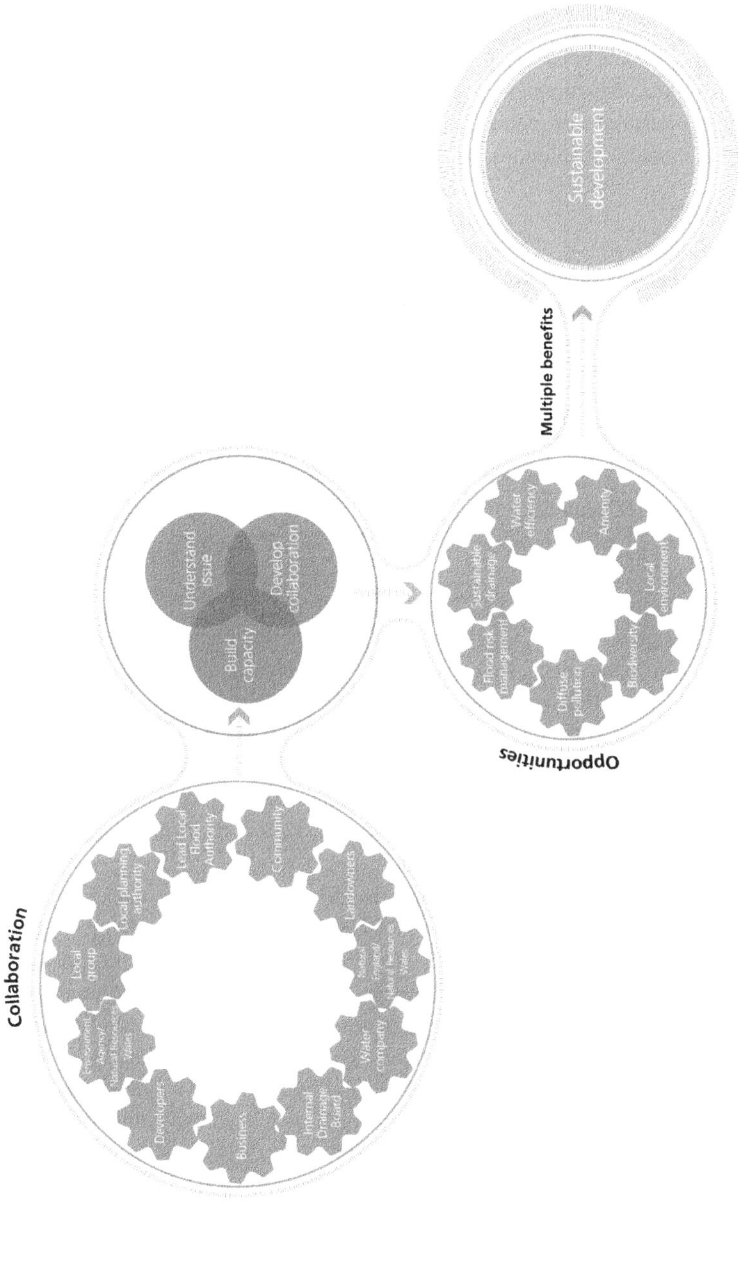

their area. This underpins collaboration by sharing objectives and builds capacity to deliver multiple benefits and strive for sustainable development.

The catchment-based approach provides a mechanism for facilitating integrated water management. Through engagement in catchment partnerships, local planners can influence what happens in other parts of the catchment, which has a direct impact on development and growth in their local area. As well as managing river flows and surface water to reduce flood risk, improvements to water quality and the local environment can unlock development potential. Where development growth is constrained by the availability of water resources or ineffective wastewater disposal or treatment, catchment partnerships can help to deliver innovative solutions, providing a wide range of benefits, for example

- reduced flood risk
- reduced pollution
- improved natural and built environment
- improved public amenity and recreational opportunities
- increased biodiversity
- additional/alternative water resources
- improved wastewater management.

The overall outcome is a more sustainable urban development, through environmental, social and economic improvements. Combining NFM techniques with SuDS can also help improve the natural environment and deliver healthy ecosystems and economies of scale.

5.5.1 Ecosystem services and natural capital

A significant step forward was achieved at the start of the current century with the publication of the global Millennium Ecosystem Assessment. This has provided the foundation for valuing the services provided by nature to humanity (e.g. Hansen and Pauliet, 2014) and clearly illustrates the multifunctionality of natural systems. With the growth in interest in greening urban areas, using NbSs as infrastructure and understanding the value and multiple benefits as a result, termed 'ecosystem services' (ESs), there is now a massive expansion in understanding and application of the ESs concepts. Box 5.5 summarises natural capital (NC) and ESs concepts, which are also further considered in Chapter 8.

Now vital to effective and efficient urban drainage, this development can be tracked from the use of 'Best Management Practices', 'source controls', 'sustainable drainage', blue-green infrastructure' and other terms in the latter part of the twentieth century (Fletcher *et al.*, 2014) to the present NbSs. The path of these developments in the UK is outlined in Ashley *et al.* (2014). There are numerous other drivers related to 'nature' that are influencing planning and development that have relevance to urban drainage. For example, the concept of 'biodiversity net gain' is being applied in the UK to developments and defined by CIRIA guidance as 'development that leaves biodiversity in a better state than before, and an approach where developers work with local governments, landowners and other stakeholders in order to support their priorities for nature conservation' (Baker *et al.*, 2019). Surrounding this and other such principles and processes is the need to improve carbon, energy and waste management and to understand how NbSs can contribute to this (House of Lords, 2022).

NC and ESs valuation are now providing the means to assign monetary value to nature. Given that, globally, most decisions made about development and associated drainage systems are based on a

Box 5.5

Natural capital and ecosystem services (Horton *et al.*, 2019)

Capitals-based accounting mirrors standard financial accounting. There are several 'capitals', which together can be considered to underpin society's health, wealth and wellbeing.

- Natural capital (NC), which is defined as 'the elements of nature that directly and indirectly produce value or benefits to people, including ecosystems, species, fresh water, land, minerals, the air and oceans, as well as natural processes and functions' (Natural Capital Committee, 2014).
- Social capital, which encompasses 'the relationships within and between organisations (including businesses), communities, groups of stakeholders and other networks' (UKWIR, 2017; based on IIRC, 2013).
- Manufactured capital, which is human-created, production-oriented equipment and tools (e.g. water and wastewater treatment works, pipes, office buildings).
- Human and intellectual capital, which relates to people and knowledge (e.g. labour, education, research and development, innovation and processes).
- Financial capital, which includes the pool of funds available to an organisation – for example, debt and equity finance.

NC supports a range of ecosystem services (ESs), which are the benefits provided by ecosystems that contribute to making human life both possible and worth living (UK NEA, 2011). These are generally split into four categories, as follows.

- Provisioning services – goods or products that people consume or are used in the production of other goods, such as crops, fruits, fibre, timber, fish and natural medicine.
- Regulating services – benefits derived as a result of an ecosystem control of natural processes, such as air quality maintenance, water quality and flows, pollination, flood protection, climate regulation and erosion control.
- Cultural services – non-material benefits, such as recreation, spiritual values and aesthetic enjoyment.
- Supporting services – natural processes that maintain the production of all other ecosystem services, such as habitat provision, nutrient cycling, soil formation and water cycling.

Collectively, these can be utilised to define the many benefits provided by nature and are the foundation for toolkits, tools and toolboxes used to ensure that natural systems are appropriately included in accounting procedures. An example is provided in Chapter 8.

measure of value for money, this has, for the first time, provided the economic data for nature to be appropriately included in cost–benefit or other monetary-based assessments. From a development planning perspective, Van Oijstaeijen *et al.* (2020) review some of the available support tools for planning investments in NbSs. Pointing out that there is a risk that monetisation can detract from the intrinsic and non-monetisable value of environments and ecosystems, it is concluded that toolkits for analysing the value of nature are likely to be essential and, despite the limitations of extant tools, are a useful support in the decision-making processes.

NC and ESs developments have had profound implications for urban planning and development worldwide, and NC accounting now has a central place in ICM. Kaval (2019) reviews applications of ES and looks back over some 25 years of ICM in various countries, finding that there has been 'an explosion' in interest and inclusion of such valuations in plans, but most of these focus only on a narrow group of ESs considered most relevant for the case examined.

Valuation of NbSs is further considered in Chapter 8.

5.5.2 Improving stormwater management by way of the planning system

Stormwater can impact the functioning of combined sewer systems (see Chapter 3), increasing flows to treatment and affecting the ability to achieve compliance with regulatory obligations. Urban development increases surface water runoff entering the sewer systems, thereby increasing the risk of sewer flooding and pollution. To enable growth that is more sustainable it will be necessary in most instances to reduce surface water flows entering sewers, primarily by reducing connected impermeable surfaces (see Chapter 7). Reducing surface water inputs to combined sewers will create capacity to help accommodate planned growth without detrimentally impacting assets, customer services or the environment. The development and urban planning processes are key to how water is managed on surfaces (White, 2010). The form, layout and function of drainage systems depend on city planning mores, which in response to planning visions and cultures, as urban areas change from informal settlements with open drains, as are common around the world, to planned cities where drainage is invariably hidden from sight. With the growth of urban areas, city sprawl, densification and new ideas about 'compact urbanism' there are seemingly constant challenges to extant and new drainage networks (Bibri, 2020). Urban planning mores can be likened to changing fashions over the centuries, with drainage culture reflecting planning (e.g. Juuti and Katko, 2005).

Effective planning processes can help ensure that diffuse pollution from runoff is minimised. Individually, each source of diffuse pollution might not be significant. However, when these accumulate, particularly in an urban area, they can be very problematic and difficult to manage or remove. Well-designed urban areas and developments can help reduce diffuse pollution by integrating SuDS and appropriate treatment (Bradley *et al.*, 2022; Woods-Ballard *et al.*, 2015).

New developments in many countries need to meet water consumption requirements imposed by legislation and regulations. This reflects the value placed on water resources and the potential challenge of water scarcity. However, the requirements on water use can be achieved, in part, through water efficiency measures such as installing dual-flush toilet cisterns, aerated taps and showerheads, lower-volume baths and water-efficient white goods. Approaches to harvesting rainwater for non-potable uses such as toilet flushing and garden water can also help achieve water savings while also managing surface water at source. This all forms an integral part of integrated water management and can be driven by requirements in national regulations and local planning frameworks.

Managing water in an environmentally sensitive way can help enhance biodiversity and improve the health of an urban environment. Planners have an opportunity to recognise the role that cities and local authorities have in stemming biodiversity loss. Tools such as the City Biodiversity Index (or Singapore Index) can be used as self-assessment tools to benchmark and monitor the progress of their biodiversity conservation efforts (Chan *et al.*, 2014; SCBD, 2023).

Planners and developers also have an opportunity to help local authorities meet the objectives of their green infrastructure plans and strategies. For example, providing wildlife corridors and connections between green spaces can help create cleaner air. However, it is important that any green infrastructure designed to improve habitat diversity introduces appropriate native plants into the local ecosystem, in line with requirements from Natural England around green infrastructure and biodiversity net gain (Baker *et al.*, 2019; Natural England, 2023).

Considering flood risk strategically and early in the planning process is essential as it enables flood risk management to become an integral part of each neighbourhood's vision for sustainable development, thus providing greater certainty to all involved, particularly developers. This helps facilitate more effective use of resources. Through integrating development, regeneration and flood risk management, developers can make better investment decisions, thereby delivering improved outcomes with multiple benefits. The National Planning Policy Framework planning practice guidance (MHCLG, 2021) and CIRIA's guidance on delivering better water management through the planning system (Bide and Coleman, 2019) explain how development and flood risk can be effectively delivered.

Wastewater conveyance and treatment can be a significant and often overlooked constraint on urban development. Collaboration between municipal planners and developers, water and wastewater service providers and environmental regulators through the development planning process should identify potential challenges and solutions.

REFERENCES

Ainger C and Fenner R (eds) (2016) *Sustainable Water*. ICE Publishing, London, UK.

Allen M (2008) *Cleansing the City*. Ohio University Press, Athens, OH, USA.

AQEG (Air Quality Expert Group) (2018) *Impacts of Vegetation on Urban Air Pollution*. Defra, London, UK. See https://uk-air.defra.gov.uk/assets/documents/reports/cat09/1807251306_180509_Effects_of_vegetation_on_urban_air_pollution_v12_final.pdf (accessed 29/07/2023).

Ashley RM, D'Arcy B, Wilson S *et al.* (2014) Looking at SuDS "from both sides now – from up and down – but still somehow it's SuDS illusions many recall". In *CIWEM Urban Drainage Group Conference*, Blackpool, November.

Ashley RM, Horton B, Walker L *et al.* (2022) The benefits of nature-based systems in a changing and uncertain world. *Proceedings of the Institution of Civil Engineers – Engineering Sustainability* **176(2)**: 61–71, 10.1680/jensu.21.00102.

Baker J, Hoskin R and Butterworth T (2019) *Biodiversity Net Gain – Good Practice Principles for Development*. CIRIA C776a. Construction Industry Research and Information Association, London, UK.

Bell S (2018) *Urban Water Sustainability*. Routledge, Abingdon, UK.

Bibri SE (2020) Compact urbanism and the synergic potential of its integration with data driven smart urbanism: an extensive interdisciplinary literature review. *Land Use Policy* **97**: 104703, 10.1016/j.landusepol.2020.104703.

Bide P and Coleman A (2019) *Delivering Better Water Management through the Planning System. Part E. Characteristics of Good Local Policy*. CIRIA C787A. Construction Industry Research and Information Association, London, UK.

Bradley J, Haygarth P, Stachyra K and Williams P (2022) *Using SuDS to Reduce Phosphorus in Surface Water Runoff*. CIRIA C808. Construction Industry Research and Information Association, London, UK.

Bridges TS, Smith JM, King JK *et al.* (2021) *International Guidelines on Natural and Nature-Based Features for Flood Risk Management*. US Army Engineer Research and Development Center, Vicksburg, MS, USA. See https://ewn.erdc.dren.mil/?page_id=4351.

Bridges TS, McKee Smith J, King JK *et al.* (2022) Coastal natural and nature-based features: international guidelines for flood risk management. *Frontiers in Built Environment* **8**: 904483, 10.3389/fbuil.2022.904483.

Catchment Based Approach (2023) Working together to improve the water environment. https://catchmentbasedapproach.org/ (accessed 29/07/2023).

cbec and Environment Agency (2017) *Natural Flood Management Toolbox*. cbec eco-engineering, Inverness, UK. See https://catchmentbasedapproach.org/wp-content/uploads/2018/08/EA-NFM-Toolbox-Final-Draft.compressed.pdf (accessed 29/07/2023).

Chan L, Hillel O, Elmqvist T *et al.* (2014) *User's Manual on the Singapore Index on Cities' Biodiversity (also known as the City Biodiversity Index)*. National Parks Board, Singapore. See https://www.cbd.int/authorities/doc/Singapore-Index-User-Manual-20140730-en.pdf (accessed 29/07/2023).

City of Bradford Metropolitan District Council (2019) Public Health – Joint Strategic Needs Assessment: People are living their lives well and ageing well. https://jsna.bradford.gov.uk/People%20are%20living%20their%20lives%20well%20and%20ageing%20well.asp (accessed 03/08/2023).

CMO (Chief Medical Officer for England) (2022) *Chief Medical Officer's Annual Report 2022: Air pollution*. Department of Health and Social Care, London, UK. See https://assets.publishing.service.gov.uk/government/uploads/system/uploads/attachment_data/file/1124738/chief-medical-officers-annual-report-air-pollution-dec-2022.pdf (accessed 29/07/2023).

Collins R, Johnson D, Crilly D *et al.* (2020) Collaborative water management across England – an overview of the Catchment Based Approach. *Environmental Science & Policy* **112**: 117–125, 10.1016/j.envsci.2020.06.001.

CRCWSC (CRC for Water Sensitive Cities) (2022) *Urban Water Guide*. Australian Water Partnership, Canberra, Australia.

Dadson SJ, Hall JW, Murgatroyd A *et al.* (2017) A restatement of the natural science evidence concerning catchment-based 'natural' flood management in the UK. *Proceedings of the Royal Society A* **473**: 20160706, 10.1098/rspa.2016.0706.

Daly D, Archbold M and Deakin J (2019) Progress and challenges in managing our catchments effectively. *Biology and Environment: Proceedings of the Royal Irish Academy* **116B(3)**: 157–166, 10.3318/bioe.2016.16.

Defra (Department for Environment, Food and Rural Affairs) (2009) *The Independent Review of Charging for Household Water and Sewerage Services*. Defra, London, UK. See https://assets.publishing.service.gov.uk/government/uploads/system/uploads/attachment_data/file/69459/walker-review-final-report.pdf (accessed 29/07/2023).

Defra (2011) *Water for Life*. CM 8320. HMSO, London, UK. https://assets.publishing.service.gov.uk/media/5a7c7b64e5274a559005a299/8230.pdf (accessed 5 December 2023).

Defra (2012) *Statement of Obligations*. Defra, London, UK. See https://www.gov.uk/government/publications/statement-of-obligations (accessed 29/07/2023).

Defra (2013) *Catchment Based Approach: Improving the Quality of Our Water Environment*. Defra, London, UK. See https://assets.publishing.service.gov.uk/government/uploads/system/uploads/attachment_data/file/204231/pb13934-water-environment-catchment-based-approach.pdf (accessed 29/07/2023).

DLUHC (Department for Levelling Up, Housing, Communities) (2023) *National Planning Policy Framework*. See https://assets.publishing.service.gov.uk/government/uploads/system/uploads/attachment_data/file/1005759/NPPF_July_2021.pdf (accessed 29/07/2023).

EC (European Commission) (2019) *Summary Report of the BLUE2 Project "Study on EU Integrated Policy Assessment for the Freshwater and Marine Environment, on the Economic Benefits of EU Water Policy and on the Costs of its Non-implementation"*. Directorate-General for Environment, European Commission, Publications Office, Brussels, Belgium. See https://data.europa.eu/doi/10.2779/87717 (accessed 29/07/2023).

EC (European Community) (2000) Directive 2000/60/EC of the European Parliament and of the Council establishing a framework for Community action in the field of water policy. *Official Journal of the European Communities* L327/1. See https://www.eea.europa.eu/policy-documents/directive-2000-60-ec-of (accessed 29/07/2023).

Fenner R, Sykes J and Ainger C (2022) *Sustainable Infrastructure*, 2nd edn. ICE Publishing, London, UK, 10.1680/si.66717.001.

Fletcher TD, Ahuster W, Hunt WF and Ashley R (2014) SUDS, LID, BMPs, WSUD and more – the evolution and application of terminology surrounding urban drainage. *Urban Water Journal* **12(7)**: 525–542, 10.1080/1573062X.2014.916314.

Halcrow (2013) *Drainage Strategy Framework. Good Practice Guidance Commissioned by the Environment Agency and Ofwat*. Halcrow Group, Swindon, UK. See https://www.ofwat.gov.uk/wp-content/uploads/2015/12/rpt_com201305drainagestrategy1.pdf (accessed 29/07/2023).

Hansen R and Pauliet S (2014) From multifunctionality to multiple ecosystem services? A conceptual framework for multifunctionality in GI planning for urban areas. *AMBIO* **43**: 516–529, 10.1007/s13280-014-0510-2.

Heijs J (2023) Stormwaters' inclusion in Three Waters reforms will worsen flood risks. *Newsroom*. 2 Feb. See https://www-newsroom-co-nz.cdn.ampproject.org/c/s/www.newsroom.co.nz/news/jan-heijs-stormwaters-inclusion-in-three-waters-reforms-will-worsen-flood-risks?amp = 1 (accessed 29/07/2023).

Helsingborg (2021) *Voluntary Local Review of Helsingborg*. See https://sustainabledevelopment.un.org/content/documents/27959Helsingborg_VLR_2021.pdf (accessed 29/07/2023).

HM Government (2011) *Water for Life*. The Stationery Office, London, UK. See https://assets.publishing.service.gov.uk/government/uploads/system/uploads/attachment_data/file/228861/8230.pdf (accessed 03/08/2023).

Horton B, Digman CJ, Ashley RM and McMullan J (2019) *B£ST Guidance – Guidance to Assess the benefits of blue and green infrastructure using B£ST*. CIRIA W047b. Construction Industry Research and Information Association, London, UK. See http://www.susdrain.org/resources/best.html (accessed 11/07/2021).

House of Lords (2022) *Nature-Based Solutions: Rhetoric or Reality?* Science and Technology Select Committee. HL Paper 147. See https://committees.parliament.uk/publications/8646/documents/87644/default/ (accessed 24/22/2023).

Howe C and Mitchell C (eds) (2012) *Water Sensitive Cities*. IWA Publishing, London, UK.

IIRC (International Integrated Reporting Council) (2013) *The International <IR> Framework*. IIRC, Delaware, USA. See https://www.integratedreporting.org/wp-content/uploads/2015/03/13-12-08-THE-INTERNATIONAL-IR-FRAMEWORK-2-1.pdf (accessed 03/08/2023).

Jessel E (2022) Explainer: What is 'nutrient neutrality' and why are nutrients stopping housebuilding? *Architects' Journal*, 12 May. See https://www.architectsjournal.co.uk/news/explainer-what-is-nutrient-neutrality-and-why-is-it-stopping-housebuilding (accessed 29/07/2023).

Juuti PS and Katko TS (eds) (2005) *Water, Time and European Cities: History matters for the Futures*. See http://www.watertime.net/docs/WP3/WTEC.pdf (accessed 23/11/2023).

Kaval P (2019) Integrated catchment management and ecosystem services: a twenty-five year overview. *Ecosystem Services* **37**: 100912, 10.1016/j.ecoser.2019.100912.

Khreis H, Ramani T, de Hoogh K *et al.* (2019) Traffic-related air pollution and the local burden of childhood asthma in Bradford, UK. *International Journal of Transportation Science and Technology* **8(2)**: 116–128.

Laville S (2022) Citizen scientists to monitor English and Welsh rivers in £7m scheme. *The Guardian* 14 Sept. See https://www.theguardian.com/environment/2022/sep/14/citizen-scientists-to-monitor-english-rivers-in-7m-scheme (accessed 29/07/2023).

Natural Capital Committee (2014) *The State of Natural Capital: Restoring our Natural Assets*. Second report to the Economic Affairs Committee. See https://www.gov.uk/government/collections/natural-capital-committee-documents (accessed 29/07/2023).

Natural England (2023) Introduction to the Green Infrastructure Framework – Principles and Standards for England. https://designatedsites.naturalengland.org.uk/GreenInfrastructure/Home.aspx (accessed 29/07/2023).

Newman R, Ashley R, Molyneux-Hodgson S and Cashman A (2011) Managing water as a socio-technical system: the shift from 'experts' to 'alliances'. *Proceedings of the Institution of Civil Engineers – Engineering Sustainability* **164(ES1)**: 95–102, 10.1680/ensu.1000032.

OECD (Organisation for Economic Co-operation and Development) (2018) *Financing Water – Investing in Sustainable Growth. Policy Perspectives*. OECD Environment Policy Paper No. 11. OECD, Paris, France. See https://www.oecd.org/water/Policy-Paper-Financing-Water-Investing-in-Sustainable-Growth.pdf (accessed 29/07/2023).

ONS (Office for National Statistics) (2022) How the population changed in Bradford: Census 2021. https://www.ons.gov.uk/visualisations/censuspopulationchange/E08000032/ (accessed 29/07/2023).

Public Health England (2020) *Improving Access to Green Space. A Review for 2020*. PHE Publications, London, UK. See https://assets.publishing.service.gov.uk/government/uploads/system/uploads/attachment_data/file/904439/Improving_access_to_greenspace_2020_review.pdf (accessed 29/07/2023).

SCBD (Secretariat to the Convention on Biological Diversity) (2023) City Biodiversity Index (or Singapore Index). https://www.cbd.int/subnational/partners-and-initiatives/city-biodiversity-index (accessed 29/07/2023).

Scottish Government (2001) *Planning Advice Note 61: Planning and Sustainable Urban Drainage Systems*. Planning and Architecture Division. The Scottish Government, Edinburgh, UK. See https://www.gov.scot/publications/pan-61-sustainable-urban-drainage-systems/ (accessed 29/07/2023).

Stantec (2011) *Comparing the Arrangements for the Management of Surface Water in England and Wales to Arrangements in Other Countries*. MHW UK. See https://copenhagenwater.files.wordpress.com/2013/11/comparing-arrangements-for-the-management-of-surface-water-in-eng-and-wlaes-to-arrnagements-in-other-countries-incl-netherlands.pdf (accessed 03/08/2023).

TCPA (Town and Country Planning Association) (2023) *Flood Risk and Coastal Erosion*. TCPA, London, UK. See https://tcpa.org.uk/wp-content/uploads/2022/12/PPG-Flood-Risk-update-FAQs-January-2023.pdf (accessed 29/07/2023).

UK Government (2018) *Further Evaluation of Partnership Funding*. Final Report FD2702. Defra, London, UK.

UK Government (2022) Draft river basin management plans consultation: summary of responses. https://www.gov.uk/government/consultations/river-basin-planning-draft-river-basin-management-plans/public-feedback/draft-river-basin-management-plans-consultation-summary-of-responses (accessed 29/07/2023).

UK NEA (UK National Ecosystem Assessment) (2011) UK National Ecosystem Assessment. http://uknea.unep-wcmc.org/#:~:text=The%20UK%20National%20Ecosystem%20Assessment%20(UK%20NEA)%20was%20the%20first,and%20reported%20in%20June%202011 (accessed 29/07/2023).

UKWIR (United Kingdom Water Industry Research) (2017) *Implementing Ecosystem Services and Natural and Social Capital Accounting Approaches – Phase 1*. Report Ref. No. 17/CL/04/15. United Kingdom Water Industry Research Ltd, London, UK. See https://ukwir.org/implementing-ecosystem-services-and-natural-and-social-capital-accounting-approaches-phase-1 (accessed 03/08/2023).

Van Oijstaeijen W, Van Passel S and Cools J (2020) Urban green infrastructure: a review on valuation toolkits from an urban planning perspective. *Journal of Environmental Management* **267**: 110603, 10.1016/j.jenvman.2020.110603.

Warner JF and Weigel H (2021) Displacement induced by climate change adaptation: the case of 'climate buffer' infrastructure. *Sustainability* **13(16)**: 9160, 10.3390/su13169160.

Washbourne C-A and Wansbury C (eds) (2023) *ICE Manual of Blue-Green Infrastructure*. ICE Publishing, London, UK.

Welsh Government (2019) *Sustainable Drainage (SuDS) Statutory Guidance*. Welsh Government, Cardiff, UK. See https://www.gov.wales/sites/default/files/publications/2019-06/statutory-guidance.pdf (accessed 29/07/2023).

White I (2010) *Water and the City*. Routledge, Abingdon, UK.

Wood-Ballard B, Wilson S, Udale-Clarke H *et al.* (2015) *The SuDS Manual.* CIRIA C753. Construction Industry Research and Information Association, London, UK. See http://www.susdrain.org/resources/SuDS_Manual.html (accessed 29/07/2023).

Wren E, Barnes M, Kitchen A *et al.* (2022) *The Natural Flood Management Manual.* CIRIA C802. Construction Industry Research and Information Association, London, UK.

WWF (2021) *Helsingborg – Net Zero, One Bite at a Time.* WWF. https://wwfint.awsassets.panda.org/downloads/wwf_us_case_2021_helsingborg_v3.pdf (accessed 06/12/2023)

Xu X, Huang A, Belle E, de Frenne P and Jia G (2022) Protected areas provide thermal buffer against climate change. *Science Advances* **8(44)**: eabo0119, 10.1126/sciadv.abo0119.

Ashley R, Smith B, Shaffer P and Caffoor I
ISBN 978-0-7277-4178-3
https://doi.org/10.1680/icehudp.41783.145

Chapter 6
Rainfall estimation and prediction for urban drainage

Alma Schellart, Adrian Saul, Andrew Bailey and Richard Ashley

6.1. Introduction

Precipitation is the 'catch-all' term for rainfall, snow, hail, sleet and drizzle that falls or condenses on the ground. This chapter focuses principally on rainfall, as this is the most relevant to urban drainage, and on UK practice. The occurrence of rainfall is highly variable in both space and time, which makes it challenging to measure and to predict. Rainfall is the key driver in generating runoff and flows in urban drainage systems. Rainfall measurements should be seen as samples of the actual process. Rainfall is measured using either networks of point measurements (e.g. rain gauges) or spatial measurements (e.g. weather radar or satellites). Rainfall estimation is the process of determining the best estimate of the actual rainfall occurring over an area based on rainfall measurements. Rainfall prediction is a term that can mean different things: it could refer to short-term rainfall forecasting, for example 'nowcasting', which involves predicting actual rainfall volumes and intensity some hours into the future, based on rainfall estimates and nowcasting models. Alternatively, it could mean predicting the current or future probability of occurrence of rainfall events of certain intensities and durations based on historical rainfall estimations; the rainfall from these specified events is defined as 'design rainfall'.

Hence many different types of rainfall data, rainfall estimation methods, short-term rainfall forecasts and longer-term rainfall predictions are used for different purposes.

6.2. Rainfall measurement

There have been significant developments in rainfall measurement technology since the 1990s, and how this is used in urban drainage. Conventionally, tipping bucket rain gauges have been used, but since 2010 rainfall radar has become increasingly common. As rainfall is highly variable in both space and time, it is useful to distinguish between point source rainfall measurements, which monitor rainfall at a single location, and spatial methods, which measure rainfall over larger areas. Examples of point source rainfall measurement devices are tipping bucket rain gauges and weighing rain gauges, which capture actual rain at a single location, and disdrometers, which measure reflections of rainfall at a single location. Examples of spatial rainfall monitoring techniques are weather radar, satellite and microwave links, which do not capture actual rainfall but are instead remote sensing techniques. The World Meteorological Organization has carried out an extensive comparison of the most common brands of rainfall intensity gauges and provides information about the pros and cons of each measurement method (Vuerich *et al.*, 2009). It is important to note that the quality of the rainfall estimation is determoned not only the precision and reliability of the instruments used, but also by how the instruments are sited, maintained and calibrated. Willems and Einfalt (2021) provide extensive practical advice on the selection of the type of instrument, the installation and calibration aspects to be considered, and the measurement data processing and adjustment needs. CIWEM (2016) provides extensive information on UK industry good practice on rainfall modelling measurements for modelling the hydraulics of urban drainage systems.

6.2.1 Tipping bucket rain gauges

Tipping bucket rain gauges are still the most used rain gauges for short-term surveys and are likely to be the cheapest. These rain gauges collect the rainfall in a circular collection cylinder of known plan area from where it is funnelled into a system of two buckets that can rotate about a central pivot (Figure 6.1). The tipping mechanism holds the first bucket in place until a set volume of rain is collected. This volume can correspond to 0.1 mm, 0.2 mm or 0.254 mm (0.1 inch) of rain falling over the plan area of the collection cylinder. The volume collected is sufficient to create a moment about the pivot such that the first bucket tips downwards and the second bucket moves into the position to collect the rainfall. This second bucket then fills with rainfall which again causes the first bucket to tip, and so on. Once a bucket has tipped the collected rain drains away from the bottom of the gauge. Either the time at which each individual tip occurred can be recorded, or the data logger can be set such that the number of tips in a certain time interval is recorded.

Non-heated tipping bucket rain gauges are not able to record solid precipitation, such as snow or hail, as this collects on top of the funnel and is only recorded once it starts melting. To overcome this problem, heated tipping rain gauges can be used, but these have more power requirements.

Tipping buckets tend to be used as part of a short- or long-term flow survey, carried out for urban drainage model verification exercises. There are several UK industry guides: the *Guide to Short Term Flow Surveys of Sewer Systems* (WRc, 1987), the *Code of Practice for the Hydraulic Modelling of Urban Drainage Systems* (CIWEM, 2017), and the *Rainfall Modelling Guide* (CIWEM, 2016). These provide best practice information on the siting of tipping bucket rain gauges and the density of gauges to ensure representative data collection across different types of catchment. The selection of a rain gauge location should also consider other factors such as health and safety, the opportunity for vandalism and the avoidance of sheltering or rain shadows. UK practice is suitable for most climates, although, as above, where snowfall is the primary form of precipitation, cold climate procedures may be more reliable. In some

Figure 6.1 Left: tipping bucket rain gauges installed on a flat roof. Right: inside view of a tipping bucket rain gauge, showing the bucket-pivot mechanism (courtesy of A. Schellart)

Table 6.1 Recommended number of rain gauges for different terrains (CIWEM, 2016)

Type of terrain	Typical number of rain gauges
Flat	1 + 1 per 4 km^2
Average	1 + 1 per 2 km^2
Mountainous	1 + 1 per 1 km^2

climates, extremes of weather in the form of gales or hurricanes may require the use of non-ground devices (e.g. Koistine *et al.*, 2004).

Rainfall patterns vary spatially across a catchment, and it is important that the rain gauge density is sufficient to measure any such variation. CIWEM (2016) recommends the density as shown in Table 6.1 for planning short-term flow surveys. Achieving this density of rain gauges in an urban area can be challenging, as sites need to be protected from vandalism, with the rain gauge not overshadowed by buildings or trees. The duration of rainfall monitoring for short-term flow surveys tends to be several months, as there should have been enough rainfall captured to wet the catchment (CIWEM, 2016). The measurements recorded for the different rain gauges should be sufficiently consistent so that rainfall at any point in the catchment can be determined by interpolation. Periods of lying snow should be omitted and there should be some variability in duration and intensity of events captured.

For model verification purposes it is also necessary to have a record of rainfall in the days prior to the events that are used in the verification exercise, such that it is possible to calculate the antecedent rainfall. Hence rainfall data should be collected for a number of days (often up to one month) prior to the start of the installation of the flow survey equipment.

In the UK, as elsewhere, it has been recognised that in order to ensure that computational models of drainage systems (Chapter 11) can be verified, there is a need to undertake long-term flow surveys to measure the different types of rainfall and the associated runoff responses that occur in different seasons. For example, high-depth, low-intensity and long-duration events in the winter can give very different drainage system responses (flows, depths and velocities) to high-intensity, short-duration events in the summer. An example case study assessing the merits of short- and long-term flow surveys is described by Terry and Margetts (2003). However, increasingly the monitoring of urban drainage systems is aimed at both flows and water quality (e.g. Nickel *et al.*, 2021) (Chapter 9).

6.2.2 Weighing rain gauges

Weighing gauges collect the rainfall and then weigh it. As these rain gauges do not use any moving mechanical parts in the weighing mechanism, they require less maintenance than tipping bucket rain gauges. These gauges can measure solid as well as liquid precipitation, or a mixture of both, as the solid precipitation does not require melting. Some weighing gauges work with a siphon or pumping system to empty them, others need to be emptied manually.

6.2.3 Disdrometers (rainfall drop counters)

Disdrometers are a 'non-catching' type of rain gauge. Instead, they record information about the size of raindrops or particles of solid precipitation, which is then integrated into rainfall volumes over time.

Optical disdrometers use one or two laser beams to detect passing particles of precipitation. The fall speed of the particles is also detected, which means the type of precipitation can be detected. Disdrometers tend to be used by meteorological offices to distinguish between types of precipitation – for example, to corroborate rainfall radar or non-heated rain gauge readings. These optical devices are not generally used alone for precipitation measurements for urban drainage systems.

Impact disdrometers are another type of rainfall drop counter, whereby the impacts of raindrops on a membrane are registered. In some systems the movement of the membrane generates elastic waves to the sensor plate and onto a piezoelectric sensor; other systems detect the amplitude and analyse the frequency spectrum generated by particles hitting the membrane (Vuerich *et al.*, 2009).

6.2.4 Radar

Radar is a 'spatial' way of estimating rainfall, as opposed to rain gauges and disdrometers, which only record rainfall at a single point. Radar is an echo-sounding system which works by transmitting an electro-magnetic wave and receiving the returned echo. For detailed description of how a weather radar works, and the UK weather radar network, see the UK Meteorological Office's *Factsheet 15 – Weather radar* (Met Office, 2020a). An extensive review of emerging developments and applications of weather radar in urban hydrology, and practical recommendations for the application of radar rainfall data in urban hydrology, are presented by Thorndahl *et al.* (2017).

Radars do not provide a direct measurement of the rainfall itself; rather, by analysing the returned echoes and checking these for errors, the returned echoes can be converted into rainfall intensity.

Sources of error in radar rainfall are described in, for example, UK Meteorological Office *Factsheet 15*, with Harrison *et al.* (2006, 2009) providing detail on the methods used. Analysis of radar data specifically for urban applications is described in Willems and Einfalt (2021).

In summary, the ability of a radar to pick up different types of precipitation accurately is dependent on a combination of several factors, such as the power of the transmitter, the wavelength, and whether the radar has a single or dual polarisation signal or a Doppler signal. Weather radar can be divided into several different categories according to the wavelength of the radar: the most used worldwide (in order of decreasing wavelength) are S-, C- and X-band radars. The more powerful the transmitter signal is, the larger the radar needs to be. Portable radars generally have a less powerful transmitter and hence a smaller range. Systems with a longer wavelength (S-band) have a longer range, but are less sensitive, systems with a shorter wavelength (X-band) are more sensitive, with C-band systems somewhere in the middle. The shorter wavelength systems are also more vulnerable to attenuation, whereby the radar signal cannot penetrate through heavy rainfall. The reach of a radar is not only dependent on the transmitter strength and the wavelength, but is also limited by the curvature of the earth. Even if the radar signal is sent out at a low angle, 0.5 degrees, after 300 km the height of the beam would be around 8 km above the earth's surface and thus mostly in or above the clouds and not able to pick up precipitation (see illustrations in e.g. UK Meteorological Office *Factsheet 15*).

The UK Meteorological Office operates a C-band weather radar network throughout the British Isles, including one located in Jersey (Met Office, 2020a). The data from each station are transferred to a central processing unit where they are corrected for various errors. Subsequently, the data are converted from radar reflectivity measurement to rainfall intensity and from polar to Cartesian grid coordinates.

The analysis of UK and European radar data is carried out using the 'Nimrod' system, developed by the UK Meteorological Office (Met Office, 2003). The development of the system was described by Golding (1998) and Harrison *et al.* (2006, 2009). Although the resolution at which the data are provided by Nimrod is 1 km^2, the data are a composite of readings from the different radars in the network. Each radar has an individual range of 255 km and the further from the centre of each radar, the radar signal is spread over a larger area and thus data become less reliable. When using radar data for an urban area it is therefore important to minimise the distance to the nearest radar to improve reliability of data.

There have been attempts to compare and contrast the measurement of rainfall using radar with that using rain gauges; for example, Schellart *et al.* (2012) assessed the influence of the rainfall measurement technique and the spatial variability of rainfall on the prediction of sewer flow at a small (11 km^2) urban scale, as outlined in Box 6.1.

Harrison *et al.* (2009) described the application of rain gauge and radar data using a sewer network model of a larger urban catchment of area 170 km^2 and a population of 600 000. The catchment was modelled using the data from ten rain gauges and from 4601 km^2 radar pixels. The results at two locations and for one storm event highlighted that the use of the radar data in the model showed a slightly better fit to monitored flow data.

Rico-Ramirez *et al.* (2015) created an error model to describe the uncertainties in rainfall radar data, and propagated this through to an urban drainage model, comparing simulated flows with measured flows. The radar rainfall error model was able to explain 55% of the uncertainties in the simulated flow volumes, indicating that there are additional sources of uncertainty in the simulated flows that must be considered, such as the uncertainty in the urban drainage model structure, the uncertainty in the urban drainage model calibrated parameters, and the uncertainty in the measured sewer flows (Chapter 11). Many efforts have therefore focused on the 'merging' of radar and rain gauge data, to improve the accuracy of radar estimates while benefitting from the wider spatial range of measurements possible from rainfall radar; for example, Wang *et al.* (2015). CIWEM (2016) provides detailed guidance on using radar data, and merging of radar and rainfall data for short-term flow surveys for verification of

Box 6.1

Comparative study of radar and rain gauge measurements of rainfall for urban drainage modelling (Schellart *et al.*, 2012)

For a small urban catchment in the UK (11 km^2), rainfall data were collected by four tipping bucket rain gauges and by the UK Meteorological Office's C-band radar network. These different sources of rainfall data have individually been used as inputs to a previously verified InfoWorks CS hydraulic network model (Chapter 11) to simulate flows and combined sewer overflow (CSO) spills in a combined sewer network. The simulation results were then compared with a new set of flow monitoring data. The study highlighted that both types of rainfall measurement had merits, but that in a comparison of the simulated and measured flow rates in the system, neither rainfall measurement system was consistently better than the other; and how good the fit of the outputs was was a function of the storm pattern, intensity and duration. No consistent trends were observed. Some errors specific to radar data, such as wind displacement, were found to be specifically pronounced at very small urban scales (e.g. 1–4 km).

urban drainage models. Be aware that the nearest radar in any network might not be working at the time of a sewer flow survey, so checks should be made on any scheduled routine maintenance of the nearest radar. In the UK, radar data are also calibrated against rain gauge data by the UK Meteorological Office, but the nearest Meteorological Office rain gauge may be some distance away from the urban catchment studied, hence it is good practice always to include at least some rain gauges in the catchment for local checking of the radar data.

6.2.5 Other remote sensing techniques for spatial rainfall monitoring

A relatively small number of studies have investigated other remote sensing techniques for spatial urban rainfall monitoring. One such technique to use microwave links – that is, radio connections from cellular networks (mobile phone networks), which are naturally more dense in urban areas. These radio waves are attenuated by raindrops. Pastorek *et al.* (2023) provide a review of research carried out to date on the use of microwave links for rainfall estimations in small urban catchments and describe techniques to reduce systematic errors in rainfall estimates from commercial microwave links. They conclude that optimised rainfall estimates using microwave links perform only slightly worse than benchmark rain gauge data collected using a rain gauge density of one rain gauge per 0.5–1 km^2, especially for events with high spatial variability, and a potential improvement when compared with rain gauge density of one rain gauge per 20–25 km^2.

Pistocchi *et al.* (2019) and Quaranta *et al.* (2022) used satellite rainfall data for an EU-wide review of water quality, making an EU-wide, high-level estimation of the amount of wastewater entering receiving waters through combined sewer overflows (CSOs). These studies were carried out as part of the review of the Urban Wastewater Treatment Directive, and the use of satellite data enabled the policy review to cover the whole EU, including all areas not monitored by radar or rain gauges.

6.3. Rainfall estimation, prediction and forecasting in space and time

Rainfall runoff processes on impermeable areas in urban drainage catchments occur much more rapidly than in natural river catchments. Urban catchments also tend to be more heterogeneous, with a mixture of impermeable areas (such as roads and roofs) and more permeable areas (parks, gardens and other green infrastructure, including sustainable drainage systems). Hence urban hydrological processes can be highly variable in space and time (see Chapter 7). As described in Cristiano *et al.* (2017), in urban areas the interactions between rainfall variability, catchment heterogeneity and hydrological response remain poorly understood.

Urban drainage models also come in a wide variety, from simple lumped catchment models estimating runoff peaks (e.g. rational method) to highly complex and detailed one-dimensional to two-dimensional models simulating urban drainage network hydraulics and overland flows (see Chapter 11). Hence great care should be taken to select rainfall inputs with temporal and spatial resolutions that are suitable for the intended runoff model (e.g. Cristiano *et al.*, 2019).

6.3.1 Rainfall estimation

As described in Section 6.2, rainfall can either be measured directly at a single point (using different types of rain gauges), or it can be inferred from spatial remote sensing techniques (such as radar). The 'true' ground rainfall is therefore always an approximation, and the process of deriving the best estimate for the 'true' ground rainfall is called rainfall estimation.

When using point source measurements, the assumption is made that the rainfall recorded at a single point is valid for the whole catchment, or that a spatially distributed rainfall pattern can be derived from several point source measurements. When using remote sensing techniques, assumptions are made to derive the ground rainfall based on the parameters measured (e.g. returned echoes in the case of radar). It is good practice to evaluate data derived from remote sensing techniques by comparing with rain gauge measurements, as these provide a more accurate measurement of the actual precipitation that reached the ground. However, if this is not done carefully then considerable spatial sampling errors can be introduced and the comparisons can become meaningless or, worse, misleading (Villarini and Krajewski, 2008).

Several techniques can be used to derive spatial rainfall from point source rainfall measurements, such that a spatially varying rainfall input can be assigned to runoff surfaces. In urban drainage this is commonly done using the simple Thiessen or Voronoi polygons methods (CIWEM, 2016), but more sophisticated isohyetal methods or geostatistical models, such as kriging, can also be used. For detailed explanations, refer to standard hydrology textbooks, for example Shaw *et al.* (2011). Figure 6.2 shows an example of Thiessen polygons.

In urban areas, which have a high degree of imperviousness and fast runoff response, it is important to capture high spatial–temporal rainfall variability (Cristiano *et al.*, 2019), and significant errors can occur when rainfall measurements with low spatiotemporal resolutions of rainfall are used, especially for convective events which exhibit high spatial variability (Maier *et al.*, 2020). The latter provide an overview of studies on the spatial variability of rainfall in urban areas; several studies have confirmed that even on a sub-kilometre scale, spatial variability can be as much as 100% within 1 km^2.

Schellart *et al.* (2012) simulated rainfall runoff, flow peaks and CSO spills in a combined sewer system for an 11 km^2 catchment that included a 4 km^2 urbanised area. The catchment was covered by four rain gauges and 25 radar pixels of 1 km^2. Using Thiessen polygons (Figure 6.2) they distributed the rain gauge data or the radar pixels as rainfall input and found considerable differences in simulated flow

Figure 6.2 Example of Thiessen polygons in use (adapted from Schellart *et al.*, 2012)

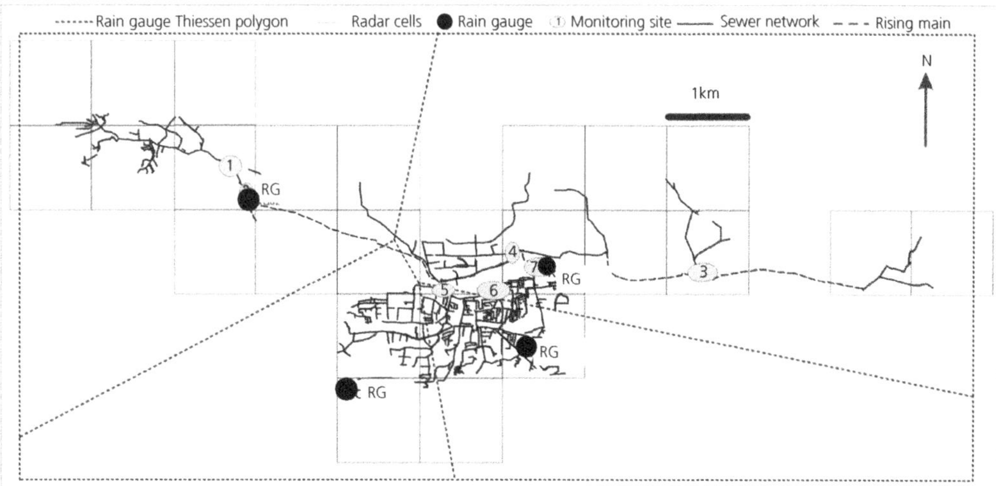

peaks due to spatial representativeness error in the rainfall estimates. They found differences between –65% and +178% in simulated CSO spill volumes at 5 min resolution (between 500 m^3 and 3000 m^3 absolute difference), and even greater differences at lower temporal resolutions.

Recent developments have focused on techniques for merging radar and rain gauge data in order to get the best possible estimates of true rainfall. This can combine the advantages of both techniques and overcome the weaknesses in each of the individual techniques (see Ochoa-Rodriguez *et al.*, 2019).

6.3.2 Rainfall prediction based on historic data: design rainfall

The probability of the occurrence of rainfall events needs to be estimated in order to design urban drainage systems to a specific design life and risk of failure. Predictions of the probability of occurrence of rainfall events are made through statistical analysis of historic rainfall data. This introduces uncertainties as climate changing drivers mean that predictions from historic data are not necessarily valid into the future. What rainfall data predictions should we use? It is incumbent on drainage system designers and analysts to ensure that the most up-to-date design rainfall is used for the locality being considered.

Design rainfall is often expressed through intensity–duration–frequency (IDF) curves (Figure 6.3), which can then be transformed into time-varying rainfall, such as single event hyetographs or design time series of rainfall, using various different methods. IDF curves show estimates of the amount (intensity) of rainfall for different rainfall event durations; for different probability of exceedance (frequency, often expressed as 'return period'). IDF curves are derived using several steps. First, historical rainfall data are gathered and processed into either an annual maximum series (by selecting the maximum value of rainfall of a certain duration (e.g. 24 h) that occurred in each water-year in the time series), or a partial duration series (peaks over threshold) that includes every independent event above a threshold. Then, a statistical distribution is fitted to the annual maximum series or the partial duration series, for each rainfall duration of interest. A variety of statistical distributions can be used for this. Then, the quantiles of the distributions are used to determine the rainfall intensity (or depth) associated with different frequencies (or return periods) for each duration. IDF curves are specific to regions and

Figure 6.3 Example of intensity–duration–frequency (IDF) curves

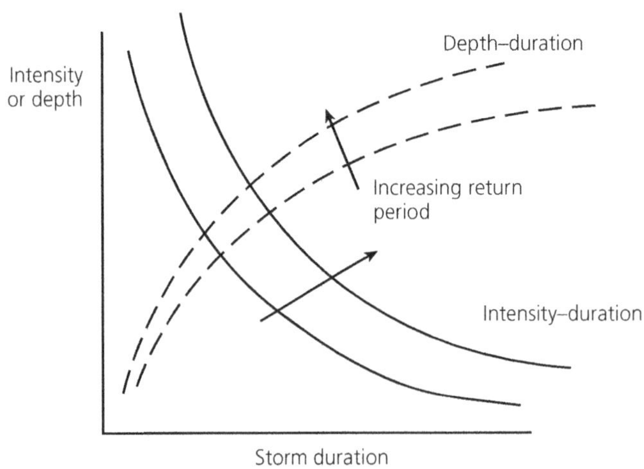

depend on the local climate. A review of rainfall frequency estimation methods used in nine countries is given in Svensson and Jones (2010a).

IDF curves are derived from historical data and assume the climate is stationary, hence they do not necessarily give a representative prediction of future rainfall, due to the effects of climate change on future regional precipitation patterns (e.g. Milly *et al.*, 2008). Ntegeka and Willems (2008) analysed a 107-year rainfall time series in Belgium and found significant multidecadal oscillations in rainfall extremes. High extremes were clustered in the 1910s to 1920s, the 1960s and the 1990s. This highlights the difficulties of analysing changes in climate using relatively short duration datasets, as well as potential bias in design rainfall events derived from datasets comprising only several decades.

An extensive review on incorporating non-stationarity from climate change into IDF curves is presented in Schlef *et al.* (2023). This is a fast-moving field of research, and it has to be kept in mind there is great uncertainty associated with predicting the probability of the occurrence of extreme rainfall events based on relatively short (e.g. several decades) historic datasets. Whenever a new 'record breaking' rainfall event is added to a time series, the estimated return period of the largest events in the series can change drastically. This has been illustrated by Vesuviano *et al.* (2021), who recalibrated the most recent UK depth–duration–frequency (DDF) model by adding two recent extreme events, and this reduced the return period of each event from approximately 1000 years to approximately 140 years.

IDF curves represent point rainfall measurements. If it is assumed that a design rainfall event occurs over the whole of an urban area, the amount of rainfall is likely to overestimated. Traditionally, areal reduction factors (ARFs) have been used to try to mitigate this problem and convert point rainfall intensity derived from an IDF curve into catchment-wide rainfall intensity. Different methods have been proposed to derive ARFs. Svensson and Jones (2010b) reviewed the methods to derive ARFs. These fall into two distinct groups: the traditional data intensive empirical methods, and the less computationally intensive analytical methods. However, the latter give some cause for concern due to uncertainty related to assumptions made which do not entirely truly describe the real rainfall process. Now that there is the increasingly widespread availability of spatial radar rainfall data, and as longer historical rainfall radar datasets are being built up, the use of ARFs is coming under increasing scrutiny because ARFs do not properly represent the true properties of extreme rainfall (a critical review can be found in Wright *et al.*, 2014).

IDF curves only provide a 'block' rainfall event – for example, a uniform rainfall intensity for a certain duration of time. The actual occurrence of such a 'block' rainfall event would be impossible. Many methods have therefore been derived to transform such a uniform block rainfall event into a hyetograph (time-varying rainfall), such as the alternating block method, the Chicago method (Keifer and Chu, 1957), or the UK summer and winter event profiles defined in the Wallingford Procedure (DoE/HR Walliingford, 1983).

Many urban drainage design and simulation software packages already incorporate different design rainfall events and event profiles. For example, in InfoWorks ICM (Autodesk, 2023) one can choose between: Australian (1987) rainfall, Australian (2016) rainfall, Chinese rainfall, Euler type II rainfall, Flood Estimation Handbook (FEH) rainfall, FEH 2013 rainfall, French rainfall (Desbordes), Hong Kong rainfall (5th edition), Hong Kong rainfall (historic), Malaysia rainfall, QM French rainfall, ReFH rainfall, SCS/NRC rainfall, and United Kingdom rainfall (Flood Studies Report; FSR).

A general drawback of design rainfall events derived through IDF/DDF curves, with or without adding ARFs and creating different shaped event profiles, is that these remain estimates for the probability of occurrence of single rainfall events. In many cases, time series rainfall would be needed instead of single events, such as when designing drainage systems that rely on filling and emptying of storage, or when looking at longer-term system performance, such as assessing cumulative pollution in receiving water (Chapter 11). Statistical time series rainfall may be defined as a sequence of historical rainfall events that are statistically representative of the annual or long-term pattern of rainfall at a given location. CIWEM (2016) describes the development of different stochastic rainfall series generation tools. There have been many reported applications of such series, and the use of rainfall time series as opposed to single events has been shown to be important (e.g. Quinn *et al.*, 2021).

Around the world there are different approaches to defining rainfall estimates to be used in the design of urban drainage systems. There are international sources of information, including the International Water Association's specialist group on rainfall (IWA, 2023). These should be considered when deciding on appropriate rainfall for the analysis of urban drainage systems. In the following, the approach currently used for UK applications is outlined.

6.3.2.1 UK approach to design rainfall

DDF curves were originally derived for all locations in the UK, following a method known as the Wallingford Procedure (DoE/HR Wallingford, 1983), which is based on the UK *Flood Studies Report* (FSR) (NERC, 1975). The Wallingford Procedure was based on two parameters – the depth of rain in 60 min for a return period of 5 years (M5-60 parameter) and the ratio between M5-60 and M5-2 days (*r* parameter). These parameters were plotted on maps for the whole of the UK and are presented within the original Wallingford Procedure documents (DoE/HR Wallingford, 1983). The values of 'M5-60' and '*r*', together with statistical equations, commonly referred to as growth factors, are used to derive the rainfall depth for durations from 5 min to 48 h, for return periods from 1 year to 100 years, and including ARF for catchments from 1 km^2 to 100 km^2.

In order to derive time-varying design rainfall events, in the UK, synthetic 'peaked' design rainfall profiles have been statistically derived for both winter and summer rainfall. The synthetic profile was shown to vary with season and, consequently, the profiles have been developed for both summer and winter rainfall patterns. The Wallingford Procedure describes the derivation of these synthetic design rainfall profiles for use in urban drainage models. It is good practice to complete a hydraulic analysis of system performance for both summer and winter profiles for a range of return periods and durations to identify the critical seasonal profile and hence the poorest (critical) performance of the system. It is also important to realise that even when there have been measurements made of both flows and rainfall, these might not necessarily have been from rainfall events that are representative of the selected design conditions. Hence model verification might only be valid for lesser rainfall events (Chapter 11).

In 1999 the UK Institute of Hydrology produced the *Flood Estimation Handbook* (FEH) (Faulkner, 1999), which superseded the FSR. The FEH includes additional hourly data, which was more than three times the amount of data available for the FSR. The FEH revisited much of the original base statistics and produced similar DDF curves, but allowed these to be applied at a 1 km grid resolution. This provided for more regional variation and allowed the influence of local topographic factors to be considered. Further refinement to the DDF model has since taken place, resulting in the FEH13 (Stewart *et al.*, 2013), which is aimed at improving the rainfall frequency estimates for rare events with a duration over 1 h. Prosdocimi *et al.* (2017) provide an overview of advantages and disadvantages of the FSR and FEH and compares both against a new DDF model based on generalised extreme value (GEV) for rainfall frequency estimation for short-duration events. The new GEV model provided satisfactory results, and although no sub-hourly data were used to develop the FEH it still gives reasonable estimates. The FSR seems to overestimate rainfall for short return periods but shows less bias for long-return-period events. Although all models could only be compared against relatively scarce reliable long data records of sub-hourly data, the absence of such datasets is a major obstacle to developing a new DDF model for the whole of the UK.

6.4. Prediction of changing rainfall patterns due to climate change

Climate change impacts sewerage and drainage systems due to changes in rainfall distribution and patterns, catchment wetness and drought (Chapter 7). These changes can lead to a range of challenges, such as increased risk of flooding and CSO spills due to increased rainfall intensity, significant changes in the amount of infiltration inflow into the system, increased numbers of blockages and sedimentation problems due to prolonged dry spells. These changes could lead to a range of problems, such as deteriorating quality of the receiving watercourse, increased risk of human interaction with harmful pathogens, and increased damage and deterioration of the sewer system itself (Chapters 2, 3 and 10).

Throughout the world, climate change predictions are continuously updated and refined. It is therefore essential to check the latest predictions made for the region of interest. A major challenge is that climate change predictions tend to be made at spatial and temporal resolutions far lower than those needed for analysing urban drainage systems. Attempts have been made to utilise spatial and temporal downscaling techniques to predict the probability of higher spatial and temporal resolution future rainfall events, as global circulation models and regional climate models do not provide a suitably high spatial and temporal resolution. However, these downscaled results tend to be highly uncertain; an overview and evaluation of methods is given in Willems *et al.* (2012).

6.4.1 UK climate projections

The latest climate change projections for the UK are described on the UK Climate Projections (UKCP) website (Met Office, 2023). Three sets of climate projections have been made for the UK so far: UKCP02 released in 2002, UKCP09 released in 2009, and UKCP18 is the most recent for this book.

UKCP18 (Met Office, 2018) uses emissions scenarios called representative concentration pathways (RCPs), which are used in the Intergovernmental Panel on Climate Change's (IPCC) Fifth Assessment Report, whereas UKCP09 was based on scenarios from the IPCC's Fourth Assessment Report. UKCP18 uses five greenhouse gas emission scenarios: RCP2.6, RCP4.5, RPC6.0, RCP8.5 and SRES A1B.

UKCP18 includes an update to probabilistic projections (UKCP Probabilistic), describing 20-, 50- and 100-year return periods of daily maximum surface air temperature, and 1-day and 5-day accumulated precipitation at 25 km resolution (Met Office, 2020b). These probabilistic projections are based on 360 climate model simulations. During the twenty-first century the uncertainty in the projections grows; for example, by 2070 for a 50-year return period under RCP8.5, UK averages of the 10th and 90th percentiles are 40 mm and 58 mm for 1-day winter precipitation (Rx1day) and 102 and 142 mm, respectively, for 5-day autumn precipitation (Rx5day). For further spatially distributed analysis – for example, by users interested in higher spatial resolution projections – UKCP18 also provides 28 global projections (UKCP Global), 12 European regional projections (UKCP Regional) and 12 UK convective-permitting projections (UKCP Local, only based on RCP8.5 and two epochs 2030s and 2070s). All these products, however, include more limited representations of uncertainty compared with UKCP Probabilistic.

Specifically for urban drainage purposes, the UKCP18 local projections from the convective-permitting model are useful as they provide 2.2 km spatial resolution projections, although they are only at hourly resolution (Met Office, 2019). The most recent guidance created specifically for UK water and sewerage companies and flood risk management authorities is presented in Dale (2021). This guidance is based on outputs from the FUTURE-DRAINAGE project, where the UKCP18 high-resolution 2.2 km data have been used to develop rainfall uplifts for the high greenhouse gas emission scenario RCP8.5. These uplifts supersede those provided in the UKWIR (2017) report *Rainfall Intensity for Sewer Design*, as the UKCP18 high-resolution data are based on a revised climate model with improved physics and improved underlying statistical methods. Dale (2021) includes links to shape files to download the UKCP 2.2 km convection permitting projections. These uplifts have been developed against a baseline period of 1981–2000 and hence are appropriate for use with FEH13 depth duration frequency (DDF) data, now that FEH22 has been released.

The UKCP18 projections therefore also supersede the UKCP09 and UKCP02 projections. UKCP09 predictions included a statistical weather generator that could be used to produce design rainfall into the future at a grid resolution of 5 km × 5 km (or multiples thereof), with an hourly resolution. However, the UKCP09 weather generator has been discontinued, as UKCP18 provides data from a physically-based modelling system that can be better evaluated against real-world observations (Met Office, 2018).

This creates problems for the design of many types of urban drainage systems, particularly sustainable drainage systems (SuDS), in which several hydrological processes need to be taken into account. This means that, for the design of SuDS, the use of future time series rainfall is more appropriate than the use of single design events (e.g. Stovin *et al.*, 2023).

6.5. Rainfall forecasting: short term (hours)

There is interest in the use of short-term rainfall forecasts for applications such as urban flood warning systems and real-time control of sewer systems.

The use of successive radar rainfall images provides an opportunity to track the patterns and movements of storms. These patterns can be used to predict the future movement of storms and thus the subsequent storm profile that is likely to occur over a particular pixel into the future. This prediction technique is termed 'nowcasting' and can provide forecasts for very short periods (no more than 6 h, down to only minutes in cases of convective rainfall). For longer forecasting periods, the rainfall distribution must be seen as a result of the changes in the atmospheric conditions, and numerical weather prediction (NWP) models would be needed (Golding, 1998). However, NWP models tend to be far more computationally intensive.

A method for combining radar-based forecasts (nowcasts) and NWP forecasts was developed at the UK Meteorological Office in collaboration with the Australian Bureau of Meteorology. This stochastic precipitation forecasting system is known as the Short-Term Ensemble Prediction System (STEPS) and merges precipitation forecasts from a nowcasting system with downscaled deterministic NWP forecasts (Bowler *et al.*, 2006). An example of the application of probabilistic radar rainfall nowcasting techniques together with numerical weather models for the prediction of flow in urban catchments was presented by

Ligouri *et al.* (2012), in which it was concluded that the approach was able to predict, with reasonable accuracy, the rainfall associated with frontal systems, but that convective storms (thunderstorms) and mixed convective and frontal storms could not be accurately predicted.

A main challenge in the use of short-term rainfall forecasts is finding a balance between the run times of the urban drainage model that takes the forecast as input (this would likely need to be a simplified conceptual model, rather than, for example, a full two-dimensional flood spreading model) and the rate of loss of accuracy of the rainfall nowcast (which can be as little as a few minutes in the case of convective rainfall).

The use of rainfall nowcasts is still full of challenges, due to the heterogeneous quality of the nowcast, and the rapidly deteriorating accuracy of the nowcast with time. A recent overview and a case study on the potential application of nowcasts in real-time control of urban drainage systems to reduce CSO spills are presented in Van der Werf *et al.* (2023). The authors concluded that significant risks are associated with using real nowcast data to control urban drainage systems, and that operation would sometimes be significantly worse were the nowcast to be used. They also investigated how the accuracy of the nowcast might be used to quantify risks associated with the use of the nowcast for system control. Knowledge gaps in short-term rainfall forecasting, as well as runoff prediction, particularly exists in the 'global south', where the climate and thus the predominant rainfall patterns, land use and urban drainage systems are quite different from the 'global north'. Chitwatkulsiri *et al.* (2022) review the use of the TITAN (thunderstorm identification, tracking, analysis and nowcasting) rainfall model (Dixon and Wiener, 1993) for real-time flood forecasting in Ramkhamhaeng Polder, Bangkok, Thailand. The authors found that 61.2% of 116 recent flash flood events were successfully predicted, with accuracy high enough for appropriate pre-warning.

6.6. Use of rainfall estimates, predictions and forecasts in urban drainage models

Chapter 11 considers modelling for urban drainage systems. Many different types of rainfall data, rainfall estimates, future rainfall predictions and short-term rainfall forecasts are being used in urban drainage applications for various reasons.

Depending on the application, it is important to consider what spatial and temporal resolution of the rainfall data is needed to get meaningful results, whether single-event or longer-term rainfall data are needed; and whether the data are used for design purposes, model calibration verification purposes or for real-time control purposes.

As the climate continues to change around the world, it is important for anyone interested in urban drainage systems to keep pace with the acquisition of related data and growth in knowledge about rainfall, other forms of precipitation and how best to use this information in analysis. There is a tendency for non-specialists to request simple rules for uplifting rainfall intensity, volumes and duration for future robustness of infrastructure, and this tendency is exacerbated by media reporting information about storms being 'once in a lifetime' or the 'hundred-year event'. As Milly *et al.* (2008) state: 'In a non-stationary world, continuity of observations is critical'. For urban drainage, it is the rainfall that is perhaps the most significant component of the system in which continuous observations are going to be essential into the future.

REFERENCES

Autodesk (2023) Autodesk InfoWorks ICM: Plan and manage flood control, stormwater and sewer systems. https://www.autodesk.co.uk/products/infoworks-icm (accessed 16/03/2023).

Bowler NE, Pierce CE and Seed AW (2006) STEPS: a probabilistic precipitation forecasting scheme which merges an extrapolation nowcast with downscaled NWP. *Quarterly Journal of the Royal Meteorological Society* **132(620)**: 2127–2155, 10.1256/qj.04.100.

Chitwatkulsiri D, Miyamoto H, Irvine KM, Pilailar S and Loc HH (2022) Development and application of a real-time flood forecasting system (RTFlood System) in a tropical urban area: a case study of Ramkhamhaeng Polder, Bangkok, Thailand. *Water* **14(10)**: 1641, 10.3390/w14101641.

CIWEM (Chartered Institution of Water and Environmental Management) (2016) *Rainfall Modelling Guide 2016*. Urban Drainage Group, CIWEM, London, UK. See https://www.ciwem.org/assets/pdf/Special%20Interest%20Groups/Urban%20Drainage%20Group/CIWEM-UDG-Rainfall-Guide-2016.pdf (accessed 21/02/2023).

CIWEM (2017) *Code of Practice for the Hydraulic Modelling of Urban Drainage Systems*. Urban Drainage Group, CIWEM, London, UK. See https://www.ciwem.org/special-interest-groups/urban-drainage-group (accessed 16/03/2023).

Cristiano E, ten Veldhuis M-C and van de Giesen N (2017) Spatial and temporal variability of rainfall and their effects on hydrological response in urban areas – a review. *Hydrology and Earth System Sciences* **21(7)**: 3859–3878, 10.5194/hess-21-3859-2017.

Cristiano E, ten Veldhuis M-C, Wright DB, Smith JA and van de Giesen N (2019) The influence of rainfall and catchment critical scales on urban hydrological response sensitivity. *Water Resources Research* **55(4)**: 3375–3390, 10.1029/2018WR024143.

Dale M (2021) *Future Drainage. Guidance for the Water and Sewerage Companies and Flood Risk Management Authorities: Recommended Uplifts for Applying to Design Storms*. FUTURE-DRAINAGE, University of Newcastle, Newcastle, UK. See https://artefacts.ceda.ac.uk/badc_datadocs/future-drainage/FUTURE_DRAINAGE_Guidance_for_applying_rainfall_uplifts.pdf (accessed 14/02/2023).

Dixon M and Wiener G (1993) TITAN: Thunderstorm Identification, Tracking, Analysis, and Nowcasting – a radar-based methodology. *Journal of Atmospheric and Oceanic Technology* **10(6)**: 785–797, 10.1175/1520-0426(1993)010<0785:TTITAA>2.0.CO;2.

DoE (Department of the Environment)/HR Wallingford (1983) *Design and Analysis of Urban Storm Drainage. The Wallingford Procedure. Volume 1: Principles, Methods and Practice*. Standing Technical Committee Report No. 28. Department of the Environment, National Water Council, Hydraulics Research Ltd, Wallingford, UK. See https://eprints.hrwallingford.com/37/ (accessed 27/02/2023).

Faulkner D (1999) Rainfall Frequency Estimation. *Flood Estimation Handbook*, vol. **2**. Institute of Hydrology, Wallingford, UK.

Golding BW (1998) Nimrod: a system for generating automated very short-range forecasts. *Meteorological Applications* **5(1)**: 1–16.

Harrison D, Driscoll SJ and Kitchen M (2006) Improving precipitation estimates from weather radar using quality control and correction techniques. *Meteorological Applications* **7(2)**: 135–144.

Harrison DL, Scovell RW and Kitchen M (2009) High-resolution precipitation estimates for hydrological uses. *Proceedings of the Institution of Civil Engineers – Water Management* **162(2)**: 125–135, 10.1680/wama.2009.162.2.125.

IWA (International Water Association) (2023) International Group on Urban Rainfall. Working Group of the IWA/IAHR Joint Committee on Urban Drainage. https://igur.org/ (accessed 20/07/2023).

Keifer CJ and Chu HH (1957) Synthetic storm pattern for drainage design. *ASCE Journal of the Hydraulics Division* **83(HY4)**: 1–25.

Koistine J, Michelson DB, Hohti H and Paura M (2004) Operational measurement of precipitation in cold climates. In *Weather Radar. Physics of Earth and Space Environments* (Meischner P (eds)). Springer, Berlin, Germany. https://doi.org/10.1007/978-3-662-05202-0_3.

Ligouri S, Rico-Ramirez MA, Schellart ANA and Saul A (2012) Using probabilistic radar rainfall nowcasts and NWP forecasts for flow prediction in urban catchments. *Atmospheric Research* **103**: 80–95, 10.1016/j.atmosres.2011.05.004.

Maier R, Krebs G, Pichler M, Muschalla D and Gruber G (2020) Spatial rainfall variability in urban environments – high-density precipitation measurements on a city scale. *Water* **12**: 1157, 10.3390/w12041157.

Met Office (2003) *Met Office Rain Radar Data from the NIMROD System*. NCAS British Atmospheric Data Centre. See http://catalogue.ceda.ac.uk/uuid/82adec1f896af6169112d09cc1174499 (accessed 19/10/2022).

Met Office (2018) *UKCP18 Guidance: UKCP18 for UKCP09 users*. Met Office, Exeter, UK. See https://www.metoffice.gov.uk/binaries/content/assets/metofficegovuk/pdf/research/ukcp/ukcp18-guidance-ukcp18-for-ukcp09-users.pdf (accessed 14/02/2023).

Met Office (2019) *UKCP18 Factsheet: UKCP Local (2.2 km) Projections*. Met Office, Exeter, UK. See https://www.metoffice.gov.uk/binaries/content/assets/metofficegovuk/pdf/research/ukcp/ukcp18-factsheet-local-2.2km.pdf (accessed 24/10/2022).

Met Office (2020a) *Factsheet 15 – Weather radar*. National Meteorological Library and Archive, Met Office, Devon, UK. See https://www.metoffice.gov.uk/binaries/content/assets/metofficegovuk/pdf/research/library-and-archive/library/publications/factsheets/factsheet_15-weather-radar-2020_temp.pdf (accessed 21/02/2023).

Met Office (2020b) *UKCP Additional Land Products: Probabilistic Projections of Climate Extremes*. Met Office, Exeter, UK. See https://www.metoffice.gov.uk/binaries/content/assets/metofficegovuk/pdf/research/ukcp/ukcp-probabilistic-extremes-report.pdf (accessed 21/02/2023).

Met Office (2023) UK Climate Projections (UKCP). https://www.metoffice.gov.uk/research/approach/collaboration/ukcp (accessed 14/02/2023).

Milly P, Betancourt J, Falkenmark M (2008) Stationarity is dead: whither water management? *Science* **319(2008)**: 573–574, 10.1126/science.1151915.

NERC (Natural Environment Research Council) (1975) *Flood Studies Report*. NERC, London, UK.

Nickel JP, Sacher F and Fuchs S (2021) Up-to-date monitoring data of wastewater and stormwater quality in Germany. *Water Research* **202**: 117452, 10.1016/j.watres.2021.117452.

Ntegeka V and Willems P (2008) Trends and multidecadal oscillations in rainfall extremes, based on a more than 100-year time series of 10 min rainfall intensities at Uccle, Belgium. *Water Resources Research* **44(7)**: W07402, 10.1029/2007WR006471.

Ochoa-Rodriguez S, Wang L-P, Willems P and Onof C (2019) A review of radar-rain gauge data merging methods and their potential for urban hydrological applications. *Water Resources Research* **55(8)**: 6356–63911, 10.1029/2018WR023332.

Pastorek J, Fencle M and Barreš V (2023) Uncertainties in discharge predictions based on microwave link rainfall estimates in a small urban catchment. *Journal of Hydrology* **617(C)**: 129051, 10.1016/j.jhydrol.2022.129051.

Pistocchi A, Dorati C, Grizzetti B, *et al.* (2019) *Water Quality in Europe: Effects of the Urban Wastewater Treatment Directive. A Retrospective and Scenario Analysis of Dir. 91/271/EEC*. Publications Office of the European Union, Luxembourg, 10.2760/303163.

Prosdocimi I, Stewart EJ and Vesuviano G (2017) A depth–duration–frequency analysis for short-duration rainfall events in England and Wales. *Hydrology Research* **48(6)**: 1624–1638, 10.2166/nh.2017.140.

Quaranta E, Fuchs S, Liefting HJ, Schellart A and Pistocchi A (2022) A hydrological model to estimate pollution from combined sewer overflows at the regional scale: application to Europe. *Journal of Hydrology: Regional Studies* **41**: 101080.

Quinn R, Rougé C and Stovin V (2021) Quantifying the performance of dual-use rainwater harvesting systems. *Water Research X* **10**: 100081, 10.1016/j.wroa.2020.100081.

Rico-Ramirez MA, Liguori S and Schellart ANA (2015) Quantifying radar-rainfall uncertainties in urban drainage flow modelling. *Journal of Hydrology* **528**: 17–28, 10.1016/j.jhydrol.2015.05.057.

Schellart ANA, Shepherd WJ and Saul AJ (2012) Influence of rainfall estimation error and spatial variability on sewer flow prediction at small urban scale. *Advances in Water Resources* **45**: 65–75, 10.1016/j.advwatres.2011.10.012.

Schlef KE, Kunkel KE, Brown C *et al.* (2023) Incorporating non-stationarity from climate change into rainfall frequency and intensity-duration-frequency (IDF) curves. *Journal of Hydrology* **616**: 128757, 10.1016/j.jhydrol.2022.128757.

Shaw EM, Beven KJ, Chappell NA and Lamb R (2011) *Hydrology in Practice*, 4th edn. Spon Press, London, UK.

Stewart EJ, Jones DA, Svensson C *et al.* (2013) *Reservoir Safety – Long Return Period Rainfall.* Defra and Environment Agency, London, UK. See https://www.gov.uk/flood-and-coastal-erosion-risk-management-research-reports/reservoir-safety-a-new-model-to-estimate-long-term-period-rainfall (accessed 28/11/2023).

Stovin V, Quinn R and Rouge C (2023) Study of continuous simulation supporting multiple design criteria for sustainable drainage systems. *Journal of Sustainable Water in the Built Environment* **9(3)**.

Svensson C and Jones DA (2010a) Review of rainfall frequency estimation methods. *Journal of Flood Risk Management* **3**: 296–313, 10.1111/j.1753-318X.2010.01079.x.

Svensson C and Jones DA (2010b) Review of methods for deriving areal reduction factors. *Journal of Flood Risk Management* **3**: 232–245, 10.1111/j.1753-318X.2010.01075.x.

Terry D and Margetts J (2003) Derby Case Study: runoff modelling issues. In *WaPUG Autumn Conference*, Blackpool, UK. See http://ftp2.ciwem.org/2003/2003-11/A2003_margetts_%20terry_paper.pdf (accessed 21/02/2023).

Thorndahl S, Einfalt T, Willems P, *et al.* (2017) Weather radar rainfall data in urban hydrology. *Hydrology and Earth System Sciences* **21(3)**: 1359–1380. https://doi.org/10.5194/hess-21-1359-2017.

UKWIR (UK Water Industry Research) (2017) *Rainfall Intensity for Sewer Design.* 15/CL/10/16-1. UKWIR Ltd, London, UK. See https://ukwir.org/eng/reports/15-CL-10-16-1/129641/Rainfall-Intensity-for-Sewer-Design-Technical-Guide (accessed 17/08/2023).

Van der Werf JA, Kapelan Z and Langeveld JG (2023) Predictive heuristic control: inferring risks from heterogeneous nowcast accuracy. *Water Science Technology* **87(4)**: 1009–1028, 10.2166/wst.2023.027.

Vesuviano G, Stewart E, Spencer P and Miller JD (2021) The effect of depth-duration-frequency model recalibration on rainfall return period estimates. *Journal of Flood Risk Management* **14(2)**: e12703. https://doi.org/10.1111/jfr3.12703

Villarini G and Krajewski WF (2008) Empirically based modelling of spatial sampling uncertainties associated with rainfall measurements by rain gauges. *Advances in Water Resources* **31**: 1015–1023, 10.1016/j.advwatres.2008.04.007.

Vuerich E, Monesi C, Lanza LG, Stagi L and Lanzinger E (2009) *WMO Field Intercomparison of Rainfall Intensity Gauges.* Instruments and Observing Methods Report No. 99. World Meteorological Organization, Geneva, Switzerland. See https://library.wmo.int/doc_num.php?explnum_id=9422 (accessed 21/02/2023).

Wang LP, Ochoa-Rodriguez S, Van Assel J *et al.* (2015) Enhancement of radar rainfall estimates for urban hydrology through optical flow temporal interpolation and Bayesian gauge-based adjustment. *Journal of Hydrology* **531(2)**: 408–426, 10.1016/j.jhydrol.2015.05.049.

Willems P and Einfalt T (2021) Sensors for rain measurements. In *Metrology in Urban Drainage and Stormwater Management: Plug and Pray* (Bertrand-Krajewski J-L *et al.* (eds)). IWA Publishing, London, UK, 10.2166/9781789060119.

Willems P, Olsson J, Arnbjerg-Nielsen K *et al.* (2012) *Impacts of Climate Change on Rainfall Extremes and Urban Drainage Systems.* IWA Publishing, London, UK. See http://library.oapen.org/handle/20.500.12657/25812.

WRc (Water Research Centre) (1987) *Guide to Short Term Flow Surveys of Sewer Systems.* WRc, Swindon, UK. See https://www.thenbs.com/PublicationIndex/documents/details?Pub=WRc&DocID=253789 (accessed 30/07/2023).

Wright DB, Smith JA and Baeck ML (2014) Critical examination of area reduction factors. *Journal of Hydrologic Engineering* **19(4)**: 769–776, 10.1061/(ASCE)HE.1943-5584.0000855.

Ashley R, Smith B, Shaffer P and Caffoor I
ISBN 978-0-7277-4178-3
https://doi.org/10.1680/icehudp.41783.161

Chapter 7
Urban hydrology

Andrew Bailey, Richard Ashley, Adrian Saul, John Blanksby and Jamie Margetts

7.1. Introduction

Chapter 3 introduced the rainfall–runoff processes and here only the catchment-based processes are considered further. The emphasis here is on analysis and modelling of relevance to urban drainage systems (see also Chapter 11), as this is the most helpful way to understand the sources, performance, design and management of urban drainage systems for practitioners.

Hydrological processes generate the flows that discharge to drainage systems and are the backbone of models used in design and operation. Despite their importance, these processes are not fully understood, especially the details of runoff mechanisms from permeable surfaces and infiltration. Much of the research and development work of the past few decades has been aimed at understanding the processes well enough that these can be included more precisely in computational models for urban drainage (e.g. McCulloch, 2022). The uncertainties relate to the process by which rainfall is transformed into either moving surface water or soil moisture. In a practical sense, the layout of the surfaces – the complex pattern of urban permeable and impermeable areas, flow pathways and formal and informal drainage systems – makes precise understanding difficult. Even in a densely built urban catchment, MacDonald *et al.* (2022) found no clear relationship between rainfall, runoff and antecedent soil moisture conditions. Evaporation and evapotranspiration of surface moisture further complicates the mechanics of the behaviour of water in the hydrological cycle.

Hydrological accounting needs to be able to apportion the splits between the various processes that rainfall is divided into; an accounting procedure termed a 'water balance' (e.g. EC, 2015). This chapter is not a treatise on hydrology as such, instead concentrating on the aspects important for urban drainage systems and their analysis. More details of rural and urban hydrology can be found in other publications (e.g. Mansell, 2003).

7.2. Urban hydrological processes

Not all the rainfall falling on the surface is translated into runoff – that is, is 'effective rainfall'. The other pathways are termed 'losses' for urban drainage analytical purposes. Figure 7.1 illustrates the hydrological cycle and aspects of the water balance globally (Mansell, 2003).

There are numerous processes important for determining the amounts and rates of runoff and infiltration from urban surfaces. In the simplest form these relate to the following.

Changing relatively slowly

- the relative topography
- vegetative type and cover

Figure 7.1 The hydrological cycle and water balance (adapted from Mansell, 2003)

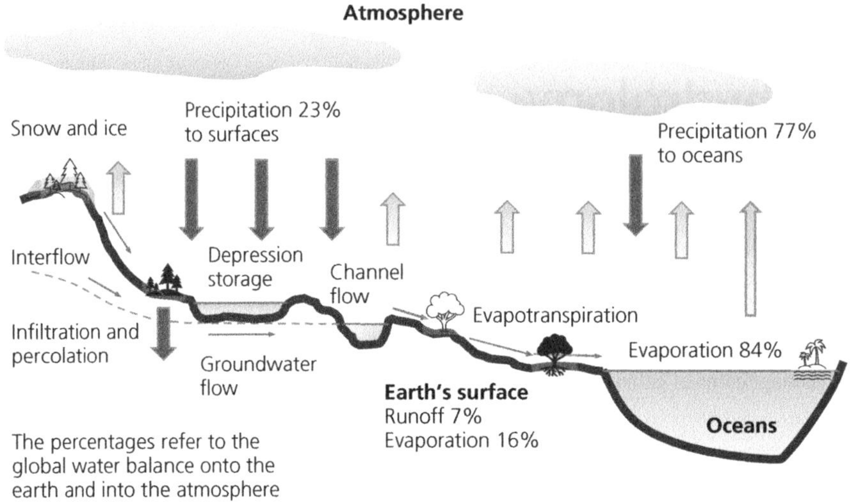

- the degree of impermeability of the catchment: that is, the proportion that is paved (or sealed)
- the nature of the sealed surfaces: smooth, rough, pitted.

Where change is frequent and different for every rainfall event

- 'antecedent conditions' and 'catchment wetness': how wet the catchment is at the start of a storm and how this changes during an event
- soil moisture and what capacity there is for the soil to absorb rainfall.

Rainfall–runoff processes comprise 'losses' from interception, detention, infiltration, evaporation and evapotranspiration. Of these, the infiltration process may result in a significant portion of the rainfall losses, where rainfall percolates into the ground and enters various soil and groundwater stores. Subsequently, this flow may be moved through the ground as part of groundwater transfer processes and may reappear elsewhere in the catchment, and it can sometimes infiltrate into drainage systems. These processes are important for two main reasons. First, they influence the degree of saturation of the ground, and this results in a change in the antecedent wetness condition, thereby affecting subsequent runoff generation. Second, due to cracks, fractures and joints in pipes, flow may enter the below-ground drainage system in the form of an infiltration inflow (either as constant baseflow controlled by the water table or as a flow that varies in response to rainfall). This infiltration inflow is usually termed rainfall-induced infiltration and generally the time of response is slower than the more immediate runoff from impervious surfaces (see Box 7.1 in Section 7.2.3). These two components of infiltration are difficult to define individually and quantify (e.g. Vellacott and Gorst, 2014).

Significant advances in understanding and the analysis of hydrological processes in relation to flooding and drainage systems have occurred in the past 50 years. Many of these advances are attributable to the collection and collation of significant amounts of data. In countries such as the UK, the wealth of data collected over a long period of time enabled standardised approaches to be developed. The *Flood Studies*

Report in the UK, published in 1975, provided a breakthrough in estimating flows from catchments and included procedures for estimating these for urbanised catchments in a supplementary report (NERC, 1975). Since then, the *Flood Estimation Handbook* (FEH) has largely replaced the original methods (CEH, 2023a), together with bespoke computer software. Much of the analytical procedures are empirical and derived from statistical analysis of data records, resulting in 'lumped models' of individual or groups of processes whereby several phenomena are grouped together in a single or few empirical transfer functions. This means the input, which may be flow for example, is assumed to be modified by the group of processes using one or more calibration functions to estimate the output.

The changing climate is also modifying hydrological processes, not least because warmer weather, often accompanied by longer periods between rainstorms, can lead to dryer ground conditions, allowing more rapid infiltration to occur. In larger rural catchments this is expected to lead to less flood volume and lower peak flows, but in urban areas the increased rainfall intensity will lead to greater runoff volumes and higher flood peaks (Sharma *et al.*, 2021). Predicting what might happen in the future is difficult and Sharma *et al.* set out how future rainfall and hydrology might be considered using alternative approaches based on, for example, surface air temperature or moisture levels. As yet there is no consensus on how best to estimate what the hydrology might be in the future or the best approach to analysis for flood estimation or control.

Although many aspects of urban hydrology are grounded in scientific process understanding, urban hydrology is largely an empirical science. Consequently, it is important, as with rainfall estimation (Chapter 6), to understand and access the local hydrological information whenever a study is underway. For example, in the USA, the Flood Frequency Reports (USGS (2023)) are used for flood frequency analysis, and information provided by the US Environmental Protection Agency (USEPA (2023a) is used in the management of flood risk through green infrastructure.

7.2.1 Rainfall/runoff losses

In what follows, the processes whereby rainfall results in runoff are outlined with a focus on urban drainage analysis and modelling (e.g. Fletcher *et al.*, 2013). More detailed treatises on urban hydrology, hydrological theory and analytical processes may be found elsewhere – for example, Mansell (2003). As this is a rapidly developing field of endeavour it is important that urban drainage practitioners access the most up-to-date understanding and analytical processes (by way of, e.g., CEH, 2023b; USEPA, 2023b).

Unsurprisingly, much of the effort to understand the rainfall–runoff processes has been directed to the longer duration events and river flood risk management. There are significant differences in the processes of importance to urban drainage compared with those relevant to river flooding. Many of the greatest impacting rainfall events for urban drainage systems are of much shorter duration than those that cause the greatest flooding from rivers (Chapter 6) and occur on a much smaller scale. Only in the past few decades have shorter duration events and consequential flood risks in urban areas gained appropriate attention (e.g. NIC, 2022). This is together with the other impacts of urban drainage systems on, for example, watercourse pollution and the use for irrigation (Chapter 3, Chapter 8 and Fletcher *et al.*, 2013). With the growing use of nature-based systems (NbSs) and attempts to move towards integrated urban drainage and water management (Chapter 12), the relative importance of getting the hydrology right in any analysis is even higher than it used to be (e.g. Rentachintala *et al.*, 2022). The changes due to climate, behaviours in urban areas, lifestyles and the propensity for urban living all emphasise this importance as there are now more urban areas that have formal and informal sealed surfaces (e.g. Alamdari and Hogue, 2022).

7.2.1.1 Topography

The rainfall converted to runoff is generally subject to some forms of loss. The Wallingford Procedure for Europe (Kellagher, 2001) defines these as initial losses, from the quantity of rainfall required to wet surfaces and just to cause overland runoff, and continuing losses, which occur throughout a rainfall event. A classification system (Terry and Margetts, 2004) subdivides storage losses into three components

- micro-topography: storage due to surface roughness
- meso-topography: storage due to puddles or surface ponding
- macro-topography: storage due to larger scale surface topography and percentage connectivity to the drainage system, permeability, gully performance, etc.

The degree of initial loss will depend on the characteristics of the surface (surface micro-topography), the catchment slope and the antecedent conditions. Continuing losses can then occur through the remainder of the event, through processes such as evaporation, interception or evapotranspiration from vegetation, or infiltration into the ground through permeable surfaces (or cracks or joints in impermeable surfaces). This is demonstrated in Figure 7.2 using blocks of rainfall/runoff.

Figure 7.2 Illustration of the variability in rainfall losses for a runoff event (adapted from Kellagher, 2001)

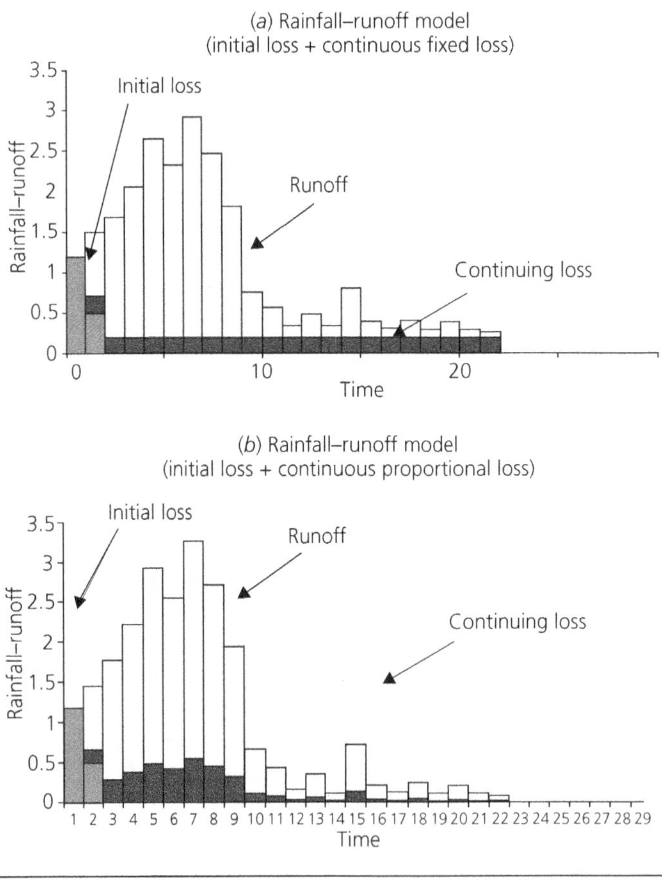

In the simplest form, initial losses and continuing losses can be specified in an analysis by setting an absolute value (i.e. mm) to be assumed lost at the start of the rainfall event, or a percentage to be lost (or added) as the rainfall event continues (continuous loss throughout the event). Continuous losses are represented in two different ways, either as a fixed constant loss over the duration of the event, or as a variable loss, whereby the percentage runoff changes in response to variations in the catchment wetness throughout a storm. The way in which these losses are described is based on the type of analysis being used and is discussed more fully in Section 7.2.3 for the various runoff models in use.

7.2.1.2 Wetting and interception (micro-topography)

At the start of the storm event some rainfall is initially lost as the catchment first becomes wet. The magnitude of the loss is a function of the catchment surface cover, and it is usually small and typically taken as *c.* 0.2–0.5 mm of rainfall for analysis of runoff for temperate climates.

7.2.1.3 Depression storage (meso-topography)

Depression storage is the filling of the voids on the catchment surface. Hence the accurate representation of depression storage requires that the micro- and meso-topography (cracks, roughness, puddles, depressions) are adequately described and considered within the analysis. The depression storage can be assumed as a fixed value for the particular surface being considered. For example, Terry and Margetts (2004), in a UK case, used fixed percentage runoff models with initial loss values ranging from 0.2 mm (accounting for wetting and interception of rainfall for the bulk of an impermeable surface) to 12 mm (for a second surface) in order to calibrate a delayed impermeable paved surface response from an old industrial development site. Different types of approach are in use now as computational models have been further developed, as explained in Section 7.2.3.

7.2.1.4 Evaporation and evapotranspiration

Evaporation and transpiration loss can be considered as either a constant loss that occurs continuously (i.e. mm/day) to a rainfall record, or as a potential evaporation rate that may be obtained directly from meteorological records (e.g. Blyth *et al.*, 2019). There is significant uncertainty with this type of application, as evaporation differs over time and between locations based on vegetation cover, soil type factors, shade temperature, relative humidity and wind speed. Consequently, the loss due to evapotranspiration is often assumed as a constant value in any analysis, *c.* 0.3 mm/h, and is taken as a continuous loss over the duration of the storm event (e.g. Rocha *et al.*, 2022).

The influence of evapotranspiration in the intervening period between events is also important because it has a major influence on the drying of the catchment surface and hence of the antecedent conditions on the surface prior to the next event. Historically, evaporative processes were often deemed to be insignificant, such as in the original UK Wallingford Procedure (item 1 in Box 7.1, Section 7.2.3; DoE and HR Wallingford, 1983) due to the events of interest being considered at the time as short-duration events only (Chapter 6). However, more recent analysis using time-series rainfall and infiltration modelling have demonstrated the importance of ensuring that evaporation is considered in more detail. Evaporation is now considered an important variable and further research is required to improve understanding of how evaporation should be measured and modelled in urban catchments. With the changing climate worldwide, it would be expected that evaporation will increase, and evaporative processes will play a much greater part in the hydrological processes important for urban drainage.

Evaporation and evapotranspiration are even more important now than in the past. The increasing use of NbSs, which include both green and blue infrastructure (Chapter 8), is resulting in more drainage infrastructure on surfaces exposed to sunlight and heat and consequentially greater rates of evaporation and transpiration (e.g. Wright *et al.*, 2021). This will be amplified due to the pace of climate change, and there will also be consequences for and due to possible vegetation change during the lifetime of any drainage infrastructure.

7.2.1.5 Infiltration

Infiltration into the ground mainly occurs through the pervious surfaces of the catchment. The representation of this process in runoff analysis varies significantly depending on the runoff model being used. The percentage permeability of a catchment should be considered in detail to ensure that an appropriate model is selected (e.g. Piadeh *et al.*, 2022; USEPA, 2023c). Examples of the approach to and modelling of infiltration losses and flows are given in Section 7.2.3.

7.2.2 Practical application of rural and urban hydrology for urban drainage

There are numerous applications and bespoke derived models for estimating the various drainage processes of relevance to urban drainage, as illustrated in Figure 7.1. Many of these have been incorporated into computational software packages. Computer models are available in many cases at no cost, or as commercial packages; two examples are presented in Table 7.1. Many are nation specific and derived using parochial data, albeit these may also be applied elsewhere when appropriately modified. It is important for the urban drainage professional to understand and utilise the models and associated

Table 7.1 Two examples of hydrological models used for apportioning flows and volumes for various drainage systems processes

Model	Function	Availability
SWAT+	A small to river catchment scale model that simulates the quality and quantity of surface and ground water, accounting for the environmental impact of land use, land management practices and climate change. Used internationally in assessing soil erosion prevention and control, non-point source pollution control and catchment management.	The US Soil and Water Assessment Tool Plus (SWAT+) is a public domain model jointly developed by the USDA US Department of Agriculture's Agricultural Research Service and Texas A&M AgriLife Research, part of the Texas A&M University System. https://swatplus.gitbook.io/docs/. See also Francesconi *et al.* (2016) which includes applications to ecosystems services.
ReFH2	An event-based rainfall runoff method for estimating hydrographs for floods of a specified frequency. Used to model urbanised catchments as both a rural area and an urban area. The urban area is modelled as an impervious extent and a pervious (green space) extent. The direct runoff from both the impervious and the pervious extents is routed through an enhanced unit hydrograph model.	Based on the original *Flood Studies Report* and Kjeldsen *et al.* (2013) for mixed urban areas, the revitalised flood hydrograph (ReFH) model uses 'lumped' hydrological models; developed by the UK Centre for Ecology & Hydrology (CEH) and Wallingford HydroSolutions Ltd. Theory and equations are public domain and the procedure has been implemented into many small catchments and urban drainage network models. https://refhdocs.hydrosolutions.co.uk/The-ReFH2-Model/Urban-Model/

computational packages best suited to the application context. These are developing continually, and recourse should be made to local and key sources, ensuring provenance is secure and reliable; for example, CIWEM (2017) and CIWEM (2021a) for UK applications. Allitt (2002) reviews the then current hydrology models in use, and CIWEM (2021b) outlines how to select the most appropriate software for urban drainage models.

Where there are mixed-surface catchments comprising large rural areas, large rivers and interacting sewer systems, including combined sewer overflows (CSOs), then combinations of models are required in an integrated modelling approach (Chapter 11). Box 7.1 summarises the main hydrological models that have been and are currently being used in UK urban applications (adapted from Foster and Allitt, 2018). A case study included in Foster and Allitt (2018) illustrates the complexities of selecting the appropriate hydrological models for the relevant parts of the catchment and integrated study.

In the model development process, it has been traditionally presumed that 'standard' computational models required only verification, a process whereby limited data are collected and used to compare with the constructed model's predictions. Adjustments to the model were made to ensure more accurate representation of the observations only by checking the basic data used in the model. Calibration is a different process and is where some of the data are used to develop the model through a calibration of the parameters, especially hydrological, and additional data are then used to confirm that the calibrated model provides a robust prediction. Further discussion of verification and calibration of models is provided in Chapter 11.

7.2.3 Runoff volume and continuing losses

As highlighted earlier, following the removal of initial losses, there are several approaches being used to determine the proportional conversion of rainfall to runoff. These account for continuing losses during the rainfall event, through either a fixed or a variable percentage reduction, with most approaches considering the impact of antecedent conditions on the percentage of the rainfall that runs off. The methods available, including those in Box 7.1, are listed in CIWEM (2021a), together with recommendations for usage in relation to analysis and modelling, with CIWEM (2017) providing guidance on the latter for UK applications. The continuing discussion here is generic and relates to various approaches, rather than a specific methodology, although UK practice is highlighted.

7.2.3.1 Simple fixed percentage runoff (item 2 in Box 7.1)

In this, a fixed percentage of the net rainfall is used, which then becomes runoff following the consideration of any initial losses through depression storage (e.g. Innovyze, 2023a; see also Chapter 11). This value is specified in the analysis and does not take account of antecedent conditions. The value specified, often a percentage, is fixed for the duration of the event, representing continuous losses. This model is simple to use, but it is less precise for analysing runoff from pervious surfaces where there is a variation in response during a rainfall event due to vegetation or slope (e.g. Tabacchi *et al.*, 2000), or where antecedent conditions have a significant impact on the amount of runoff produced for any given event.

The assumption of a fixed percentage of runoff is most applicable for impermeable surfaces where there is a known percentage runoff (PR), which is unlikely to vary significantly during the event, such as high-quality roads and roofs, generating close to 100% runoff. It is less applicable for poor-quality paved areas, with lower PR due to cracks, joints and large depressions or ponds, as these may produce varying degrees of runoff through increasing activation and connectivity under extreme event conditions (Terry and Margetts, 2004). Analysis of runoff from surfaces in poor condition, with low PR from low return

Box 7.1

The main hydrological models in use for urban areas in the UK (adapted from Foster and Allitt, 2018)

1. **The 'Wallingford Procedure'** provides an average percentage runoff (PR), initially for the catchment and subsequently for individual contributing areas. The 'PR Equation' has variables for soil type, annual rainfall and percentage impermeability. At very low impermeability this gave problems which led to the '10 metre rule' (see Section 7.2.3.2). The PR remains constant throughout the storm event and is identical for all storm events.
2. **Fixed runoff** methodology has a fixed PR from each type of surface, including permeable surfaces. The PR from each surface can be set differently; paved and roof areas are frequently set at between 85% and 100% and permeable surfaces between 0% and 35%. PR remains constant throughout the storm event and is identical for all storm events.
3. **New UK runoff** methodology has been the most widely used in the UK. It has a fixed PR from impervious surfaces, and a variable PR from permeable surfaces. A small number of variables can be used to adjust the runoff from the permeable surfaces, both in terms of magnitude and duration. As permeable runoff can vary between storms, 'model verification' has been replaced by 'model calibration'. Calibration of the model for long-duration time series of rainfall ensures robustness for a range of events. However, when used with synthetic design storms, exceptionally large and false flooding volumes can be predicted.
4. **Ground infiltration module**: Rainfall-induced infiltration (RII) can either be caused by very slow surface runoff or by means of shallow groundwater responding to rainfall and infiltrating into the sewer. Some sewers return rapidly to baseflow while others can take up to a week. Modelling RII is problematic even when using the ground infiltration module (GIM), which is a 'black box' process with multiple variables.
5. **UKWIR runoff methodology** was developed to overcome the problems associated with the 'New UK' method when simulating with synthetic design storms. This method uses more variables than the New UK method and uses an alternative soil classification.
6. **ReFH/ReFH2** runoff methodology is used for large permeable areas either within or immediately adjacent to urban areas. The method is suitable for 'small' areas (<0.5 km^2) draining to small watercourses. It is not a suitable method for large watercourses.

period events, can result in significant uncertainty for representing what happens during more extreme events, due to the likely changes in PR when all surfaces are potentially saturated to a higher depth – that is, the cracks and fissures can become effectively submerged.

Standard runoff coefficients for different surface types are presented within the various industry standards (e.g. CIWEM, 2021a) or specified by drainage authorities. Typically, these values could range from 20% for large open areas with limited connectivity to local drainage systems, to 40–60% for poor-quality surfaces (e.g. paved areas in old industrial sites that have been subject to differential settlement), and even up to almost 100% for high-quality paved surfaces drained by a high density of gullies.

7.2.3.2 The UK runoff models (items 3, 5 and 6 in Box 7.1)

There have been continuous developments in the understanding of and the best way to represent runoff in analysis and modelling in the UK, summarised by Kellagher (2014a). As defined in various chapters of this book, the Wallingford Procedure (DoE and HR Wallingford, 1983) has been referenced regarding the management of urban runoff. This was the standard runoff model in UK urban drainage for much of the past 30 years, subsequently being replaced in the UK by fixed PR models, the UK Wallingford Procedure (new/variable) PR model, the Wallingford Ground Infiltration Model and, more recently, the UK Water Industry Research (UKWIR) runoff equation (Kellagher, 2014b). There are also alternative models in use and CIWEM (2021a) lists these.

The development of the earlier Wallingford Procedure models was based on in-sewer measurements of rainfall responses for 510 storm events across 17 trial catchments in the UK (Allitt, 2002). Considerable variability was observed across the sites in terms of their characteristics (area, percentage impermeability and PR produced), highlighting the importance of the many different factors contributing to runoff generation.

The model had limitations in that for very low values of the main parameters, negative values of PR could be predicted, indicating that there is no runoff but, in many cases, runoff had been observed to occur. Such negative values caused problems with the application of the model, especially for semirural areas or partially separate areas, where sub-catchments containing large amounts of permeable area have been predicted to produce little runoff, despite an obvious impermeable contribution. This is not surprising given that many of the catchments used to develop the model originally had more than 40% impermeability.

Titterington *et al.* (2014) introduced the UKWIR runoff model, developed to rectify some of the problems in the earlier relationships. Significant in these was low predicted runoff and underprediction of the true runoff, which led to the practice of artificially removing permeable areas from models through the '10 metre rule' (Chapman, 2009; WaPUG Committee, 2009). This led to models not representing the connected area 'on the ground' and was considered to be a 'fix'. However, Allitt (2002) notes that the '10 metre rule' had then fallen into disuse with applications of many commercial models, resulting in the runoff model often not being applied correctly. As with all modelling using 'opaque' and attractive graphical user interfaces, users can erroneously assume that detailed understanding of the science and its application in the model is not necessary, only how to 'feed in the numbers'. This 'blind' use of standard model packages is further discussed in Chapter 11.

The new runoff model was developed in 1987 by the (then) Institute of Hydrology and is described in detail by Osborne (2009). The model addresses several of the limitations of the original Wallingford (old/fixed) PR model and, specifically, the ability to allow the calculation of a variable PR for permeable surfaces. This represents the changing wetness of permeable surfaces during longer duration events and is more applicable to catchments with low impermeability or low runoff from soil surfaces as the model treats impermeable and permeable surfaces separately.

Urban drainage analysts were initially slow to utilise this 'new' model due to a lack of experience and uncertainty as to how to derive design values for antecedent catchment wetness (Osborne, 2000). Nevertheless, this model became the preferred approach in the UK as it better represented the delayed responses observed in many catchments associated with rainfall responsive runoff from permeable surfaces. Furthermore, the ability to represent continuously varying catchment wetness makes it more applicable to time series or multipeaked events where increased runoff is observed on the second or third rainfall peaks, as illustrated in Figure 7.3, in which the model results for the second flow peak are inaccurate (Allitt, 2002).

Figure 7.3 Model verification (observed against modelled results) for flows (m^3/s) in a trunk sewer where the variable PR model has not been applied (reproduced from Allitt, 2002 with permission of CIWEM)

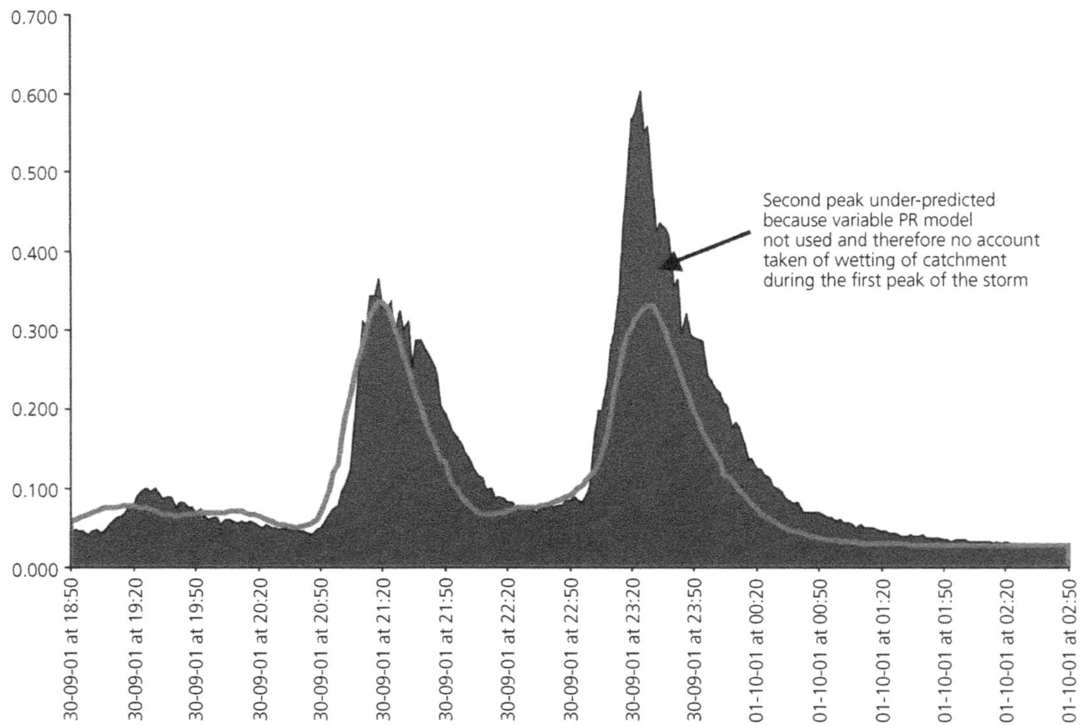

Due to recent advances in long-term monitoring, ongoing model calibration and real-time forecasting of sewer performance, the ability to use models continuously to simulate the network performance for long time series of rainfall data is important. Due to the ability to vary catchment wetness, the new PR model made this possible (Osborne, 2009; Titterington, 2014). While this enhanced model allowed responses from permeable surfaces to be better represented, it had drawbacks. Due to all the permeable area being included within the model, it was found that excessive runoff from permeable surfaces could be generated under extreme event conditions, especially where verification of the model had been made only for low return period events. Squibbs and Jack (2003) gave examples demonstrating the problems from extrapolating calibrated parameters to extreme event conditions. This is because the model allowed catchment wetness to increase, so that an unrealistic 100% PR could be generated where there are large permeable areas.

Part of the problem was reliance on verifying computational models against short-term flow survey data. This resulted in calibration of models and the flows from permeable surfaces to represent the conditions for a particular season, which are not necessarily representative of what is likely to be observed in other seasons due to differences in the prevailing hydrological processes (Terry and Margetts, 2004). Ideally, verification of runoff responses, particularly from permeable surfaces, should be undertaken using long-term flow survey data covering a range of seasons and antecedent conditions. However, budget and programme pressures often override other considerations, preventing such detailed investigations. As most modelling is to investigate high-flow conditions, and observed high return period rainfall event data are rarely available, it is imperative to assess the robustness of the model predictions of runoff in extreme events (Chapter 11). This highlights the importance of ensuring that all urban drainage analysts understand the limitations of the runoff model and the data used in the verification of an urban drainage model.

7.2.3.3 The UKWIR runoff model (item 5 in Box 7.1)

As outlined above, there were significant limitations to the runoff models being used for UK practice up until the mid-2010s. This was partly due to lack of scientific knowledge and a lack of suitable recorded data to create any alternative formulations. These limitations led to the emergence of ReFH2 (Table 7.1) and a water industry initiative to tackle the problems of the 'new PR/UK' model particularly related to runoff from mixed permeable and impervious areas and the need for accurate models for continuous flow simulation. The UKWIR (UK Water Industry Research) runoff model is illustrated in Figure 7.4 and is the current standard being used in UK practice for urban runoff (Kellagher, 2014a; Titterington *et al.*, 2014), although the new PR equation (item 3 in Box 7.1) is still in use as an alternative. ReFH2 (item 6 in Box 7.1) is being used where there are watercourses in catchments, complementary to the UKWIR runoff model for the drainage systems, although the former might not be as reliable for continuous simulation (e.g. Siniscalchi *et al.*, 2019).

7.2.4 Runoff routing models

Following application of the losses, the runoff flow remaining is routed to the drainage system, representing flows over the surface and within very localised, unmodelled, drainage systems. This includes a time delay and attenuation prior to entering the formal drainage system. Routing models and coefficients are included within the runoff models to represent this variability. These generally take the form of simple representative 'reservoir-based tank' models, rather than complex kinematic wave solutions, due to the computational demands required by larger-scale models.

Figure 7.4 Diagrammatic illustration of the UKWIR runoff model (Titterington *et al.*, 2014)

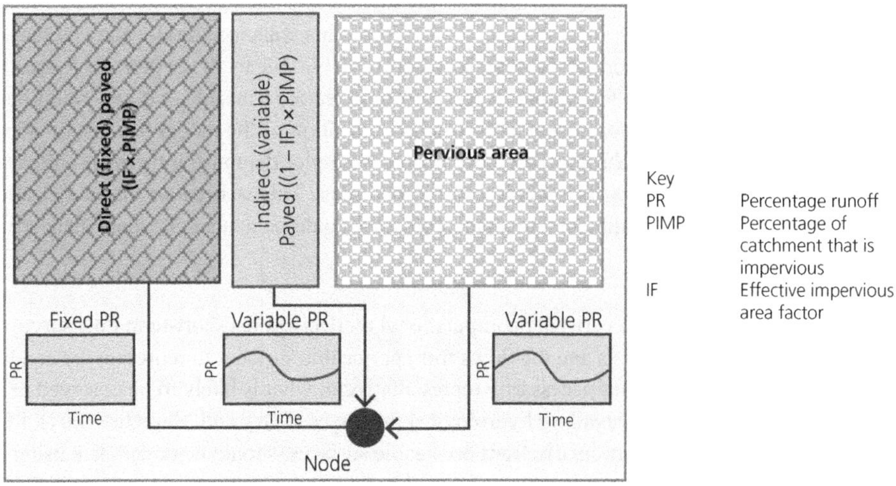

There are numerous ways in which flows on surfaces are included in models, a process termed 'routing'. There are simple models and more complex models (e.g. Szelag *et al.*, 2022). When the aim was to obtain adequate representation of surface flow movements into buried piped systems, it was often sufficient to simplify these processes, assuming a uniformity of flow from individual connected (mainly impermeable) surfaces, making assumptions about large-scale aggregations. Changes in the objectives for developing models have meant that representation of surface flow routing needs to be far more detailed and reliable at small scale. This was initially due to the need to model localised surface flooding even at low-flow depths, and now to be able effectively to represent the routing and detention of flow around and through NbSs.

The original Wallingford Procedure (item 1 in Box 7.1) suggested a non-linear reservoir routing model, in which attenuation relationships were developed through statistical analysis of a number of catchments across Europe. Nine standard unit hydrographs were implemented based on different slopes and paved area per road gulley; however, these were deemed to be too inflexible for the many varied types of situations that were encountered in initial modelling studies. In response to this, a double linear reservoir routing model was developed, in which a routing coefficient was developed from measured data from 20 small catchments. This linked the routing and attenuation of runoff to rainfall intensity, contributing area and surface slope.

Due to the empirical nature of this routing model, there were limits to its application. The minimum slope was 0.002 m/m, and area that could be modelled ranged from 1000 m^2 to 10 000 m^2. Due to these contributing area-based limitations, large catchment routing models have since been developed that better represent large drainage catchments, up to 10 ha, with significant areas of unmodelled sewer.

More recently, with the advent of accurate digital mapping of terrain, it has been possible to develop more sophisticated models that can better determine flow pathways even in areas where there are small differences in surface level (e.g. Néelz and Pender, 2009). The outcome from this is that fully contiguous modelling of flows on surfaces and the interactions with the below- and above-ground systems is becoming feasible. More details are given in Chapter 11.

7.2.5 Ground infiltration models

The infiltration of groundwater into sewer networks can be a significant issue in many catchments and is typically identified where long drain-down times are observed in flow monitoring sites following a significant rainfall event. These flows may not all be as the result of groundwater infiltration and may also be due to unaccounted for land drainage, waterways directed into sewer systems, attenuation tanks and sustainable drainage systems (SuDS). Interpreting flow data to determine the sources is extremely difficult and costly. Such flows cannot be replicated using the standard surface runoff models identified and discussed in the previous sections. Instead, specific infiltration models are generally applied in tandem with a surface runoff model in order to represent such flows. For the UK, this is usually using the 'ground infiltration module' (GIM; item 4 in Box 7.1) within the InfoWorks ICM commercial modelling platform (Innovyze, 2023b). Although this is also being used in other countries (e.g. the USA: City of Columbia, 2020). Alternatively the Green–Ampt (Green and Ampt, 1911) or SCS (Rallison and Miller, 1981) models are predominantly utilised in the public domain Storm Water Management Model (SWMM) modelling package (USEPA, 2023d).

As an illustration of the significance of this issue, Figure 7.5 shows flows modelled for a rural catchment using the GIM. It shows that more than half of the peak flows during a 30-year design storm event are estimated to be due to groundwater infiltration. Figure 7.6 shows the implications from this for CSO spills above 50 m^3 using a 7-week time series of rainfall and for four catchments. Groundwater infiltration is predicted to increase the spills dramatically, hence getting the representation correct when designing control measures is imperative. Using the GIM enables calibration that includes soil and groundwater storage components, for which the parameters can be adjusted to match the observed hydrographs.

Caution must be taken when using the GIM as the significant range of model parameters can lead to an over-parameterised fit against a single event that does not properly represent other observed events. Common practice is to review the full period of data from the flow survey being used for verification, in order to give a 'best overall fit' to the observed data. This is a limitation as the model is not physically based.

Figure 7.5 Modelled results for a rural catchment where half of the predicted flows are due to groundwater infiltration (adapted from Vellacott and Gorst, 2014)

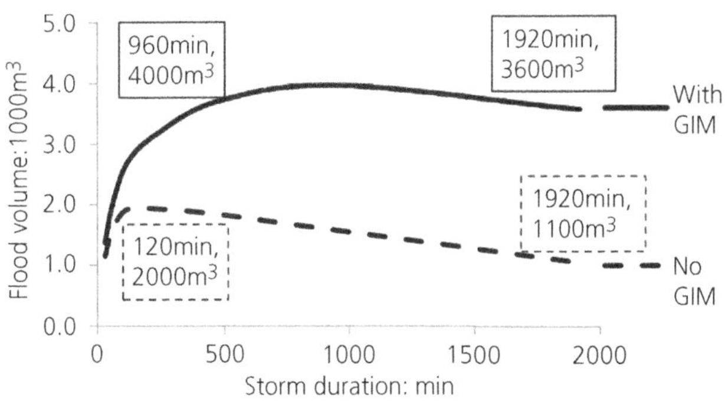

Figure 7.6 Implications of groundwater infiltration leading to additional combined sewer overflow (CSO) spills (adapted from Vellacott and Gorst, 2014)

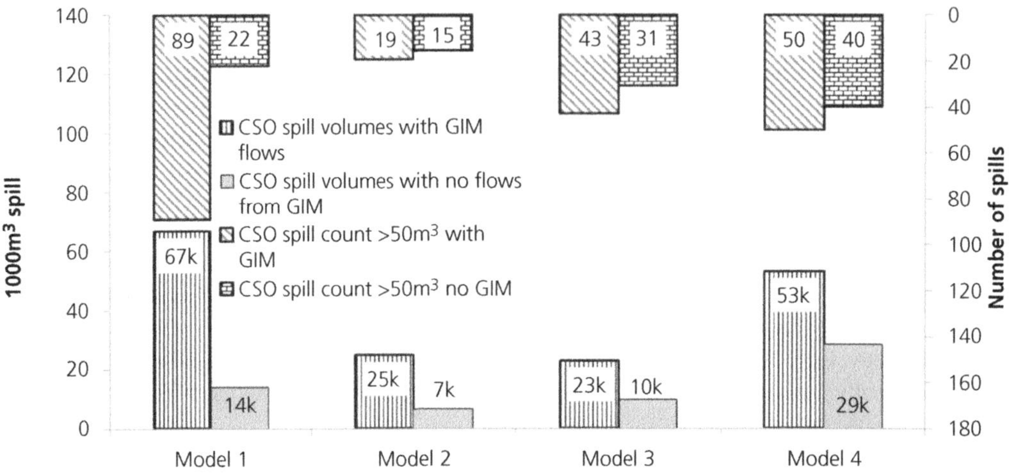

Alternative approaches include the use of the historical Horton infiltration model (Horton, 1940). This has been used successfully to verify a model for two-dimensional (2D) pluvial flood flows from a semirural catchment (e.g. Hall *et al.*, 2015). However, the application of alternative infiltration models is limited in the literature. This is mainly due to a lack of guidelines for the application of these models to the verification of urban drainage models, as well as the limitations of their predictions when extrapolated to high return period storms for design not being understood. Therefore, the application of any of these infiltration models as part of a model verification should be undertaken with caution. This is particularly important when the model is to be used for assessment of volumes, such as in the case of CSO spill durations or in the design of a storage tank. In these cases, over-prediction of infiltration flows due to the limitations of the model could result in significant over-design of new assets.

The increasing use of one-dimensional/2D models for the interactive processes between surface and below-ground drainage systems has highlighted the need to ensure that all the hydrological processes are properly included, and for better models to be developed. Martínez *et al.* (2021) pointed out the deficiencies in many model studies in not properly accounting for these processes, when developing coupled modelling of rainfall–runoff infiltration and sewers. A novel infiltration module algorithm based on the Green–Ampt method (Green and Ampt, 1911) was used together with SWMM to improve simulation of historical flooding in Greenfield, Glasgow, Scotland. Use of the Green–Ampt method in the 2D model had a direct impact on the overland flood depths and the effects of the soil infiltration capacity on flows.

7.3. The need for better runoff models

Given the scale of the climate crisis, hydrology is at the centre of any attempts to understand the consequences for future impacts, especially for urban areas. The rapid pace of climate changes is such that any attempt to design, by way of modelling, urban drainage systems that can deliver the required service throughout their lifetime of use must consider the very high uncertainties in the understanding of the hydrological processes and hence predictions of performance of any assets (e.g. Sharma *et al.*, 2021).

Few, if any, publications attempt to assign any uncertainty to the hydrological process analyses used (see also Chapter 11).

In urban hydrology there has been an increase in the routine use of long-term flow monitoring and real-time forecasting, coupled with a growth in 2D modelling and the use of remote sensing data to build models. For example, Van Zijl *et al.* (2020) used remote sensed data together with 2D modelling to determine the hydrological impacts from development in a suburb of Cosmo City, South Africa, showing the effects of soil sealing to be responsible for increased runoff and reduced evaporative processes, and to determine the source of groundwater flooding. Nevertheless, the relative precision in such analyses is low due to the limited current understanding of the hydrological processes involved, and the paper did not point out the significant uncertainties involved in the study.

Irrespective of the challenges of the changing climate, and despite recent advances as outlined in this chapter, there is still a need for runoff models that give better representation of conditions across all seasons and antecedent wetness conditions, and in ways more suitable for the new generation of models. As most of the models available are based on a mix of physical determinants and parameters used to define uncertain processes, there is a need to ensure that calibration is carried out methodically and in full knowledge of the science and unknown processes. Clear guidance, such as provided by CIWEM (2017, 2021a), is helpful, as are ongoing advancements in knowledge, particularly to do with permeable surface runoff. However, current model calibration is based on a relatively small number of parameters representing a multitude of complex processes.

The increasing use of NbSs and associated multifunctional land use has also increased the need for more accurate representation of the hydrological processes involved, especially how these might vary into the future (Chapter 8). This issue has been exacerbated in recent years by the need to produce integrated models through, for example, the Surface Water Management Plan process in England, to investigate urban flooding from all potential sources. The increasing use of NbSs such as SuDS has led to urban drainage models representing small watercourses or inflows from rural areas, where flows from permeable surfaces may comprise a greater portion of the flow in the system. There have been successful uses of forms of the rational method for localised NbSs, including by Miller *et al.* (2023), who developed an 'adapted nature-based solutions rational method' modified using the FEH index of flood attenuation in an application to a 74 km^2 catchment at Calthorpe Park in England, part of the Birmingham drainage system. The study, looking at distributed NbSs, demonstrated the importance of understanding the effects of spatial distribution on flows and showed that local SuDS measures would be the most effective at attenuating downstream flood risks. Despite this, there is a need for better representation of the hydrology of distributed NbSs where the measures are spread across catchments. Baker *et al.* (2021) show how few published studies are sufficiently rigorous or scientific when considering the effects of tree planting on hydrological processes and consequent runoff. The studies examined rarely tested model results against observations. Given the drive for tree planting across the world, to assist with climate change mitigation and flood risks, there is an urgent need to develop the knowledge in this area to justify the costs and land take.

The drive for efficiencies in asset management has made over-design due to runoff model uncertainty increasingly unacceptable. While there is a move towards a risk-based approach to analysis, including sensitivity testing of uncertain or potentially variable parameters, there is a need for better representations of the hydrological processes in order to increase model accuracy and user confidence.

REFERENCES

Alamdari N and Hogue TS (2022) Assessing the effects of climate change on urban watersheds: a review and call for – future research. *Environmental Reviews* **30(1)**: 61–71, 10.1139/er-2021-0003.

Allitt R (2002) Rainfall, runoff and infiltration re-visited. *Proceedings of the WaPUG Spring Meeting*. 2002, Coventry, UK. See http://ftp2.ciwem.org/2002/2002-05/allitt%20S2002.PDF (accessed 31/07/2023).

Baker HJ, Hutchins MG and Miller JD (2021) How robust is the evidence for beneficial hydrological effects of urban tree planting? *Hydrological Sciences Journal* **66(8)**: 1306–1320, 10.1080/02626667.2021.1922692.

Blyth EM, Martinez-de la Torre A and Robinson EL (2019) Trends in evapotranspiration and its drivers in Great Britain: 1961 to 2015. *Progress in Physical Geography: Earth and Environment* **43(5)**: 666–693, 10.1177/0309133319841891.

CEH (Centre for Ecology & Hydrology) (2023a) Flood Estimation Handbook and Flood Studies Report. https://www.ceh.ac.uk/services/flood-estimation-handbook (accessed 31/07/2023).

CEH (2023b) About the UK Centre for Ecology & Hydrology (UKCEH). https://www.ceh.ac.uk/ (accessed 31/07/2023).

Chapman R (2009) *The Wallingford Procedure Percentage Runoff Equation*, version 3. WaPUG User Note 9. Wastewater Planning Users Group, CIWEM, London, UK. See https://www.ciwem.org/assets/pdf/Special%20Interest%20Groups/Urban%20Drainage%20Group/WAPUG_User_Note_09%20(1).pdf (accessed 31/07/2023).

City of Columbia (2020) Clean Water 2020 Program. *Sewer System Hydraulic Model Report*. August 2020. See https://cleanwater2020.com/docs/default-source/consent-decree/sewer-system-hydraulic-model-report-aug-2020.pdf?sfvrsn=d102416d_4 (accessed 31/07/2023).

CIWEM (Chartered Institution of Water and Environmental Management) (2017) *Code of Practice for the Hydraulic Modelling of Urban Drainage Systems*. Urban Drainage Group, CIWEM, London. UK. See https://www.ciwem.org/assets/pdf/Special%20Interest%20Groups/Urban%20Drainage%20Group/Code%20of%20Practice%20for%20the%20Hydraulic%20Modelling%20of%20Ur.pdf (accessed 31/07/2023).

CIWEM (2021a) *Integrated Urban Drainage Modelling Guide. Appendix H – Hydrology*. Urban Drainage Group, CIWEM, London. UK. See https://www.ciwem.org/assets/uploads/Appendix_H_-_Hydrology.pdf (accessed 31/07/2023).

CIWEM (2021b) *Integrated Urban Drainage Modelling Guide*. Urban Drainage Group, CIWEM, London. See https://www.ciwem.org/assets/uploads/IUD_1.pdf (accessed 31/07/2023).

DoE (Department of the Environment) and HR Wallingford (1983). *Design and Analysis of Urban Storm Drainage: The Wallingford Procedure*. Standing Technical Committee Report No. 28. Department of the Environment, National Water Council, Hydraulics Research Ltd, Wallingford.

EC (European Commission) (2015) *Guidance Document on the Application of Water Balances for Supporting the Implementation of the WFD*. European Commission Technical Report – 2015 – 90. Office for Official Publications of the European Communities, Luxembourg. See https://circabc.europa.eu/sd/a/820ec306-62a7-475c-8a98-699e70734223/Guidance%20No%2034%20-%20Water%20Balances%20Guidance%20%28final%20version%29.pdf (accessed 31/07/2023).

Fletcher TD, Andrieu H and Hamel P (2013) Understanding, management and modelling of urban hydrology and its consequences for receiving waters: a state of the art. *Advances in Water Resources* **51**: 261–279, 10.1016/j.advwatres.2012.09.001.

Foster K and Allitt R (2018) Combining the worlds of rural and urban hydrology. *Proceedings of the CIWEM UDG Conference*, Blackpool, UK. See http://ftp2.ciwem.org/2018/2018-Autumn/Paper%2011%20%20Combining%20the%20worlds%20of%20rural%20and%20urban%20hydrology%20-%20%20Kerry%20Foster%20-%20Richard%20Allitt.pdf (accessed 31/07/2023).

Francesconi W, Srinivasan R, Pérez-Minana E, Willcock SP and Quintesro M (2016) Using the Soil and Water Assessment Tool (SWAT) to model ecosystem services: a systematic review. *Journal of Hydrology* **535**: 625–636, 10.1016/j.jhydrol.2016.01.034.

Green WH and Ampt G (1911) Studies of soil physics, part I – the flow of air and water through soils. *Journal of Agricultural Science* **4**: 1–24, 10.1017/S0021859600001441

Hall A, Bailey A and Burt K (2015) A collaborative approach to solving pluvial and sewer flooding utilising 2D overland flow modelling to develop a sustainable solution. *Proceedings of the CIWEM UDG Spring Conference*. Birmingham, UK. CIWEM. See http://ftp2.ciwem.org/2015/2015-Spring/6%20Paper%207%20A%20Collaborative%20approach%20to%20solving%20pluvial%20floodinghellip;%20Presenter%20Alastair%20Hall,%20Clear%20Environmental%20Consultants.pdf (accessed 31/07/2023).

Horton RE (1940) An approach towards physical interpretation of infiltration capacity. *Proceedings of the Soil Science Society of America* **5**: 399–417, 10.2136/sssaj1941.036159950005000C0075x.

Innovyze (2023a) Storm, Sewer, and Flood. InfoWorks ICM. https://help-innovyze.refined.site/space/XPRESOURCE/14221333/Storm%2C+Sewer%2C+and+Flood (accessed 31/07/2023).

Innovyze (2023b) Ground Infiltration Model. https://help2.innovyze.com/infoworksicm/Content/HTML/ICM_ILCM/Ground_Infiltration_Model.htm (accessed 31/07/2023).

Kellagher R (2001) *The Wallingford Procedure for Europe – Best Practise Guide to Urban Drainage Modelling*. HR Wallingford Ltd, Wallingford, UK. See https://eprints.hrwallingford.com/461/ (accessed 31/07/2023).

Kellagher R (2014a) *Development of the UKWIR Runoff Model: Main Report*. UK Water Industry Research. London, UK. See https://ukwir.org/water-research-reports-publications-viewer/6dcaa6c1-baff-4a6f-a064-5d1b26fb29ec

Kellagher R (2014b) *A Guide to the Use of the UKWIR Runoff Equation*. UK Water Industry Research, London, UK. See https://ukwir.org/water-research-reports-publications-viewer/62a08379-83c8-4121-8aa0-018530e80801 (accessed 31/07/2023).

Kjeldsen TR, Miller JD and Packman JC (2013) Modelling design flood hydrographs in catchments with mixed urban and rural land cover. *Hydrology Research* **44(6)**: 1040–1057, 10.2166/nh.2013.158.

McCulloch C (2022) Dramatic growth of hydrological science in the UK in the 1960s and early 1970s. *Hydrology Research* **53(5)**: 754, 10.2166/nh.2022.005.

MacDonald N, Redfern T, Miller J and Kjeldsen TR (2022) Understanding the impact of the built environment mosaic on rainfall-runoff behaviour. *Journal of Hydrology* **604**: 127147, 10.1016/j.jhydrol.2021.127147.

Mansell MG (2003) *Rural and Urban Hydrology*. Thomas Telford Publishing, London, UK.

Martínez C, Vojinovic Z, Price R and Sanchez A (2021) Modelling infiltration process, overland flow and sewer system interactions for urban flood mitigation. *Water* **13(15)**: 2028, 10.3390/w13152028.

Miller JD, Vesuviano G, Wallbank JR, Fletcher DH and Jones L (2023) Hydrological assessment of urban nature-based solutions for urban planning using Ecosystem Service toolkit applications. *Landscape and Urban Planning* **234**: 104737, 10.1016/j.landurbplan.2023.104737.

Néelz S and Pender G (2009) *Desktop Review of 2D Hydraulic Modelling Packages*. Environment Agency, Bristol, UK.

NERC (Natural Environment Research Council) (1975) *Flood Studies Report*. NERC, London, UK. https://www.ceh.ac.uk/services/flood-estimation-handbook (accessed 31/07/2023).

NIC (National Infrastructure Commission) (2022) *Reducing the Risk of Surface Water Flooding*. NIC, London, UK. See https://nic.org.uk/studies-reports/reducing-the-risks-of-surface-water-flooding/surface-water-flooding-final-report/ (accessed 31/07/2023).

Osborne M (2000) Runoff models – lessons from study audits. In *WaPUG Spring Meeting*, Coventry, UK. See http://www.ciwem.org/knowledge-networks/groups/wapug/events/past-events–presentations.aspx (accessed 31/07/2023).

Osborne M (2009) *A New Runoff Volume Model*, version 3. WaPUG User Note 28. Wastewater Planning Users Group, CIWEM, London, UK. See https://www.ciwem.org/assets/pdf/Special%20Interest%20Groups/Urban%20Drainage%20Group/WAPUG_User_Note_28.pdf (accessed 31/07/2023).

Piadeh F, Behzadian K and Alani AM (2022) A critical review of real-time modelling of flood forecasting in urban drainage systems. *Journal of Hydrology* **607**: 127476, 10.1016/j.jhydrol.2022.127476.
Rallison RE and Miller N (1981) Past, present, and future SCS runoff procedure. In *Proceedings of the International Symposium on Rainfall-Runoff Modeling*. Mississippi, USA.
Rentachintala LGNP, Muni Reddy MG and Mohapatra PK (2022) Urban stormwater management for sustainable and resilient measures and practices: a review. *Water Science & Technology* **85(4)**: 1120–1140, 10.2166/wst.2022.017.
Rocha AD, Vulova S, van de Tol C, Forster M and Kleinschmit B (2022) Modelling hourly evapotranspiration in urban environments with SCOPE using open remote sensing and meteorological data. *Hydrology and Earth System Sciences* **26**: 1111–1129, 10.5194/hess-26-1111-2022.
Sharma A, Hettirachchi S and Wasko C (2021) Estimating design hydrologic extremes in a warming climate: alternatives, uncertainties and the way forward. *Philosophical Transactions of the Royal Society A* **379(2195)**: 20190623, 10.1098/rsta.2019.0623.
Siniscalchi F, Gianniou P and Quail T (2019) Integrated catchment modelling: an application of the ReFH2 runoff model. In *Proceedings of the CIWEM UDG Conference*, Scotland. See http://ftp2.ciwem.org/2019/2019-Scotland/Paper%207%20ICS_An%20application%20of%20ReFH2%20model%20by%20RPS.pdf (accessed 31/07/2023).
Squibbs G and Jack A (2003) The impact of permeable runoff on CSO discharges. In *WaPUG Scottish Meeting 2003*, Dunblane, UK. www.ciwem.org/groups/wapug
Szelag B, Lagód G, Musz-Pomorska A *et al.* (2022) Development of rainfall-runoff models for sustainable stormwater management in urbanized catchments. *Water* **14(13)**: 1997, 10.3390/w14131997.
Tabacchi E, Lambs L, Guilloy H *et al.* (2000) Impacts of riparian vegetation on hydrological processes. *Hydrological Processes* **14(1617)**: 2959–2976, 10.1002/1099-1085(200011/12)14:16/17%3C2959::AID-HYP129%3E3.0.CO;2-B.
Terry D and Margetts J (2004) Will we ever understand and model surface wetting and drying? In *WaPUG Autumn Conference 2004*. Blackpool, UK. See http://www.ciwem.org/knowledge-networks/groups/wapug/events/past-events–presentations.aspx (accessed 31/07/2023).
Titterington J, Wisdish A, Nicholls L and Ravnkilde K (2014) Urban runoff – using the UKWIR runoff equation. In *Proceedings of the CIWEM UDG Autumn Conference*. See http://ftp2.ciwem.org/2014/2014%20Autumn/Paper%207.pdf (accessed 31/07/2023.)
USEPA (US Environmental Protection Agency) (2023a) *Manage Flood Risk*. See https://www.epa.gov/green-infrastructure/manage-flood-risk (accessed 31/07/2023).
USEPA (2023b) Urbanization - Stormwater Runoff. https://www.epa.gov/caddis-vol2/urbanization-stormwater-runoff (accessed 31/07/2023).
USEPA (2023c) Infiltration Models. https://www.epa.gov/water-research/infiltration-models (accessed 31/07/2023).
USEPA (2023d) Storm Water Management Model (SWMM). https://www.epa.gov/water-research/storm-water-management-model-swmm (accessed 31/07/2023).
USGS (US Geological Survey) (2023) Flood Frequency Reports. https://www.usgs.gov/streamstats/science/flood-frequency-reports (accessed 31/07/2023).
Van Zijl G, van Tol J, Bouwer D, Lorentz S and le Roux P (2020) Combining historical remote sensing, digital soil mapping and hydrological modelling to produce solutions for infrastructure damage in Cosmo City, South Africa. *Remote Sensing* **12(3)**: 433, 10.3390/rs12030433.
Vellacott F and Gorst A (2014) Modelling ground infiltration: approaches and impacts. In *Proceedings of the CIWEM UDG Conference*, Spring. See http://ftp2.ciwem.org/2014/2014-Spring/UDG9_%20Modelling%20Ground%20Infiltration_Presentation%20by%20F%20Vellacott.pdf (accessed 31/07/2023).

WaPUG Committee (2009) *Runoff Equations and Catchment Data*, version 2. WaPUG User Note 21. Wastewater Planning Users Group, CIWEM, London, UK. See https://www.ciwem.org/assets/pdf/Special%20Interest%20Groups/Urban%20Drainage%20Group/WAPUG_User_Note_21.pdf (accessed 31/07/2023).

Wright MSP, Santelmann MV, Vaché DB and Hulse DW (2021) Modeling the impact of development policies and climate on suburban watershed hydrology near Portland, Oregon. *Landscape and Urban Planning* **214**: 104133, 10.1016/j.landurbplan.2021.104133.

Ashley R, Smith B, Shaffer P and Caffoor I
ISBN 978-0-7277-4178-3
https://doi.org/10.1680/icehudp.41783.181

Chapter 8
Managing surface water using natural systems

Bridget Woods-Ballard, Steve Wilson and Richard Ashley

8.1. Introduction to a more sustainable approach to urban drainage

The development of understanding regarding the performance and needs for surface water drainage over the past few decades has been outlined in Chapter 3. The development of ideas for this is illustrated in Figure 8.1. From managing the problems to exploiting the opportunities, surface water drainage has developed into an area where it is no longer 'drainage' in terms of the original definition of getting water away rapidly and safely. The value of urban runoff to provide outcomes supporting adaptation to climate change and contributing to place-making and improved quality of life in towns and cities is becoming recognised by policymakers and practitioners alike.

The increasing awareness of the unaffordable economic costs and impacts from using traditional buried piped drainage (Chapters 2, 3 and 12; De Sousa *et al.*, 2012) has raised interest in the potential multiple opportunities afforded by nature-based systems (NbSs). This approach takes a different perspective from that of problem-centred traditional drainage, and in so doing creates a more sustainable drainage system. What was until recently an innovative approach, NbSs are rapidly becoming mainstreamed and increasingly are the first option proposed for surface water management (OECD, 2015).

Figure 8.1 Diagrammatic illustration of the development of stormwater management for urban drainage systems (see Box 8.1 for abbreviations)

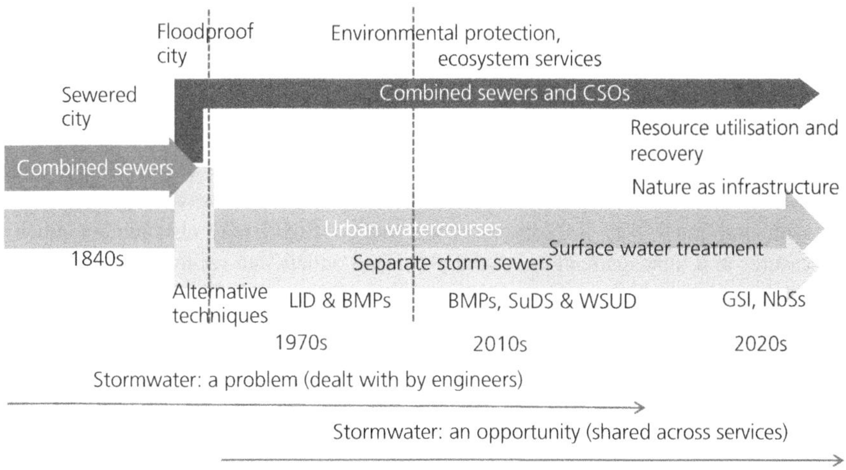

Figure 8.2 A surface drainage system using blue and green spaces, contrasted with an equivalent traditional buried piped system (Richard Ashley and iStock/Inzyx)

All forms of water are useful and opportunities for exploiting this are common in urban systems, especially utilising urban runoff as a resource. Figure 8.2 illustrates the contrast between using sustainable drainage systems (SuDS) and traditional buried grey infrastructure.

There is currently a plethora of terms in use that encompass 'nature-based' infrastructure. Even more so than when the earlier review of non-piped drainage systems was published (Fletcher *et al.*, 2014). Box 8.1 sets out some terminology and relationships to surface water systems. Here we use NbSs and SuDS, with the latter specifically related to stormwater systems and NbSs encompassing the wider nature-based systems (e.g. Xing *et al.*, 2017).

The use of natural drainage systems is not new. Storage areas and open channels on the surface have been used throughout time for human settlements in locations without formal sewerage infrastructure. Wetland treatment beds have been used to provide natural treatment of wastewater in remote areas. It is the growing acceptance that traditional pipework is fundamentally unsustainable and that there is a need for a move towards using natural, vegetated surface systems for all surface water management that is novel, especially in countries such as the UK. However, the extent of existing buried piped systems and associated assets in industrialised countries requires integrated approaches whereby SuDS are used in conjunction with traditional grey infrastructure. In practice, SuDS may also require pipes, linking individual measures in a 'management train' and providing outlets into receiving water bodies.

There are many guides for using and designing SuDS (in the UK, Woods-Ballard *et al.*, 2015) and for retrofitting SuDS (Digman *et al.*, 2012). The use of SuDS in the context of blue-green infrastructure (BGI) is set out in a companion book (Jones *et al.*, 2023). Many projects now consider SuDS to be useful for adapting existing urban drainage systems to cope with climate and other changes (e.g. Ashley *et al.*, 2023; Kabisch *et al.*, 2017).

Box 8.1

Definitions of nature-related systems that can be used for and supported by surface water

Non-piped drainage systems are no longer the primary province of engineers and interest a wide range of disciplines – for example, landscape architects and urban designers. Each of these has a particular perspective on these systems and hence a particular definition of what they are. Desired outcomes and drivers also vary and it is important to understand this for effective communication. The definitions here are individual and alternatives may readily be found.

Nature-based solutions (NbSs) are defined by the European Commission as 'Solutions that are inspired and supported by nature, which are cost-effective, simultaneously provide environmental, social and economic benefits and help build resilience' (Network Nature, 2021).

As with many such definitions, terms can be misleading. Use of 'solutions' implies a once and for all solving of a problem. 'Systems' would be better, and it also avoids the problem-centric perspective. NbSs should be about opportunities.

Similarly, 'sustainable drainage systems' (SuDS) may or may not be any more or less sustainable than an alternative and requires detailed analysis of the technical, economic, social and environmental performance (as explained in Chapter 12) in order to demonstrate how 'sustainable' they may be in a given case.

Blue-green infrastructure (BGI) is also the focus of aspects of NbSs and many SuDS. Arguably a subset of NbSs, Washbourne and Wansbury (2023) define BGI as 'the managed network of terrestrial and water spaces found across our urban and rural landscapes that help deliver socio-economic and ecological benefits supporting ecosystem functions and societal wellbeing.'

Green infrastructure (GI) is a subset of BGI encompassing only green infrastructure, but ironically relying on blue, in the form of surface and sub-surface water, to support ecological health. The term 'green stormwater infrastructure' (GSI) is used in the USA.

Other terminology of relevance includes water-sensitive urban design (WSUD), low-impact development (LID), stormwater control measures (SCMs), source control, Best Management Practices (BMPs). Each of these includes elements of NbSs. See Fletcher *et al.* (2014) for a review of the development of terminology in use in relation to urban drainage.

Box 8.2

A SuDS approach (adapted from Defra, 2021)

A SuDS approach

- mimics natural drainage systems: delivering surface water management that recognises the value of rainfall and runoff as a resource
- manages surface water flooding and the increased rates, volumes and frequency of runoff from developments
- captures and treats the pollutants in urban runoff
- uses components in combination as a 'management train': integrating these throughout the development and its landscape to help create healthy and resilient spaces for people and habitats for wildlife
- manages runoff close to its source, and prioritises components that lie on the surface and incorporate vegetation
- meets the requirements for delivering multiple benefit SuDS over the lifetime of the development, by planning for a changing climate and provision of appropriate long-term maintenance.

'SuDS' is the collective term that has been used for drainage systems in England and Wales (Flood and Water Management Act 2010; Woods-Ballard *et al.*, 2015) to manage urban runoff on the surface: maximising infiltration opportunities, reducing runoff rates and volumes to greenfield equivalents (or ideally below) to mimic natural catchments, optimising pollution removal using natural treatment processes, and enhancing biodiversity and amenity benefits through the use of features that provide habitat, visual, amenity and recreational value. Not all SuDS provide added value. Some, including manufactured and proprietary systems, such as buried plastic crates, have a place in the effective control and management of stormwater, but do not bring added benefits to nature, other than, in some cases, infiltration into groundwater.

The place of SuDS within urban public open spaces, often in residential and commercial conurbations, provides a strong statement on the shift towards finding more sustainable and resilient ways of supporting development and, with appropriate education and awareness-raising (Chapters 4 and 5), can and should support a greater public interest in and commitment to the water cycle and environment. These benefits will, however, be dependent on the quality of the systems that are implemented and their design and effective integration with the development and surrounding landscape. Inadequate planning and engagement, flawed designs, poor construction practices and insufficient maintenance can all affect public perception and acceptance – and reduce the quality of the local environment. The inherent visibility of these systems drives the high levels of potential environmental benefit that they can offer in delivering more sustainable ways of providing urban drainage. It also means that high-quality and reliable workmanship and maintenance are essential. Figure 8.3 illustrates a high-quality scheme in a social housing area in Germany that, despite being easily accessible, suffers no vandalism.

The principles and objectives of any SuDS design must encompass hydraulic (runoff flows and volumes), water quality, amenity and biodiversity criteria, as set out in Box 8.2 and outlined in Table 8.1. These principles and objectives are overlapping and in some cases interdependent. In characterising and defining the components for any scheme, planners and designers should also consider opportunities for buffering and adapting to climate change, at the same time minimising health and safety risks. This usually also leads to more hydraulically robust systems. Designers should also take full consideration of ease of construction, long-term maintenance, and the carbon footprint and other impacts of the proposed scheme (e.g. Sagrelius *et al.*, 2020 and Chapter 12). Key principles of SuDS design to meet all these objectives are to maintain the surface features at a shallow depth, avoiding the deep and steep slopes that

Figure 8.3 A high-quality SuDS scheme in Germany (Richard Ashley)

Table 8.1 SuDS design objectives

Category	Design objective
Water quantity	Manage the quantity of surface runoff to minimise flood risk and erosion and maintain and protect the natural water cycle
Water quality	Manage the quality of runoff to support and protect natural systems
Amenity	Create and sustain better places for people
Biodiversity	Create and sustain a better natural environment

are typical of poorly designed SuDS features, often referred to as 'bomb craters', and to avoid 'end-of-pipe' SuDS, such as an 'estate pond'.

The surface water management philosophy when using SuDS should be based on managing both runoff and pollution as close to the source as possible (source control). The prevention of pollution (by either removing the pollutant sources or changing the pathways through which flows pass into the environment) and reduction of runoff (by either harvesting and using the rainwater or preventing surface runoff by making the urban surface less impervious) provide the highest level of 'source control'. In fact, the benefits of green roofs are underestimated. They protect against extremes of temperature and solar impacts, enhancing the thermal properties of buildings, and can attenuate as much as 50% of rainfall volumes annually.

Where polluted runoff cannot be prevented, drainage components should be used in a management train to infiltrate (soak) and slow, store and treat the runoff. NbSs are known to be effective at controlling polluted runoff (e.g. Matos and Roebeling, 2022), although few studies provide details and most are interested in water quantity rather than quality. Grass drainage channels (swales) that run alongside a road, individual property soakaways or a basin that manages runoff from an adjacent car park all deliver control of runoff at source. However, infiltration rates change within the landscape, which can often be a challenge for designers (Chapter 7).

Upstream source control acts to 'protect' downstream features by trapping debris and sediments and controlling runoff rates and volumes. Without these, maintenance requirements and costs associated with downstream systems are likely to be disproportionately greater. The SuDS design criteria from this perspective are given in Woods-Ballard *et al.* (2015).

It may not always be cost-effective to design 'at-source' or 'near-source' components for the more extreme rainfall events, and runoff might need to be collected and routed downstream to larger storage systems, which might be designed to provide hydraulic control for many small upstream sub-catchments. Wherever possible, conveyance routes should be on the surface and designed to provide further hydraulic control and treatment. Overland flood flow routes will also be required to convey and control floodwater safely during the most extreme events (i.e. in excess of those for which the system is designed). These are collectively assessed under the concept of 'designing for exceedance' as discussed in Chapter 3 (Figure 3.2) and good practice for this is set out in Digman *et al.* (2014a, 2014b) and Diaz-Nieto *et al.* (2012; 2016).

8.1.1 Designing for a range of flows and maximum benefits

Designing for flows on the surface can be more complicated than when designing buried piped infrastructure, as outlined in Chapters 3 and 11.

Standards for the hydraulic design of drainage systems are set out in codes of practice as summarised in Chapters 2 and 3. There are also local regulations and ordnances to be followed (Chapter 5). British standards, such as BS 8582 (BSI, 2013a), set out how to design to achieve required performance, and Water UK (2022a, 2022b) sets out how to build to an adoptable standard in England. *Homes for a Healthy Life* (Design for Homes, 2023) is an example of how SuDS can be an essential part of the planning and design of housing areas for healthy living.

Many standards and guides to good practice of relevance to SuDS and NbSs have been developed internationally, including BS ISO 24536:2019 and PD ISO/TR 24539:2021 (BSI, 2019, 2021). As developments in this area are advancing rapidly, it is recommended that when planning to use SuDS or NbSs, a review of current and locally relevant material is made, especially for water-sensitive urban design (WSUD).

Often the best way to comply with standards and regulations is to follow industry guidance documents or manuals. The UK Construction Industry Research and Information Association (CIRIA) has produced many manuals and guidance documents of relevance to SuDS and NbSs (CIRIA, 2023a). Here, the second edition of *The SuDS Manual* (Woods-Ballard *et al.*, 2015) is used to show how designs can follow good practice to achieve both a safe environment and one that uses surface water to maximise the benefits to society. Designs may be used as part of new developments or as modifications to an urban area to improve the performance of the drainage system through retrofitting (e.g. Digman *et al.*, 2012).

Typically, standards focus on the design events characterised by some measure of risk, usually expressed as rainfall event return periods (Chapter 6), together with some expectations that larger events will be managed at least so as to minimise damage, as is expected in the National Planning Policy Framework (NPPF) for England (Chapter 5). As an example, in England SuDS have typically been designed to comply with the non-statutory technical standards for sustainable drainage systems (Defra, 2015), which require that no flooding will occur outside of the defined drainage system for 1 in 30-year rainfall events and no internal flooding for 1 in 100-year events. Thus risk acceptance in this context is a planning need. Normally, 1 in 100-year events are used for drainage systems (and required in local guidance). Elsewhere, expected standards will be set nationally. As explained in Chapter 5, because of the relatively rapid rate of change, and implications of this for infrastructure systems, policy, regulations, standards, guidance and the precise specification of performance needs must be agile enough to cope (e.g. BSI, 2023). It is essential to review the latest local specifications and ensure that these are used in designs. In the UK, for example, the National Infrastructure Commission (NIC, 2023) provides guidance to government for standards for all types of infrastructure.

There is reliance on emergency and civil contingency planning for 'extreme' rainfall, by way of regional resilience fora. In Chapter 3, the concept of exceedance drainage (i.e. for events greater than designed for) is explained, and Figure 3.2 illustrates where this fits into the range of flow conditions and types of rainfall event to be considered in designing SuDS and other drainage features. It is essential to understand how the entire range of rainfall runoff events (including everyday rainfall) needs to be considered in designing and operating urban drainage systems. The change in paradigm now is that virtually every type of event affords opportunities, as Figure 3.2 illustrates.

Many of the benefits brought by using SuDS are in domain 1 of Figure 3.2, that is, they accrue because of the lowest magnitude 'everyday rainfall' events; although, in common with other forms of drainage, SuDS are primarily designed for domain 2 (see also Chapter 11), the so called 'design' events.

8.2. Sustainable drainage systems (SuDS)

Sustainable drainage comprises a range of surface features which can be supplemented, where necessary (e.g. in constrained sites), by subsurface pipework or proprietary drainage facilities providing conveyance, storage and/or treatment opportunities. Each feature can be uniquely designed for site-specific application, and this provides a high level of flexibility, allowing SuDS implementation for any development site, including retrofitting.

Typical SuDS components are illustrated in Woods-Ballard *et al.* (2015), the SuDS Sector Guidance for London (Mayor of London, 2023) and Jones *et al.* (2023). Elsewhere, there are clear illustrations for stormwater control measures (SCMs) (the USA, e.g. USEPA, 2023a) and WSUD components (Australia, e.g. Sharma *et al.*, 2019).

The second edition of *The SuDS Manual* (Woods-Ballard *et al.*, 2015) explains how SuDS can be used on perceived 'problematic' sites and also on flood plains. For example, where a site is contaminated, there may be opportunities to infiltrate below the contamination layer, or else the drainage system can be sealed from the layer at the surface. Queen Mary's Walk, Llanelli, Wales is an example of an underdrained swale on poor/contaminated ground. The swale, lined to prevent infiltration into the ground, attenuates and slows flow before discharging back to the combined sewer at a controlled rate. Where sites need to be very densely developed and there is very little opportunity to utilise public open space, then green or blue roofs or roof water collection systems together with permeable car parking surfaces can provide cost-effective and beneficial drainage. Where sites lie over impermeable soils, this may rule out infiltration, but by maximising the multifunctional use of public open space, land take for surface storage systems can be minimised. On steep sites, conveyance routes can follow the site contours or systems can be effectively 'stepped' to minimise high velocities and system scouring.

For drainage systems to provide the greatest benefits, site-specific design objectives must be clearly stated, within the context of those set out in Table 8.1, and criteria set that meet regulatory and local community requirements. Effective and sustainable surface water management strategies should be embedded within planning policy from national through regional to local level, and stakeholder relationships and engagement processes will be fundamental. This is further explored in Chapters 4 and 5.

8.3. Introduction to the design of SuDS and NbSs

Here the design is summarised, as full details are provided in Woods-Ballard *et al.* (2015). Charlesworth and Booth (2017) also provide details and international case study examples. With a long history of utilisation of best management practices (BMPs) and low-impact development (LID), in the USA there are numerous sources that can be accessed, including WEF/ASCE (1998), USEPA (2023a), the latter focused on pollution prevention. Around the world there will be developed standards and practice guides, and these should be accessed locally where a scheme is being designed. These will often use different terminology for SuDS or NbSs, such as WSUD (e.g. Sharma *et al.*, 2019), and have varying objectives, as illustrated in Box 8.3. Ongoing and former international projects have and are developing new ideas and standards related to NbSs, including SKINT (Pötz and Bleuzé, 2012), BEGIN (2023a) and ClimateADAPT (2023). These demonstrate how rapidly knowledge and practice are advancing in this arena.

Box 8.3

Water sensitive urban design (WSUD) objectives (National Water Commission, 2010, cited in Sharma *et al.*, 2019)

- 'minimise impacts on existing natural features and ecological processes/minimise impacts on natural hydrological behaviour of catchments
- protect water quality of surface and groundwaters
- minimise demand on the reticulated water supply system
- improve the quality of and minimise polluted water discharges to the natural environment
- incorporate collection treatment and/or reuse of runoff, including roof water and other stormwater
- reduce runoff and peak flows from urban development
- reuse treated effluent and minimise wastewater generation
- increase social amenity in urban areas through multipurpose green space, landscaping, and integrating water into the landscape to enhance visual, social, cultural, and ecological values
- add value while minimising development costs (e.g. drainage infrastructure costs)
- account for the nexus between water use and wider social and resource issues
- harmonise water cycle practices across and within the institutions responsible for waterway health, flood management, pollution prevention, and protection of social amenity.'

Box 8.3 illustrates the slightly different emphasis of WSUD, at least in Australia, compared with SuDS in the UK. In many places water quality improvement and protection is the objective, rather than flood risk reduction. SuDS, as a form of urban drainage, aim first at controlling flows and water quality. Other uses of SuDS or NbSs may prioritise other aspects of performance, such as amenity, and can be influenced by local drivers or funding approaches. As we are concerned here with urban drainage, the primary aims considered are for water quantity and quality control. See Washbourne and Wansbury (2023) for a town-planning focus for using BGI. Elsewhere, integrated approaches to water management are being used as the focus (Chapter 12), and it is important to decide on the main objective of any development where NbSs are being used, especially to satisfy funder requirements (e.g. Ashley *et al.*, 2023).

The design of SuDS is based on the key principles set out in Box 8.2, together with the objectives in Table 8.1. The main SuDS concepts are summarised in the following.

8.3.1 Hydraulic (water 'quantity') objectives

Ideally, sustainable development and drainage should have no or insignificant negative impact on the environment and preferably be beneficial. The hydraulic focus for SuDS must therefore be on maintenance or improvement of the pre-development water balance and runoff characteristics. Where benefits other than flow controls are of interest, there may be opportunities to change the water regime, retaining water on site, supporting ecosystems or increasing amenity and/or biodiversity.

In terms of controlling the hydraulic characteristics of the surface water runoff, the effectiveness of the drainage scheme can be evaluated using four criteria.

1. Preventing surface runoff from the site for small rainfall events or an initial depth of rainfall, often referred to as 'everyday rainfall'.
2. Controlling runoff volumes by prioritising rainwater harvesting and infiltration.
3. Where outfalls are required, prioritising discharges to surface waters over combined sewers.
4. Controlling runoff rates.

These are set within the context of the four domains approach explained in Chapter 3, which include exceedance drainage (Figure 3.2). In the past, the focus of drainage professionals has tended to be on limiting peak runoff rates, or volumes, alone. However, the emphasis is now on the need to adhere to the requirements above, each an important element of the design of any SuDS scheme. Delivery mechanisms for key runoff management characteristics are described in the following sub-sections.

8.3.1.1 Preventing runoff from impervious surfaces for small rainfall events/an initial depth of rainfall

Small rain events (everyday rainfall) falling on greenfield sites (domain 1 in Figure 3.2) will tend to be intercepted by vegetation, used to replenish soil moisture storage, stored in natural depressions, evapotranspired, and/or infiltrated into the ground. In temperate climates like the UK, at least half of all rainfall events will cause no measurable runoff. In contrast, rain falling on impermeable surfaces will almost always cause runoff, even at very low intensities, with little, if any, infiltration occurring.

Small receiving streams that take runoff from impermeable areas can be very flashy (i.e. subject to regular, sudden rapid flows) causing unwanted scouring and sedimentation, with associated changes to both the morphology and ecology. Soil sealing, or compaction, with reduced infiltration, particularly over large areas, can affect base flows of local streams and rivers and, potentially, groundwater recharge processes (Chapter 7).

If the development site and its drainage system are designed to prevent runoff from small events or an initial depth of rainfall (e.g. 5–10 mm) (an approach referred to as 'interception' or 'source control'), the adverse impacts can be reduced, and additional benefits realised, including the following.

- Reducing the additional volumes of runoff discharged from the site and using those volumes on the site for non-potable purposes or to restore natural hydrological functionality.
- Lowering the frequency and volumes of discharges of contaminated runoff to the receiving water body, thus helping to reduce total pollutant loadings.
- Where sites have to discharge to combined sewers, reducing surface water loads, which could contribute to lowering the frequency and volumes of combined sewer overflow (CSO) spills.

Regular, small or everyday rainfall events will wash many of the contaminants that accumulate on the urban surface into the drainage system. The cumulative effect of all these small events is likely to be a significant proportion of the total pollutant load from the area (Figure 11.6). Source control ensures such events are retained and contaminants treated as far as practicable using natural processes (e.g. WEF/ASCE, 1998).

Infiltration systems, vegetated surfaces (e.g. green roofs, swales, detention basins, bioretention zones), other permeable surfaces (e.g. pervious block paving) and rainwater harvesting systems can all be designed to deliver on a requirement for interception/source control. In some cases, it is possible to deliver more than 5 mm onsite retention by using a combination of components, even in sites underlain by clayey soils with relatively poor infiltration. Provision of interception (particularly if distributed across a site) will also reduce the volume of attenuation storage required. Figure 8.4 shows a well-established green roof.

Figure 8.4 A green roof (Robert Bray Associates/Adelina Iliev Photography)

8.3.1.2 Limiting peak runoff rates

The proportion of rainfall that runs off urban surfaces and the rate at which this runoff occurs is significantly higher than from an equivalent greenfield area (Chapter 7). This is the result of the reduction in permeability of the surface; reduction in vegetated surfaces, which intercept and slow runoff; and changes from natural drainage channels (with high hydraulic roughness) to smooth pipes or concrete channels which are designed to convey flows as rapidly as possible to receiving water bodies.

Sustainable drainage should aim to reduce development runoff rates to rates more similar to greenfield conditions for a range of different scales of event. In practice, to match all points on the flow frequency curve (Figure 11.7) would be extremely difficult, requiring a complex level of hydraulic control, so in England, the 1 in 1-year and 1 in 100-year rainfall events are often specified, occasionally with the addition of the 1 in 30-year event – particularly if the outfall discharges to a separate surface water sewer (which may have been designed for a 30-year standard of service). There is a range of SuDS components which can contribute to the required reduction in peak flows during extreme events, as shown in Box 8.4.

Attenuation storage using SuDS can be provided in various ways, either with surface features or below ground, with multiple benefits maximised through the use of surface, vegetated systems.

On the surface

- in ponds or wetlands, above the permanent water level, within contoured storage zones
- in basins (areas that are normally dry and temporarily store water)
- in surface storage zones (basins), above areas of car park, play areas or other public open spaces
- in swales, that can be designed to provide storage as well as convey water
- on roofs, as part of a green, or blue roof, within the roof void, or below hard surfaces on podium decks
- in bioretention areas – these can provide some storage above, below and within the treatment system.

Box 8.4
SuDS techniques

- **Infiltration**: the net rainfall intensity can effectively be reduced by infiltrating runoff at a constant rate throughout a rainfall event. This is dependent on the maintenance of adequate infiltration capacity during wet weather conditions. Results from observed infiltration tests on site need to be used with added factors of safety.
- **Rainwater harvesting**: provided the system is designed so that there is always storage available during a rainfall event. Either the harvesting tank can be oversized to provide the required attenuation storage, or additional storage (with controlled draw down) can be added above the storage required for supply purposes, or smart systems can be deployed that empty tanks in advance of extreme weather (e.g. Liang *et al.*, 2019).
- **Slow conveyance systems**: conveyance routes using, for example, wet or dry swales or geocellular boxes will have a higher hydraulic roughness than piped systems, transferring flows more slowly downstream.
- **Attenuation and hydraulic control**: the principle of reducing peak runoff rates is known as attenuation. Runoff can be captured and stored within dry or wet depressions, or within granular fill or other structurally formed subsurface cavities beneath hard or soft permeable surfaces. By controlling the flow at the outfall to set thresholds, the (attenuation) storage will fill temporarily during design events and be emptied in a controlled manner, such that the rate of discharge from the storage is below the runoff rate.

Just below the surface

- within a permeable pavement sub-base
- within open filter drains
- within granular, geocellular or other storage layers below impermeable surfaces, such as car parks
- within granular, geocellular or other storage layers below sports pitches and games areas.

Underground

- within oversized pipes
- as attenuation tanks.

Systems may potentially be designed to use BGI for smaller events (to deliver the multiple benefits), and 'grey' subsurface infrastructure for more extreme flow management where space is constrained.

Attenuation storage can be provided either on-line or off-line as for traditional below-ground systems (Chapter 3), although the integration of storage with treatment and other multiple-benefit functionality will drive the use of on-line components.

8.3.1.2 Flow control

Where shallow surface systems are used with source control, simple flow controls such as weirs and orifice plates are usually appropriate (left and right images in Figure 8.3). These can be low cost and easy to maintain. The use of source control (that acts to trap debris and sediments upstream of storage systems) also minimises the risk of the flow controls becoming blocked. An overflow route should always be provided around a flow control so that blockage does not result in unacceptable flooding (this will require consideration of overland flow or exceedance routes).

Where site and regional controls are necessary, other more substantive flow controls such as vortex devices may need to be considered. These are generally more expensive and more maintenance intensive than the simple controls (e.g. CIRIA, 2023b). The advantages of vortex flow controls are much less where systems are on shallow sloping areas and there is very low differential head of water when the storage and flow controls are in operation.

8.3.1.3 Limiting runoff volumes

Most urban surfaces restrict natural infiltration and the trapping of rainfall (and subsequent evapotranspiration) by vegetation and soils (Chapter 7). The total volume of runoff from an extreme rainfall event from a paved area is typically between two and 10 times the runoff volume from the same area in an undeveloped state. SuDS should aim to reduce runoff volumes by using a SuDS approach that prioritises the use of rainwater harvesting, infiltration and BGI within drainage designs. This approach will help the drainage to reflect a more natural response, while protecting the natural hydrological processes (e.g. evapotranspiration, soil moisture replenishment and groundwater recharge) to which the site contributes. Use of the runoff as a non-potable resource (e.g. toilet flushing, irrigation) will support local water resource resilience and improve the capacity of the green infrastructure (GI) to be self-sustaining during dry periods.

8.3.2 Water quality objectives

Pollution arising from (urban) land-use activities that are dispersed across a catchment or sub-catchment and do not arise as a process effluent, municipal sewage effluent or an effluent discharge from buildings is termed 'diffuse' (Chapter 3). Urban sources account for many important diffuse pollutants, and urban water bodies are often badly polluted as a consequence. Further discussion on sources and nature of pollutants in runoff is provided in Chapter 3.

Stormwater runoff can vary widely in quality, and the range of pollutants and the loads reaching drainage systems will be dependent on: the local hydrology and climatic conditions; the size and type of urban area; land use; and the types and frequencies of activities taking place (e.g. vehicle washing, street cleaning, pesticide application, etc.). As illustrated in Chapter 3, Figure 3.5, pollutants comprise a wide variety of particles and dissolved substances. The UK's 2nd edition of *The SuDS Manual* provides details of types of pollutants and their potential impacts in annexes to Chapter 25 (Woods-Ballard *et al.*, 2015). WEF/ASCE (1998) and the BMP database (Clary *et al.*, 2022) provide pollutant information as well as data for the performance of SuDS. Information on diffuse pollution from surface runoff is outlined in Chapter 3, and in Lundy and Wade (2013) and D'Arcy (2013).

The impacts on receiving water bodies from these various and diverse substances conveyed in runoff are considered as 'diffuse pollution' in that they arise from widespread and diffuse sources (see also Chapter 3). Many of these sources are classed as 'priority substances' (PSs) and therefore are priority for control. EU states have been compiling inventories of emissions of 45 PSs, as required by the Environmental Quality Standards Directive (EC, 2017). There is a subset of priority hazardous substances (PHSs) also identified, including cadmium and its compounds and mercury and its compounds. All member states were meant to stop any discharge of PHSs by 2020 under the Water Framework Directive. Nevertheless, few have yet to quantify diffuse emissions originating from the wide variety of sources. Consequently, most of the measures identified in relation to chemical pollution are too general, with unquantified outcomes, rather than substance or source specific (EC, 2015), hence deciding on controls and treatment is difficult. However, as regards PSs and PHSs in stormwater, the focus has been on: (*a*) the assumption that, over a 20-year time period, many combined sewers will be separated and the problems will then be confined to flows to treatment plants; (*b*) what needs to be done to uprate treatment plants to manage new pollutants, including PSs and PHSs (e.g. Anon, 2012).

The move towards the use of separate sewerage systems worldwide in the mid-twentieth century (Chapter 3) has resulted in major pollution of receiving waters as surface water drainage systems were presumed 'clean' and therefore subject only to minor constraints on flows and volumes. It is not only inadvisable but also impossible to rely on sewer separation to manage diffuse pollution. Hence, SuDS and other surface water treatment systems will need to be employed to control, at least in part, pollutants in surface runoff, including PSs and PHSs.

Biswal *et al.* (2022) make the case for using NbSs for simultaneous surface water pollution control and air pollution control, based on studies in Singapore. Air phytoremediation by way of roadside vegetation is considered an effective option for reduction of particulate matter ($PM_{2.5}$ and PM_{10}) and gaseous pollutants (nitric oxide and carbon monoxide) in urban air. They call for better guidelines for the design of NbSs to help deal with pollution in the air, on land and in water.

Lundy and Wade (2013) review the occurrence and potential for pollution control by using SuDS, based on data from Scottish catchments. It is important to include commercial and industrial yards and any agriculture in the catchment in any analysis as many PSs and PHSs will originate from these areas. There are also risks from spillages on highways and from displaced loads and accidents, as well as from railways and other infrastructure (e.g. Osborne and Montague, 2005). These require buffer zones to detain any runoff.

There are two main components to dealing with runoff pollution – management and prevention – at source and otherwise. The objectives of SuDS in protecting the quality, ecology and usability of any receiving water body should be twofold

- minimise the risks of pollution of receiving water bodies from regular or occasional discharges of urban runoff (chronic pollution)
- act as a buffer for accidental (episodic or irregular) spills to prevent their direct discharge (acute pollution).

8.3.2.1 Managing runoff pollution using SuDS

The utilisation of SuDS components in series (the SuDS management train) can progressively reduce the contaminant concentrations (and total loads) in the surface runoff using natural treatment processes (see Jones *et al.*, 2023). Processes by which pollutants are removed or lost from runoff include

- sedimentation: reduction of flow velocities for particle settlement
- filtration and biofiltration: by trapping in soil matrix, plants or in or on geotextiles
- adsorption: binding of pollutants to surfaces
- biodegradation: by way of the natural biological processes of breakdown or utilisation of compounds
- volatilisation: transfer of organic compounds from solution to emission into the atmosphere

- precipitation: chemical reactions enable metal and other compounds in solution to be settled by forming particulates
- uptake by plants: nutrients in particular are required for plant metabolism
- nitrification: oxidised by bacteria in the ground to form nitrate, a plant nutrient
- photolysis: breakdown of pollutants by exposure to ultraviolet light.

The processes listed above function with varying degrees of effectiveness depending on the nature of the inflows and the configuration of the SuDS component. Regular monitoring and synthesis of studies of treatment efficiencies for urban runoff are published from data collected in the USA in the BMP database (Clary *et al.*, 2022). These illustrate the wide variability in performance of natural SuDS, and the more predictable pollutant removal performance of proprietary SuDS devices.

Volumetric losses from SuDS (e.g. by way of infiltration and evapotranspiration) are also recognised as key mechanisms for reducing total pollutant loads from the site.

Wade and Garcia-Haba (2013) review the performance of SuDS in managing diffuse pollution and the potential for accruing multiple benefits. There are many examples of the pollutant removal performance of SuDS (e.g. Woods-Ballard *et al.*, 2015) and BMPs (from the international stormwater database: Clary *et al.*, 2022). The control of PSs from the use of SuDS, including metals, was the subject of the ScorePP EU programme in the early 2000s (Eriksson *et al.*, 2011). This developed an international risk prioritisation approach.

Within SuDS components, runoff is slowed and/or stored to promote the settling of sediments and associated contaminants and the opportunities for natural treatment processes, including uptake by plants. The detained solids and adsorbed substances within the SuDS due to the treatment processes may require management. (Allen *et al.* (2017) provide a matrix for evaluating the sediments and trace metals detained in SuDS, to help in deciding if and when these need to be dealt with.)

It is the differences in the purification potential of the various SuDS components in a train that make the overall system robust and the treatment train concept so important. This makes SuDS far more effective at treating pollution than a single engineered treatment device such as an oil separator. The range of drainage components that can be considered provides a toolbox with which drainage specialists and others can design appropriate and functional schemes to meet site-specific objectives.

The design approach is based on a 'train' of SuDS and the performance of individual units is not necessarily representative of how the SuDS will perform when in series, as successive inflows between individual units will be more dilute than when a SuDS component is used in isolation.

The number of drainage components required in a train should reflect the pollution risk at a particular site. This risk will be related to the level of hazard (dependent on the site characteristics and use) and the consequences of any pollution, which usually relates to the sensitivity and vulnerability of the receiving water body. The probability of pollution occurring is also a consideration when assessing risk, especially with accidental spillages.

For a particular site, for any one event, the runoff pollution concentrations will reflect the extent of contaminant deposition on the surface since the previous rainfall, the elapsed period of time since rainfall and the duration and intensity of the event. Studies have shown the early runoff in a storm is usually the most polluted, and that small, frequent events – particularly those during the summer months – tend to contribute the highest pollution load, often when the receiving watercourse is at its most vulnerable due to low base flows (Figure 11.6). By using the ‘interception’ principle described in Section 8.3 above, small or everyday rainfall events are automatically retained by the drainage system, producing zero runoff, and thereby protecting the receiving waterbody from pollution. This design philosophy, together with the prioritisation of rainwater harvesting and infiltration, also promotes reduced runoff volumes – which has been demonstrated to have a significant impact on total pollutant loads from a site.

When rainfall events exceed the ‘interception’ depth, runoff may occur, and the drainage system must provide treatment of this runoff to reduce contaminant levels as far as practicable. In the UK, the event for which treatment is generally designed for is about the 1 in 1-year return period event. Events larger than this will tend to have sufficient dilution both within the drainage system and in the receiving watercourse, and effective treatment is therefore less crucial.

There are four key mechanisms by which treatment is promoted within individual SuDS

- filtration
- infiltration
- detention
- retention in permanent open water.

These are considered further in what follows.

There is a range of materials that can be used as filtration media to treat runoff. Permeable pavements and trenches use gravels which promote adsorption, precipitation and biofiltration of pollutants on the gravel surfaces. Precipitation here does not refer to rainfall precipitation, but to the chemical process of precipitation. Source control features should be designed to trap and retain the coarse sediment fraction and its associated contaminants effectively. Bioretention areas, swales, filter strips and other vegetated surfaces use grasses and other plants to the same effect. Wetlands and pond systems use aquatic vegetation, and sand filters are also in use.

Where water is filtered through material on the surface – that is, exposed to ultraviolet light and with wetting and drying cycles – this can assist with managing bacteria and pathogen levels, which is important where surface water is to be used directly or indirectly for, for example, irrigation (e.g. Shen *et al.*, 2020).

Infiltration is a form of filtration that occurs vertically through the soil matrix, where runoff travels towards hydrogeological flow paths or groundwater bodies. To promote effective treatment, the soil matrix must be sufficiently fine that adsorption, precipitation and biofiltration processes are initiated (e.g. Allen *et al.*, 2017). Where runoff can find high-flow pathways through, for example, fractured rock or fissured limestone then treatment is not likely to be effective.

Detention basins tend to have a wide base that can be used as an effective filtration device for small events if correctly designed. At higher flows, where the basin begins to fill, retaining and delaying the event will promote the settling out and retention of coarse sediments and associated contaminants.

Where a system has a permanent body of open water (the permanent pond volume), this volume will remain in the system between rainfall events. Its size and the downstream flow control, together with characteristics that influence mixing processes within the pond (e.g. shape, baffles), will dictate the length of time that runoff will remain in the pond before being discharged. Capture of a depth of rainfall of 10–15 mm (approximately 90% of events) is usually considered appropriate for treatment. A reasonable period of retention should allow even fine silts to fall out of suspension, which is important for pollution control, as well as promoting uptake of contaminants by plants and treatment by way of in situ microbial activity. Maintenance of retention ponds and sedimentation management are of particular importance. If left to build up in ponds for a long period of time, the sediment will change state and then be classed as special waste in countries such as the UK, which may then require costly disposal at special waste management sites.

High-quality treatment systems (which could be a treatment train) often use dry vegetative source control treatment to achieve the capture of coarse sediments and the effective treatment of hydrocarbons, but also include a feature at the end of the system that is permanently wet to facilitate final 'polishing', including the removal offine silts, prior to discharge. This ideal may not be practicable for small or high-density sites, which occasionally may need to rely on subsurface treatment systems (see Woods-Ballard *et al.*, 2015).

Despite the information outlined above, there is still limited knowledge and reliable and repeatable evidence regarding the water quality treatment of SuDS and NbSs. With climates changing (Fowdar *et al.*, 2021; Hulsmann *et al.*, 2015), there are many additional risks likely to influence both point and diffuse sources of pollution through increased flushing of pollutants during peak flows and/or reduced dilution during environmental flows (e.g. Wilby *et al.*, 2006). Repeated (and increased) flood inundation will remobilise and redistribute large amounts of pollutants from both urban surfaces and in-pipe sources. This could cause much more widespread and extreme acute pollution effects on receiving waters and ecosystems and to human life (e.g. Butterworth *et al.*, 2011).

8.3.2.2 Pollution prevention

Pollution prevention strategies are methods used to manage pollutants at their source. Strategies work in a variety of ways. Many are commonsense housekeeping or site management practices and are relatively inexpensive and easy to implement. For example, dust and debris can be swept or vacuumed from a work area and disposed of elsewhere instead of being washed into a drainage system or SuDS component. In some cases, pollution prevention strategies are sufficient to remove the pollution risk entirely, depending on the source, and these are set out in Woods-Ballard *et al.* (2015).

Pollution prevention strategies comprise

- avoid and minimise: not using (or limiting the use of) materials and activities that generate pollutants
- prevent: not allowing pollutants to mix with rainfall
- capture: trapping pollutants in runoff before they enter the drainage system or receiving water body.

Much of what is required for prevention entails engagement with and education of potential polluters. This includes all activities that could impinge on surface water runoff quality – that is, everyone in a catchment. Engagement is considered in Chapter 4.

SuDS that manage water at or near the surface provide greater visibility and opportunities for visibly monitoring the quality of runoff and any potential pollution risks.

Potential problems arising from separate surface water drainage systems, which can convey washed off pollutants rapidly to receiving waters, are exacerbated by wrongly connected foul drainage entering surface water systems. Worldwide, this is a major problem and often occurs due to poorly constructed or careless plumbing in all parts of drainage systems, as discussed in Chapters 2 and 3. Pollution prevention strategies include the need to investigate and eliminate these wrong connections.

It is important also to take a risk-based approach using, for example, BS ISO 31000:2018 (BSI, 2018) and to anticipate potentially sudden or catastrophic events (e.g. Gilbertson *et al.*, 2011).

8.3.3 Operation and maintenance of SuDS and NbSs

Responsibilities for ensuring that SuDS continue to function and perform as intended may fall on various stakeholders. These responsibilities may also change during the lifetime of the system due, for example, to personnel, institutional, regulatory and environmental changes and performance expectations changing. Hence it is important to define clearly the operational and maintenance requirements from the outset of the project, normally during the planning stages, and revise these into the future as necessary.

In a UK example, Dolowitz *et al.* (2018) explain how the policy of integrating SuDS in all new developments in Glasgow, Scotland, has seen many NbS measures being built in preference to piped systems to control flooding. However, there was a number of constraints to the use of NbSs in terms of how the responsibilities for operation and maintenance were decided. Key organisations managing infrastructure, including the local authority and the public water service provider, have strictly defined which SuDS they would maintain and fund, and this has constrained the use of certain types of NbSs in this project.

Careful consideration of the maintenance implications of the design is required from an early stage. This will require consultation with the final owner or operator of the scheme during the planning and design stages. Ensuring future ease of maintenance begins with appropriate design. A well-managed construction process that protects the functionality of the final SuDS components will also be crucial in ensuring early maintenance activities are as envisaged and extensive remedial operations are not needed. It is important to ensure that the following characteristics and principles (adapted from the Landscape Institute interim technical guidance note on the management and maintenance of SuDS landscapes (Landscape Institute, 2016)) are considered appropriately and incorporated into the site SuDS management plan

- soil selection and preparation
- plant selection and likely development
- safe and efficient access for management and maintenance (to meet hydraulic, water quality, amenity and biodiversity design objectives)
- inspection points that are easy to access
- appropriate sediment management to prevent clogging and reduce the need for costly maintenance downstream
- bypass systems or temporary drainage structures for use if required during maintenance/management
- disposal areas for organic arisings (green waste) and sediments.

Different types of SuDS/NbSs have different operational requirements, most of which differ significantly from the maintenance of traditional drainage. Operation of SuDS, although different, is no more and often substantially less difficult than for traditional drainage systems, albeit that maintenance of the latter is often overlooked. Many SuDS maintenance activities are linked to the provision of multiple benefit objectives, such as water quality, amenity and biodiversity provision – that do not form part of traditional drainage design. This should be taken into account. The main differences lie between landscaped, vegetative surface SuDS and engineered solutions, such as permeable surfaces, geocellular tanks, treatment channels and other proprietary systems.

Operation of landscape SuDS usually requires routine landscape maintenance practices familiar to any landscape contractor. It may even be part of normal public park or open space maintenance processes. Maintenance of engineered SuDS usually requires machinery, such as road sweepers for permeable pavements or 'gulley suckers' for silt traps, etc. Engineered source control systems such as treatment channels can usually be cleaned out manually.

Up until recently all the published information on the operational or maintenance requirements for SuDS in the UK has been largely based on theoretical assumptions, supported by information from international experience. Growing experience of maintenance over the past 20 years in the UK and worldwide has shown that many of the theoretical assumptions are pessimistic and that required maintenance is lower than anticipated (when the design is good). There are now numerous sources of information for maintenance, including, for UK applications, Woods-Ballard *et al.* (2015) and Wilson and Davies (2017), and for US applications, City of Portland (2020). There are guides for BMP maintenance for homeowners, such as Harrisburg (2017), and although terminology may vary internationally, the principles are globally relevant, albeit climate variability and risks from non-water-related sources may be different. Many municipalities provide guidance and templates (e.g. CIRIA, 2023c). The philosophy or guiding principle of landscape architect Tom Liptan is 'design to maintain' (Liptan and Santen, 2017).

There will be situations where vandalism occurs, or the base of an infiltration basin silts up, requiring remedial works. However, this will not occur in all systems and the best way to allow for this if adopting many SuDS is to apply a contingency cost provision to each scheme.

All SuDS should have a management plan defined and approved at the design stage. This should be a living document that evolves with feedback from inspection and evaluation of expected against required maintenance activities for the site.

8.4. The value of nature as infrastructure

Systems that include green, blue or natural features and are on the surface of the land can provide numerous benefits to humans and nature other than just for controlling water. All or most infrastructure projects occupy land, and even those that are buried have a 'footprint'. There are thus numerous opportunities for including blue, green or natural components that can add benefits to people, flora and fauna, often with little added costs. Even when there is no explicit consideration of the implications for water systems, these systems will be affected by any form of change to the land, rural or urban.

The concept of 'nature as infrastructure' is being taken up widely (e.g. EESI, 2019; Roy, 2018), which not only exhorts the use of NbSs as infrastructure, but also emphasises the need actively to manage nature for this purpose. Unlike grey infrastructure, there is no option to 'fit and walk away', leaving the infrastructure to manage itself for decades without intervention. In contrast, after being built in 1888, the Dundee interceptor sewer (Figure 3.13) would have required little, if any, maintenance to ensure functionality over its continuing lifetime and certainly not any consideration of the seasonal effects on performance. NbSs will function differently at different times of the year and functioning will change as the climate changes (e.g. Fowdar *et al.*, 2021). It is therefore important to 'manage expectations from seasonal transitions to the ongoing operation and maintenance necessary to keep them running smoothly so the owner understands the nature of this working landscape from a holistic viewpoint' (Liptan and Santen, 2017). They will also not be 'oven-ready' once built, as they will take time to establish and develop the ecological systems that will bring many of the benefits (e.g. Perales-Momparler *et al.*, 2014).

There is evidence around the world that the use of 'green infrastructure' in various forms is increasing, along with requisite investments. For example, the projects that have used GI across the EU, and funding sources, have been reviewed by HKV and RPA (2014) in a study on economic and social benefits of environmental protection and resource efficiency, showing the increasing investment in GI especially in regard to flood protection.

> Overall, it was found that green infrastructure projects have a high proportion of capital costs which are incurred up-front, with benefits that are delivered into the future. As a result, the creation, restoration, and maintenance of green infrastructure projects often require substantial investment. This highlights the importance of taking account of the future flow of benefits when appraising green infrastructure projects, but also identifies that funding opportunities may be important in encouraging uptake of green infrastructure projects.

In other industrialised countries (e.g. in the USA and China), NbSs have been and are becoming the preferred approach. In the developing world, there are problems with inadequate infrastructure for the management of solid wastes and sanitation, inhibiting the effective uptake of NbSs (e.g. Ashley and Horton, 2022). Not all the initiatives have considered the multiple benefit opportunities, although many have linked the use of NbSs to ecosystem services (ESs), capitals accounting and the Sustainable Development Goals (SDGs) (UN, 2021).

Those involved in the management of urban drainage have become familiar with using analytical models to quantify the physical performance of a drainage system (NbSs or grey infrastructure). Even the professionals who traditionally are concerned with the quality and aesthetics of places respect the numerical quantification of costing any enhancements. A lot of the added benefits from nature as infrastructure are less tangible than the predicted results from traditional models of drainage performance. These provide the value of the reduction of flood risks to properties and places, or for the reductions in water pollution. Many of the added benefits provided by NbSs cannot readily be quantified simply in physical or numerical terms. Therefore, there are challenges in assessing the value of the benefits brought from nature as infrastructure, especially for the professionals who are more familiar with traditional ways of quantifying the value of changes to systems, such as cost–benefit analysis.

There are well-established economic approaches to valuing 'intangibles' used as part of environmental or ecological economics, especially for putting a value on scarce resources that are usable for natural infrastructure systems (e.g. USEPA, 2023b); seen in this perspective as 'non-market goods'. In Chapter 5, ESs and the Millennium Ecosystem Assessment were introduced. This has provided an internationally agreed foundation for valuing the benefits provided by nature (e.g. TEEB, 2023; UN, 2023). The understanding and definition of the multiple benefits provided by nature is an evolving field of endeavour, but currently has followed the ESs reductionist, 'financially cost everything' principles of neo-liberal economics. Most decision-makers are of the view that if it does not have an economic value, then it has less of a role in influencing how options for a new or modified drainage system are to be assessed. For many, in contrast, the put-a-value-on-nature is anathema, and they believe nature should intrinsically have value not set in human economic terms (e.g. Erickson, 2022; Monbiot, 2020). In practice, however, decision-makers remain steadfast in wanting the 'best value for money' for infrastructure investments, be they natural or traditional. This is further considered in Chapter 12 on sustainable water systems.

Even best value for money valuations are complex and challenge decision-makers to address questions such as: 'whose or what value?'. For example, Drupp *et al.* (2018) consider inequality in access to and use of NbSs: 'From a sustainability perspective, economic valuation should not only seek to determine a society's willingness to pay for environmental goods to devise an efficient allocation of scarce resources but should also account for distributional effects to ensure justice.'

In Chapter 3, aspects of the way in which decisions should be made about the beneficiaries for flood risk management, balanced with environmental impacts, were considered, based on Semple (1991). This included aspects of equity and need. Although there are often explicit considerations given to benefiting the most vulnerable in society when allocating flood risk reduction funding (e.g. Environment Agency, 2020), the concepts of 'fairness' or 'justice' are used only rarely and remain difficult to incorporate into decision-making for infrastructure investments (e.g. Sayers, 2017).

With the foregoing caveats for how nature has been valued as 'non-market' goods, the following considers the added, or multiple, benefits of SuDS as NbSs. These values are commonly classified in terms of two areas

- the value to ecosystems by way of encouraging living systems, expressed in terms of biodiversity
- the other aspects of value to humans, collectively termed amenity, although benefits in this area cover much more than amenity alone.

There is no standardised agreed methodology for valuing the ESs of catchment systems. Kaval (2019) reviews the various approaches in use, highlighting boundary problems in applications (including temporal boundaries, and between services, catchments and governance jurisdictions).

As stated earlier, for many projects, the objectives may be to focus on delivering infrastructural changes for the primary objectives. In addition, this may also bring about value for one or other of the added benefit categories above. For example, the upgrading of the roads in and around the Canal Road corridor in Bradford, England, has provided an opportunity to use BGI and SuDS for the associated drainage and to daylight part of the existing Bradford Beck, a watercourse that runs alongside the road (BEGIN, 2023b), as illustrated in the case study in Chapter 14.

Capturing the value from using NbSs in a form that is understandable to decision-makers is necessary to support the development of an evidence base and informed decision-making. Most decisions are made based on evidence provided in a business case (e.g. Ashley *et al.*, 2023). The components and definitions being used to make a business case for using SuDS or NbSs accounting for the potential multiple benefits are outlined in the following sections.

8.4.1 The capitals

'Capital describes any resource or asset that stores or provides value to people' (Natural Capitals Coalition, 2019). From this, four capitals have been defined for accounting purposes (see also Chapter 5).

- natural capital: the renewable and non-renewable natural resources that provide benefits to people
- social capital: the cultural networks that facilitate cooperation within and among groups
- human capital: the knowledge, skills, competencies and attributes in individuals that contribute to improved performance and wellbeing
- produced capital: the human-made goods and financial assets that are used to produce goods and services of use to society; sometimes known as 'manufactured capital', together with 'financial capital'.

Natural capital is of particular relevance for NbSs, and is the stock of natural assets, such as water bodies, habitats and soils. These assets support and provide a flow of ESs – for example, clean water. The stocks and flows in turn provide a range of economic, social and environmental benefits to people. Restoring and enhancing natural capital can therefore lead to an increase in each of the four capitals above.

A 'capitals approach' is broader than considering ESs alone in that it considers the stocks of assets as well as the flows, is more amenable to taking an integrated approach and fits with well understood business accounting processes (Eftec, 2021). Thus when making a business case for using NbSs, basing this on the capitals can provide an understandable and comparative source of information for decision-makers (Ashley *et al.*, 2023; BEGIN, 2023c). In England, both the government and major infrastructure providers (particularly water companies) are following a 'capitals' approach to account for nature and social value (e.g. Conlan *et al.*, 2017; HM Government, 2023). This is a rapidly developing area and aims to provide a marketable value for nature to attract private investment and minimise public investments. Projects elsewhere are also seeking cost-beneficial incentives for investing in nature, such as the European 'Woodlands for Water' initiative (Forest Research, 2023), set within the ESs context, and whereby woodland investment can be linked directly with the value of improving runoff quality and for flood risk reduction.

NbSs and social capital are an interesting concept, particularly with regard to the younger generation. There is much to be said for 'social capital'. In so-called developed countries, what people value is their surroundings, such as open areas and green space. However, establishing social capital is neither quick nor easy. It requires trust, common rules, exchange and connectedness, which can only be achieved when communities feel engaged, are treated with honesty and respect and can influence what happens; but high social capital results in cooperation and collective action, thus meaningful engagement is crucial, as in the Living with Water programme in Hull, Yorkshire (Ashley *et al.*, 2023; BEGIN, 2023a).

The capitals approach may not be the only one being used to make the case for NbSs. Alternative approaches are being developed and should be considered when valuing NbSs. For example, there are projects developing marketable cases for investing in NbSs, such as Invest4nature (2023), although most approaches use non-market valuation (e.g. the INFFEWS tool for water-sensitive cities; Pannell, 2020).

The two added benefit categories of biodiversity and amenity are considered next, then Section 8.4.4 outlines current approaches to valuing the overall multiple benefits from using SuDS and NbSs.

8.4.2 Biodiversity net gain

Biodiversity encompasses the number, abundance and distribution of all species of life on earth. It includes the diversity of individual species, the genetic diversity within species and the range of habitats that support them. Biodiversity also includes humans and our interactions with the environment (e.g. Dale *et al.*, 2011). Locally, biodiversity reflects the character of the plants and wildlife that share the space in which humans live, work and play. Ecology is the study of plants (flora) and animals (fauna) and the relationships between these and the physical environment. An ecosystem comprises both a biological community and its physical environment (Dale *et al.*, 2011). This is the key component of natural capital, as defined in Section 8.4.1.

Designing for biodiversity can bring new life and opportunities into urban areas for plants, people and wildlife, and this enhancement to environmental value or natural capital provides clear economic as well as social benefits that align with the capitals in Section 8.4.1. The contribution of biodiversity to development that is more sustainable and improves the quality of spaces is recognised throughout local, national and international policies. For many years, the emphasis has been on preserving, protecting, conserving or restoring biodiversity despite development pressures and activities (e.g. Royal Society, 2023; UNEP *et al.*, 2015).

More recently the vision has been to add to existing biodiversity, ensuring that there is a gain in biodiversity for each development; an objective or outcome termed ‘biodiversity net gain’ (BNG) (Baker *et al.*, 2019; Defra, 2023). Development planning approval in England is now subject to demonstrable BNG (Knight-Lenihan, 2020). Mitigation by, for example, offsetting, is still an option, whereby biodiversity impacts or losses from a development project are compensated for by gains at another site. This raises questions about how to assess net gain against net loss (e.g. Cuckston, 2019).

Although the term BNG is now established for use in England, elsewhere this definition is seldom used. A new version of the global biodiversity framework, coordinated by the Worldwide Convention on Biological Diversity, is expected in the first half of the 2020s decade. A framework that nations have signed up to will need to be adhered to by way of national biodiversity strategies and action plans (Convention on Biological Diversity, 2023). NbSs are and will be a key component of delivering strategies and action plans (Brorken *et al.*, 2022).

Good practice SuDS design can ensure that urban space is used cost-effectively to maximise wildlife and landscape benefits while at the same time delivering efficient drainage of urban surfaces. Wildlife will always inhabit the areas used by people whether the environment is designed to be hospitable for animals and plants or not; even where the conditions that enhance wildlife are neglected. However, the design of landscape features that support diverse habitats and associated ecosystems will provide a healthy and stimulating environment that can add significant value to urban living. SuDS provide opportunities to create ecologically rich green (vegetated and landscaped) and blue (water) corridors in developments – connecting people to their environment and enriching biological value by connecting networks of habitats and ecosystems together. As many SuDS utilise and also enhance GI provision, they provide significant opportunities to add to urban habitats and support ecosystems, especially where there are interconnected networks of green corridors and 'nodes' where wildlife can travel and feel secure (e.g. Washbourne and Wansbury, 2023).

Biodiversity benefits can be maximised through the inclusion of open water, green corridors or wetland mosaics (ideally with a range of water depths and inundation frequencies) to create a range of habitat types; by using varied but always local planting, and by designing ecologically sensitive maintenance plans and waste management strategies. Ideally, systems should provide the right conditions to support diverse, self-sustaining and resilient ecosystems.

Biodiverse environments are invariably linked with high-quality and highly valued open space. By integrating surface drainage systems within urban space it is possible to recreate lost habitats (albeit different habitats) and start to link natural areas with 'blue-green' networks, providing linked corridors, feeding, breeding and resting grounds for local species. Such networks can potentially provide a significant contribution to local biodiversity goals and will increase resilience to climate change impacts. The future potential of an area for wildlife and habitat development can be evaluated, starting with the current habitat characterisation and considering what species of flora and fauna might be able to exploit the new habitat created by SuDS. This may be based on standardised techniques, such as the natural vegetation classification methodology (JNCC, 2023).

ESs concepts can be used to estimate the value and hence the importance of ecology, ensuring diverse natural communities based on an ecological assessment of existing conditions and expected enhancements because of the development (e.g. Biodiversity in Planning, 2023; BSI, 2013b).

Designing SuDS for biodiversity is dependent on meeting objectives around managing water quantity and quality. It also requires the integration of knowledge and practices from a wide range of disciplines, including spatial and urban planning, landscape design and ecology with engineering drainage design good practice. Integrated design requires an understanding of both ecological principles and the specific design actions needed to provide a supportive habitat for animals and plants. Table 8.2 provides examples of objectives and indicators for delivering enhanced biodiversity using SuDS (see also Washbourne and Wansbury, 2023; Woods-Ballard *et al.*, 2015).

8.4.3 Amenity

Amenity is 'a useful or pleasant facility or service', which includes both tangible and intangible benefits. The tangible benefits can be measured in terms of use, such as numbers of people; the less tangible benefits are not measurable in physical terms, but can be experienced as pleasurable or aesthetically appealing experiences. In terms of the capitals in Section 8.4.1, amenity comprises both social and financial capital. Having a pleasant urban space, for example, includes regulating temperature (e.g. urban heat islands), a major consideration for using WSUD in Australian cities. Temperatures can be reduced by several degrees over 24-h periods by using BGI, reducing mortality rates among the most vulnerable (e.g. ClimateADAPT, 2023).

Table 8.2 Biodiversity principles and examples of indicators

Biodiversity objective	Example indicator
Support and promote natural local habitat	The number, type, extent, connectivity and diversity of habitats delivered by SuDS
Contribute to the delivery of local biodiversity objectives	The habitats delivered by the SuDS scheme that meet biodiversity plans or equivalent objectives (including green infrastructure provision)
Contribute to biodiversity value through SuDS design	The number/area of drainage components that add biodiversity value to the site
Deliver diverse and resilient ecosystems	The range of habitat types delivered by the SuDS scheme, and the likely resilience of the ecosystems and supporting habitats to potential disturbance from climate change and other potential impacts

This definition of amenity can be used for defining the multifunctional opportunities associated with SuDS designs and can link with 'place-making' (Chapter 5), used in describing the quality of space in urban design. Amenity also covers 'needs' and 'liveability', which are associated with wellbeing and factors that are essential to and can improve the quality of life for inhabitants, as illustrated in Table 8.3, for the water-sensitive city. As considered in Chapter 12, liveability may be seen as the 'here and now' and functionality aspects of sustainability.

Over a century of living with subsurface drainage networks and surface flood 'defence' has led to a lack of societal recognition and understanding of the impact of development on the receiving environment and, in particular, flood and water quality risks. As climate change drives increasing rainfall intensities and consequently drainage systems become increasingly vulnerable to exceedance, living with and sustainably managing water locally, within communities, provides more tangible benefits than uneconomic approaches to increasing drainage capacities.

By integrating drainage systems, on the surface, within developments, through WSUD (Chapter 12), SuDS and drainage good practice, there is a demand for an increased environmental awareness and levels of acceptance from communities (Chapter 4). Particularly within urban areas, water has been so efficiently removed from the urban landscape that there is, unsurprisingly, a real fear of the potential associated safety risks of 'water on the surface'. Such fears can only be managed through community education and participation, high-quality drainage design and appropriate system maintenance. Concerns tend to be highest when systems look unattractive or are poorly established, where open water quality is poor, or where maintenance is inappropriate. Systems are also believed to be a greater hazard when the open water is hidden or is not in full public view. These factors also lead to poor public acceptability of the drainage system and often to sites being at higher risk of vandalism (Chapter 4).

Amenity criteria need to, among other aspects of SuDS design, include health and safety through appropriate design; minimising slopes adjacent to open water, keeping water depths shallow and providing safety benching at the margins. In accordance with health and safety legislation, systems should be safe for both construction and maintenance work and during their ongoing operation. Any feature that

Table 8.3 Urban water societal needs and the contribution that SuDS can provide (adapted from Ashley *et al.*, 2013)

Needs category		Urban water societal need	Description	SuDS potential contribution
Existence	Physical and material needs	Potable water	Safe, secure supply of water for consumption	Harvesting from source or storage components; replenishing groundwater
		Non-potable water	As above for all other purposes	As above, but lower levels of treatment
		Public health	Protection from contaminated water, controlled microclimates, public places promoting physical and mental health	Water quality improvements; greening and heat island mitigation; parks and green infrastructure benefits
		Public safety	Protection of people from hazards such as flooding	A primary function of all SuDS; landscape SuDS also reduce crime levels
		Property protection	As above for property	As above
		Economic activity	Industries and jobs that rely on water and related services	Landscape SuDS create and sustain jobs, tourism etc.; source controls make water available near places of need
Relatedness	Social interaction and inter-personal relationships	Recreation	Places for play, sport and leisure	Many SuDS provide green or blue areas that facilitate these activities
		Social cohesion	Safe and secure places for social interaction and human connectedness with people and nature	Local and or source control SuDS especially can be community managed; landscape SuDS can connect people with ecosystems
	Societal environmental inter-relationships	Beauty	Aesthetic urban environments	Landscape SuDS are aesthetically attractive
		Comfort	Pleasant micro-climate and landscape for thermal comfort	Landscape SuDS reduce urban heat and also absorb greenhouse gases
		Ecological health	Clean and healthy ecosystems with no negative impact on other ecosystems	Landscape SuDS create new ecosystems and sustain these; by providing treatment pollutants are managed

(*continued on next page*)

Table 8.3 (*Continued*)

Needs category		Urban water societal need	Description	SuDS potential contribution
Growth	Societal self-esteem and self-actualisation	Identity	Harmony with culture and tradition; belonging. Proud association with urban water systems	Not an attribute typically found in the UK other than for industrial or commercial developments
		Purpose and ambition	Progress towards a shared vision of a water-sensitive future	This varies around the world. The various UK countries each have visions for a more joined-up approach to water management (Chapters 5 and 12), but SuDS can provide key opportunities and incentives for managing water more collectively
		Control and independence	Choice and influence on decision-making about water infrastructure and services	Around the world there are various governance and institutional arrangements for water management (Chapter 5); most welcome the use of SuDS/NbSs, although the form of use varies
		Equity and social justice	Equal opportunities to access the benefits of urban water systems	Landscape SuDS, where best on public land to provide this? (introduction to Section 8.4)
		Intergenera-tional equity	Preserve the ability of future generations to meet their water needs	SuDS are more flexible and adaptable to future changes and can therefore be modified easily if needed (Chapter 12)

may constitute a hazard (e.g. a drop of >2 m) should be reconsidered and, if still required, fenced appropriately.

Surface drainage systems need careful integration with development 'space' to maximise the benefits of potential multifunctionality, to maximise the amenity and aesthetic value of water, to maximise contributions to community or planning objectives for an area, and to add economic value (Chapter 5). There are acknowledged and measured health benefits of urban green space (and in particular water space), and physical benefits associated with mitigating local temperature increases associated with urban heat island effects.

It is fundamental that, alongside good design and maintenance regimes, individuals and communities are educated as to their potential impact on the water environment, how drainage systems function and mitigate these impacts, particularly during periods of extreme rainfall, and the implications of potential future damage and poor performance. By also tasking communities with some responsibilities with respect to longer-term maintenance, a more sustainable future for drainage can potentially be realised. Amenity design principles, together with associated criteria, are set out in Table 8.4.

Urban environments are constantly changing and are never complete (Chapter 5). Living with and helping to shape change in surface water systems is relatively easy when SuDS are used. Changes can come about due to environmental factors such as climate change and also from changes in planning and public demands and expectations. It is often straightforward to modify a SuDS component in response to change by adding an additional area or component or modifying part of the system. SuDS have been shown to have greater flexibility than underground piped drainage systems due to their relative ease of modification, and therefore provide greater resilience (Chapter 12). Hence, adapting to future uncertainty is also easier with SuDS. SuDS will contribute to ensuring that landscape value can be maintained or even enhanced when change happens and that the other amenities described in this section and Chapter 12 will continue to be supported.

Table 8.4 Amenity principles together with examples of indicators that may be used to assess how well the principles are being achieved

Amenity principles	Example indicators
Use water as a resource	The proportion of runoff that is 'used' as a resource, e.g. for domestic use, for landscape irrigation or horticulture, as part of a community recreational feature, etc.
Maximise multifunctionality	The numbers and variety of additional and multifunctional uses for the SuDS scheme, e.g. for recreational areas, for car parking, for traffic management, for exceedance flow management, etc.
Enhance visual character	The proportion of the drainage system that is designed to be visually attractive, add visual value to the development, support local heritage and landscape character, etc.
Accommodate change	The proportion of the drainage system that is designed with future climate change or development change impacts accounted for. The proportion of the drainage system that will contribute to the development's climate resilience, e.g. through reducing the heating/cooling needs of buildings, through shade provision, etc. and is readily amenable to modification
Design for safety	The consideration of safety within the design of each SuDS component and for the system as a whole. Reduced numbers of traffic-related accidents. As a minimum, demonstration that the health and safety principles set out in the health and safety framework have been followed effectively
Raise awareness	The extent of interpretation boards, signposting, community awareness, school involvement, visitor provision is planned for/initiated as part of the development submission

An example of the contribution SuDS can make to community amenity through place-making and improved environments is illustrated in Figure 8.5 in Rotterdam, where water squares are serving to provide water storage during exceedance flows and recreational spaces when dry.

Figure 8.5 does not show the more typical SuDS that help to green areas, as illustrated in Figure 8.6.

Figure 8.5 A water square in Rotterdam (Richard Ashley)

Figure 8.6 A typical retrofit SuDS in Haringey, London (Robert Bray Associates/Maple Photography)

8.4.4 Valuing the multiple benefits from using SuDS and NbSs

As outlined above, value is 'in the eye of the beholder' and can mean many things to many beneficiaries. Ultimately, the entire global society benefits where infrastructure that is more sustainable is used (Chapter 12). However, investors (public or private) are interested in the value they gain from this. For many investors, wider societal benefits may be an objective, whereas for others, especially institutional service providers, such as the private water companies in England, they are constrained by the licences they hold to operate and need to deliver a direct and clear outcome that deals with a problem. Grey infrastructure provides a clear investment path and a clear and focused outcome. NbSs, in contrast, in providing multiple benefits, deliver a much more complicated set of outcomes, even when these deal with the original outcome. Thus, for NbSs, often the primary beneficiaries are not necessarily the investor, but other stakeholders (Ashley *et al.*, 2023). To demonstrate this, the costs and benefits of using SuDS/NbSs need to be determined using agreed approaches. Costs are considered in Section 8.4.5, here the benefits are presented.

There are several approaches to categorising and valuing the multiple benefits that can be gained from using SuDS and NbSs. As outlined in the preceding sections, these usually utilise ES classifications in assigning financial value to the benefits. It is first necessary to estimate the direct or indirect physical elements of the benefits when these are tangible and can be measured or modelled (Chapter 11). Then an economic or financial value can be determined based on databases or established procedures. Table 8.5 lists a number of categories that can contribute to value, and how SuDS can contribute to these.

The categories in Table 8.5 are examples of benefits in use for various assessment approaches and tools. These utilise differing categories and take different approaches to the assessment of benefits. Some of the analytical tools in use are outlined in Chapter 11. As an example, Box 8.5 shows how the B£ST tool developed by CIRIA (2023d) is used to outline the approach being used in assessing and valuing the multiple benefits. B£ST may be used for both grey infrastructure and SuDS/NbSs. The valuation process is the same, although the grey infrastructure will have fewer benefit categories.

In a comparative evaluation of the benefits of CSO spill reduction compared with the investment costs in Yorkshire, traditional grey infrastructure investments were found to solve the problems and cost the least, whereas investments in the SuDS alternatives brought many more added benefits but cost up to three times more. These multiple benefits were mainly to property values, recreation and community activities – that is, to social capital (Ashley *et al.*, 2018). As the water service provider's main function was to solve the CSO problems, any investment in wider societal benefits would require difficult justification of the significantly greater expenditure needed and may require partnership working.

A major challenge is that many of the benefits do not accrue to the investors as indicated above, instead going to elevated property values, health benefits and other non-water-related societal benefits (e.g. Ashley *et al.*, 2018; Zhou *et al.*, 2013). Chini *et al.* (2017) reviewed the motivations for using NbSs in 27 applications in the USA. In each plan, multiple benefits were expected from the use of NbSs. Of the cities, only five explicitly included water quality improvements due to CSO reductions without this being mandated, although there were nine cities with an 'EPA consent decree' reason for improving water quality (legal mandates stating that the city is lawfully bound to decrease CSO discharge volumes). Most benefits accrued to 'cleaning up waterways' and to 'greening aesthetically pleasing areas'.

Table 8.5 Examples of benefit categories potentially supported by SuDS (see Ashley *et al.*, 2013)

Benefit category	Examples	Physical characteristic examples
Rainwater harvesting and urban agriculture opportunities	Direct collection of rainwater to use for domestic and other purposes saves water, as well as potentially providing essential irrigation resources and long-term viability for urban plants and crops	Volume collected for use/volume of potable water saved
Recreational opportunities	SuDS provide a wide range of green and blue spaces that can be used, for example, for walking, cycling, informal play or organised sports and games	Numbers of people using new spaces or using enhanced existing spaces
Parks, other public spaces/aesthetic and visual amenity	SuDS can provide opportunities within the public realm or parks that may overlap with recreation above but also provide pleasant and aesthetic places to be. Below-ground SuDS, such as infiltration basins or geocellular systems, can keep spaces open, as these cannot be built over	Numbers of people with an enhanced view of green and blue spaces
Investment enhancement	Using SuDS to collect surface water to irrigate the green areas and creating attractive places encourages and supports inward investment and often attracts tourists	Amount of added investment as a result of added blue and green spaces
Biodiversity and ecology (ecosystem)	Landscape SuDS help to support flora and fauna and here SuDS amenity and biodiversity value come together	Numbers of additional species as a result of the new spaces
Jobs and labour productivity enhancement	Productivity is enhanced in attractive environments, such as business parks with green spaces, which can help reduce staff recruitment costs. Ponds and landscape SuDS provide job opportunities for maintenance that last indefinitely, with additional jobs also being created by the added tourism from pleasant urban landscapes	In Philadelphia, for example, it was estimated in 2009 that 170 permanent jobs would be created from 25% of the city being retrofitted with SuDS (Stratus Consulting, 2009)
Air quality improvements	SuDS using blue and green areas, including grass and trees, provide significant air quality improvements by, for example, trees 'scrubbing' fine particulates from urban streets, resulting in reduced accident and emergency hospital admissions for respiratory based conditions	Toronto's urban forest comprising some 10 million trees removes over 1400 t of air pollutants annually (City of Toronto, 2013; Pugh *et al.*, 2012)

Table 8.5 *(Continued)*

Benefit category	Examples	Physical characteristic examples
Traffic calming/ management/ car parking	SuDS, such as rain gardens and bioretention systems, can provide horizontal constraints in roads, discouraging driving at excess speeds. Spaces between SuDS components or the components themselves can also provide parking spaces for cars and bicycles	Numbers of vehicles calmed by highway constrictions
Noise reduction	SuDS and associated trees and grassed areas can provide noise-absorbent barriers and surfaces in noisy urban areas. Green roofs can also provide sound insulation within buildings	Volumetric noise reduction levels
Aesthetics and quality of places	SuDS provide aesthetic value and green/ blue space and contribute to biodiversity, albeit there is subjectivity in assessing the quality of places	Guidance as to how best to provide and assess this, especially for large-scale developments, from Landscape Institute (2016)
Land and property values	Landscape SuDS add value to land and property nearby. In all studies, certain types of open space enhancement and preservation have contributed to higher housing prices	The increased value of property on markets
Tourism	SuDS in themselves may provide interest for tourists, especially where they are a novelty, as in Malmo in Sweden, but SuDS are most likely to be part of creating attractive places that will appeal to tourists	Increased numbers of tourists
Energy use reduction	SuDS require far less energy use in all stages of the supply chain and life cycle than conventional drainage, and harvesting water at source also saves energy	There are some 100 million mature trees around residences in the USA. These trees save about $2 billion annually in reduced energy costs (Fazio, 2023) by protecting from winds in winter and giving shade in the summer and as a renewable source of fuel. De Sousa *et al.* (2012) considered the relative merits of using porous pavements, trees and bioretention SuDS for CSO control in New York and concluded that there was a reduction of some 314 kWh/megalitre in energy needs for the wastewater treated at the downstream plant when using SuDS

(continued on next page)

Table 8.5 (*Continued*)

Benefit category	Examples	Physical characteristic examples
CO_2 and other greenhouse gases reduction	Plants and soils take in and store CO_2 and other greenhouse gases; hence where SuDS use plants this potential can be exploited	On average one tree absorbs some 25 kg of carbon dioxide a year (Ecotree, 2023)
Temperature extreme buffering	Green and blue infrastructure buffers and moderates temperatures, which will become increasingly important as the climate changes and urban areas get hotter in future	The use of SuDS in Australia as part of water-sensitive urban planning and design (e.g. Rosetti, 2013) has been shown to reduce urban heat island temperatures by judicious location to maximise their effectiveness (see also Gill *et al.*, 2013)
Community cohesion	SuDS can help bring communities together. By increasing opportunities for human interaction and creating a more enjoyable environment, people are more likely to feel they belong to the community and take a greater pride in their neighbourhood. This is especially the case if the community has been involved in the SuDS design process and residents have ownership of the ongoing maintenance (even if only in part)	An example of this is the award winning permeable paved scheme in the Home Zone, The Dings, Bristol, where the community came together to decide on what they wanted (CIRIA, 2023a)
Public education	Using green and blue spaces as part of the management of the water cycle provides many opportunities to support education both formally in schools and in communities	Can be measured by the new numbers of pupils/students accessing these spaces

There are also issues with determining 'whole-life value' for both grey and green infrastructure. As stated earlier, NbSs do not provide immediate benefits following construction, but take some time to establish. There is therefore is a 'run-in' period when benefits start to accrue. NbSs also usually need more maintenance than grey infrastructure – for example, to mow grassed areas and clear litter. Grey and green systems also have differing requirements for major upgrading, depending on the nature of the NbSs. Without continuing verification and interventions as needed, NbSs might not provide the expected added benefits, such as aesthetic value, into the unknown future. As the future is uncertain, responsibilities might change over time and different operators may become responsible for maintaining functionality. This is especially true where the systems are on private land or operated by, for example, a licensed contractor. It is apparent from the neglect of maintenance of private piped drainage (Chapter 2) systems that this can be a problem for grey infrastructure, and even more so when NbSs are used. Values may also change in the future. What is seen as an important added benefit from NbSs now might not be so important in the future. It is essential that when planning and designing NbSs (and is also true for grey infrastructure, but largely ignored), some formal futures assessment is undertaken if whole-life value is to be properly determined (Ashley *et al.*, 2018). This is considered further in Chapter 12.

There is thus a need for 'value integration'. Value integration means the coming together of different values of different stakeholders, for the benefit of society, making our world more secure, smart, shared, sustainable and satisfying (BEGIN, 2022). Each stakeholder will have individual objectives, constraints and limitations on their actions and ability to utilise the multiple benefits. These need to develop and use different types of business cases for funding (Ashley *et al.*, 2023). It is therefore important to understand and work within these differing needs, perspectives and frames (e.g. Wild *et al.*, 2017).

As this type of valuation process is in its early stages of development, it is a rapidly advancing area. New and modified analytical processes and tools, now mainly linked to national databases by way of a geographic information system (GIS), are being developed and it is important to access the latest ideas for this when undertaking valuations of either grey or nature-based infrastructure systems. For UK applications see, for example, Valuing Nature (2023). There are numerous other sources worldwide such as WEF (2023) and the White House (2022) that can be accessed depending on context and objectives.

Box 8.5

How the Benefits Estimation Tool (B£ST) is used to assess the value of infrastructure

The core of the valuation process is the impact pathway, shown in Figure 8.7 for the ecosystems category. The scheme options can be analysed, and physical performance determined in terms of water quantity and quality aspects. The wider aspects of performance, such as enhancements to biodiversity, or other benefit categories (Table 8.5) can also be quantified. Then, the economic value of these changes can be determined using standard, usually national, agreed information, as shown in Figure 8.7.

Figure 8.7 Impact pathway

In B£ST, as an example, amenity values are consistently among the most valuable benefits, even for water management schemes (Ashley *et al.*, 2023). This is because people value blue and green spaces, and any additions will usually increase property values. For example, a Land Trust (2018) study found a price premium of 5.4% for every 100 m (within 500 m) that a house is closer to a new city park. The Green City, Clean Waters programme in Philadelphia (Stratus Consulting, 2009) found a property premium of 1.7% to 12.7% (best estimate 10.3%) associated with proximity to green surface water infrastructure features. Thus it is possible to assign a financial value to this benefit. Of course, the value accrues to property owners, rather than necessarily the scheme investors. This raises various issues, not least whether the property owners should contribute directly to scheme costs (Ashley *et al.*, 2018).

8.4.5 Costs of SuDS and NbSs

The wide application and numerous examples now available demonstrate that SuDS/NbSs can be both cheaper and costlier to implement than traditional grey systems. However, once benefits are considered, the overall net present value of using NbSs is invariably better than for traditional approaches, even when costs are higher. As with benefits, costs should be assessed for the whole of the life of the assets (Woods-Ballard *et al.*, 2015). Costs can be considered in terms of the following, discounted over the lifetime (adapted from Sayers *et al.*, 2022)

- capital costs (CapEx): the capital expenditure associated with initial implementation of the measure (as part of a scheme or alone)
- operational cost (OpEx): the (annual) operational expenditure for operation, maintenance and periodic renewal
- carbon cost: including construction (embodied) and operational carbon, with carbon emissions valued using standard national guidance
- support costs: annual costs that are not directly attributable to a scheme, such as improvements of existing functionality and payment for additional benefits, during the scheme lifetime.

There are numerous other costs associated with the above, including

- on-costs: due to, for example, increased treatment costs associated with increased conveyance of flows through the drainage network
- land purchase costs: which can be significant for NbSs, especially on private land. There may also be a need to block land for future expansion of a scheme – that is, for adaptation purposes
- construction disruption: some measures, such as piped drainage improvements, are often associated with significant disruption during construction
- transfer payments: related to taxation transfers, which may or may not be applicable
- residual asset values and benefits: returns on investments within the appraisal period will continue to accrue benefits within the design life. After this, the asset may continue to perform and accrue benefits
- optimism bias: there is a tendency for cost assessments and benefit assessments to be over-optimistic. Most jurisdictions have rules and recommendations for managing this (see Woods-Ballard *et al.*, 2015)
- affordability: investments are assumed to take place from day 1 and then further adaptation will occur in response to future change
- resources and capacity building: especially important for NbSs, in which every stakeholder must come to terms with the approach.

These cost categories are usually the most significant and some may be applicable to both grey and nature-based systems.

The relative feasibility of using SuDS is dependent in part on whether schemes are retrofitted or newly built. SuDS schemes for new developments are generally less capital intensive, can have lower operational costs if managed well, and are inherently more flexible in that they can delay the expansion of existing grey infrastructure. Figure 8.8 shows capital costs and Figure 8.9 annual operational costs for SuDS in the UK determined from a survey of new developments (Ashley *et al.*, 2017).

There is a wide range of costs for the various reported GI projects that consider SuDS compared with a traditional approach, making comparisons and generalisation incoherent. For example,

Figure 8.8 Capital cost ranges for SuDS usage for (primarily) development schemes (Ashley *et al.*, 2017)

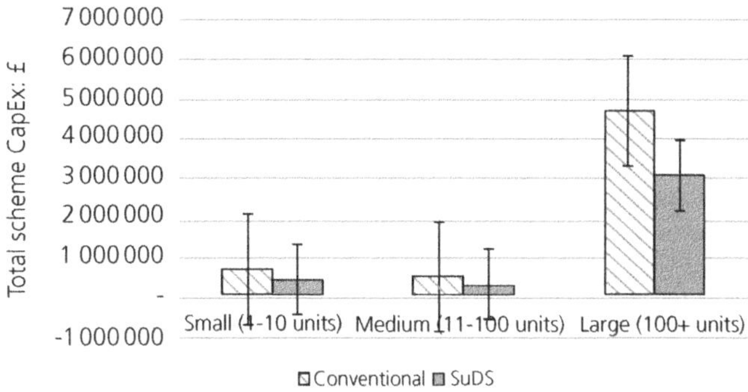

Figure 8.9 Annual operational costs of SuDS compared with traditional drainage schemes (Ashley *et al.*, 2017)

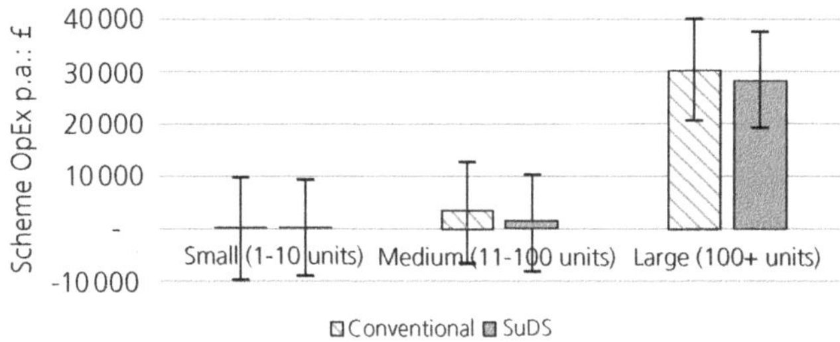

Wright *et al.* (2016) demonstrate that costs per cubic metre of stormwater runoff removed from a network in the USA can range from \$3 to \$600 depending on the SuDS component and site-specific conditions, and payback periods ranged from 3 years to never. Other studies have considered the value for money (e.g. Johnson and Geisendorf, 2019) of SuDS in terms of scale of application.

In situations where retrofit SuDS can result in eliminating an existing underground conveyance network, cost savings can be substantial. By contrast, in the short term, retrofit SuDS may only be feasible when they can be competitive with options to maintain or upgrade such infrastructure, although they will provide many added benefits.

Financial viability is possible under the right conditions even when retrofitting. In the example from Philadelphia mentioned earlier, the city's water department set out a plan to spend 67% of its water quality improvement budget until 2026 on GI. Although nominally charged with minimising CSO spills, the city adopted an assessment methodology that allowed inclusion of the wider multiple benefits (Dolowitz *et al.*, 2018). The city's municipally-owned water department was able to cooperate closely with other planning-related departments to identify and implement some of the multiple benefits of using GI, all of which accrue to the duties of the Mayor's Office. The availability of land, due to vacant lots and decreasing house prices, provided the opportunity for multiuse land utilisation to be implemented at minimum cost.

The importance of land expense and availability can be seen in the Glasgow Surface Water Management Project. Scholz *et al.* (2006) identified feasible scheme types based on land value, with all SuDS interventions with an above-ground footprint being not feasible for land more expensive than £200 per m^2 (prices relevant at the time). Interventions with no footprint, such as permeable paving and underground storage, were also included in the analysis. A secondary impact of adding green spaces is that neighbourhoods which utilise SuDS have experienced increased property values and gentrification, not always a desirable outcome in places such as Glasgow (e.g. Eckart *et al.*, 2017).

There is a perception that the costs (capital and operational) of using grey infrastructure are well understood and relatively easy to access. This will depend on national sources and availability plus what has been included and what has not in the cost estimation (see the bulleted lists above). For example, Webber *et al.* (2019) compare the costs of grey piped drainage with those of SuDS. They provide costs of upgrading a network of pipes in the UK to cope with climate change conditions, based on cost estimates given by, for example, the Environment Agency (EA, 2015) and utilising data from Babtie (2003). The raw costs of constructing pipelines and ancillary structural assets in the UK are provided in the Spon (2023) price book series, although interpretation of the civil engineering and other costs requires specificity of scheme details and an itemised bill of quantities to be meaningful.

Several other examples of costing SuDS in the UK are available, including most recently the London Strategic SuDS Pilot Study (Arcadis, 2020). Earlier UK estimates are provided by Royal Haskoning (2012) and specifically for roads by SCOTSNET (2010).

The costs of SuDS and piped drainage systems aimed at addressing changing flood risks in England were compared by Sayers *et al.* (2022). Table 8.6 summarises the cost information and sources used for this.

Despite the above examples, it is difficult to generalise about the costs of SuDS, due to the variabilities of units used, location, scale, etc. For example, Naumann *et al.* (2011) show how there is a wide range of per hectare costs for their 127 reviewed European GI projects, with capital costs ranging from €250 to almost €1 million per hectare. Not all these projects were to deal with storm or surface water management, and unlike piped drainage systems, included a wide range of cost elements. Only a few of the 127 projects sought to use GI as a substitute for grey infrastructure. Most projects aimed primarily to meet nature conservation objectives and/or to create and maintain green space for recreational, amenity and landscape purposes. The largest element of costs for six of the largest projects related to land management and restoration works, comprising 43% of the overall total cost. The largest continuing costs related to the management and maintenance of land and buildings, 68% of the total, and project management and administration, 25%.

Maintenance and operational costs are also difficult to assess and hence predict for deciding on the best SuDS/NbSs option. Lack of experience of NbS use in many parts of the world makes estimation difficult. There are databases and published data, for example, Clary and Piza (2017), Clary *et al.* (2022) and Weiss *et al.* (2012), but these may not be applicable in other contexts.

Table 8.6 Average costs of SuDS and piped drainage used in the assessment of adaptation to changing flood risks in England (Sayers *et al.*, 2022)

Type of SuDS	Cost per urban property protected: £ (2021 prices)		Carbon cost per property protected, per £ spent on SuDS: £ (2021 prices)
	CapEx	OpEx (per year)	
New build	4678–11 523	34.7–35.6	0.000209
Retrofit bioretention	496	2.0	
Retrofit street trees	4 728	46.8	
Retrofit green roof	56	0.9	
Retrofit planters	755	7.6	
Piped drainage			
Design standard for piped drainage	Cost per km: £ (2021 prices)		Carbon cost per property protected, per £ spent on SuDS: £ (2021 prices)
New build (1 in 30 years)	709 284	4.79	0.000240
New build (>1 in 30 years)	709 284		
Retrofit (1 in 30 years)	1 107 803		
Retrofit (≥1 in 30 years)	8 560 293		

Costs for new build SuDS are based mainly on Ashley *et al.* (2017), supported by Ossa-Moreno *et al.* (2017) and Arcadis (2020). Whole-life cost of carbon emissions is based on the £ spent (tonnes) assuming 98 300 $kgCO_2e$ per unit of area managed by SuDS (Stantec, 2022) and SCOTSNET (2010) which gives an average of 0.4 $kgCO_2e$ per £ invested in SuDS.
For piped drainage, retrofit costs are based on increasing the capacity of existing drainage of which there is on average some 14.9 m of existing pipework per UK property in urban areas. Sourced from Babtie (2003), EA (2015), Webber *et al.* (2019) and Spon (2023).

8.4.6 Comparing the costs and benefits of SuDS and NbSs

In most applications an economic case will need to be made for the use of either new or retrofit SuDS or NbSs. There are various ways of doing this and different business cases in use (see examples in Thurston, 2012). The most common types of business case are (Ashley *et al.*, 2023)

- strategic: making the case for change and alignment with a strategic objective
- economic: identifying the proposal that delivers best value for society
- commercial: showing the proposal is viable commercially
- financial: demonstrating affordability and funding to stakeholders
- management: demonstrating arrangements for delivery, maintenance, monitoring and evaluation.

There is a need to engage with decision-makers to ensure that the appropriate infrastructure system is utilised, be it grey or nature based, and there may be several parties involved, each with their own needs and objectives, hence a compelling business case may often be needed for each stakeholder. Of course, motivations other than the clear value of benefits to costs may prevent selection of the most effective or more sustainable option (see Chapter 12). The BEGIN (2023c) EU project provides guidance on developing a compelling business case for using NbSs, which needs to be undertaken in partnership with all stakeholders. The guidance makes the following recommendations.

1. Define the vision and expected outcomes for the project.
2. Identify and assess all relevant impacts of each option.
3. Develop an evidence-based BGI business case for selected options.
4. Communicate clearly and in an appropriate manner throughout the process.

The foregoing sections provide support for how to obtain evidence for performance of any proposed NbS (and grey infrastructure for comparison) and the cost and value of each option considered. Whole-life costs and benefits may be compared in order to arrive at a net present value discounted over the lifetime of any option being considered. There is a range of alternative economic approaches for valuation, see examples from the decades-long use of SuDS in the USA, in Garmestani *et al.* (2012), in which a variety of economic analyses have been used by different decision-makers. The GI being implemented in Portland, Oregon, USA (City of Portland, 2020), is seen as providing future headroom to the existing combined drainage system. It forms part of the solution to deliver the current requirements and will help to reduce the risk of flooding and lengthen the asset life of the sewers. As an example, in the Tabor to the River scheme (City of Portland, 2012), changing the scheme from pipes only to pipes together with green infrastructure reduced the costs from $144 million to $86 million, a saving of some $60 million.

Another example of the wider economic benefits from NbSs is illustrated in Box 8.6.

Overall, it is concluded that in the studies examined, the use of NbSs has been economically preferable to traditional piped drainage systems. In many cases, the least-cost option has not been selected; instead, communities have preferred alternatives that provide greater overall benefits, especially longer term.

Economic information, in whatever form is appropriate, may be used in helping to decide which option is preferable. Other aspects, not only the business cases (there will need to be several as a range of stakeholders/funders are involved), will inform the final decision (Chapters 4 and 12).

There are rapid advancements in this area and care should be taken to ensure that the latest knowledge is being utilised – for example, Greater Manchester *et al.* (2023); Nature Conservancy (2023); Network Nature (2023); White House (2022).

Box 8.6

Thornton Creek Water Quality Channel, Seattle, USA (adapted from Seattle Public Utilities, 2009)

Thornton Creek receives and treats runoff from 680 acres of highly developed land, most of which was formerly untreated before flowing to Lake Washington. The highly urban nature created water quality and flow control problems, degrading the lake's function as an aquatic habitat. By integrating with the adjacent private development, the use of NbSs has improved the balance between urbanisation and environmental sustainability as part of a larger strategy to revitalise the area and facilitate economic development. The functional requirements included creating a welcoming public open space that would integrate well with surrounding private commercial and residential development and provide numerous environmental benefits. The design effectively uses limited space to achieve multiple community goals by integrating public open space and a highly technical water quality facility. Water serves as the unifying element, meandering through the site; and what was once a sterile, paved parking lot is now a unique 2.7 acre urban park and public open space amenity in the heart of a dense urban neighbourhood. The development and construction costs of $14.7 million are believed to have generated over $200 million in adjacent private development.

8.5. Outlook for NbSs

The upsurge in nature-based approaches is extending to all walks of life (e.g. Discover Wildlife, 2023). Identified for particular uses, such as flood risk reduction, as well as for encouraging wildlife in urban areas, reducing urban heat islands and many other benefits, NbSs appear to be widely accepted by local communities and seen by most professionals as an obvious means of providing infrastructure that will deliver many needs. In many countries, such as the UK and other industrialised parts of the world, the past few decades of experience with the use of surface-based systems for urban drainage in conjunction with buried piped systems has helped practitioners gain confidence in the use of SuDS (UK), BMPs/LID/SCMs (USA) and WSUD (Australasia). Nevertheless, there are many examples of poorly performing systems, as illustrated in Figure 8.10, and poor construction or maintenance appears to be a major issue even where these approaches have been in use for some years. The major change has been the shift from BMPs, SuDS and WSUD to NbSs, where nature itself is seen as infrastructure.

Delivery of effective and multifunctioning NbSs is not as straightforward as for grey infrastructure systems or for BMPs/SuDS, etc. Many professionals and other stakeholders are still learning what NbSs are and how to plan, design, operate and maintain them. To be effective, all aspects of the policies, regulations, standards and professional practices need to be reviewed and modified to account properly for the greater complexity of NbSs as infrastructure. In Philadelphia, for example, at the start of the Green City process in the early 2000s, the city ordinances effectively prohibited the use of GI. The staff charged with planning and delivery were untrained and, in some cases, hostile to switching from buried piped drainage. Since then, ordinances have been modified and staff appropriately trained and engaged to deliver one of the most successful 'green' CSO control programmes in the world (Green Philly, 2020).

Figure 8.10 A not so permeable pavement (Richard Ashley)

This contrasts with London, where aspects of the water cycle are managed by so many organisations that the integration of approaches and collaboration required for effective uptake of NbSs is fraught with problems. In addition, as the land use planning functions are separated from responsibilities for water supply, drainage, sewerage and main rivers, this inhibits institutions from utilising the most effective technologies and seizing opportunities for multifunctional infrastructure to manage the CSOs (e.g. Loftus and March, 2019). This case is further considered in Chapter 12.

There are clear incentives and benefits for using NbSs for urban drainage and other purposes, but there is a danger of seeing this approach as the answer to all aspects of urban water management, especially in the face of climate change needs. Seddon *et al.* (2020) point out the danger of utilising 'low-grade' NbSs as a panacea. This includes afforestation with non-native monocultures, which would be a maladaptation, missing the opportunity for multifunctional, diverse and more resilient natural systems. They also identify the need for clear measurable and repeatable evidence of the many performance benefits of NbSs (e.g. Fowdar *et al.*, 2021), which is ultimately needed to develop effective planning, design and selection processes to achieve objective targets.

Hybrid, integrated blue-green-grey infrastructure systems, with the blue-green components delivering the multifunctional benefits, are seen as the best way forward, not necessarily driven by drainage or even water needs. Coping with the impacts of climate change on the entirety of our systems and services is the primary driver for uptake. Hybrid systems are promoted in cities such as Copenhagen for stormwater systems (e.g. Jørgensen *et al.*, 2022), and also for local sanitation services, with microbiological treatment and energy recovery (e.g. Capodaglio *et al.*, 2021).

The lack of, or seeming lack of, clear evidence for the overall performance of NbSs is seen by many as a major impediment to engineers when deciding on the most effective infrastructure for urban drainage. Knowledge fortunately continues to develop in this area, and it is expected that in the future the use of NbSs will become the default urban drainage option, although this will need to be seen in the context of integrated management of the water cycle as a whole (e.g. Bunclark and Hernandez, 2022; Nika *et al.*, 2020), as considered further in Chapter 12.

REFERENCES

Allen D, Haynes H and Arthur S (2017) Contamination of detained sediment in SuDS. *Water* **9(5)**: 355, doi.org/10.3390/w9050355.

Anon (2012) Priority substances: impact on water utilities. *WaterWorld.* 1 Sept. See www.waterworld.com/articles/wwi/print/volume-27/issue-4/editorial-focus/micropollutants/priority-substances-impact-on-water.html (accessed 02/08/2023).

Arcadis (2020) *London Strategic SuDS Pilot Study: SuDS CAPEX Estimation, Technical Note.* Arcadis UK, London, UK. See https://www.susdrain.org/files/resources/evidence/8_10016816_arc_xx_xx_de_rp_0004_01_suds_capex_estimation_technical_note.pdf (accessed 01/08/2023).

Ashley RM and Horton B (2022) Characterizing financing needs and financing capacities in different regions: a global perspective on water-related financing flows and drivers for investment needs. In *Financing Investment in Water Security – Recent Developments and Perspectives* (Leflaive X, *et al.* (eds)). Elsevier, Amsterdam, the Netherlands.

Ashley RM, Balmforth DJ and Horton B (2013) *Demonstrating the Multiple Benefits of SuDS – A Business Case (Phase 2). Draft Literature Review (October 2013).* Construction Industry Research and Information Association, London, UK. See https://www.susdrain.org/files/resources/ciria_guidance/ciria_rp993_literature_review_october_2013_.pdf (accessed 01/08/2023).

Ashley RM, Horton B, Lavers T and McLaughlin A-M (2017) *Sustainable Drainage Systems on New Developments: Analysis of Evidence Including Costs and Benefits of SuDS Construction and Adoption*. Final Report for the Welsh Government. Environmental Policy Consulting, Stratford-upon-Avon, UK. See https://www.susdrain.org/files/resources/evidence/analysis_of_evidence_including_costs_and_benefits_of_suds_wg_2017_.pdf (accessed 01/08/2023).

Ashley RM, Digman CJ, Horton *et al.* (2018) Evaluating the longer-term benefits of sustainable drainage. *Proceedings of the Institution of Civil Engineers – Water Management* **171(WM2)**: 57–66, 10.1680/jwama.16.00118.

Ashley RM, Horton B, Walker L *et al.* (2023) The benefits of nature-based systems in a changing and uncertain world. *Proceedings of the Institution of Civil Engineers – Engineering Sustainability* **176(2)**: 61–71, 10.1680/jensu.21.00102.

Babtie (2003) *An Investigation into the Costs of Sewer Flooding Alleviation Schemes*. Report for Ofwat. Babtie Group, Glasgow, Scotland.

Baker J, Hoskin R and Butterworth T (2019) *Biodiversity Net Gain. Good Practice Principles for Development*. CIRIA C776F. Construction Industry Research and Information Association, London, UK.

BEGIN (2022) *Value Integration for Delivering Blue and Green Infrastructure: Balancing Technical, Spatial, Social and Sustainability Values*. Policy Brief from the Interreg North Sea Region BEGIN project. See https://northsearegion.eu/media/19995/final_kuitert-van-buuren_-policy-brief_value-integration_2022.pdf (accessed 01/08/2023).

BEGIN (2023a) BEGIN – Blue Green Infrastructures through Social Innovation. https://northsearegion.eu/begin/ (accessed 01/08/2023).

BEGIN (2023b) *Bradford – Shipley Canal Road Corridor and Harrogate Road/New Line improvement*. See https://northsearegion.eu/begin/bgi-pilot-projects/bradford/ (accessed 01/08/2023).

BEGIN (2023c) *Social Innovations for Delivering Blue and Green Infrastructure: Developing a Compelling Business Case*. See https://northsearegion.eu/media/18718/begin_business-case_brief.pdf (accessed 01/08/2023).

Biodiversity in Planning (2023) Background to ecological assessment. https://www.biodiversityinplanning.org/about/what-is-an-ecological-assessment/#:~:text=An%20ecological%20assessment%20is%20an,a%20suitably%20qualified%20professional%20ecologist%20 (accessed 01/08/2023).

Biswal BK, Bolan N, Zhu Y-G and Balasubramanian R (2022) Nature-based Systems (NbS) for mitigation of stormwater and air pollution in urban areas: a review. *Resources, Conservation & Recycling* **186**: 106578, 10.1016/j.resconrec.2022.106578.

Brorken C, Hugé J, Dahdouh-Guebas F *et al.* (2022) Monitoring biodiversity mainstreaming in development cooperation post-2020: exploring ways forward. *Environmental Science & Policy* **136**: 114–126, 10.1016/j.envsci.2022.05.017.

BSI (British Standards Institution) (2013a) BS 8582:2013. Code of practice for surface water management for development sites. BSI, London, UK.

BSI (2013b) BS 42020:2013. Biodiversity. Code of practice for planning and development. BSI, London, UK.

BSI (2019) BS ISO 24536:2019. Service activities relating to drinking water supply, wastewater and stormwater systems. Stormwater management. Guidelines for stormwater management in urban areas. BSI, London, UK.

BSI (2021) PD ISO/TR 24539:2021. Service activities relating to drinking water supply, wastewater and stormwater systems. Examples of good practices for stormwater management. BSI, London, UK.

BSI (2023) *Agile Standards*. BSI, London, UK. See https://www.bsigroup.com/globalassets/localfiles/en-gb/knowledge-services/agile-standards_ks.pdf (accessed 01/08/2023).

BSI (2018) BS ISO 31000:2018. Risk management. Guidelines. BSI, London, UK.

Bunclark L and Hernandez IM (2022) Scientific mapping of research on nature-based solutions for sustainable water management. *Water Resources Management* **36**: 4499–4516, 10.1007/s11269-022-03242-w.

Butterworth J, McIntyre P and da Silva Wells C (eds.) (2011) *SWITCH in the City*. IRC International Water and Sanitation Centre, The Hague, the Netherlands. https://ruaf.org/assets/2019/11/SWITCH-in-the-city-Putting-urban-water-use-to-the-test.pdf (accessed 06/12/2023).

Capodaglio AG, Bolognesi S and Cecconet D (2021) Sustainable, decentralized sanitation and reuse with hybrid nature-based systems. *Water* **13**: 1583, 10.3390/w13111583.

Charlesworth SM and Booth CA (eds) (2017) *Sustainable Surface Water Management*. Wiley, Chichester, UK.

Chini CM, Canning JF, Schrieber KL, Peschel JM and Stillwell AS (2017) The Green Experiment: cities, green stormwater infrastructure, and sustainability. *Sustainability* **9**: 105, 10.3390/su9010105.

CIRIA (Construction Industry Research and Information Association) (2023a) Susdrain – The community for sustainable drainage. https://www.susdrain.org/ (accessed 01/08/2023).

CIRIA (2023b) Component: Vortex flow control. Susdrain. https://www.susdrain.org/delivering-suds/using-suds/suds-components/inlets-outlets-and-control-structures/vortex-control-systems (accessed 01/08/2023).

CIRIA (2023c) *Example of a SuDS Maintenance Plan: Sustainable Drainage and Landscape Management Plan for Robinswood Primary School*. CIRIA, London, UK. See https://www.susdrain.org/files/resources/SuDS_manual_output/paper_rp992_23_example_suds_maintenance_plan.pdf (accessed 01/08/2023).

CIRIA (2023d) *Benefits Estimation Tool*. CIRIA, London, UK. See https://www.susdrain.org/resources/best.html (accessed 01/08/2023).

City of Portland (2012) Tabor to the River Rain Garden Partnerships. https://www.portlandoregon.gov/bes/article/425334 (accessed 05/08/2023).

City of Portland (2020) *City of Portland Stormwater Management Manual*. Bureau of Environmental Services, Portland, OR, USA. See https://www.portland.gov/bes/stormwater/swmm (accessed 29/11/2023).

City of Toronto (2013) *Sustaining and Expanding the Urban Forest: Toronto's Strategic Forest Management Plan*. City of Toronto, Parks, Forestry and Recreation, Urban Forestry. Toronto, Ontario, Canada. See https://www.toronto.ca/data/parks/pdf/trees/sustaining-expanding-urban-forest-management-plan.pdf (accessed 01/08/2023).

Clary J and Piza H (2017) *Cost of Maintaining Green Infrastructure*. ASCE/EWRI, Reston, VA, USA.

Clary C, Jones J, Leisenring M, Hobson P and Strecker E (2022) *International Stormwater BMP Database. 2020 Summary Statistics*. Water Research Foundation, Alexandria, VA, USA. See https://www.waterrf.org/system/files/resource/2020-11/DRPT-4968_0.pdf (accessed 01/08/2023).

ClimateADAPT (2023) Water sensitive urban and building design. https://climate-adapt.eea.europa.eu/en/metadata/adaptation-options/water-sensitive-urban-and-building-design (accessed 01/08/2023).

Conlan K, Cryle P, Dickie I *et al.* (2017) *Implementing Ecosystem Services and Natural and Social Capital Accounting Approaches – Phase 1*. Report Ref 17/CL/04/15. UK Water Industry Research Ltd, London, UK. See https://ukwir.org/implementing-ecosystem-service-and-natural-and-social-capital-accounting-approaches-phase-1-0 (accessed 01/08/2023).

Convention on Biological Diversity (2023) Preparations for the Post-2020 Biodiversity Framework. https://www.cbd.int/conferences/post2020 (accessed 01/08/2023).

Cuckston T (2019) Seeking an ecologically defensible calculation of net loss/gain of biodiversity. *Accounting, Auditing & Accountability Journal* **32(5)**: 1358–1383.

Dale K, Thomson C, Kelly J, Hay D and MacDougall K (2011) *Delivering Biodiversity Benefits through Green Infrastructure*. CIRIA C711. Construction Industry Research and Information Association, London, UK.

D'Arcy B (2013) *The Development of a Strategic Approach to Managing Diffuse Pollution*. PhD thesis, University of Abertay, Dundee, UK.

Defra (Department for Environment, Food and Rural Affairs) (2015) *Sustainable Drainage Systems. Non-statutory Technical Standards for Sustainable Drainage Systems*. See https://assets.publishing.service.gov.uk/government/uploads/system/uploads/attachment_data/file/415773/sustainable-drainage-technical-standards.pdf (accessed 01/08/2023).

Defra (2021) *Recommendations to Update Non-Statutory Technical Standards for Sustainable Drainage Systems (SuDS). Final Report.* Defra, London, UK. See https://randd.defra.gov.uk/ProjectDetails?ProjectID=20287&FromSearch=Y&Publisher=1&SearchText=WT15122&SortString=ProjectCode&SortOrder=Asc&Paging=10#Description (accessed 01/08/2023).

Defra (2023) Understanding biodiversity net gain. https://www.gov.uk/guidance/understanding-biodiversity-net-gain (accessed 01/08/2023).

Design for Homes (2023) *Building for a Healthy Life*. Design for Homes, Saffron Walden, UK. See https://www.designforhomes.org/project/building-for-life/ (accessed 01/08/2023).

De Sousa M, Montalto FA and Spatari S (2012) Using life cycle assessment to evaluate green and grey combined sewer overflow control strategies. *Journal of Industrial Ecology* **16(6)**: 901–913, 10.1111/j.1530-9290.2012.00534.x.

Diaz-Nieto J, Lerner DN, Saul AJ and Blanksby J (2012) GIS water balance approach to support surface water flood-risk management. *Journal of Hydrologic Engineering* **17(1)**: 55–67, 10.1061/(ASCE)HE.1943-5584.0000416.

Diaz-Nieto J, Lerner DN, Saul AJ and Blanksby J (2016) Least cost path analysis to identify retrofit surface water conveyance solutions. *Journal of Hydrologic Engineering* **21(3)**: 04015071, 10.1061/(ASCE)HE.1943-5584.0001315.

Digman CJ, Ashley R, Balmforth D, Stovin V and Glerum J (2012) *Retrofitting to Manage Surface Water.* CIRIA C713. Construction Industry Research and Information Association, London, UK.

Digman CJ, Ashley R, Hargeaves P and Gill E (2014a) *Managing Urban Flooding from Heavy Rainfall – Encouraging the Uptake of Designing for Exceedance. Case Study Report.* CIRIA C738c. Construction Industry Research and Information Association, London, UK.

Digman CJ, Ashley R, Hargeaves P and Gill E (2014b) *Managing Urban Flooding from Heavy Rainfall – Encouraging the Uptake of Designing for Exceedance – Success Factors and Lessons.* CIRIA C738b. Construction Industry Research and Information Association, London, UK.

Discover Wildlife (2023) How nature could help fix our flooding and sewerage problems. https://www.discoverwildlife.com/how-to/wildlife-gardening/nature-based-solutions-how-nature-can-help-prevent-flooding/ (accessed 01/08/2023).

Dolowitz DP, Bell S and Keeley M (2018) Retrofitting urban drainage infrastructure: green or grey? *Urban Water Journal* **15(1)**: 83–91.

Drupp MA, Meya JN, Baumgärtner S and Quaas MF (2018) Economic Inequality and the Value of Nature. *Ecological Economics* **150**: 340–345, 10.1016/j.ecolecon.2018.03.029.

EA (Environment Agency) (2015) *Cost Estimation for SUDS – Summary of Evidence*. Environment Agency, Bristol, UK. See https://assets.publishing.service.gov.uk/media/6034ee6c8fa8f54334a5a6a9/Cost_estimation_for_SUDS.pdf (accessed 01/08/2023).

EC (European Commission) (2015) *Communication from the Commission to the European Parliament and the Council. The Water Framework Directive and the Floods Directive: Actions Towards the 'Good Status' of EU Water and to Reduce Flood Risks*. EC, Brussels, Belgium. See https://eur-lex.europa.eu/legal-content/EN/TXT/PDF/?uri = CELEX:52015DC0120&from=ET (accessed 01/08/2023).

EC (2017) Surface water. https://environment.ec.europa.eu/topics/water/surface-water_en#intro (accessed 01/08/2023).

Eckart K, McPhee Z and Bolisetti T (2017) Performance and implementation of low impact development – a review. *Science of the Total Environment* **607–608**: 413–432, 10.1016/j.scitotenv.2017.06.254.

Ecotree (2023) How much CO_2 does a tree absorb? https://ecotree.green/en/how-much-co2-does-a-tree-absorb#answer (accessed 17/12/2023).

EESI (Environmental and Energy Studies Institute) (2019) *Nature as Resilient Infrastructure: An Overview of Nature-Based Solutions*. Fact Sheet. September. See https://www.eesi.org/files/FactSheet_Nature_Based_Solutions_1016.pdf (accessed 01/08/2023).

Eftec (2021) *Natural Capital Accounting Practices in the UK*. Briefing note in partnership with COWI. See https://www.eftec.co.uk/sites/default/files/pdf/NCA%20Practices%20in%20the%20UK%20Briefing%20Note%20-%20eftec%20in%20partnership%20with%20COWI.pdf (accessed 01/08/2023).

Environment Agency (2020) *Calculate GiA funding for FCERM projects 2020*. External guidance: LIT 58360. HM Government, UK. See https://www.gov.uk/government/publications/calculate-gia-funding-for-fcerm-projects-2020 (accessed 01/08/2023).

Erickson JD (2022) *The Progress Illusion – Reclaiming our Future from the Fairy Tale of Economics*. Island Press, Washington DC, USA.

Eriksson E, Revitt DM, Ledin A, *et al.* (2011) Water management in cities of the future using emission control strategies for priority hazardous substances. *Water Science & Technology* **64(10)**: 2109–2118. Doi: 10.2166/wst.2011.797

Fazio JR (ed.) (2023) *How Trees Can Save Energy*. Tree City USA Bulletin, No.21. Arbor Day Foundation, Nebraska City, NE, USA. https://www.arborday.org/trees/bulletins/coordinators/resources/pdfs/021.pdf (accessed 17/12/2023).

Fletcher TD, Shuster W, Hunt WF *et al.* (2014) SUDS, LID, BMPs, WSUD and more – The evolution and application of terminology surrounding urban drainage. *Urban Water Journal* **12(7)**: 525–542, 10.1080/1573062X.2014.916314.

Fowdar H, Payne E, Schang C, *et al.* (2021) How well do stormwater green infrastructure respond to changing climatic conditions? *Journal of Hydrology* **603(A)**: 126887, 10.1016/j.jhydrol.2021.126887.

Garmestani AS, Clements J, Pratt J and Hair L (2012) The economics of green infrastructure and low-impact development practices. *In Economic Incentives for Stormwater Control* (Thurston HW (ed.)). CRC Press, Boca Raton, FL, USA, pp. 101–126.

Gilbertson A, Gibb A, Gilbertson A and Kappie G (2011) *Guidance on Catastrophic Events in Construction*. CIRIA C699. Construction Industry Research and Information Association, London, UK.

Gill SE, Rahman MA, Handley JF and Ennos AR (2013) Modelling water stress to urban amenity grass in Manchester UK under climate change and its potential impacts in reducing urban cooling. *Urban Forestry & Urban Greening* **12**: 350–358.

Greater Manchester *et al.* (2023) *The IGNITION Project: Building a Business Case for NBS*. See https://www.greatermanchester-ca.gov.uk/media/4142/ignition-fact-files-building-a-business-case-final.pdf (accessed 01/08/2023).

Green Philly (2020) Eco-Explainer: What are Combined Sewer Systems? https://www.thegreencities.com/water/eco-explainer-what-are-combined-sewer-systems/ (accessed 01/08/2023).

Harrisburg (2017) *Homeowners Guide to Stormwater BMP Maintenance*. Harrisburg, Pennsylvania, USA. See https://www.phrc.psu.edu/assets/docs/publications/SW_Booklet_2017.pdf (accessed 01/08/2023).

HKV and RPA (2014) *Study on Economic and Social Benefits of Environmental Protection and Resource Efficiency Related to the European Semester*. (ENV.D.2/ETU/2013/0048r) Final Report prepared for DG Environment. See https://ec.europa.eu/environment/integration/green_semester/pdf/RPA%20Final%20Report-main%20report.pdf (accessed 01/08/2023).

HM Government (2023a) *Nature Markets: A Framework for Scaling Up Private Investment in Nature Recovery and Sustainable Farming*. HMG, London, UK. See https://assets.publishing.service.gov.uk/government/uploads/system/uploads/attachment_data/file/1147397/nature-markets.pdf (accessed 01/08/2023).

Hulsmann A, Grützmacher G, Van den Berg G *et al.* (eds) (2015) *Climate Change, Water Supply and Sanitation*. IWA Publishing, London, UK.

Invest4nature (2023) The economics of nature-based solutions. https://invest4nature.eu/ (accessed 01/08/2023).

JNCC (Joint Nature Conservation Committee) (2023) National Vegetation Classification (NVC). https://jncc.gov.uk/our-work/nvc/ (accessed 01/08/2023).

Johnson D and Geisendorf S (2019) Are neighborhood-level SUDS worth it? An assessment of the economic value of sustainable urban drainage system scenarios using cost-benefit analyses. *Ecological Economics* **158**: 194–205, 10.1016/j.ecolecon.2018.12.024.

Jones J, Woods-Ballard B, Duffy A and Walsh CL (2023) Sustainable drainage systems (SuDS). In *ICE Manual of Blue-Green Infrastructure* (Washbourne C-L and Wansbury C (eds)). ICE Publishing, London, UK, pp. 67–87.

Jørgensen G, Fryd O, Lund AA, Andersen PS and Herslund L (2022) Nature-based climate adaptation projects, their governance and transitional potential – cases from Copenhagen. *Frontiers in Sustainable Cities* **4**: 906960, 10.3389/frsc.2022.906960.

Kabisch N, Korn H, Stadler J and Bonn A (eds) (2017) *Nature-Based Solutions to Climate Change Adaptation in Urban Areas*. Springer, Cham, Switzerland. See https://doi.org/10.1007/978-3-319-56091-5_1 (accessed 05/08/2023).

Kaval P (2019) Integrated catchment management and ecosystem services: a twenty-five year overview. *Ecosystem Services* **37**: 100912, 10.1016/j.ecoser.2019.100912.

Knight-Lenihan S (2020) Achieving biodiversity net gain in a neoliberal economy: the case of England. *Ambio* **49**: 2052–2060, 10.1007/s13280-020-01337-5.

Landscape Institute (2016) *Landscape and Visual Impact Assessment (LVIA)*. See https://www.landscapeinstitute.org/technical-resource/landscape-visual-impact-assessment/ (accessed 01/08/2023).

Land Trust (2018) *The Economic Value of our Green Spaces*. Land Trust, Warrington, UK. See https://thelandtrust.org.uk/wp-content/uploads/2018/02/The-economic-value-of-our-green-spaces.pdf (accessed 01/08/2023).

Liang R, Di Matteo M, Maier HR and Thyer MA (2019) Real-time, smart rainwater storage systems: potential solution to mitigate urban flooding. *Water* **11**: 2428, 10.3390/w11122428.

Liptan WT and Santen JD (2017) *Sustainable Stormwater Management*. Timber Press, Portland, OR, USA.

Loftus A and March H (2019) Integration for whom? Financialisation and the Thames Tideway Tunnel. *Urban Studies* **56**: 2280–2296, 10.1177/0042098017736713.

Lundy L and Wade R (2013) *A Critical Review of Urban Diffuse Pollution Control: Methodologies to Identify Sources, Pathways and Mitigation Measures with Multiple Benefits*. Centre of Expertise for Waters (CREW), Aberdeen, UK. See https://nora.nerc.ac.uk/id/eprint/502911/1/CREW%20Stage1Report_FINAL.pdf (accessed 02/08/2023).

Matos FA and Roebeling P (2022) Modelling impacts of nature-based solutions on surface water quality: a rapid review. *Sustainability* **14**: 7381, 10.3390/su14127381.

Mayor of London (2023) SuDS Sector Guidance. https://www.london.gov.uk/programmes-and-strategies/environment-and-climate-change/climate-change/surface-water/suds-sector-guidance (accessed 02/08/2023).

Monbiot G (2020) How did wildlife groups start collaborating in the destruction of nature? *The Guardian*. 24 Jun. See https://www.theguardian.com/commentisfree/2020/jun/24/how-did-wildlife-groups-start-collaborating-in-the-destruction-of-nature- (accessed 02/08/2023).

Natural Capitals Coalition (2019) *What is a Natural Capital Approach?* See https://naturalcapitalcoalition.org/wp-content/uploads/2019/06/NCC-WhatIs-NaturalCapitalApproach-FINAL.pdf (accessed 02/08/2023).

Nature Conservancy (2023) *Insights into Practice: The Business Case for Nature-Based Solutions*. See https://www.nature.org/content/dam/tnc/nature/en/documents/NBSFactSheet.pdf (accessed 02/08/2023).

Naumann S, Davis M, Kaphengst T, Pieterse M and Rayment M (2011) *Design, Implementation and Cost Elements of Green Infrastructure Projects*. Final report to the European Commission, DG Environment. Ecologic institute and GHK Consulting. See https://www.ecologic.eu/11382 (accessed 02/08/2023).

Network Nature (2021) *Nature-Based Solutions in Light of the Pandemic. Outcomes*. Network Nature. See https://networknature.eu/sites/default/files/uploads/nn-s1-outcomes-website-final.pdf (accessed 02/08/2023).

Network Nature (2023) *Nature-Based Solutions: Taskforces – Welcome Package*. See https://networknature.eu/sites/default/files/images/Welcome_package_for_Horizon_NBS_Task_Forces.pdf (accessed 02/08/2023).

NIC (National Infrastructure Commission) (2023) Impartial, expert advice on major long term infrastructure challenges. https://nic.org.uk/ (accessed 02/08/2023).

Nika CE, Gusmaroli L, Ghafourian M *et al.* (2020) Nature-based solutions as enablers of circularity in water systems: a review on assessment methodologies, tools and indicators. *Water Research* **183**: 11598, 10.1016/j.watres.2020.115988.

OECD (Organisation for Economic Co-operation and Development) (2015) *Water and Cities – Ensuring Sustainable Futures*. OECD Publishing, Paris, France. See https://www.oecd.org/env/water-and-cities-9789264230149-en.htm (accessed 02/08/2023).

Osborne M and Montague K (2005) *The Potential for Water Pollution from Railways*. CIRIA C643. Construction Industry Research and Information Association, London, UK.

Ossa-Moreno J, Smith KM and Mijic A (2017) Economic analysis of wider benefits to facilitate SuDS uptake in London, UK. *Sustainable Cities and Society* **28**: 411–419, 10.1016/j.scs.2016.10.002.

Pannell DJ (2020) *Benefit: Cost Analysis and Strategic Decision Making for Water-Sensitive Cities*. Cooperative Research Centre for Water Sensitive Cities, Melbourne, Australia. See https://watersensitivecities.org.au/wp-content/uploads/2020/05/INFFEWS-BCA-and-Strategic-DM-2020-02-5.pdf (accessed 02/08/2023).

Perales-Momparler S, Hernández-Crespo C, Vallés-Moran F *et al.* (2014) SuDS efficiency during the start-up period under Mediterranean climatic conditions. *Clean Soil Air Water* **42**: 178–186, 10.1002/clen.201300164.

Forest Research (2023) Payments for Ecosystem Services (Forest for water). https://www.forestresearch.gov.uk/research/payments-for-ecosystem-services-forest-for-water/pesfor-w-objectives/ (accessed 02/08/2023).

Pötz H and Bleuzé P (2012) *Urban Blue-Green Grids*. Atelier Groenblauw, Delft, the Netherlands. See (for) https://www.naibooksellers.nl/green-blue-grids-manual-for-resilient-cities-groenblauwe-netwerken-handleiding-voor-veerkrachtige-steden.html?:_store = english&:_from_store = default (accessed 02/08/2023).

Pugh TAM, MacKenzie AR, Davies G and Whyatt D (2012) A futures-based analysis for urban air quality remediation. *Proceedings of the Institution of Civil Engineers – Engineering Sustainability* **165(ES1)**: 21–36.

Rosetti J (2013) *Valuation of Australia's GI: Hedonic Pricing Model using the Enhanced Vegetation Index*. Honours thesis, Department of Economics, Monash University, Melbourne, Australia. See https://datainspace.org/wp-content/uploads/2017/04/Joe-Rossetti-2013-Thesis-1.pdf (accessed 02/08/2023).

Roy D (2018) *The Multiple Benefits of Natural Infrastructure*. International Institute for Sustainable Development. https://www.iisd.org/articles/insight/multiple-benefits-natural-infrastructure (accessed 02/08/2023).

Royal Haskoning (2012) *Costs and Benefits of Sustainable Drainage Systems*. Haskoning UK Ltd, London, UK. See https://silo.tips/download/costs-and-benefits-of-sustainable-drainage-systems (accessed 02/08/2023).

Royal Society (2023) Reversing biodiversity loss. https://royalsociety.org/topics-policy/projects/biodiversity/ (accessed 02/08/2023).

Sagrelius PO, Blecken G, Hedström A, Ashley R and Viklander M (2020) Environmental impacts of stormwater bioretention systems with various design and construction components. *Journal of Cleaner Production* **359**: 132091, 10.1016/j.jclepro.2022.132091.

Sayers PB (2017) Evolution of strategic flood risk management in support of socialj, ecosystem health, and resilience. *Oxford Research Encyclopedia*: Natural Hazard Science. Oxford University Press, Oxford, UK, 10.1093/acrefore/9780199389407.013.85.

Sayers P, Eccleston P, Horton, *et al.* (2022) *Surface Water: Future Risk and Investment Needs*. Report prepared for the National Infrastructure Commission. Sayers and Partners, UK. See https://nic.org.uk/app/uploads/NIC-Sayers-and-Partners-Surface-Water-Future-Risk-and-Investment.pdf (accessed 02/08/2023).

Scholz M, Corrigan NL and Yazdi SK (2006) The Glasgow sustainable urban drainage system management project: case studies (Belvidere hospital and Celtic FC stadium areas). *Environmental Engineering Science* **23(6)**: 908–922, 10.1089/ees.2006.23.908.

SCOTSNET (2010) *SuDS for roads. Whole life cost and whole life carbon toolkit*. See https://fidic.org/sites/default/files/CALC%2028%20-%20SUDS%20WLC%20WLC%20-%20final.pdf (accessed 02/08/2023).

Seattle Public Utilities (2009) *Thornton Creek Water Quality Channel*. See https://www.seattle.gov/util/cs/groups/public/documents/webcontent/spu01_006146.pdf (accessed 02/08/2023).

Seddon N, Chausson A, Berry P *et al.* (2020) Understanding the value and limits of nature-based solutions to climate change and other global challenges. *Philosophical Transactions of the Royal Society B* **375**: 20190120, 10.1098/rstb.2019.0120.

Semple NG (1991) Principles of environmental protection. *Proceedings of the Institution of Civil Engineers* **90(4)**: 837–851. https://www.icevirtuallibrary.com/doi/abs/10.1680/iicep.1991.15385.

Sharma AK, Gardner T and Begbie D (eds) (2019) *Approaches to Water Sensitive Urban Design*. Elsevier, Amsterdam, the Netherlands. See https://www.sciencedirect.com/book/9780128128435/approaches-to-water-sensitive-urban-design#book-description (accessed 02/08/2023).

Shen P, Deletic A, Bratieres K and McCarthy DT (2020) Real time control of biofilters delivers stormwater suitable for harvesting and reuse. *Water Research* **169**: 115257, 10.1016/j.watres.2019.115257.

Spon (2023) *Spon's Civil Engineering and Highway Works Price Book 2023*. Edited by @AECOM. CRC Press, Abingdon UK. See https://www.routledge.com/Spons-Civil-Engineering-and-Highway-Works-Price-Book-2023/AECOM/p/book/9781032331751 (accessed 02/08/2023).

Stantec (2022) *Storm Overflows Evidence Project*. HM Government, London, UK. See https://www.gov.uk/government/publications/storm-overflows-evidence-project (accessed 01/08/2023).

Stratus Consulting (2009) *A Triple Bottom Line Assessment of Traditional and GI Options for Controlling CSO Events in Philadelphia Watersheds*. Office of Watersheds, City of Philadelphia Water Department, Philadelphia, PA, USA. See https://www.epa.gov/sites/default/files/2015-10/documents/gi_philadelphia_bottomline.pdf (accessed 02/08/2023).

TEEB (The Economics of Ecosystems & Biodiversity) (2023) https://teebweb.org/ (accessed 02/08/2023).

Thurston HW (ed.) (2012) *Economic Incentives for Stormwater Control*. CRC Press, Boca Raton, FL, USA.

UN (United Nations) (2021) *UN-Water, 2020: Summary Progress Update 2021 – SDG 6 – Water and Sanitation for All*. United Nations, Geneva, Switzerland.

UN (2023) System of Environmental Economic Accounting. https://seea.un.org/ecosystem-accounting (accessed 02/08/2023).

UNEP, Convention on Biological Diversity, WHO (2015) *Connecting Global Priorities: Biodiversity and Human Health. A State of Knowledge Review*. World Health Organization, Geneva, Switzerland.

USEPA (United States Environmental Protection Agency) (2023a) National Menu of Best Management Practices (BMPs) for Stormwater-Pollution Prevention and Good Housekeeping. https://www.epa.gov/npdes/national-menu-best-management-practices-bmps-stormwater-pollution-prevention-and-good (accessed 02/08/2023).

USEPA (2023b) Environmental Economics. https://www.epa.gov/environmental-economics (accessed 02/08/2023).

Valuing Nature (2023) The Valuing Nature Programme. https://valuing-nature.net/about (accessed 02/08/2023).

Wade R and Garcia-Haba E (2013) *Critical Review of Mitigation Measures, from Source to End of Pipe, for Diffuse Pollution Prevention and an Assessment of their Multiple Benefits*. Centre of Expertise for Waters (CREW), the James Hutton Institute, Aberdeen, UK. See https://www.crew.ac.uk/sites/www.crew.ac.uk/files/documents/CREW%20Stage2Report_FINAL.pdf (accessed 05/08/2023).

Washbourne C-L and Wansbury C (eds) (2023) *ICE Manual of Blue-Green Infrastructure.* ICE Publishing, London, UK.

Water UK (2022a) *Sector Guidance in Relation to the Adoption of Sewerage Assets by Sewerage Companies in England.* Approved version 2.2. Water UK, London, UK. See https://www.water.org.uk/wp-content/uploads/2022/10/Sewerage-Sector-Guidance-v2-update-2.pdf (accessed 05/08/2023).

Water UK (2022b) *Design and Construction Guidance for foul and surface water sewers offered for adoption under the Code for adoption agreements for water and sewerage companies operating wholly or mainly in England ('the Code').* Sewerage Sector Guidance Appendix C. Approved Version 2.2. Water UK, London, UK.

Webber JL, Fu G and Butler D (2019) Comparing cost-effectiveness of surface water flood management interventions in a UK catchment. *Journal of Flood Risk Management* **12(Suppl. 2)**: e12523, 10.1111/jfr3.12523.

WEF (World Economic Forum) (2023) *The Chairperson's Guide to Valuing Nature. White Paper.* WEF, Geneva, Switzerland. See https://www3.weforum.org/docs/WEF_The_Chairpersons_Guide_2023.pdf (accessed 02/08/2023).

WEF/ASCE (1998) *Urban Runoff Quality Management.* WEF Manual of Practice No. 23/ASCE Manual and Report on Engineering Practice No. 87. Water Environment Federation/American Society of Civil Engineers, Reston, VA, USA.

Weiss PT, Gulliver JS and Erickson AJ (2012) Costs and effectiveness of stormwater management practices. In *Economic Incentives for Stormwater Control* (Thurston HW (ed.)). CRC Press, Boca Raton, FL, USA, pp. 9–44.

White House (2022) *Opportunities to Accelerate Nature based Solutions: A Roadmap for Climate Progress, Thriving Nature, Equity, & Prosperity.* Report to the National Climate Task Force. The White House, Washington, DC, USA. See https://www.whitehouse.gov/wp-content/uploads/2022/11/Nature-Based-Solutions-Roadmap.pdf (accessed 02/08/2023).

Wilby RL, Orr HG, Hedger M, Forrow D and Blackmore M (2006) Risks posed by climate change to the delivery of Water Framework Directive objectives in the UK. *Environment International* **32(8)**: 1043–1055, 10.1016/j.envint.2006.06.017 PMID: 16857260.

Wild T, Henneberry J and Gill L (2017) Comprehending the multiple 'values' of green infrastructure – valuing nature-based solutions for urban water management, from multiple perspectives. *Environmental Research* **158**: 179–187, 10.1016/j.envres.2017.05.043.

Wilson S and Davies O (2017) *Maintenance of SuDS.* Susdrain factsheet. Construction Industry Research and Information Association, London, UK. See https://www.susdrain.org/files/resources/fact_sheets/05_17_fact_sheet_maintenance.pdf (accessed 02/08/2023).

Woods-Ballard B, Wilson S, Udale-Clark H *et al.* (2015) *The SUDS Manual*, 2nd edn. CIRIA C753. Construction Industry Research and Information Association, London, UK.

Wright TJ, Liu Y, Carroll NJ, Ahiablame LM and Engel BA (2016) Retrofitting LID practices into existing neighborhoods: is it worth it? *Environmental Management* **57**: 856–867, 10.1007/s00267-015-0651-5.

Xing Y, Jones P and Donnison I (2017) Characterisation of nature-based solutions for the built environment. *Sustainability* **9**: 149, 10.3390/su9010149.

Zhou Q, Panduro TE, Thorsen BJ and Arnbjerg-Nielsen K (2013) Adaption to extreme rainfall with open urban drainage system: an integrated hydrological cost-benefit analysis. *Environmental Management* **51(3)**: 586–601, 10.1007/s00267-012-0010-8.

Ashley R, Smith B, Shaffer P and Caffoor I
ISBN 978-0-7277-4178-3
https://doi.org/10.1680/icehudp.41783.229

Chapter 9
Sensing and control in urban drainage systems

Andy Nichols, Kieran Williams and Hannah Casswell

The monitoring of urban drainage systems varies depending on regional asset investment programmes, which drive efforts to manage flood risk, reduce pollution, and enhance operational performance and national guidance (Ofwat, 2019). Assets such as wastewater pumping stations, combined sewer overflows (CSOs) and treatment works have been monitored on a routine basis for some time (Berge *et al.*, 2014; Gruber *et al.*, 2005; Vanrolleghem and Lee, 2003), but monitoring of the entire sewer network is not commonly conducted on a permanent, continuous basis. Instead, the focus tends to be on surveys of key assets driven by the need for rehabilitation, or by economic or legislative needs. This approach can be attributed to several factors.

- Sewer networks are difficult to monitor due to their buried nature, often under roads, or their remote location.
- The flow medium is usually not homogenous and makes the science of measurement problematic. Debris, sanitary material, faeces, silt from highways, industrial discharges and the misuse of the sewer system means that site-specific monitoring applications can be diverse and challenging.
- The hydraulic variation is wide-ranging and complex, from zero flows in dry conditions to inundation creating backwater effects and reverse flows.
- It is a dangerous working environment requiring trained and competent site operatives able to work within confined spaces. Temporal access restrictions can occur, for example, where specified ambient gas detection limits are exceeded or above certain flow depths, which can often necessitate night works during periods of lowest dry weather flow conditions. Many site operatives can be required where access to the flow is challenging, deep and/or requires traverses.
- Qualified and competent staff are required for processing data and undertaking quality assurance/quality control (QA/QC) using specific industry-accepted software exported in formats that can be readily understood by modelling systems.
- The cost of monitors and communications technology is often dictated by performance criteria, accreditation and certain regions stipulating the requirement for explosive atmospheres (ATEX) certified equipment.
- The monitoring of sewer water quality has historically been expensive and time consuming and typically completed by the physical collection of sewage samples and subsequent accredited laboratory analysis.

Despite these challenges, recent advancements in sensing and wireless communications technology have paved the way for the development of in-sewer monitors that can collect, process and transmit flow and condition data in near real time. These monitors have the potential to revolutionise the way we manage urban drainage systems, providing valuable insights into their performance and enabling proactive and data-driven decision-making. Water companies are increasingly investing in monitoring solutions for drainage systems, and this is recognised by the formation of several focused groups, such as the Instrumentation, Control and Automation (ICA) specialist group of the International Water Association (IWA), the Water Industry Process Automation and Control group (WIPAC), and the sensor-focused non-profit Sensors for Water Interest Group (SWIG).

This chapter will explore the key issues associated with monitoring urban drainage systems, and the latest advances in monitoring instrumentation and operation. The move towards real-time monitoring and management will be discussed, while examining the implications of these developments for the future of urban drainage system monitoring. With the rapid progression of technology and the growing need for reliable and efficient sewer management, this chapter provides a comprehensive overview of the current state and future prospects of sensors in the water industry.

9.1. Application areas

Water sensors are being increasingly utilised in the various components of urban drainage systems. In drainage channels, sensors can be used to monitor water levels and flow rates to give early flood warnings and optimise water management. In sustainable drainage systems (SuDS), sensors can help to monitor water quality and flow rates, ensuring the system is functioning as intended. In sewers, sensors can be used to detect blockages and monitor flow conditions to improve system performance. This section explores the specific applications of water sensors in each component of urban drainage systems, highlighting the benefits of using these sensors for efficient and effective water management.

9.1.1 Rivers and channels

Rivers and channels play a vital role in the water industry as they provide important water resources for communities, industries and the environment. To be able to manage and protect these resources effectively, it is crucial to monitor their health and understand their changing conditions. Sensors have an important role in this process, allowing the gathering of valuable data on various parameters that impact the quality and health of rivers and channels.

Some of the key water quality parameters that sensors can measure include pH, temperature, dissolved oxygen, ammonia and turbidity. These parameters provide insight into the state of the water body and the wellbeing of aquatic life, which is critical for maintaining a healthy ecosystem, and will be discussed further in Section 9.6. For example, changing dissolved oxygen and pH levels in water can indicate the presence of pollutants from CSO discharges and misconnections, harmful chemicals from industrial discharges or other environmental issues.

In addition to water quality parameters, sensors can also be used to monitor the flow rate and volume of water in rivers and channels (Godley, 2002). This information is critical for effective water management and ensuring that resources are used sustainably (O'Grady *et al.*, 2021). Understanding changes in flow rate and volume can help water managers make informed decisions about water usage and distribution, reducing the risk of water scarcity or overuse.

9.1.2 Sustainable drainage systems (SuDS)

Sensors play a vital role in the monitoring and management of SuDS. SuDS are designed to manage rainwater and prevent overloading of traditional drainage systems, which can lead to problems such as flooding and water pollution from the excessive operation of CSOs (Gimenez-Maranges *et al.*, 2020; Joshi *et al.*, 2021). Use of SuDS is increasing in the UK (Melville-Shreeve *et al.*, 2018). The use of sensors in SuDS can help to ensure that the interaction between above-ground drainage and the below-ground wastewater conveyance system is functioning effectively and efficiently (Cotterill and Bracken, 2020).

Some of the key aspects of SuDS that can be monitored using sensors include the following.

- Water levels: Sensors can be used to measure the height of water in different parts of the SuDS system, which can help to prevent overloading and manage the discharge of water (see Section 9.3).
- Rainfall: Sensors can be used to measure rainfall intensity and volume, which can be critical in predicting the potential impact of a heavy rain event on the SuDS system.
- Flow rates: Sensors can be used to measure the flow of water through the SuDS system, which can help to manage the discharge of water and prevent overloading (see Section 9.4).
- Water quality: Sensors can be used to monitor parameters such as pH, temperature, dissolved oxygen, ammonia and turbidity, which can provide valuable information about the health of aquatic life and the overall state of the water body (see Section 9.6).
- Evapotranspiration: Sensors can be used to monitor volume loss due to evapotranspiration.

By using sensors in SuDS, it is possible to gain a near real-time understanding of the performance of the system, which can help to prevent problems such as flooding and water pollution. This can also help to maintain healthy aquatic ecosystems and ensure that water resources are managed sustainably.

9.1.3 Sewers

Sensors play a crucial role in the efficient and effective management of sewer systems. Their applications span across various aspects of sewer system monitoring, including flow rate, water level and temperature. The following are the key benefits of using sensors in sewer systems.

- Capacity management: Sensors can help in monitoring the flow rate and water level of the sewer system. This is essential in determining the system's capacity in preventing overflows and ensuring self-cleansing velocities are maintained to minimise sedimentation and blockage occurrence. This information can also be used to optimise the system's operation and improve its overall efficiency.
- Early problem detection: The data collected by the sensors can be used to detect potential issues in the sewer system, such as blockages, infiltration, exfiltration, or changes in flow patterns. Early detection of such problems can help to mitigate larger, more serious issues that could potentially overwhelm the system leading to urban flooding or CSO activity.
- Wastewater quality monitoring: Sensors can be used to monitor the quality of wastewater, including the presence of hazardous and toxic pollutants. This information can be used to ensure that the sewer system is operated in a safe and sustainable manner, and to prevent or quantify the release of hazardous pollutants into the environment and the potential threat to public health, aquatic life and ecological harm.
- Compliance with regulations: Monitoring wastewater quality with the help of sensors can help organisations comply with regulations, standards and legal obligations related to the treatment and disposal of wastewater.

9.2. Barriers to change

The water industry has historically been slower to adopt sensing technologies in drainage systems than in water distribution systems. This has been due to several factors, including cost of technologies, accuracy of sensors, intermittent versus continuous flows and lack of regulatory pressure to monitor sewer flows.

However, in recent years, advances in sensor technology and data analytics have made it more feasible and cost-effective to monitor urban drainage flows, while legislation such as the Water Framework Directive (Environment Agency, 2016) gives impetus for implementing sensing regimes. Despite this, there are still some existing barriers that hinder the uptake of water sensors in urban drainage systems.

9.2.1 Safety

Regulations relating to explosive atmospheres are essential to the safety of equipment used in hazardous sewer environments. In the UK, The Dangerous Substances and Explosive Atmospheres Regulations 2002 (DSEAR; HMG, 2002) (which implement EU Directive 99/92/EC, the 'ATEX' directive (EC, 1999)) classifies places where explosive atmospheres might occur according to the severity of the risk. The places with the highest risk are classified as zone 0, where a flammable mixture is present continuously or frequently, or zone 1, where it is likely to occur occasionally in normal operation. Most sewer systems are classified as zone 1, but some manufacturers find it cost-effective to meet the more stringent zone 0 standards.

To meet the ATEX standards, manufacturers must adopt either the intrinsic safety (IS) or explosion proof device (EXD) approach. These standards ensure that the equipment is designed and tested to withstand potential explosive conditions and prevent ignition. The rigorous test programme is conducted by approved test houses.

It is important to note that these regulations apply not only to new equipment, but also to existing equipment that is being used in hazardous environments. Manufacturers must demonstrate that their products have undergone and passed the required tests and that they comply with the ATEX standards. This helps to ensure that equipment used in sewer environments is safe and that the risk of accidents or incidents from use of the equipment is minimised.

In addition to the ATEX regulations, reference should also be made to EU Directive 2014/34/EU (EC, 2014), which covers equipment and protective systems in potentially explosive atmospheres. This directive provides a harmonised set of safety requirements that are applicable to the entire EU and helps to ensure a high level of safety for workers and the public.

9.2.2 Certification of performance

In the UK, the Environment Agency has taken a proactive approach to ensuring consistent standards of environmental measurements through the creation of its monitoring certification scheme (MCERTS). The scheme outlines a comprehensive framework for environmental measurements to be made in compliance with the agency's quality requirements and performance standards, which are based on various international and national standards, including CEN, ISO and others. MCERTS covers a vast array of monitoring, sampling and inspection activities, including continuous emissions monitoring systems (CEMS), continuous ambient air monitoring systems (CAMS), portable emissions monitoring equipment and continuous water monitoring equipment (CWMS). The performance standards for each of these monitoring systems can be viewed in detail on the MCERTS website (Environment Agency, 2014).

One of the specific applications of MCERTS is in the sewer flow monitoring field. The scheme has already been applied to flow measurement at wastewater treatment plants, with a standard accuracy of ±8% required (Environment Agency, 2022). However, the non-homogeneous nature of effluent in sewer networks creates unique challenges for reliable monitoring. To address these challenges, the US Environmental Protection Agency has established a similar process to MCERTS, known as the Environmental Technology Verification Program (ETV). The ETV (USEPA, 2022) aims to verify the performance of innovative technologies with the potential to enhance environmental protection and human health. Through the verification and promotion of innovative technologies, ETV is working towards bringing new environmental technologies to the domestic and international marketplaces.

The MCERTS programme offers several key benefits to the environment and to industry. First, it ensures that all environmental measurements are carried out to a consistent standard, providing reliable data on environmental quality and trends. This helps regulators and industry to make informed decisions about how to reduce environmental impacts and improve environmental protection. Second, the MCERTS programme provides a level of assurance for industry and allows companies to demonstrate their commitment to environmental protection and high-quality environmental management. Finally, the MCERTS programme helps to promote innovation in environmental monitoring technologies, as companies strive to meet the standards set by the Environment Agency. By promoting innovative technologies and practices, MCERTS helps to ensure that the UK remains at the forefront of environmental protection and management.

9.3. Depth sensing

Measurement of flow depth in a sewer network provides a critical understanding of the hydraulic loading on the system and its proximity to failure, which can occur as a result of a surcharge. The continuous measurement of water levels, particularly the rate of change, can serve as an early warning indicator of impending failure.

Conventionally, the measurement of flow depth has been conducted using pressure transducers placed at the bottom of the flow or ultrasonic depth monitors placed above the water level. Recent advancements have led to the development of new technologies that may be suitable for this application, including inductance, capacitance and radar-based approaches (Rak *et al.*, 2023). These new technologies provide innovative solutions to the challenges faced in measuring sewer flow depth and ensuring accurate data.

9.3.1 Visual depth gauges

Perhaps the simplest method for measuring the depth of a fluid is a visible scale that penetrates the flow vertically and can be read by eye. This scale may be permanently fixed in place (gauge board), most often at river banks or channels, or it may be used as a mobile inspection method (gauge rod) that is placed into the flow temporarily in order to record an instantaneous depth (see Figure 9.1). The limitation of a visual approach is that it requires human presence on site to collect the measurement accurately, and to input the data correctly into a database.

In instances when it is only necessary to record the maximum water level reached during an event at a critical location in the system, a maximum water level device can be used. These are usually tubular devices with the maximum level reached during a storm being recorded by a colour marking. The measuring cylinder contains a graduated rod onto which a transparent self-adhesive colour band is fixed. The rising water in the measuring cylinder rinses the colour out reliably, up to the respective water level.

Figure 9.1 Examples of (left) a gauge rod and (right) a gauge board

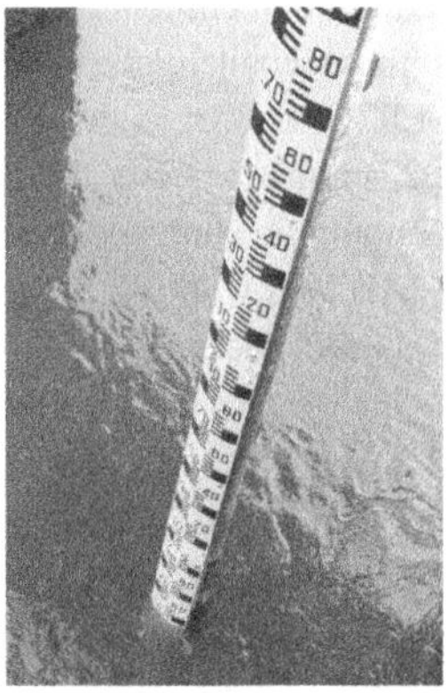

A sharp dividing line displays the highest water level that occurred during an event. The exchange of the colour band is easily done by loosening the upper cylindrical head screw. There is a straining device at the base of the cylinder that acts as a filter for coarse dirt and as a damping device preventing influence by the wash of waves in the measuring cylinder. These devices are low cost and simple to install but are limited for application to multiple storm events as they require a manual reset. Data are also typically recorded manually, although camera-based systems may be used to record the colour demarcation.

9.3.2 Conductance-based depth measurement

Discrete conductance gauges may be used to detect when a certain level in a drainage system occurs. The basic principle is that the sewer effluents act as the conductance medium between two metal electrodes, and when the water depth meets the electrodes, an electrical current is allowed to pass between them (Shi *et al.*, 2021). This confirms that the water level is equivalent to where the electrodes have been placed. Multiple pairs of electrodes may be used to monitor the occurrence of different discrete flow depths. An example of such a system is the gulley monitor, which is used to establish whether a gulley is empty, full or somewhere between. The fluid's conductivity is not a requirement for measurement; however, the limitation is that the difference in water level between two measurement points cannot be determined, thus preventing the monitoring of changes in level over time.

The same principle can be applied to monitor sewer flow depth at a larger number of vertical positions in a drainage system. Power requirements for this type of system are minimal, but the measurement is binary (wet or dry) at a finite number of depths. Similar monitors could be installed to the weir of CSO structures to identify the onset of spill flow and its duration.

Like the discrete conductance measurement, continuous conductance techniques involve measuring the conductance between two electrodes placed vertically in a fluid. An initial calibration is performed to establish the conductance values for several different depths of water. Once the calibration constants are known, the instantaneous water depth can be calculated from the measured conductance. In this manner, the exact water depth can be monitored over time, to a resolution according to the analogue to digital converter used to record the data. The electrodes can also be made to conform to the inner curvature of a sewer pipe to avoid ragging and blockage. The limitation of this system is that the calibration constants must remain constant, so any change in conductivity of the fluid (due to temperature, salinity, chemical content, etc.) could result in errors. For this reason, the approach is commonly used in laboratory experiments (where the sensors are usually termed 'wave monitors' or 'wave gauges'), but is less commonly used in the field, where conditions are changeable (Rak *et al.*, 2023).

A more advanced conductance-based technique for measuring depths of flow and sediment in drainage systems is described in Section 9.9.3.

9.3.3 Wave-based depth measurement

Acoustics have been used to record water depth for many years and there is currently a wide range of off-the-shelf ultrasonic monitors available to use (Rak *et al.*, 2023). The working principle is usually 'time of flight', whereby a short ultrasonic pulse is emitted downwards towards the water surface and the reflected pulse is then recorded. The time delay between transmission and reception is used to calculate the distance to the surface by multiplying by sound speed (often itself calculated by measuring the ambient temperature of the air to ensure accuracy across all seasons). Provided the distance from the sensor to the bottom of the pipe/channel is known, then the depth can be calculated. Some systems measure the phase difference between the sent and received signals to improve the accuracy of the measurement.

Radar depth sensors function in the same way as acoustic depth sensors. A wide variety of commercial radar depth sensors is available on the market, each offering unique features and specifications to suit different applications. Despite their accuracy in determining water depth, radar depth sensors do have certain limitations. One of the main drawbacks of radar depth sensors is that the signal footprint is relatively large, making it difficult to measure small waves. However, the large footprint of the signal also allows for an accurate representative average depth to be achieved, even in conditions with significant wave action.

9.3.4 Pressure-based depth sensing

Pressure sensors are one of the most established methods for sensing the depth of a fluid (Bergan *et al.*, 1968; Temple *et al.*, 2020). Typically, a strain gauge-based sensor is used to measure the static pressure at the base of a channel/chamber/pipe. The static pressure at a given depth is given by the product of the fluid density, fluid depth and gravitational acceleration. As gravity can be taken as constant, the flow depth can be calculated from a pressure measurement, assuming fluid density is known. Over a reasonable temperature range (0–32°C) the density of water varies between 995 and 1000 kg/m^3, so an assumed density results in only minor (up to 0.5%) errors. If the density of the flow changes substantially (e.g. substantial amounts of suspended solids) then this may need to be accounted for. Pressure sensors can give a very accurate and reliable depth measurement, but they are required to be placed within the flow, which can lead to ragging and blockage, and they may be affected by sedimentation.

9.3.5 Limit sensors

Limit sensors, such as tilt float switches, are widely used and cost-effective devices for measuring the depth of flow in sewer systems. Typically, these devices are designed with a micro switch that is protected by a rubber or plastic membrane. Rather than providing continuous absolute depth measurement, these devices are used as simple triggers to indicate when a threshold has been exceeded.

The switch typically consists of a float that is attached to the wall of a pipe, channel or manhole. As the water level rises, the float rises and activates a switch, which sends a signal to a control panel. By monitoring the water level to enable early flood warnings or inform management processes, the limit switch helps to ensure that the system operates safely and efficiently. By installing limit sensors at different levels within a sewer pipe or access chamber and providing a reference to a datum, it is possible to establish a depth hydrograph of the flow in the sewer. This enables a better understanding of the dynamics of the flow and helps to identify potential problems.

While these devices can be reliable, sudden fluctuations in debris flows can affect their accuracy. Therefore, this technology is better suited for locations with relatively calm flow conditions, such as CSO chambers.

9.4. Flow measurement

Accurate flow measurement is critical in urban drainage systems, where it is essential to monitor and manage the conveyance of wastewater and stormwater. There are various contemporary flow measurement methods available that use different technologies to measure water flow in open channels and closed pipes (Sun *et al.*, 2021). The selection of the appropriate flow measurement sensor depends on various factors, including the type of fluid, pipe or channel size, flow rate, accuracy requirements and installation conditions. In this section, the different types of flow measurement sensors typically used in the water industry will be discussed, particularly those used in urban drainage systems. This will include the principles behind their operation, advantages and disadvantages, and the applications where they are most suitable. By understanding the characteristics of each sensor type, engineers and water managers can make informed decisions when selecting the most appropriate sensor for their specific flow measurement needs.

9.4.1 Primary devices

A primary flow measurement device is a physical structure that guides the flow through a known cross-section, to produce a fixed relationship between the liquid level (or head) and the flow rate. The measurement of liquid level at a known point specific to the design and proportions of a primary device can then be converted to a flow rate. Primary devices have different properties appropriate for specific applications. It is critical that the primary device is selected carefully and appropriately designed and installed. This is to ensure that the range of flow conditions meets the minimum uncertainty threshold (i.e. when the flow is subcritical) and does not exceed the maximum measurable range.

Baffles for stilling the flow upstream of the primary device are an important element of the system design. This ensures that error is not introduced where the liquid level is being measured. Cleaning and maintenance of primary devices is essential in ensuring accurate continuous measurement. A build-up of detritus, silt, sediment, or organic matter, for example, will affect the head-flow rate curve for that structure and introduce uncertainty in the calculated flow rate. This is of greater significance for smaller primary devices or where the measurable range is narrow.

There are two major categories of primary devices: weirs and flumes. A comprehensive review of different types of flumes and weirs, including physical explanation and mathematical derivation, can be found in Clemmens *et al.* (2010). Weirs are structures that span an open channel, incorporating a specifically designed slot or notch over which liquid flows. Accurate flow measurement can be obtained where critical discharge conditions are maintained subject to well-documented weir design requirements. Weirs are

Figure 9.2 Triangular notch weir (left) and Venturi flume (right)

typically classified by their design. The most common weir designs are triangular or v-notch (Figure 9.2 left), rectangular (with and without end contractions), trapezoidal (or Cipolletti) or a compound (or hybrid) incorporating a v-notch within a rectangular weir design. Weirs are typically less costly than flumes in their design and installation and can be a good choice where the open channel flow conditions are more limited.

A flume is a specifically shaped section of an open channel that typically comprises a converging section, a throat section and a diverging section (Figure 9.2 right). The flow rate is a function of the liquid level at specific points within the flume design. This is typically upstream of the converging section although it can vary for the many types of flume designs. The key advantage of a flume over a weir is that high flow velocities induced by critical or supercritical states of flow are maintained. This is a particular benefit where self-cleansing velocities are necessary to prevent the build-up of sediment and other deposits.

The installation of flumes inserted into the bed of the sewer, together with associated depth measurement, is a long-established practice for the measurement of sewer flows. A development of this type of approach is a fully portable, precalibrated and easily installed flow metering device (often called an insertion meter), capable of accurate measurement in all sewer and stormwater flow conditions, including partially filled or full pipe, reverse flow and submerged flow.

9.4.2 Area–velocity flow measurement

Submerged area–velocity (AV) flow meters incorporate sensors that measure both the level of the water and the velocity through the water column. Typically, the characteristics of the pipe or channel, including the relative positioning of the sensor within the cross-sectional area, are programmed into the associated flow meter, and these measurements are then resolved to determine the flow rate. Continuous Doppler AV sensors typically measure the velocity in a cone of ultrasound reflection inclined in front of the sensors. This gives a good representation of average velocity until depths become high (Larrarte *et al.*, 2008). Gated Doppler sensors measure velocity at fixed positions (known as bins) along sound beams emitted by the sensors. This generates a cross-sectional map of the velocity field. However, this technology is costly and might not work as well at low depths as the transducers need to be fully submerged.

9.4.3 Surface velocity measurement

Non-contact sensors that measure surface velocity provide an alternative to submerged AV sensors. This technology is best suited to shallow, fast flows, where access to the flow presents a health and safety risk or where access is difficult or restricted, compared with submerged AV sensors.

Flow depth and velocity is measured from above the water surface. Level is typically measured using radar or ultrasonic sensors, while velocity is measured at the flow surface or slightly below the flow surface using either radar, acoustics or laser, each based on the Doppler principle.

The operating principle for radar and acoustic systems is to project energy towards the free surface at an oblique angle and to receive the backscattered signal that has reflected from the rough water surface. The Doppler shift in the signal indicates the speed that the rough surface is advecting at. This relies on the assumption that the wave pattern on the water surface is travelling at the same speed as the fluid, which is usually, but not always, the case (Dolcetti *et al.*, 2019, 2020). Surface backscatter methods are not best suited where the flow surface is very smooth or very turbulent. Furthermore, an assumed relationship (based on empirical data) is usually used to convert the surface velocity to a cross-sectionally averaged flow velocity. The accuracy of this assumption depends on the application area, so in many sites a local calibration is required.

Laser-based flow measurement devices consist of a laser and optical sensors that measure the velocity of particles within a layer of water below the water surface, using the Doppler effect. Because the laser signal is reflected from particles within the flow, it is considered a more accurate measurement of the true fluid velocity when compared with radar and acoustic methods that use the reflection from surface waves.

Surface backscatter and laser-based methods are not recommended where there is a variable velocity profile across the flow width. The devices are typically more expensive than submerged area–velocity sensors. One example of successful application of radar sensors is in the use of flumes at inlet works downstream of screens and in environments with shallow flow. However, it is important to consider carefully the specific needs and conditions of a particular location before choosing a non-contact flow monitoring device.

9.4.4 Electromagnetic sensing

Electromagnetic flow meters (e.g. Figure 9.3) measure the flow rate by detecting the velocity of a conductive liquid in a pipe. They consist of two main components: an electrode pair and a transducer. The electrodes generate an electromagnetic field within the pipe, which creates an electrical current in the liquid proportional to the flow rate. The transducer detects this current and converts it into a flow rate measurement. This is a well-established method based on fundamental principles (Shercliff, 1962).

These meters are typically used in full pipe flow applications, where the pipe is full of the liquid being measured. In the case of sewer flows, they are mainly relevant in rising mains, where the pipe is full. Electromagnetic flow meters are known for their high accuracy and low maintenance requirements (Wu *et al.*, 2018). However, they have limitations such as the requirement for a conductive liquid and the need for a constant flow rate for accurate measurements. Additionally, they require mains power and field calibration is recommended, even though they are calibrated by the instrument manufacturer.

Figure 9.3 Electromagnetic flow meter

9.4.5 Tracer detection/dilution gauging

The methodology of dilution in environmental flow measurements relies on introducing a tracer substance into the flowing water and then assessing the degree of dilution of the tracer solution (Swarnkar *et al.*, 2022). There are various types of tracer substances used, including salt, fluorescents and radioactive substances. The latter two types of tracer require advanced measurement techniques, whereas the application of low concentrations of salt can be easily measured using a standard conductivity meter.

There are two main techniques used in dilution flow measurement, the constant rate injection method and the slug injection method. The constant rate injection method involves the continuous injection of a tracer solution at a constant concentration into the flow stream during a storm event. In contrast, the slug injection technique requires the input of a single dose of tracer solution into the sewer system at a location upstream from the measurement point, to ensure thorough mixing of the tracer. Both techniques require the application of complex mathematical equations to determine the flow rate, but these are well understood and widely used.

9.5. Condition assessment

Proactive maintenance regimes, which are capable of averting service disruptions and mitigating costly failures, such as sewer flooding, require accurate knowledge of drainage asset condition. Historically, the condition of urban drainage assets has been assessed either visually (where possible) or by way of an observed impact on flow conditions. More recently, technological methods have been developed to enable objective measurement of the condition of sewerage systems. These novel approaches have enabled the collection of more reliable and robust data, which has proved to be instrumental in formulating effective and targeted maintenance strategies. This section looks at the various technological methods for condition assessment of urban drainage systems, their advantages, and the challenges they present.

Figure 9.4 Robotic inspection of sewer pipes

9.5.1 Robotic closed-circuit television inspection

Robotic closed-circuit television (CCTV) inspection of sewer pipes is a method of using a remotely controlled robot with an integrated camera system to inspect the condition of sewer pipes from the inside (Haurum and Moeslund, 2020), see Figure 9.4. The robot is inserted into the sewer pipe through an access point and navigates through the pipe, transmitting high-quality images and video to an operator who can view the footage in real time. This method allows the operator to identify any blockages, cracks, corrosion or other defects in the sewer pipe, which can then be repaired or replaced before they lead to more serious problems such as sewer overflows or structural failure. Robotic CCTV inspections can be used for both routine and emergency inspections and can cover long distances in a relatively short time. This technology allows for a more efficient and cost-effective way of inspecting and maintaining sewer pipes, reducing the need for excavation and the associated costs and disruptions to the surrounding community. State-of-the-art methods employ artificial intelligence to recognise features and defects automatically (Kumar, 2020).

9.5.2 Acoustic pipe inspection

Innovative acoustic-based systems have been designed to evaluate the condition of pipes in a sewer network (Yu *et al.*, 2021). The systems employ advanced audio signal processing techniques to detect and diagnose various types of pipe defect, including blockages, lateral connections and cracks, among others. One such system involves the deployment of a highly sophisticated acoustic sensor into the sewer network using a telescopic rod (Bin Ali *et al.*, 2011; Plihal *et al.*, 2016). The sensor generates audio signals in the form of chirps that are then reflected back by any defects in the pipes. These reflections are captured by an array of microphones and processed using complex mathematical algorithms to assess the overall condition of the pipes.

One of the key advantages of this type of system is its ability to survey pipes over long distances and around bends, making it ideal for assessing the state of small and complex sewer networks. The system offers a rapid, reliable and cost-effective way of evaluating the condition of each pipe in the sewer network. In addition, repeated surveys of the same parts of the system can be used to monitor deterioration over time, including the impacts of climate change. The use of this technology provides a step change in the way conditions of sewer networks are surveyed and assessed, providing valuable information to help prioritise maintenance and rehabilitation efforts.

9.5.3 Manhole zoom camera

Manhole zoom cameras (MZCs) are specialised cameras designed to inspect the interior of manholes, allowing utilities to assess the condition of pipes, structures and other assets located underground. MZCs are equipped with advanced lighting and zoom capabilities, allowing operators to see clearly and document the condition of underground assets. The cameras can be connected to remote control systems, allowing operators to manoeuvre the camera remotely within the manhole, providing a comprehensive view of the entire area (Jorge *et al.*, 2020).

The use of MZCs has numerous benefits in the water industry. By providing a visual inspection of underground assets, utilities can quickly identify potential issues, such as cracks or corrosion, before they become more significant problems. This allows water companies to take proactive measures to address these issues before they result in costly repairs or service disruptions. MZCs also help improve safety for workers as they reduce the need for manual inspections of manholes, which can be dangerous and time consuming. In addition, MZCs provide accurate and reliable data that can be used to prioritise maintenance and upgrades, reducing costs and improving overall operational efficiency. The high-resolution images captured by MZCs can also be used for reporting and documentation purposes, providing a comprehensive record of the condition of underground assets.

MZCs present a distinct advantage over robotic CCTV cameras, which have to travel along the pipe and may become lodged and are limited by the inability to see around pipe bends. Images can be geotagged, and new research shows that accurate feature sizes can be extracted from the recorded images. The camera will not detect objects hidden behind other objects and, while most of the products are waterproof, they are not capable of collecting clear images under water (for typical wastewater or stormwater). A combination of acoustic inspection and MZCs may present an attractive solution, with rapid detection of features afforded by the acoustic system and rapid investigation of the detected features using the camera.

9.5.4 Surveying critical infrastructure

Maintenance and inspection of large-diameter pipes or conduits carrying sewage is a part of good practice to ensure a safe, efficient service for all customers. Quite often large-diameter conduits are not easily accessible, have inherent health and safety risks and are not readily appraisable by current technology. Unless there is an exceptional circumstance, physical 'walk through' inspection should be avoided, and remote inspection systems should be employed. However, with the advent of the ever-increasing size and depth of sewerage infrastructure, such as the Lee Tunnel in London (6.9 km of 7.2 m diameter, depth of between 75 m and 80 m) and the Thames Tideway Tunnel in London (25 km of 7.2 m diameter, depth of between 30 m and 70 m), the available technology and methods of surveying and investigating large-diameter conduit critical infrastructure is limited. Note: These assets are referred to as conduits because 'tunnel' is a form of construction.

Large-diameter conduit critical infrastructure assets are by their very nature considered high risk and are likely to be strategically important to the operation of the wastewater network. The consequence of failure could cause significant disruption for the delivery of service as well as the potential for reputational damage and significant unplanned capital expenditure.

To mitigate the risk of failure it is important to understand how the condition of these assets changes under their operating environment, so an inspection regime is necessary to inform an asset condition profile. It is important to understand the rate of deterioration such that long-term investment plans can be formulated. Given the strategic nature of these assets, ideally they should be surveyed once every 15 years.

Within the UK there are limited contractors who can offer a service for surveying large-diameter conduits. Advances in the use of drone technology may offer an effective method of surveying such assets.

9.6. Water quality measurement

Water quality monitoring is critical to ensuring the safety of water resources and protecting public health, with various approaches and technologies used to assess the quality of the water (Kruse, 2018; Yaroshenko *et al.*, 2020). The UK water industry asset management plan (AMP) process (Ofwat, 2019) establishes standards for water quality, which are expressed through percentile standards (PS) and fundamental intermittent standards (FIS).

The conventional approach to monitoring sewer effluent involves collecting representative samples using automatic water samplers, followed by laboratory analysis to detect various determinants, including sanitary suites, biochemical oxygen demand (BOD), heavy metals, nutrients, bacteriological parameters, emerging substances and priority pollutants. This approach is project specific, and it can be costly both in terms of equipment and laboratory analysis. Collecting timely samples of sufficient volume while maintaining sample integrity during the chain of custody requires diligence and robust processes, systems and procedures.

Continuous water quality monitoring is well established using multiparameter water quality sondes. The range of sensors and performance continues to evolve from several established manufacturers. Sensor types are wide ranging and can include pH, temperature, dissolved oxygen, conductivity, oxygen reduction potential (redox), turbidity, ammonium, nitrate, fluorescent dissolved organic matter (fDOM), crude oil, etc. These sensors may be used in continuous mode, providing a constant stream of data, or they can be triggered on an event basis when user-defined thresholds are met. There is also the potential to utilise the sensors on these instruments as surrogates and to calibrate or reference these against other more difficult to measure parameters.

Ongoing research is focused on the development of new techniques and instruments that are based on ultraviolet-visible light absorption, infrared, fluorescence and microfluidics.

9.6.1 Temperature

In recent years, there have been significant advancements in sensors for monitoring temperature making them both cost-effective and reliable. The established nature of these technologies can be attributed to the fact that temperature serves as a fundamental reference point for accurate measurements of other key parameters, including but not limited to dissolved oxygen, conductivity, oxygen reduction potential, pH and ammonia. Some common temperature sensors are as follows.

- Thermistors: These are widely used in wastewater and stormwater flows because of their high sensitivity, accuracy and stability. Thermistors are made of semiconductor materials that change resistance with temperature. Several studies have used thermistors to measure wastewater temperature, such as Gangar *et al.* (2021).
- Resistance temperature detectors (RTDs): RTDs are another common type of temperature sensor used in the water industry. They are made of a wire or film of a metal, such as platinum, nickel or copper, that changes resistance with temperature. RTDs are accurate, stable and operate over a wide temperature range (Nuclear Regulatory Commission, 1981).
- Thermocouples: These are commonly used in wastewater and stormwater flows because of their durability and ability to withstand harsh environments. They are composed of two dissimilar metals that generate a voltage proportional to the temperature difference between the two junctions. Thermocouples are suitable for measuring the temperature of corrosive or abrasive wastewater. Several studies have used thermocouples to measure wastewater temperature, for example, Oh *et al.* (2018).
- Infrared (IR) sensors: IR sensors are non-contact temperature sensors that measure the infrared radiation emitted by objects. They are useful in applications in which direct contact with wastewater or stormwater is not possible or desirable. IR sensors are suitable for measuring surface temperature and for monitoring the temperature of large water bodies. Several studies have used IR sensors to measure wastewater temperature, such as Sobrino *et al.* (2016).

9.6.2 Dissolved oxygen

Dissolved oxygen (DO) is an essential component for all forms of aquatic life and refers to the amount of oxygen present in water. The measurement of DO is considered one of the best indicators of the overall health of a water ecosystem, as it provides important information about the availability of oxygen for aquatic life. DO levels are influenced by various factors, such as temperature, diffusion, aeration, photosynthesis, respiration and decomposition.

Some common sensor types used in the water industry (Wei *et al.*, 2019) are as follows.

- Electrochemical sensors: These sensors use a probe that contains a membrane-covered electrode. The electrode is sensitive to oxygen, and when it comes into contact with the water sample it generates a signal that is proportional to the concentration of DO. The sensor output can be displayed on a meter or data logger, allowing continuous monitoring of DO levels.
- Optical sensors: These sensors use fluorescence or luminescence to measure the amount of DO in the water sample. They work by exciting a fluorescent or luminescent material with light and measuring the resulting emission, which is proportional to the DO concentration. Optical DO sensors are robust and reliable, but they require regular maintenance to prevent fouling and maintain calibration when applied in sewer flows (Wang *et al.*, 2019).
- Amperometric sensors: These sensors measure the electric current produced by the electrochemical reaction between oxygen and an electrode. They are widely used in laboratories for analysing DO in biological and environmental samples. In environmental monitoring, amperometric sensors are used for measuring the concentration of various pollutants, such as nitric oxide in the air. They are also used in water quality monitoring for the detection of DO and other analytes.

- Galvanic sensors: Galvanic DO sensors use a galvanic cell that generates an electrical potential through a chemical reaction between two metals (usually lead and silver). Oxygen diffuses through a membrane and reacts with the lead, producing a current that is measured by the silver electrode. Galvanic sensors are simple, fast and accurate, but have a limited lifespan and can be affected by other chemicals in the water. Calibration and maintenance are necessary to ensure accuracy. Galvanic DO sensors are widely used for measuring DO in water for environmental monitoring, aquaculture and wastewater treatment.

9.6.3 Biochemical oxygen demand

Biochemical oxygen demand (BOD) is a crucial parameter in determining the quality of water and organic matter present in it. It is a measure of the amount of oxygen needed to break down the organic material in water, giving an idea of its impact on the ecosystem and the health of the water body.

Traditionally, the measurement of BOD is carried out through the collection and laboratory analysis of samples. The water samples are incubated to measure the decrease in the dissolved oxygen levels over a specified period, which is then used to determine the BOD. Specimen preservation from collection and throughout the chain of custody is essential in maintaining sample integrity and ensuring each sample is analysed prior to deviation.

However, this method is not only time consuming and costly, but also requires specialised equipment, trained personnel and a well-equipped laboratory. These factors have driven the development of new and innovative techniques for measuring BOD, such as the use of rapid and portable field sensors and the integration of automated laboratory analysis techniques (Sonawane *et al.*, 2020). These advances offer the advantage of faster, more cost-effective and more accessible ways to measure BOD, making it easier to monitor water quality and track changes over time.

Inferring BOD from other parameters has also been investigated. However, highly variable water matrices, especially within a sewer environment, make accurate measurement challenging, affecting measurement accuracy and uncertainty.

9.6.4 Ammonia

Various standard ammonia probes are now widely available, cost-effective and reliable (Li *et al.*, 2020; Lin *et al.*, 2019; Timmer *et al.*, 2005).

Ammonium can be continuously monitored using ion-selective electrode (ISE) technology. The electrode contains a sensing membrane that is selective to ammonium ions, which are commonly found in wastewater, industrial effluent and natural waters. In ISEs for ammonia, the electrode responds to the presence of ammonium in the water to generate a current proportional to the concentration of ammonia present. This electrical signal is processed and converted into a digital reading of the ammonia concentration, which is then used to provide information on the level of ammonia present in the water being monitored.

Colorimetric sensors can also be used. These use a chemical reaction to produce a colour change that is proportional to the ammonia concentration. Colorimetric sensors are simple and cost-effective, but they may be less accurate than other sensors and can be affected by environmental conditions.

Optical sensors use a fluorescent or luminescent probe that responds to changes in ammonia concentration. Optical sensors are highly specific and sensitive, but they may be affected by interferences from other substances.

Properly maintained ammonia probes can provide real-time, accurate information about the quality of sewer flow, and help inform important decisions related to water management; however, the sensors are prone to contamination and their calibration is a more frequent requirement compared with other standard sanitary parameters. The accurate measurement of pH and temperature is also critical to derive ammonia from the measured ammonium. ISE technology continues to evolve; however, the accuracy is currently not as good when compared with optical techniques.

9.7. Power and telemetry

Power and telemetry systems play a vital role in the collection, transmission and analysis of water sensor data. Power is required to operate the sensors, data loggers and associated communication systems that collect, store and transmit data. Telemetry equipment enables the data to be transmitted wirelessly over long distances. The importance of power and telemetry in the use of water sensors cannot be overstated, as they impact the reliability, frequency and timely use of the data collected.

9.7.1 Power systems

The simplest and most reliable method for providing power to sensing systems is to connect them to mains electricity. This is usually possible at key locations, such as pumping stations and treatment plants, and in some applications municipal power from street lighting circuits and the like have even been used. However, in most remote locations, battery power and/or renewable sources (Dewan *et al.*, 2014) must be relied on.

Battery life is a critical factor in reducing operating costs for sewer monitoring systems. These systems, which are typically mounted in manholes, require traffic management and a sewer team to perform even simple tasks such as battery replacement. As a result, the expected battery life and ease of maintenance of the proprietary battery pack should be considered when selecting a system. The battery pack can either be replaceable or rechargeable but is typically easily replaceable in the field. The cost of the battery pack should also be considered.

The battery life of the system is determined by the current consumption of the various components, including the main operating system circuit, the data transfer system and the sensors. The main operating system circuit should draw as little power as possible in between data samples and communications sessions, typically in the range of 1 μA. In active mode, when the system wakes up to take data readings or perform communications sessions, the current consumption can be kept as low as 100 μA. The data transfer system can have varying levels of current consumption depending on the method used, with options ranging from a direct cable connection to a GPRS connection. The consumption of the sensors will depend on the type and number of sensors being used.

It is important to consider the sampling regime of the system carefully to ensure that enough data are gathered without collecting unnecessary data, such as during dry weather flow when no relevant information can be obtained. A number of novel techniques have been developed to optimise sensor operation to reduce battery consumption (Callebaut, 2021).

There are four main areas that affect current consumption.

- Main operating system circuit – idle: In between data samples and communications sessions, the system should draw as little power as possible. Modern microprocessors can run down to as little as 1 μA in this state, switching to active mode only when it is time to do so.
- Main operating system circuit – active: In active mode, a microprocessor would periodically wake, look at the time and then go back to sleep if nothing is to be done. This can be achieved in as little as 100 μs. If the system has to perform a communications session, or take a data reading, then it will remain awake while these are carried out. Current consumption can be as low as 100 μA in this state.
- Data transfer system: When it comes to data transfer systems, the amount of current required can vary. For a direct cable connection, very little extra current is required, around 5 mA. On the other hand, a low-power radio link requires around 15 mA to run a low-power transceiver. For a GSM dial out, the average current required is around 300 mA for the duration of the call. Similarly, for a GPRS connection, the average current required is around 300 mA for the duration of the connection.
- Sensor power: The consumption will depend greatly on the type and number of sensors. In most sewer monitors, the sensors have been designed specifically for the hardware they are to interface with, although there are some standard 4–20 mA sensors available.

The selection of the appropriate battery type for a data logger is critical and is mainly determined by its capacity, which is measured in ampere hours (Ah). A 1 Ah battery is capable of supplying equipment drawing 1 ampere for 1 hour. The more the current drawn by the equipment, the shorter the battery life will be.

Battery types can be broadly categorised into two main groups: rechargeable and non-rechargeable. Rechargeable batteries are not ideal for sewer loggers as they are unlikely to have an in situ charging mechanism, such as solar panels. A rechargeable battery may be selected if the battery pack is too expensive to dispose of after its life is over, or if the situation makes solar panels viable – for example, where there is an existing above-ground kiosk or mast. Most sewer loggers have a battery life of more than 3 years, and some even surpass 5 years, so a rechargeable battery pack is not necessary unless the monitoring regime is active enough to require frequent battery replacements.

Rechargeable batteries such as Li-ion and NiMH hold their capacity only until about 0°C. In sewer monitoring, the logger is unlikely to encounter such low temperatures, but at CSOs, the system could be exposed to such temperatures, affecting the performance of the battery. Similarly, a logger that is not explosion proof would have to be mounted outside the hazardous zone, monitoring the sewer through a barrier, so is also likely to be exposed to low temperatures.

Lithium cells or batteries are widely used in modern sewer loggers as they offer a broad operating temperature range (–60 to +85°C) and have high capacities, with a single D cell providing as much as 18 Ah at 3.6 V. Lithium batteries are also relatively low cost and have low self-discharge, which is crucial in loggers that are designed to run on low current. It is important to note that even a lithium battery or cell that has never been used will lose only 20% of its capacity after 10 years, whereas other battery technologies can self-discharge within just 6 months, especially the rechargeable types.

9.7.2 Data logging and telemetry

Data logging and telemetry are techniques used in data collection and remote monitoring. While several data storage and transmission methods are available, they differ in how they transmit, process and store data, and each has its own unique advantages. When selecting a data logging system for urban drainage, it is crucial to choose a system that has been designed specifically for the sewer environment, as opposed to one that has been adapted from other logging applications. This will ensure that the operating environment is well understood and that factors such as temperature, chemical resistance and water ingress have been taken into consideration in the design.

Traditional data logging involves collecting and storing data from sensors on a local device, such as a hard disk or SD card. The sensors are typically placed into the water and connected to the logger by a cable. The data are retrieved from the device and then processed and stored locally. This can be considered advantageous because it has the following attributes.

- Cost-effective: Data loggers are relatively inexpensive and can be easily deployed in remote or hard-to-access locations without requiring any additional infrastructure.
- Self-contained: Data loggers are self-contained devices that do not require connection to the internet, making them ideal for remote and off-grid locations.
- High-resolution data: Depending on storage capacity, data loggers can collect high-resolution data at frequent intervals, providing a more detailed picture of the monitored parameters over time.
- Versatility: Data loggers can be attached to sensors to measure a wide range of parameters, making them suitable for various applications, such as environmental monitoring, industrial process control and research.

Telemetry involves transmitting data wirelessly, often in real time, to a central server where it can be processed, analysed and acted on. In telemetry systems, the data are typically transmitted over a network, such as the internet or a cellular network, to a remote server or cloud-based platform for processing and analysis. Several technologies exist to enable the wireless retrieval of sensor data (Jamshed *et al.*, 2022). However, data are often also logged locally for back-up in case transmission fails or until successful receipt of the data packet is confirmed. While telemetry systems present potential cyber-security issues, some of the benefits of telemetry are as follows.

- Real-time monitoring: Telemetry provides real-time data that can be accessed remotely, allowing for immediate analysis and decision-making.
- Scalability: Telemetry can be scaled up to handle large amounts of data from multiple locations, making it suitable for large-scale monitoring projects.
- Reduced maintenance: Telemetry eliminates the need for manual data retrieval, reducing the need for site visits and maintenance.

Overall, data logging and telemetry each have their own benefits and drawbacks, and the choice between the two depends on the specific monitoring requirements and resources available. Data logging is often more cost-effective and suitable for small-scale or short-term monitoring projects, while telemetry is more suitable for large-scale and real-time monitoring applications.

9.8. Data analytics

Data analytics refers to the use of advanced computational techniques to extract insights and information from the data collected by sensors. These insights can help water companies make more informed decisions about how to manage their resources, optimise their operations, and improve the quality of service they provide to customers (Mounce, 2020). For example, data analytics tools can be used to identify patterns and trends in the data that may indicate issues with the drainage system, such as blockages, infiltration and exfiltration. This information can be used to identify areas that require maintenance or repairs, or to optimise the system to improve its efficiency and effectiveness. Data analytics can also be used to model the behaviour of water systems, which can help water companies to predict how the system will respond to changes in weather patterns or changes in the demand for water. This can be useful for water companies in planning for future infrastructure investments or managing resources more efficiently.

For analytics to be effective, it is essential that data are reliable and accurate and that any artificial intelligence (AI) algorithms and disciplines are robust. Ensuring the accuracy and completeness of data collected in challenging environments and harsh conditions is not trivial and should not be underestimated. Noisy or incomplete datasets can be unfit for analysis by intelligent algorithms. One approach is to filter noisy data or infer missing data periods. However, this can remove or fail to detect the signatures and patterns that form the insights used for the early detection of system failure required to enable operational decision support. From an operational perspective, the result can be false positives and negatives leading to a lack of confidence, reliance and use of what is an important tool in the management of water systems. Where learning AI algorithms are used, the appropriate discipline is critical. For example, artificial neural networks (ANNs) require a stable network and range of conditions to derive sufficient training data to encapsulate the necessary range of scenarios fully. The training data can become out of date due to network changes rendering the algorithm redundant. Fuzzy logic (FL) systems are based on linguistic descriptions of complex systems and can effectively mimic human reasoning and internalise expert knowledge without the need for large datasets. This can be a very useful method when conditions are changing or when it is difficult to capture fully the full range of scenarios or system behavioural response.

9.9. Emerging technologies

Emerging water sensor technologies have the potential to revolutionise urban drainage systems by providing real-time data on water flow and sediment levels. These technologies include acoustic holography, fibre-optic flow sensing, electrical flow and sediment profiling and real-time control. By using these advanced sensors, engineers and water managers can gain more accurate and timely information about the status of drainage systems, allowing them to respond more quickly and effectively to changing conditions. This section explores the latest advancements in water sensor technologies and their potential applications within urban drainage systems.

9.9.1 Acoustic holography

Current research proposes to measure the flow conditions in sewers using acoustics (Nichols *et al.*, 2013). A transducer projects an acoustic field towards the free surface offlowing water in a pipe or open channel, see Figure 9.5. The unique dynamic pattern of waves on the surface scatters the acoustic waves, which are detected by a plurality of microphone receivers. This can enable the reconstruction of the three-dimensional (3D) shape of the water surface (Johnson *et al.*, 2022). Other research has shown that this 3D water surface shape contains information about the flow itself, because it is generated by the turbulence processes within the flow (Muraro *et al.*, 2021). Evidence suggests that it will be possible to determine the flow depth, surface flow velocity and flow rate, and also turbulence intensity, which relates

Figure 9.5 Acoustic wave monitor system developed by Nichols *et al.* (2013)

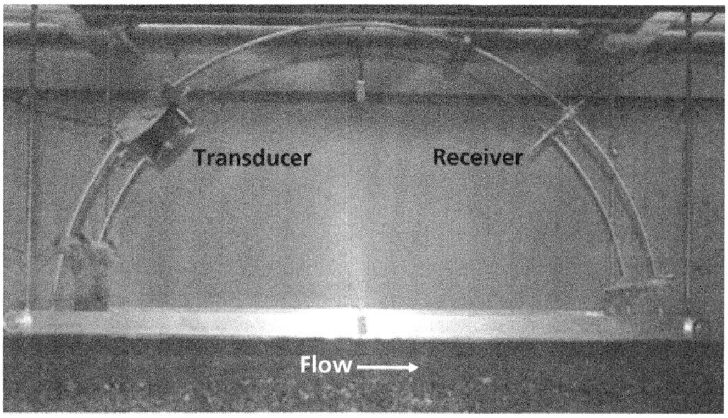

to pollutant mixing and sediment entrainment. Variants of this technique are also under investigation using microwave transducers or digital image correlation (Wu *et al.*, 2022).

9.9.2 Fibre-optic flow sensing

Fibre-optics present the opportunity to obtain a distributed measurement of flow properties along the length of a fibre-optic cable (Prisutova *et al.*, 2022). Research has shown that vibrations can cause subtle changes in the stress on a fibre-optic element, which results in measurable artefacts in the optical signal. This can be used to measure the acoustic or vibration signal along the length of the cable. Recent research has investigated the installation of a fibre-optic cable along the invert of a sewer pipe. In this case it is able to detect fluctuations in water level, which can be correlated with flow depth and flow rate along the length of the pipe. This provides a promising opportunity for widespread measurement offlow properties in a distributed network of drainage assets.

9.9.3 Electrical flow and sediment profiling

A novel sensor, currently under development for commercialisation, employs advanced conductance techniques (discussed in Section 9.3.2). It utilises an array of electrodes of increasing length to provide conductance measurements over different ranges. This enables the interface between conductive or non-conductive media to be detected (for example the sediment–water boundary and the water–air boundary). Due to its advanced capabilities, this sensor can be used to measure continuous, dynamic water level and sediment accumulation. This sensor does not require calibration, thus avoiding the limitations of the discrete and continuous methods that are currently in use. Additionally, the sensor can employ electrical impedance spectroscopy to characterise the impedance 'signature' of the surrounding medium, indicating changes in water composition or sediment structure. The sensor can be manufactured at low cost using flexible printed circuit board (PCB) technology. This means it can be adhered to the inside walls of pipes or irregular shapes with negligible obstruction to flow and negligible risk of ragging or blockage. Wang *et al.* (2020) developed a system of this kind and validated it against physical measurements and a COMSOL model (COMSOL is general-purpose simulation software based on advanced numerical methods). Figure 9.6 shows this system in development, indicating the output of the sensor and the accuracy in detecting water and sediment levels. This type of system is shown to have the potential to measure the depth and composition of settled sediment and sewer flows with high accuracy and low energy requirements.

Figure 9.6 Electrically sensed water and sediment levels, taken from Wang *et al.* (2020): (a) the fitting results for three scenarios (upper line: COMSOL modelling results; lower line: sensor measurement results); (b) the calculated depths demonstrated in the circular pipe section; (c) photograph of the setup in the experiment

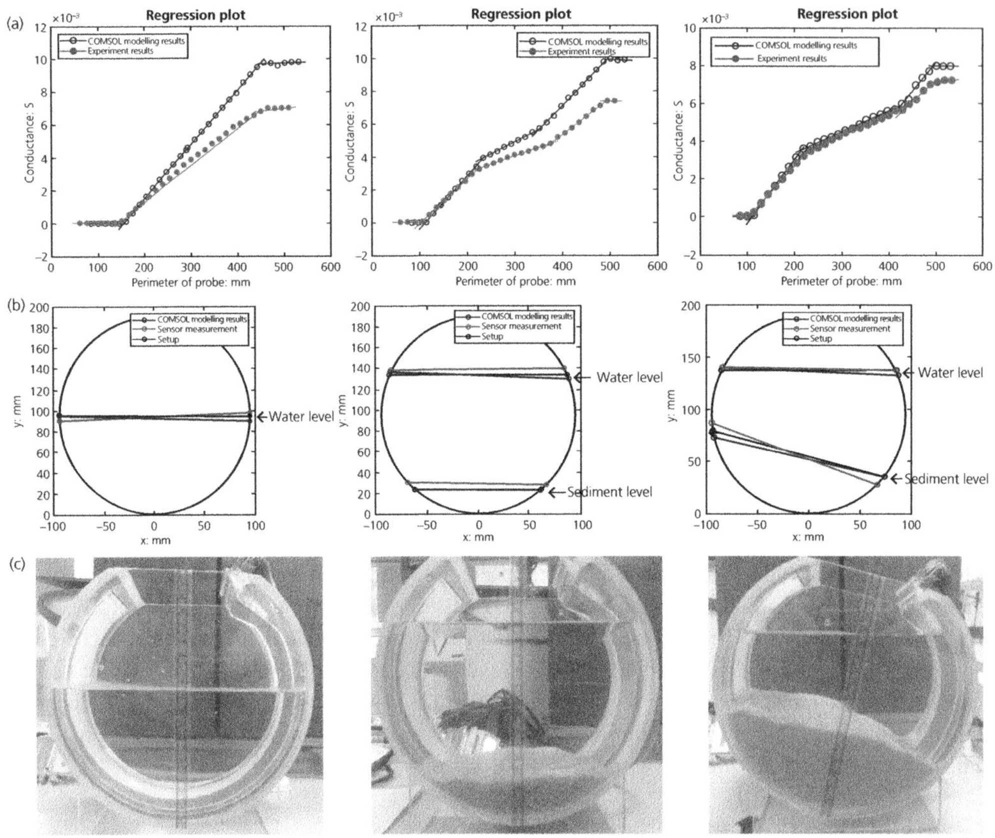

9.9.4 Active flow control

For many years, water companies have used sensor data to control the flow of clean water in water distribution systems (Creaco *et al.*, 2019). These 'smart grids' use sensors to monitor water usage and system pressure and identify areas of high demand or poor supply. The data collected from the sensors are analysed in real time, and control algorithms are used to optimise the flow in the water distribution system. This includes automatically adjusting the flow of water to different areas, identifying leaks, and ensuring that water is distributed efficiently and sustainably.

In drainage systems, active flow control, such as actuated gates, can be used to restrict the flow in order to control the levels at different points in the system, with the aim of reducing flood risk. When these active devices are automatically controlled based on sensor measurements, it is known as real-time control (RTC). RTC technologies use data and analytics to achieve operational goals through specific control objectives. This approach is becoming more readily adopted in the field of urban drainage systems to help manage and control the flow of wastewater and stormwater and mitigate flooding. This is driven by their flexibility, cost (total expenditure and/or carbon) and how readily deployable they are. RTC systems are designed to monitor the status of a network and adjust actuators to alleviate the severity and frequency of system escapes. These systems use data from sensors and analytics to achieve this. Specifically, RTC systems adjust the operation of pumps, valves and gates to utilise available dispersed storage (Ostojin *et al.*, 2017).

In traditional urban drainage systems, water is collected in underground pipes and transported to a receiving wastewater treatment works. This is a primarily gravity-driven process, but can be managed by a series of gates, pumps and valves controlled by an operator. However, during heavy rain events, the water flow can exceed the capacity of the drainage system, resulting in urban flooding or CSO activation. RTC technology can help to address this problem by intelligently adjusting the operation of the drainage system based on the network status. For example, if the water level in a specific part of the drainage system starts to rise rapidly, the RTC system can respond by opening a valve or closing a gate within the network to mobilise existing storage within a system that has available, latent capacity (Campisano *et al.*, 2013).

In addition to mitigating flooding, RTC technology/active flow control can also help to reduce the energy consumption of urban drainage systems. By optimising the operation of pumps and other equipment based on near real-time data, energy can be saved by only using the equipment when necessary. These approaches provide an economical alternative either as a standalone system or as part of an adaptive pathway to optimise the performance of another solution by maximising the additional storage capacity created. This could include SuDS or traditional engineering solutions such as using offline storage tanks or increasing the size of existing infrastructure. However, the system must be carefully understood to ensure the RTC system does not introduce any negative effects, such as increased sedimentation (Williams *et al.*, 2009).

Overall, the implementation of RTC/active flow control in urban drainage systems is helping to improve the efficiency and effectiveness of stormwater management, reducing the risk of flooding, and improving the resilience of urban infrastructure in the face of climate change and increasing urbanisation.

REFERENCES

Bergan PO, Tørum A and Trætteberg A (1968) Wave measurements by a pressure type wave gauge. In *Coastal Engineering*. ASCE, New York, NY, USA. pp. 34–51, 10.1061/9780872620131.003.

Berge SP, Lund BF and Ugarelli R (2014) Condition monitoring for early failure detection. Frognerparken Pumping Station as Case Study. *Procedia Engineering* **70**: 162–171, 10.1016/j.proeng.2014.02.019.

Bin Ali MT, Horoshenkov KV and Tait SJ (2011) Rapid detection of sewer defects and blockages using acoustic-based instrumentation. *Water Science and Technology* **64(8)**: 1700–1707.

Callebaut G, Leenders G, Van Mulders J *et al.* (2021) The art of designing remote IoT devices – technologies and strategies for a long battery life. *Sensors* **21(3)**: 913, 10.3390/s21030913.

Campisano A, Cabot PleJ, Muschalla D, Pleau M and Vanrolleghem PA (2013) Potential and limitations of modern equipment for real time control of urban wastewater systems. *Urban Water Journal* **10(5)**: 300–311, 10.1080/1573062X.2013.763996.

Clemmens AJ, Wahl TL, Bos MG and Replogle JA (2010) *Water Measurement with Flume Weirs*. Water Resources Publications, Lone Tree, CO, USA.

Cotterill S and Bracken LJ (2020) Assessing the effectiveness of sustainable drainage systems (SuDS): interventions, impacts and challenges. *Water* **12(11)**: 3160, 10.3390/w12113160.

Creaco E, Campisano A, Fontana N *et al.* (2019) Real-time control of water distribution networks: a state-of-the-art review. *Water Research* **161**: 517–530, 10.1016/j.watres.2019.06.025.

Dewan A, Ay SU, Karim MN and Beyenal H (2014) Alternative power sources for remote sensors: a review. *Journal of Power Sources* **245**: 129–143, 10.1016/j.jpowsour.2013.06.081.

Dolcetti G, Krynkin A, Horoshenkov KV and Tait SJ (2019) Effect of dispersive water waves on non-contact surface Doppler spectra in shallow river flows. *Geophysical Research Abstracts* **21**: 1-1.

Dolcetti G, Hortobágyi B, Perks M and Tait SJ (2020) The effect of surface gravity waves on the measurement of river surface velocity. In *River Flow* (Uijttewaal W, Franca MJ, Chavarrias V *et al.*, (eds)). CRC Press, Boca Raton, FL, USA, pp. 862–971, 10.1201/b22619.

EC (European Community) (1999) Directive 1999/92/EC of the European Parliament and of the Council of 16 December 1999 on minimum requirements for improving the safety and health protection of workers potentially at risk from explosive atmospheres. *Official Journal of the European Communities* **L023**.

EC (2014) Directive 2014/34/EU of the European Parliament and of the Council of 26 February 2014 on the harmonisation of the laws of the Member States relating to equipment and protective systems intended for use in potentially explosive atmospheres (recast). *Official Journal of the European Communities* **L96**.

Environment Agency (2014) Monitoring emissions to air, land and water: MCERTS. https://www.gov.uk/government/collections/monitoring-emissions-to-air-land-and-water-mcerts (accessed 02/08/2023).

Environment Agency (2016) *Water Framework Directive Assessment: Estuarine and Coastal Waters*. See https://www.gov.uk/guidance/water-framework-directive-assessment-estuarine-and-coastal-waters (accessed 02/12/2023).

Environment Agency (2022) Minimum Requirements for Self-Monitoring of Flow: MCERTS Performance Standard. https://www.gov.uk/government/publications/minimum-requirements-for-self-monitoring-of-flow-mcerts-performance-standard (accessed 20/02/2023).

Gangar T, Satyam S and Patra S (2021) Monitoring/sensing techniques to address pollutant heterogeneity assessment in wastewater. In *Microbial Ecology of Wastewater Treatment Plants* (Shah M and Rodriguez-Couto S (eds)). Elsevier, Amsterdam, the Netherlands, pp. 279–314, 10.1016/B978-0-12-822503-5.00014-X.

Gimenez-Maranges M, Breuste J and Hof A (2020) Sustainable drainage systems for transitioning to sustainable urban flood management in the European Union: a review. *Journal of Cleaner Production* **255**: 120191, 10.1016/j.jclepro.2020.120191.

Godley A (2002) Flow measurement in partially filled closed conduits. *Flow Measurement and Instrumentation* **13(5–6)**: 197–201, 10.1016/S0955-5986(02)00050-X.

Gruber G, Winkler S and Pressl A (2005) Continuous monitoring in sewer networks: an approach for quantification of pollution loads from CSOs into surface water bodies. *Water Science and Technology* **52(12)**: 215–223.

Haurum JB and Moeslund TB (2020) A survey on image-based automation of CCTV and SSET sewer inspections. *Automation in Construction* **111**: 103061, 10.1016/j.autocon.2019.103061.

HMG (HM Government) The Dangerous Substances and Explosive Atmospheres Regulations 2002. The Stationery Office, London, UK.

HSE (Health and Safety Executive)) (2023) ATEX and explosive atmospheres. https://www.hse.gov.uk/fireandexplosion/atex.htm (accessed 19/02/2023).

Jamshed MA, Ali K, Abbasi QH, Imran MA and Ur-Rehman M (2022) Challenges, applications, and future of wireless sensors in Internet of Things: a review. *IEEE Sensors Journal* **22(6)**: 5482–5494, 10.1109/JSEN.2022.3148128.

Johnson M-D, Krynkin A, Dolcetti G *et al.* (2022) Surface shape reconstruction from phaseless scattered acoustic data using a random forest algorithm. *Journal of the Acoustical Society of America* **152(2)**: 1045, 10.1121/10.0013506.

Jorge F, Costelha H and Neves C (2020) Laying ground for automated manhole inspection: a review. In *Sustainability and Automation in Smart Constructions Advances in Science, Technology and Innovation* (Rodrigues H, Gaspar F, Fernandes P and Mateus A (eds)). Springer, Cham, Switzerland, pp. 181–189, 10.1007/978-3-030-49540-4_16/.

Joshi P, Leitão JP, Maurer M and Bach PM (2021) Not all SuDS are created equal: impact of different approaches on combined sewer overflows. *Water Research* **191**: 116780, 10.1016/j.watres.2020.116780.

Kruse P (2018) Review on water quality sensors. *Journal of Physics D: Applied Physics* **51(20)**: 203002, 10.1088/1361-6463/aabb93.

Kumar SS, Wang M, Abraham DM *et al.* (2020) Deep learning-based automated detection of sewer defects in CCTV videos. *Journal of Computing in Civil Engineering* **34(1)**: 04019069, 10.1061/(ASCE)CP.1943-5487.0000866.

Larrarte F, Bardiaux JB, Battaglia P and Joannis C (2008) Acoustic Doppler flowmeters: a proposal to characterize their technical parameters. *Flow Measurement and Instrumentation* **19(5)**: 261–267, 10.1016/j.flowmeasinst.2008.01.001.

Li D, Xu X, Li Z, Wang T and Wang C (2020. Detection methods of ammonia nitrogen in water: a review. *Trends in Analytical Chemistry* **127**: 115890, 10.1016/j.trac.2020.115890.

Lin K, Zhu Y, Zhang Y and Lin H (2019) Determination of ammonia nitrogen in natural waters: recent advances and applications. *Trends in Environmental Analytical Chemistry* **24**: e00073, 10.1016/j.teac.2019.e00073.

Melville-Shreeve P, Cotterill S, Grant L *et al.* (2018) State of SuDS delivery in the United Kingdom. *Water and Environment Journal* **32**: 9–16, 10.1111/wej.12283.

Monitoring emissions to air, land and water: MCERTS. (n.d.). Gov.UK. Retrieved February 19, 2023, from https://www.gov.uk/government/collections/monitoring-emissions-to-air-land-and-water-mcerts

Mounce SR (2020) Data science trends and opportunities for smart water utilities. In *ICT for Smart Water Systems: Measurements and Data Science* (Scozzari A, Mounce S, Han D, Soldovieri F and Solomatine D (eds)). Springer, Cham, Switzerland, 10.1007/698_2020_482.

Muraro F, Dolcetti G, Nichols A, Tait SJ and Horoshenkov KV (2021) Free-surface behaviour of shallow turbulent flows. *Journal of Hydraulic Research* **59(1)**: 1–20, 10.1080/00221686.2020.1870007.

Nichols A, Tait S, Horoshenkov K and Shepherd S (2013) A non-invasive airborne wave monitor. *Flow Measurement and Instrumentation* **34**: 118–126, 10.1016/j.flowmeasinst.2013.09.006.

Nuclear Regulatory Commission (1981) *Review of Resistance Temperature Detector Time Response Characteristics. Safety Evaluation Report.* Report No. NUREG-0809. Office of Nuclear Reactor Regulation, Washington, DC, USA.

Ofwat (2019) *Business Plans.* https://www.ofwat.gov.uk/regulated-companies/price-review/2019-price-review/business-plans/ (accessed 21/02/2023).

O'Grady J, Zhang D, O'Connor N and Regan F (2021) A comprehensive review of catchment water quality monitoring using a tiered framework of integrated sensing technologies. *Science of The Total Environment* **765**: 142766, 10.1016/j.scitotenv.2020.142766.

Oh C, Ji S, Cheong Y and Yim G (2018) Applicability of electrochemical wastewater treatment system powered by temperature difference energy. *Journal of Hazardous Materials* **351**: 108–116, 10.1016/j.jhazmat.2018.02.026.

Ostojin S, Shepherd W, Williams K, Simões N and Steinhardt J (2017) CENTAUR: smart utilisation of wastewater storage capacity to prevent flooding. In *CIWEM UDG Autumn Conference and Exhibition 2017*. Blackpool, UK, 8–10 November.

Plihal H, Kretschmer F, Bin Ali MT *et al.* (2016) A novel method for rapid inspection of sewer networks: combining acoustic and optical means. *Urban Water Journal* **13(1)**: 3–14.

Prisutova J, Krynkin A, Tait S and Horoshenkov K (2022) Use of fibre-optic sensors for pipe condition and hydraulics measurements: A review. *CivilEng* **3(1)**: 85–113.

Rak G, Hočevar M, Kolbl Repinc S, Novak L and Bizjan B (2023) A review on methods for measurement of free water surface. *Sensors* **23(4)**: 1842, 10.3390/s23041842.

River Basin Management Plans updated 2022: summary programmes of measures mechanisms - GOV.UK. (n.d.). Retrieved February 19, 2023, from https://www.gov.uk/guidance/river-basin-management-plans-updated-2022-summary-programmes-of-measures-mechanisms/1-introduction#:~:text=

Shercliff JA (1962) *The Theory of Electromagnetic Flow Measurement*. Cambridge University Press, Cambridge, UK.

Shi X, Tan C, Dong F, Dos Santos EN and Da Silva MJ (2021) Conductance sensors for multiphase flow measurement: a review. *IEEE Sensors Journal* **21(11)**: 12913–12925, 10.1109/JSEN.2020.3042206/

Sobrino JA, Del Frate F, Drusch M *et al.* (2016) Review of thermal infrared applications and requirements for future high-resolution sensors. *IEEE Transactions on Geoscience and Remote Sensing* **54(5)**: 2963–2972, 10.1109/TGRS.2015.2509179.

Sonawane JM, Ezugwu CI and Ghosh PC (2020) Microbial fuel cell-based biological oxygen demand sensors for monitoring wastewater: state-of-the-art and practical applications. *ACS Sensors* **5(8)**: 2297–2316, 10.1021/acssensors.0c01299.

Sun B, Chen S, Liu Q *et al.* (2021) Review of sewage flow measuring instruments. *Ain Shams Engineering Journal* **12(2)**: 2089–2098.

Swarnkar K, Nikam V, Gupta K and Pearson JM (2022) Review of the state-of-the-art for monitoring urban drainage water quality using rhodamine WT dye as a tracer. *ISH Journal of Hydraulic Engineering*. Epub ahead of print 11 Jun 2022, 10.1080/09715010.2022.20986892.

Temple NA, Webb BM, Sparks EL and Linhoss AC (2020) Low-cost pressure gauges for measuring water waves. *Journal of Coastal Research* **36(3)**: 661–667, 10.2112/JCOASTRES-D-19-00118.1.

Timmer B, Olthuis W and van den Berg A (2005) Ammonia sensors and their applications – a review. *Sensors and Actuators B: Chemical* **107(2)**: 666–677, 10.1016/j.snb.2004.11.054.

USEPA (US Environmental Protection Agency) (2022) *Environmental Technology Verification (ETV) Program: Site Characterization and Monitoring*. See https://www.epa.gov/remedytech/environmental-technology-verification-etv-program-site-characterization-and-monitoring (accessed 02/08/2023).

Vanrolleghem PA and Lee DS (2003) On-line monitoring equipment for wastewater treatment processes: state of the art. *Water Science and Technology* **47(2)**: 1–34, 10.2166/wst.2003.0074.

Wang Q, Zhang J-M and Li S (2019) Minreview: recent advances in the development of gaseous and dissolved oxygen sensors. *Instrumentation Science & Technology* **47(1)**: 19–50, 10.1080/10739149.2018.1453835.

Wang S, Corredor Garcia JL, Davidson J and Nichols A (2020) Conductance-based interface detection for multi-phase pipe flow. *Sensors* **20(20)**: 5854, 10.3390/s20205854.

Wei Y, Jiao Y, An D *et al.* (2019) Review of dissolved oxygen detection technology: from laboratory analysis to online intelligent detection. *Sensors* **19(18)**: 3995, 10.3390/s19183995.

Williams K, Tait S and Ashley R (2009) In-sewer sedimentation associated with active flow control. *Water Science and Technology* **60**: 55–63, 10.2166/wst.2009.286.

Wu J-P, Xu K-J, Xu W and Yu X-L (2018) Transient process based electromagnetic flow measurement methods and implementation. *Review of Scientific Instruments* **89**: 095108, 10.1063/1.5049451.

Wu J, Nichols A, Krynkin A *et al.* (2022) Digital image correlation for stereoscopic measurement of water surface dynamics in a partially filled pipe. *Acta Geophysica* **70**: 2451–2467, 10.1007/s11600-022-00792-w.

Yaroshenko I, Kirsanov D, Marjanovic M *et al.* (2020) Real-time water quality monitoring with chemical sensors. *Sensors* **20(12)**: 3432, 10.3390/s20123432.

Yu Y, Safari A, Niu X, Drinkwater B and Horoshenkov KV (2021) Acoustic and ultrasonic techniques for defect detection and condition monitoring in water and sewerage pipes: a review. *Applied Acoustics* **183**: 108282.

Ashley R, Smith B, Shaffer P and Caffoor I
ISBN 978-0-7277-4178-3
https://doi.org/10.1680/icehudp.41783.255

Chapter 10
Asset management, operation and maintenance

Jeroen Langeveld, Nick Orman and Brian Smith

10.1. Introduction

Operation and maintenance are activities undertaken to ensure the performance of a drain or sewer system continues to meet the required performance. The two terms are often used interchangeably; however, in broad terms operational activities will be considered as those activities that involve no permanent physical change to the asset (e.g. adjustment of penstocks, switching of pumps). Correspondingly, maintenance activities will be considered as those that involve some form permanent or semi-permanent physical change to the asset (e.g. repair, removal of debris or sediment). The future, long-term development of urban drainage and sewer systems is discussed in Chapter 3.

Asset management is defined in BS ISO 55000:2014 as the 'coordinated activity of an organization to realize value from assets', with an asset being defined as an 'item, thing or entity that has potential or actual value to an organization' (BSI, 2014). An asset is recognised when the investment results in future economic benefit to the business. An asset may be either tangible (i.e. physical, such as sewers or pumps, and included within a company's asset register) or intangible, such as the reputation of a company. All these things can bring value to a company and need to be well managed in order to make the most of that value.

The institutional arrangements for the management of drain and sewers systems and the split between the responsibilities of property owners and public sewerage authorities varies widely across Europe and internationally. In the UK, sewerage authorities are typically responsible for any pipe in the network that is outside the boundary of a property. In other countries, the public sewerage authorities only become responsible for the larger sewers taking groups of properties, leaving the smaller sewers as the shared responsibility of the individual property owners. In the UK, road drainage is the responsibility of the highway authority, which is separate from the sewerage authority, while in much of the world these responsibilities are held by a single organisation, often the municipality. In many countries the responsibility for rivers is held by a different entity, while in some it can be part of the same organisation as the sewerage authority.

Where there are split responsibilities, specific arrangements need to be made to coordinate activities between the various entities responsible.

10.2. Asset management

Asset management is about knowing what we want to achieve with an asset and how to make it happen, in addition to assessing risks associated with that asset. It translates the objectives of the organisation into asset-related decisions, plans and activities based on a risk-based approach to ensure the performance of their assets aligns with the owner's or operator's requirements. All assets need maintaining and as such will benefit from long-term plans and strategies.

There are numerous sources of information and supporting guidance related to asset management. Of perhaps greatest significance is the BS ISO 55000 series of standards, which has a strong focus on the organisational aspects of asset management – that is, how to organise asset management. BS ISO 55000 reflects the general approaches towards asset management adopted by the urban drainage community over the past decade(s) in many countries, such as described in some detail in BS EN 752:2017 (BSI, 2017a) and in BS ISO 24516-3:2017 (BSI, 2017b). Assets are (usually) tangible things in regard to water systems; however, there are numerous non-water initiatives and sources of information that are relevant to effective management in the water domain. Increasingly these relate to the circularity of systems, including economics, energy and resources (e.g. Syed and Lawryshyn, 2020).

The Global Forum on Maintenance and Asset Management's *The Asset Management Landscape* (GFMAM, 2014) publication recognises asset management as a continual and iterative process designed to deal with complex and dynamic systems. The main feature of asset management systems is that they allow organisations to learn from their actions, plans and decisions by requiring information feedback loops. These feedback loops are typically expressed in terms of the well known 'plan-do-check-act' cycles and inform business decisions at various levels, ranging from strategic to tactical to operational. Sewer asset management is generally seen as a prerequisite for cost-effective sewer management and the main solution to deal with future, long-term challenges such as climate change, resource recovery and urban development.

Sewer infrastructure represents a major investment in essential services contributing to public health and the protection of the environment. Some assets are identified as critical infrastructure, and management activities should be set at strategic, tactical and operational levels. However, sewerage infrastructure has not been maintained over the years on a truly sustainable basis, with funding and implementation of rehabilitation programmes postponed, and a focus on short-term repairs or an allowed decrease in the level of service provided.

Optimisation will become necessary to compensate for aging and wear and tear and respond to changing societal and economic conditions. There is a need not only to focus on maintenance and rehabilitation, but also to keep future requirements and developments in mind. The systematic management of assets and a risk-based investigation should be undertaken to determine maintenance and rehabilitation priorities. The aim in the effective management of assets is to provide an appropriate service life while fulfilling given requirements in a cost-effective manner.

European Standard BS EN 752:2017 Drain and sewer systems outside buildings – Sewer system management (BSI, 2017a) specifies the functional requirements for achieving the objectives and principles for strategic and policy activities relating to planning, design, installation, operation, maintenance and rehabilitation. It recommends that an integrated approach is taken to the management of the drain and sewer systems and sets out a four-stage process that should consider all aspects of the system, including the hydraulic, environmental, structural and operational performance. The process is summarised in Figure 10.1. Procedures for the first two stages are described in more detail in BS EN 13508-1:2012 Investigation and assessment of drain and sewer systems outside buildings – General requirements (BSI, 2012).

Figure 10.1 Integrated sewer system management planning process (BS EN 752:2017; BSI, 2017a). Permission to reproduce extracts from British Standards is granted by BSI Standards Ltd. (BSI). No other use of this material is permitted. British standards can be obtained from BSI Knowledge

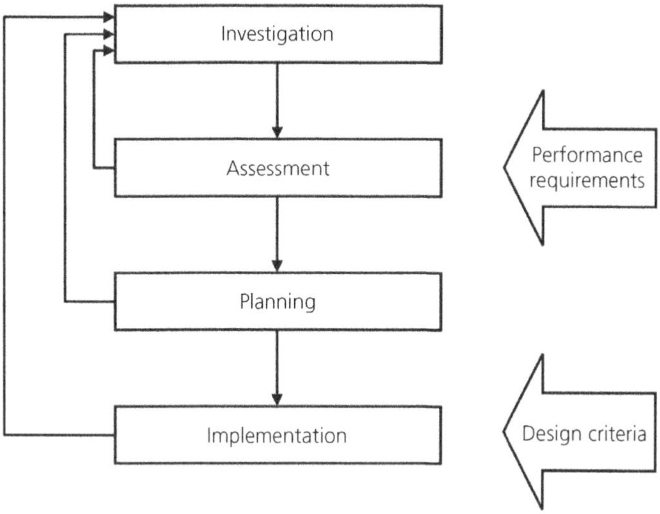

An almost identical process is described in BS ISO 24516-3:2017 Guidelines for the management of assets of water supply and wastewater systems – Wastewater collection networks (BSI, 2017b).

The processes can be applied in stages for strategic planning, tactical planning and operational planning (see Figure 10.2).

A more detailed description of this staged approach is given in BS EN 14654:2021 Drain and sewer systems outside buildings – Management and control of activities – General requirements (BSI, 2021a).

Figure 10.2 Stages in the planning process (BS ISO 24516-3:2017; BSI, 2017b). Permission to reproduce extracts from British Standards is granted by BSI Standards Ltd. (BSI). No other use of this material is permitted. British standards can be obtained from BSI Knowledge

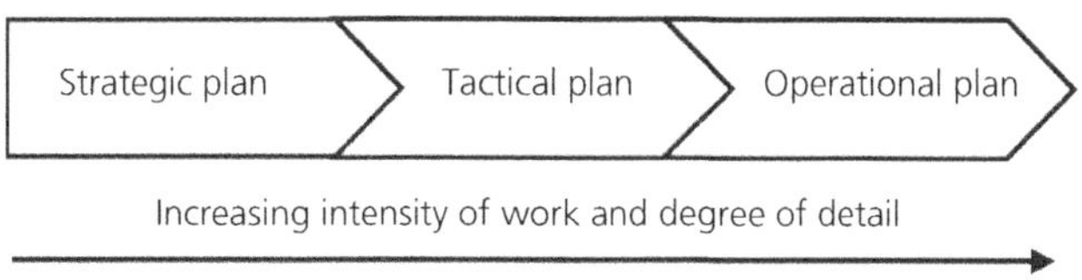

The investigation stage can use a variety of techniques, including the analysis of operational data, specific surveys (e.g. closed-circuit television (CCTV) inspection, flow surveys or water quality surveys) and hydraulic or water quality models, to measure the current and possibly the future performance of the system. The assessment stage then compares the measured performance with a set of performance requirements to identify the performance deficiencies. The third stage involves the development of an integrated sewer systems development plan to overcome the performance deficiencies. This should incorporate a rehabilitation plan, a new development plan and an operations and maintenance plan for the system. The final stage involves implementation and monitoring. The investigation stages are described in more detail in BS EN 13508-1:2012 Investigation and assessment of drain and sewer systems outside buildings – Part 1: General requirements (BSI, 2012).

The hydraulic performance of the drain and sewer system is interrelated to the flow in the rivers and coordination is therefore required with river and coastal flood planning; frameworks such as the Flood Risk Management Plans under the EU Floods Directive (EC, 2007) can provide the basis for this. Water quality is impacted by discharges from combined sewer overflows (CSOs) and surface water outfalls. Initiatives to improve river water quality include those in the River Basin Management Plans produced through the EU Water Framework Directive (EC, 2000).

10.2.1 Risk

As failure of the sewer system is irregular, the most useful measure of risk of sewer failure over time is an annualised risk. Risk is defined as the product of the likelihood of a failure occurring and its effect. The concept of risk has been used in flood management for many years (MAFF, 2001) and is now a requirement of the EU Floods Directive (EC, 2007). The concept allows a range of possibilities to be considered to mitigate a failure. Risk can be managed by reducing the likelihood of the failure, reducing the impact of the failure, or reducing both. Risks can be managed, for example, by engineered solutions, such as bigger pipes or attenuation structures, use of warning systems and measures to prevent the failure or limit its impact, such as increasing the level of the first floor of new dwellings, or improving the resilience to the failure, such as by using tiles instead of wood for flooring. When the risk is expressed in monetary terms then an economic assessment of risk management measures can be undertaken by comparing the present value of the reduction in annualised risk (e.g. in £ per year) with the cost of the proposed investment. This approach is used in the UK in the appraisal of flood and coastal erosion risk management schemes (Defra, 2009).

While it may be relatively easy to identify some assets as having a high risk of failure (e.g. large sewage pumping stations), for gravity sewers the most difficult step in setting up a proactive maintenance regime, whether it is for collapse or blockage, is to distinguish those sewers with a high failure risk from those with a low failure risk.

In determining the risk of failure there are three main considerations. The first is the likelihood that the asset failure will occur (e.g. the collapse or blockage). The second is what effect that asset failure will have on the performance or serviceability of the system. Finally, there is an assessment of the magnitude of that effect.

When the difficulty is in identifying the likelihood of failure of the asset, one approach is to prioritise purely on the magnitude of the consequence. An asset in which the consequence of failure is high is called a critical asset.

The criticality of sewers in relation to their collapse risk has been used to prioritise CCTV inspection of sewers since the early 1980s, when the first edition of the *Sewerage Rehabilitation Manual* (WAA/WRc, 1983) was introduced. In recent years, however, it has been recognised that the assessment of the consequence of collapse did not always give proper weighting to the possibility of flooding or pollution resulting from collapse. Greater consideration therefore needs to be given to the assessment, particularly of flooding impacts due to a collapse or blockage, with appropriate weight given to the magnitude of the consequence.

10.2.2 Maintenance

There are many approaches to maintenance, such as incident-based, preventive, condition-based or predictive. These can be broadly classified based on two basic types: reactive and planned. Reactive maintenance is the execution of remedial work in response to a performance failure (e.g. flooding, or fish kills due to dry weather CSO discharges) or an asset failure (e.g. a pump failure, pipe blockage or collapse). There will always be a role for reactive maintenance in any sewer system because incidents will continue to arise even when planned maintenance is undertaken. As, by definition, reactive maintenance is a response to a failure it cannot reduce the number of failures. However, poor execution of maintenance work can make the recurrence of a failure more likely. For example, poor blockage clearance practice can make another blockage more likely.

Reactive maintenance approaches are the most appropriate where the risk of failure (the likelihood and the consequence) is low. The likelihood of failure differs strongly depending on the sewer component. Figure 10.3 shows the observed failure rates of main sewers, gully pots and house connections, expressed in events per 100 km main sewer/year. A failing house connection has a very strong impact on the serviceability of one household, whereas a failing main sewer may have a much broader impact, including flooding and road collapse. Good asset management limits the risk to society to a predefined acceptable level, balancing the cost of pre-emptive inspection and, if necessary, repair or rehabilitation of a particular asset.

Figure 10.3 Failure rates of sewer components (after Post *et al.*, 2016)

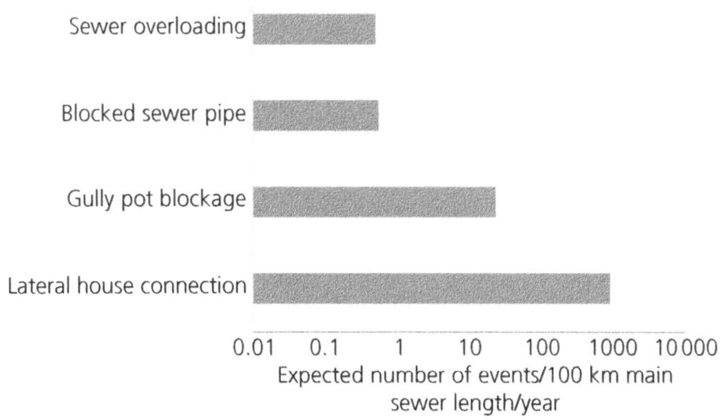

Pro-active or planned maintenance is a strategy to ensure that the necessary requirements can be economically achieved. It can be more economic than crisis maintenance and is more appropriate where the risk of failure is high. It can be carried out either at regular intervals (regular maintenance) or irregularly (in response to monitoring or inspection).

Examples of assets in which planned maintenance is preferred include sewage pumping stations, where both the consequence and likelihood of failure can be high, or a sewer collapse, which would result in extensive flooding. The risk could consider only the financial costs to the sewerage utility, but it is usually more appropriate to consider the costs to society as a whole. The cost of failure therefore includes, for example, not just the direct repair and reinstatement costs, but also the direct and indirect costs of any damage to third parties, including damage to property, loss of earnings, pollution impacts, traffic disruption, etc. When considering costs to society as a whole care must be taken to exclude costs that are merely transfers between the parties. For example, the inclusion of damage costs as well as the insurance claim would count a cost twice, because the insurance payment is just a transfer of the damage cost to another party, the insurance company.

10.3. Serviceability and performance indicators

BS EN 752:2017 and BS EN 13508-1:2012 give a clear description of the process for sewer performance assessment, which can also be applied to sustainable drainage systems (SuDS) components, such as surface water infiltration facilities, ditches and swales. BS EN 752 assesses the performance of sewers based on four types of criteria, covering hydraulic performance, environmental impact, structural condition and operational deficiencies (see Figure 10.4). The hydraulic performance is generally considered to be the basic function of sewers. However, other aspects of the performance of a drain or sewer system are also important. These can include a range of issues such as nuisance from odour or vermin and health and safety issues, for example due to defective manhole covers.

The set of performance assessments, Figure 10.4, allows us to determine the ‘serviceability’ of the system, which has been defined as the capability of a system of assets to deliver a reference level of service to customers and to the environment, both now and in the future (UKWIR, 2005).

The performance or serviceability of the system can be considered in two relevant ways: the level of service experienced by the end users, and the performance of the total system and/or each individual asset or component in the system.

The level of service is only partly determined by the actual performance of sewer systems. For example, overland flows are tolerated when the urban drainage system is hydraulically overloaded by an intense storm event exceeding the design capacity, but not when there is a storm event smaller than the design capacity. This means that the context also plays an important role in the experienced level of service.

The performance of the (individual) assets depends on the condition of the asset and the context. For example, a wide crack in the sewer (condition) will only result in infiltration of groundwater if the groundwater table is above the sewer (context). This means that in sewer asset management, information on the context is very relevant, but this is often overlooked.

Figure 10.4 Process for sewer system performance assessment (BSI, 2017a). Permission to reproduce extracts from British Standards is granted by BSI Standards Ltd. (BSI). No other use of this material is permitted. British standards can be obtained from BSI Knowledge

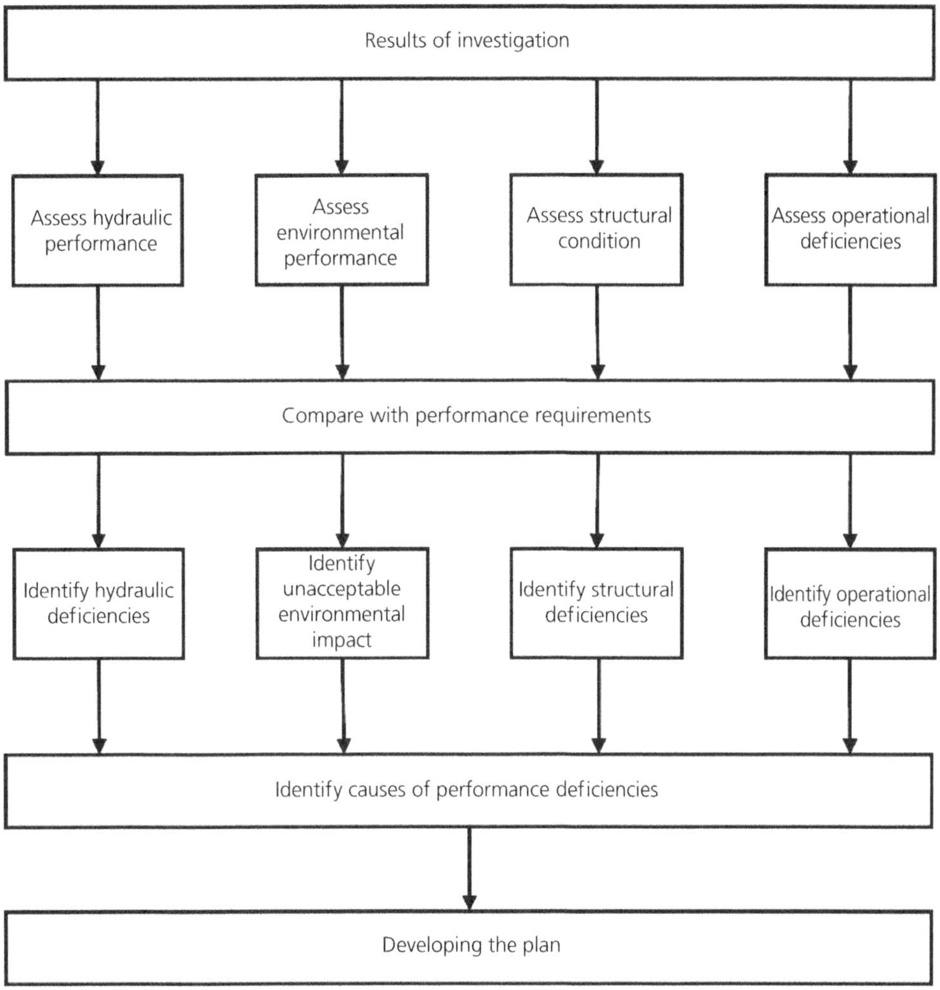

10.3.1 Levels of service

The level of service can include any performance issues as perceived by the customer or impacting on the environment. Levels of service issues may include

- flooding of buildings, land or roads (other than designated flood storage or conveyance routes)
- restricted use of sanitary facilities
- pollution of surface receiving waters
- pollution of groundwater
- obstruction of roads

- hazard to road users
- odour nuisance
- nuisance from vermin
- untidiness of vegetative treatment systems.

As failures to provide an acceptable level of service have a direct impact on customers or the environment, they are useful to regulators. However, they are typically of more limited use in the management of operation and maintenance as the perceived performance is also affected by the (often unknown) context.

10.3.2 Asset performance

Asset performance relates purely to the performance of an asset or group of assets in fulfilling their intended function (UKWIR, 2005). However, even if some parts of drainage systems are performing poorly, the sewer system robustness and redundancy limits the impact on serviceability. Only if the asset is a critical asset could its poor performance impact serviceability in a noticeable way.

Causes of poor asset performance in underground drain or sewer systems can include

- cracking, fracturing or deformation of pipes or ancillary structures
- collapse of the pipes or ancillary structures
- sedimentation of pipes or tanks
- blockage of pipes
- failure of power supply to pumps
- failure of pump control systems
- blinding of screens
- illegal connections or illicit discharges.

For non-piped urban drainage systems, asset performance issues may include

- blockage of infiltration systems
- sedimentation of ponds or channels
- death of vegetation
- untidiness or overgrowth of vegetation
- illegal connections or illicit discharges.

Asset performance indicators record the functional failure of a component of the system, even if the failure has not led to a functional failure of the whole system. They are therefore useful to operators in monitoring the effectiveness of maintenance regimes.

10.3.3 Asset condition

Asset condition is different from asset performance. It can be defined as a measure of deterioration in the physical state of the asset when compared to its new state (WRc, 2009). The effect of deteriorating condition on performance of a mechanical component has been described by Moubray (1997) (see Figure 10.5). The condition of a component can deteriorate over a long period of time without any detectable reduction in performance. The condition deteriorates, usually at an accelerating rate, until the reduction in performance of the component becomes apparent (potential failure) and further until it eventually fails (functional failure).

Figure 10.5 The effect of condition on performance (Moubray, 1997)

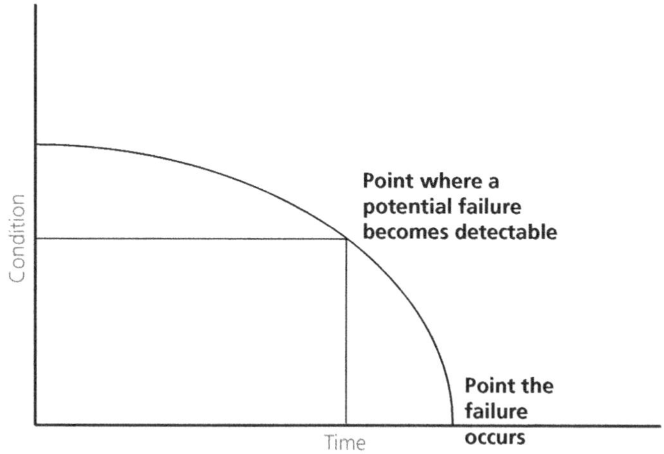

Asset condition can be assessed by inspecting the assets. Until recently, CCTV has been perceived as the one-size-fits-all inspection technique. However, research has shown that CCTV inspections are also error prone, mainly due to human errors (Dirksen *et al.*, 2013). Alternative techniques are being proposed, such as the use of laser profilers to measure the internal diameter and calculate wall losses (Clemens *et al.*, 2015), the use of acoustics for detecting any anomalies in the pipe cross-section (Plihal *et al.*, 2016), and conductivity methods for leak detection (Tuccilo *et al.*, 2011).

10.3.4 Performance indicators

Performance indicators (or serviceability indicators) can either be service indicators or asset performance indicators.

A service indicator has been defined as a quantity that may be measured, calculated or estimated in order to provide an indication of a particular aspect of the service being provided by a utility to customers and the environment. This includes compliance with statutory obligations (UKWIR, 2005).

An asset performance indicator has been defined as a quantity that reflects the performance of an asset or group of assets in fulfilling its intended function (UKWIR, 2005).

Performance indicators can either measure the level of performance by comparison with a reference standard or just measure the performance without any such reference. Where performance is measured against a reference standard, then this can either be explicit (e.g. the number of properties at risk of flooding internally more frequently than once in 10 years) or implicit (e.g. the number of sewer collapses). An extensive list of performance indicators for wastewater services has been produced by the International Water Association (IWA, 2003).

In using performance indicators to compare the performance of different systems or parts of systems, the indicators are frequently normalised based on the population served, the number of connected properties, the length of sewer, etc. The most appropriate method of normalisation will depend on the particular aspect of performance.

Performance indicators can be used for benchmarking performance between countries. However, different arrangements influence management and working practices and make direct comparisons of performance between different countries difficult. For example, a sewerage authority in a country where sewerage authorities are not responsible for smaller diameter sewers would be expected to have far fewer sewer blockages than one ina country in which sewerage authorities are responsible for sewers right up to property boundaries.

Where performance indicators are used as part of regulatory compliance measurement, the reference level of service is often compared with a fixed performance requirement.

Performance requirements should be clearly separated from design criteria. If the performance requirement is used as the basis for design, then only a small amount of deterioration will mean that the system fails to meet the performance requirements. To allow for this, design criteria should be set at a higher level than the performance requirements to allow some headroom for future growth in the system and for expected deterioration.

Traditionally, performance indicators are defined to measure the status or performance of piped systems. However, the application of SuDS and the acknowledgement that urban flooding can best be managed by using the optimal combination of above- and below-ground infrastructure and urban design requires that performance indicators are expressed to cover the whole range of serviceability indicators. The selection of appropriate performance indicators is a very important driver for asset management.

10.4. Maintaining performance and directing development

Urban drainage systems, including traditional pipe sewers and SuDS, fail as soon as the load to the system exceeds the capacity. This may be due to an increase of the load (which may be hydraulic or structural) and/or due to a decrease in the carrying capacity. Asset management is a means to maintain performance: in the short term by recognising or anticipating failures and taking direct actions (repair, cleaning), and in the long term by increasing system capacity or adapting systems to anticipated future developments, such as climate change or demographic changes. Integrated asset management should include both short-term optimisation and long-term development. The selection of appropriate performance indicators is fundamental in this respect.

10.4.1 Failure modes

Urban drainage systems can fail in a number of different ways including the following.

- Increase in flows leading to hydraulic incapacity. These can increase above design levels for a variety of reasons.
  - Development growth increases foul and surface water flows, and surface runoff from impermeable areas in new developments.
  - Urban creep is the increase in impermeable area associated with existing development (e.g. land that naturally soaks up the water removed by impermeable surfaces, such as paving a front garden for parking, the construction of extensions and conservatories and patios). Urban creep could be considered as placing additional obligations on water companies and

highway authorities without any associated increase in income. In England, development control requirements were changed in 2008 (OPSI, 2008) to limit the extent of creep.

- Misconnection of surface water to foul sewers. It has been estimated that up to 2% of properties in England and Wales are misconnected (Ellis and Butler, 2015), a number that can also be found in international literature (Schilperoort *et al.*, 2014).
- Misconnection of land drainage to the sewer system. There is no legal right of connection for land drainage to sewer systems in the UK, and sewerage undertakers have no legal powers for land drainage. However, land drainage is sometimes connected to the system and. Land drainage flows have relatively low peaks, but they often have significantly longer durations than direct surface water connections, and can have a disproportionate impact on those systems relying on storage to provide peak flow capacity.
- Infiltration: Leakage of groundwater into the sewer system through defective joints in pipes or manholes or other defects in the system or areas where there are high groundwater levels. The problem is often not just confined to the sewer system; significant infiltration can also come from building drainage systems. There can also be direct connections of groundwater to the sewer system.
- Excessive rainfall: Current predictions of the effect of climate change in much of the UK suggest an increase in the frequency of high-intensity rainfall, which will lead to deterioration in the flood performance of sewer systems. Current guidance from Defra suggests that the effect on rainfall intensities will vary across the country, but might increase by up to 40% by the 2050s and up to 50% by 2070s. The effects on peak river flows might be significantly greater (see Environment Agency, 2022).
- Changing patterns of water use in recent years, through an increased number and use of water-using appliances and changing habits.
- Changes in the use of drain and sewer systems, with an increased disposal of inappropriate materials to sewers such as fats, oils and grease (FOG), wet wipes, etc. (Chapter 2).

■ Decrease in sewer capacity caused by a chronic build-up of sediment and gross solids or other causes.

- Sediment build-up due to a lack of self-cleansing velocity. More information on sediments can be found in the IWA's *Solids in Sewers* (Ashley *et al.*, 2004).
- Sewer blockage due to gross solids, typically consisting of paper, rags, wet wipes and miscellaneous sewage litter.
- Fats, oils and grease (FOG) discharged to sewers by industrial wastewater, food catering establishments and domestic customers.
- Tree root intrusion: Root penetration can occur due to inadequate leak tightness, poor installation or pipe damage. Contrary to popular belief, the root system of a tree is not a mirror image of the branches, nor is there usually a 'tap root'. Typically, the root system of any tree is limited to the surface 600 mm of soil, by the need for roots to have both air and water. The roots extend radially in any direction for distances frequently more than the tree's height. There are many factors to be considered in understanding growth habit and root development. The rate of growth is dependent on the age of the tree. Vigorous young trees are the most likely to be capable of rapid growth or rapid root regeneration. The deeper the root system, the less rapid the growth. Other factors influencing the rate of growth are

  (*a*) ground conditions
  (*b*) temperature of the soil

(*c*) temperature of the air
(*d*) level of flows within the sewer.

The construction of a sewer can create a favourable environment for root growth at much greater depths by providing a free-draining filling material. Research carried out by IKT, Germany, showed that a tree root tip can develop a lateral pressure in excess of 12 bar and that the root is seeking a void, and not water and nutrients (Stützel *et al.*, 2004). Once in the pipe, the tree root can grow into a mass of small fibrous roots or into a tap root, obstructing the flow.

- Structural failure of the fabric of the system due to overloading of the system by traffic or overburden or to decreasing structural strength. There are several different mechanisms of structural failure of drains and sewers causing asset deterioration, including the following.

  - Ring failure, where a rigid pipe is overloaded and fractures.
  - Beam failure, where lack of support under part of a single pipe causes excessive bending moments along the length of the pipe leading to its failure. As with ring failure, migration of soil into the pipe caused by surcharge or groundwater movement can lead to further subsidence and eventual misalignment and blockage of the pipe.
  - Joint failure: Failure of joint sealing materials can allow migration of soil into the pipe leading to progressive subsidence and eventual failure. Older jointing materials, including puddle clay, lime mortar, hemp and mortar and natural rubber joints, are all subject to failure in time.
  - Buckling: Flexible pipes (e.g. plastic pipes) derive their strength from the passive side support of the pipe under loading. Any loss of side support or overloading is therefore likely to cause deformation of the pipe and, in extreme cases, to buckling of the pipe.
  - Chemical attack: Different materials can be subject to different forms of chemical attack. Metallic pipes can be subject to corrosion; cementitious pipes are vulnerable to acid attack, including attack as a result of septicity in sewage; plastic pipe can also be vulnerable to various hydrocarbons. The attack can be external, due to natural chemicals in the groundwater or from contaminated soils, or it can be internal, as a result of chemicals in the sewage.
  - Wear: Sediments (particularly mineral sediments) or debris in the sewage can cause wear of the pipe material, particularly where the flow velocity is very high (e.g. at the foot of deep drop pipes). It can also cause wear of pump impellers, which can significantly reduce the performance of the pumps.
  - Third party damage: In many locations third party damage, particularly during work on other utility services, is a major cause of sewer failure.
  - Sewer works: Maintenance and upgrading should improve the sewer structural condition and hydraulic performance. However, poorly performed sewer works are one of the main causes of sewer failure. Examples include cleaning the sewer with too high a pressure, which can result in erosion, and forgetting to remove 'temporary' concrete or other barriers placed in manholes to allow safe working downstream, etc.

- Mechanical or electrical failure of pumps, valves or screens.
  - Wear of pump impellers reduces the capacity and efficiency of pumps, and sharp edges on chipped impellers increase the probability of pump blockages, which can lead to motors overheating and sometimes burning out.
  - Changes in pressure and flow velocity, particularly rapid changes, during starting and stopping of pumps can create surge pressures, or water hammer, which can exceed the design pressures and cause fatigue in some materials. This can significantly reduce the design life of a pumping main, as well as damage pumps and valves.
  - Low pressures on the inlet to a pump can result in the formation of vapour-filled bubbles, or cavitation. This can lead to the formation of pressure waves in the liquid, which are likely to cause serious mechanical damage to the impeller.
  - Excess vibration, noise or heat can result in mechanical damage.
  - Wear on other mechanical equipment can lead to unreliability.
  - Septicity and the generation of hydrogen sulphide can result in the production of sulphuric acid, which can cause corrosion of metal parts such as valves.
  - Failure of seals on electrical equipment or breakdown of insulation can lead to electrical failure.
  - Air valves can become blocked with FOG in the sewage, making them ineffective. The resulting accumulation of air increases friction and can cause an airlock, increasing the pressures and reducing the capacity of the pumps.
  - Seizure of reflux valves can cause backflow when pumps stop, emptying the main back into the wet well.

- Exfiltration, where leakage of foul or combined sewage through a defective sewer can lead to unacceptable pollution of groundwater and aquifers. The risk is particularly high when the sewer is close to a groundwater source protection zone or water abstraction point, or when the sewage contains chemicals from industrial effluents.

10.4.2 Maintenance activities

Maintenance activities can include the following.

- Sewer cleaning.
  - Blockage removal: Blockage removal is typically the most frequent sewer maintenance activity. To prevent re-occurrence it is important that the blockage material is either removed completely from the drain or sewer or is broken up into small fragments. Dislodging the blockage without breaking it up is a major cause of further blockages. The most common methods of removal are jetting and rodding.
  - Grease removal from sewers: Blockages caused by FOG can affect quite long lengths of drain or sewer and are difficult to remove. Where the grease has totally blocked the pipe then the flow should first be restored by jetting or by rodding. However, further cleaning should

then be carried out to remove the remainder of the grease. Where grease has adhered to the wall of the pipe over a large area, removal can sometimes cause the grease to become detached in large plates which can block the sewer downstream. The use of spinning jets at right angles to the wall of the pipe can be used to ensure that grease removed from the wall is broken up into small pieces. A jetter with a suction pipe should be used to ensure that the material is removed from the sewer to prevent further blockages and consequent flooding or pollution. In large sewers where there are severe grease deposits, manual removal is sometimes the only practical option. As FOG-related blockages are known to re-occur, regular inspection is advised after grease removal.

- Tree root removal: Fine tree roots can usually be removed by jetting; however, sometimes the effect of the jetting is merely to clean the strands of root rather than remove them. Wire brushes, either rodded or winched through the sewer, can also be used to remove fine roots that are growing into the sewer. For tap roots, conventional jetting or specialised high pressure water cutting equipment may be necessary. For repeat regrowth, the opening through which the roots entered the pipe should be sealed, usually by lining.
- De-sedimentation: Settled deposits in the invert of the sewer can be removed by jetting, winching or rodding. Where sewer sediments have been in place for some time they can accrete and become difficult to remove. In small diameter sewers, a small high-pressure/low-volume jetter can be effective as for blockage removal. However, in large diameter sewers, larger HGV-mounted low-pressure/high-volume machines will be necessary. The use of vacuum extraction equipment to remove the sediments can be advantageous, as it avoids the risk of re-deposition downstream and also allows the recycling of the water. Winching was the traditional method of sediment removal but is now less frequently used. Various tools can be used with winching for sediment removal, including buckets ploughs and brushes. Skill is required to avoid damaging the sewer. Rodding is only practical for removing sediment from small drains. When removing sediment from surface water sewers discharging to a waterbody, measures should be taken to prevent the sediment-laden flow from entering the receiving water as it could cause environmental damage. Also, where a sewer is discharging to a small wastewater treatment plant the sediment load could adversely impact on the performance of the plant.
- Cleaning of pumping stations: Sediments, grease and sanitary solids collect in the wet wells of pumping stations and if not properly controlled can cause failure of the control sensors or pump blockage. Pumping station wet wells should therefore be cleaned periodically to prevent failure of the pumping station. Cleaning is generally done by vacuum extraction and a high-pressure water lance.

- Structural repair, renovation or replacement: Damaged pipes can often continue to function without further intervention; however, structural repair, renovation or even replacement is sometimes necessary. Where there are extensive structural problems, solutions should be developed that take into account other problems in the system.

  - Replacement: Where much of a length of pipe is in poor condition the traditional solution has been to replace the pipe. This still can be the best option when an increase in flow capacity is also required or where renovation of the pipe is not practicable.
  - Renovation: Where a damaged pipe is still able to carry all the imposed load it is often possible to stabilise it by sealing the defects or pipe wall. Linings can also be used to provide additional strength, either by creating a composite structure by bonding the lining to the existing structure or by designing the lining to take the whole of the imposed load.

For smaller pipes, where the existing pipe is deformed, re-rounding techniques are available. These usually involve the insertion of a tool that restores the shape of the pipe and inserts a clip to stabilise it prior to the insertion of a lining. There is a wide range of techniques available, which are described in BS EN 15885:2018 (BSI, 2018).

- Localised repairs can be undertaken by replacement of the damaged pipe or by localised lining. Localised replacement is now effected by removal of the damaged section back to a clean-cut end of pipe and insertion of a new section of pipe jointed with suitable repair couplings.

- Increasing hydraulic capacity: Where it is necessary to increase hydraulic capacity this can be done in several ways.

  - Removal of unwanted flow from the system can create additional capacity. It can be practical to disconnect surface water from a system by separation, connecting it to another piped surface water system or to a sustainable drainage system.
  - Diversion of flow or part of the flow into a neighbouring system with available capacity can also be used to create capacity.
  - Maximisation of existing capacity: Where there is available storage capacity in the system it can be practical to mobilise this additional storage in order to accommodate short-term peak flows.
  - Removal of local restrictions: In some cases the flow is restricted in only a small section. Examples include a section of pipe downstream of a major junction or a pumping station. In these cases a small improvement can substantially increase the capacity. Care should be taken to ensure that such an improvement does not simply move the problem further downstream.
  - Attenuation can be a practical solution where the hydraulic problems are caused by short-term peaks in the flow. Attenuation solutions can be in the form of on-line or off-line tanks constructed either as large chambers or as enlarged sections of sewer (tank sewers). Attenuation solutions are likely to be most cost-effective where it would otherwise be necessary to upsize a long length of sewer. Attenuation solutions are less effective when there are delayed responses in the system (e.g. due to infiltration or land drainage connections).
  - Reinforcement: Additional carrying capacity can be provided within the system by replacing a sewer with a larger one (replacement) or by laying an additional sewer in parallel (reinforcement). In either case this will increase peak flows downstream and the effect of this will need to be considered.

10.5. UK procedures supporting asset management

Drainage Area Plans (DAPs) or Sewerage Management Plans (SMPs) have provided the basis for water companies' investment planning for sewer networks since the publication of the first edition of the *Sewerage Rehabilitation Manual* (SRM) in 1983 (WAA/WRc, 1983). The SRM emphasises the importance of an integrated approach to sewerage planning, in which all the hydraulic, environmental, structural and operational problems are considered together to find the most cost-effective solutions. A key aspect of the SRM procedure is to improve capital efficiency by the developing integrated solutions that take account of all aspects of performance. The SRM procedure has been used in conjunction with the Urban Pollution Management (UPM) procedure for investigating the environmental performance of sewer networks. The UPM provides a process for the investigation of urban wet weather water quality issues and identification of cost-effective solutions for continuous and intermittent discharges. More recently, Drainage and Wastewater Management Plans have introduced the wider catchment flooding aspects in collaboration with other stakeholders.

Pressures from customers and regulators to improve the water environment, particularly through the implementation of various EU environmental directives and customer and regulatory pressures to reduce sewer flooding, led to a series of major work-streams that dominated water company capital programmes. The pressures to complete these very large capital programmes in very short timescales made taking the integrated approach very difficult, particularly as water companies and regulators are responding to different deadlines. Recognising the need for a longer-term strategic approach to water company planning, Ofwat and the Environment Agency commissioned a Drainage Strategy Framework (Halcrow, 2013). This has led to the development of drainage and wastewater management planning (Water UK, 2018).

10.5.1 The SRM procedure

The SRM is an 11-step, risk-based procedure for the planning of sewer systems. It can be used for planning at different scales, from major capital investment right down to maintenance activities. The procedure is a development from the procedure in previous editions of the SRM (WRc, 2001). The procedure is illustrated in Figure 10.6.

The four quadrants of the procedure are aligned with the four stages of the BS EN 752 integrated sewer system management planning process (see Figure 10.1), the procedure working in an anti-clockwise direction. Whereas previous editions of the SRM were applied largely at a single level, the drainage area, the current SRM procedure can be applied at different scales to suit the nature of the problem or the type of sewerage asset being considered. For example, hydraulic and structural problems in sewers might continue to be considered at the level of the drainage area, but a water quality problem in a large urban area might need to be considered on a larger scale, incorporating several drainage areas. Equally, in the application of the procedure to an asset that has no influence on other aspects of the performance of

Figure 10.6 The Sewerage Rehabilitation Manual (SRM) procedure (WRc, 2009)

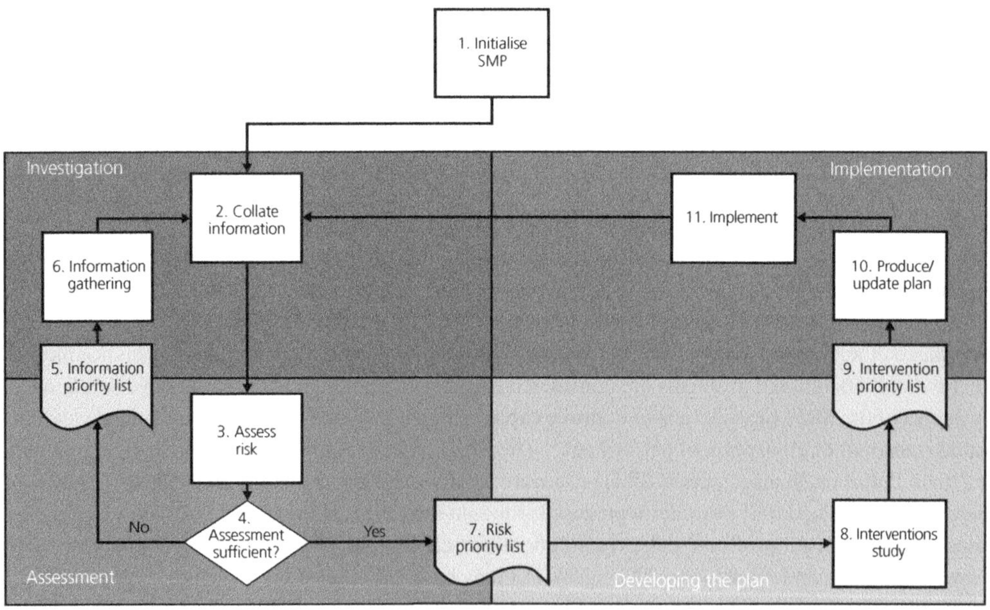

the system, such as manhole covers, it may be better to set the area as the whole area covered by the sewerage utility. The 11 steps are as follows.

Step 1 – Initiate SMP
Step 2 – Collate data
Step 3 – Assess risk
Step 4 – Assessment sufficient
Step 5 – Information priority list
Step 6 – Information gathering
Step 7 – Risk priority list
Step 8 – Interventions study
Step 9 – Intervention priority list
Step 10 – Produce/update Sewerage Management Plan
Step 11 – Implementation.

After setting the objectives, performance requirements and planning horizons in step 1, an initial assessment of current and future risks is undertaken using existing data. This risk assessment is then reviewed and if necessary further investigations are undertaken to refine the risk assessment until it is adequate for the purpose. After that a priority list of performance deficiencies is collated (step 7) and interventions are developed to deal with these deficiencies, which are then subject to economic appraisal. Finally, a priority list of interventions is produced and an SMP prepared and implemented.

There is no explicit monitoring and review stage as the intention is that the whole procedure is cyclic; so the first repeat of steps 2 and 3 in the next round of the procedure form the monitoring and review stages.

10.5.2 Capital maintenance planning common framework

In the UK, the capital maintenance planning common framework (CMPCF) (UKWIR, 2002) describes the water industry's risk-based approach to investment appraisal. This was developed in conjunction with, and approved by, the economic regulator Ofwat. The CMPCF uses an appraisal system based on the assessment of current and forecast future service failure risk, with and without investment. However, Ofwat is now focusing on asset management maturity improve understanding of how companies manage risks from assets and to consider its approach to regulation in light of that understanding (Ofwat, 2021).

CMPCF uses a cost–benefit approach to investment appraisal (Ofwat, 2008) and in the assessment of flood and coastal erosion risk management (Defra, 2009). Risk-based cost–benefit approaches can be applied to investment decisions regarding planned periodic maintenance activities such as sewer cleaning as well as the construction of sewer upgrading schemes.

The CMPCF (UKWIR, 2002) has three principal stages.

- A historical analysis of maintenance expenditure and serviceability indicators.
- A forward-looking analysis to identify future maintenance expenditure to meet regulatory objectives.
- A conclusion which takes the results of the first two stages and makes the case for future maintenance.

These principles have been incorporated into the recent editions of the SRM procedure.

10.5.3 *Urban Pollution Management Manual*

The *Urban Pollution Management Manual* (FWR, 1998) was developed as a procedure for planning works to reduce the impact of sewer systems, including wastewater treatment plants, on surface receiving waters. It complements the SRM procedure, providing more detailed guidance on these aspects.

Following publication of the *Urban Pollution Manual* in 1998, there were significant changes in legislation, including the implementation of the Water Framework Directive (EU, 2000), with its new water quality and ecological standards. There were also significant improvements in computing power and software development, resulting in new enhanced modelling tools for urban drainage planning. In response to these changes, a third edition of the manual was published in 2012 (FWR, 2012), updated to reflect the new priorities for a range of environmental issues, such as the impact of climate change and adaptation.

10.5.4 Drainage and Wastewater Management Plan procedure

The institutional arrangements for the water cycle in the UK present some challenges to the effective implementation of asset management principles across the whole water cycle. To achieve this, collaboration is required between a number of institutions.

In 2018, a collaboration between the water industry, regulators and local government developed a framework for producing Drainage and Wastewater Management Plans (DWMPs) (Water UK, 2018, 2021) covering planning for surface water management, including flood risk and water quality. DWMPs concentrate on the hydraulic and environmental performance of sewer systems, but do not address structural and operational issues. However, operational issues also need to be considered as they can be vital to the environmental and hydraulic performance of networks.

DWMPs set out how water and sewerage companies intend to extend and maintain a robust and resilient drainage and wastewater system. A DWMP takes a long-term view, setting out a planning period that is appropriate to the risks faced, but with a minimum period of 25 years (Water UK, 2021). The plan is developed in collaboration with relevant stakeholders and involves working at least three different levels: level 1 (company), level 2 (catchment) and level 3 (tactical).

The issues and difficulties that surround engagement with others are not confined to flooding. Frequent CSO activity can exert a chronic polluting effect, but CSO load is only part of the load that water companies need to control. Addressing water company commitments under the Urban Wastewater Treatment Directive for continuous and intermittent discharges plus compliance with the Water Framework Directive requires a collaborative approach to understand both the contribution of the sewer system and that of others.

Water quality is not just a water company issue. Local highway authorities discharge surface water both to sewers and direct to watercourses. Local planning authorities now regulate surface water discharges from new developments and must consider both the flood risk and water quality impacts of these discharges.

Figure 10.7 Collaborative core for integrated planning for the major stakeholders

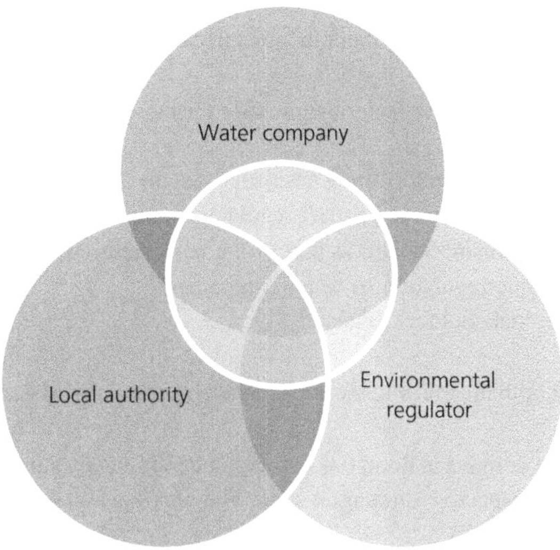

The key stakeholders are the water companies, environmental regulations and local authorities, but other stakeholders may also need to be involved. The 'systems thinking' philosophy encourages wider collaboration between these stakeholders (Figure 10.7); for example, with the regulator to maintain sustainable environmental flows within some watercourses through the discharge of good quality final effluent, and with water supply companies to balance the need for adequate assimilative capacity for dilution of discharges against the abstraction requirements for supply. Collaborative opportunities exist to explore the potential for demand efficiencies to offset growth impacts at wastewater treatment plants and to understand whether this would lead to a possible commensurate increase in blockages.

10.5.4.1 Managing uncertainties

Long-term planning will always include uncertainties. The rate and magnitude of climate change is a major uncertainty, but uncertainties also include the location and timing of new developments and new infrastructure. In the past, decisions have often been made only when the uncertainties have been resolved, leading to short-term sub-optimal decisions. Collaborative solutions bring further uncertainties with the availability and timing of funds to deliver those solutions.

The DWMP procedure proposes the use of an adaptive pathway approach, as set out in BS 8631:2021 Adaptation to climate change. Using adaptation pathways for decision making (BSI, 2021b), to manage the uncertainties, particularly those associated with climate change.

There need to be economic tools to provide a basis for deciding between pathways that provide flexibility. One such approach is real options analysis which applies probabilities to the costs and benefits of different pathways. It has been used in making decisions on climate change adaptation.

10.5.5 Fitting it all together

10.5.5.1 Linking DWMPs to other stakeholders' planning activities

DWMPs require companies to collaborate with other stakeholders during the process. However, each stakeholder still needs to produce plans to meet its objectives.

For a water company, its planning activities must provide for the following.

- Tactical operation and management of the wastewater system.
- Management of its capital programme to meet regulatory requirements in a cost-effective way.
- Providing the basis for its business plan, as part of the periodic review process that is a component of the economic regulation of water companies.
- Providing input to other stakeholders' plans.

This must apply equally for both the foul and the surface water sewer networks and wastewater treatment.

Other stakeholders who own or manage flood risk assets, such as highway authorities, need to have asset management plans for their respective drainage assets. The area for collaboration is only part of their planning activities (see Figure 10.7).

The processes used by each stakeholder to produce their own plans should be aligned to minimise any additional work required for DWMPs.

REFERENCES

Ashley RM, Bertrand-Krajewski JL, Hvitved-Jacobsen T and Verbanck M (2004) *Solids in Sewers.* IWA Publishing, London, UK.

BSI (British Standards Institution) (2012) BS EN 13508-1:2012. Investigation and assessment of drain and sewer systems outside buildings – General requirements. BSI, London, UK.

BSI (2014) BS ISO 55000:2014. Asset management – Overview, principles and terminology. BSI, London, UK.

BSI (2017a) BS EN 752:2017. Drain and sewer systems outside buildings – Sewer system management. BSI, London, UK.

BSI (2017b) BS ISO 24516-3:2017. Guidelines for the management of assets of water supply and wastewater systems – Wastewater collection networks. BSI, London, UK.

BSI (2018) BS EN 15885:2018. Classification and characteristics of techniques for renovation, repair and replacement of drains and sewers. BSI, London, UK.

BSI (2021a) BS EN 14654-1:2021. Drain and sewer systems outside buildings – Management and control of activities – General requirements. BSI, London, UK.

BSI (2021b) BS 8631:2021. Adaptation to climate change. Using adaptation pathways for decision making. Guide. BSI, London, UK.

Clemens F, Stanić N, Van der Schoot W, Langeveld J and Lepot M (2015) Uncertainties associated with laser profiling of concrete sewer pipes for the quantification of the interior geometry. *Structure and Infrastructure Engineering* **11(9)**: 1218–1239.

Defra (Department for Environment, Food and Rural Affairs) (2009) *Appraisal of Flood and Coastal Erosion Risk Management. A Defra Policy Statement.* Defra, London, UK. See https://assets.publishing.service.gov.uk/government/uploads/system/uploads/attachment_data/file/69419/pb13278-erosion-manage-090619.pdf (accessed 09/08/2023).

Dirksen J, Clemens FHLR, Korving H *et al.* (2013) The consistency of visual sewer inspection data. *Structure and Infrastructure Engineering* **9**: 214–228.

EC (European Community) (2000) Directive 2000/60/EC of the European Parliament and of the Council establishing a framework for Community action in the field of water policy. *Official Journal of the European Communities* **L327**. See https://eur-lex.europa.eu/legal-content/EN/TXT/?uri=celex%3A32000L0060 (accessed 09/08/2023).

EC (2007) Directive 2007/60/EC of the European Parliament and of the Council of 23 October 2007 on the assessment and management of flood risks. *Official Journal of the European Communities* **L288**. See https://eur-lex.europa.eu/legal-content/EN/TXT/?uri=celex:32007L0060 (accessed 09/08/2023).

Ellis JB and Butler D (2015) Surface water sewer misconnections in England and Wales: pollution sources and impacts. *Science of the Total Environment* **526**: 98–109, 10.1016/j.scitotenv.2015.04.042.

Environment Agency (2022) Flood risk assessments: climate change allowances. https://www.gov.uk/guidance/flood-risk-assessments-climate-change-allowances#peak-rainfall-intensity-allowance

FWR (Foundation for Water Research) (1998) *Urban Pollution Management Manual – A Planning Guide for the Management of Urban Wastewater Discharges During Wet Weather*, 2nd edn. FWR, Marlow, UK.

FWR (2012) *Urban Pollution Management Manual*, 3rd edn. Foundation for Water Research, Marlow, UK. http://www.fwr.org/UPM3/.

GFMAM (Global Forum on Maintenance and Asset Management) (2014) *The Asset Management Landscape*, 2nd edn. GFMAM. See www.gfmam.org/sites/default/files/2019-05/GFMAMLandscape_SecondEdition_English.pdf (accessed 05/08/2023).

Halcrow (2013) *Drainage Strategy Framework. Good Practice Guidance Commissioned by the Environment Agency and Ofwat*. Halcrow Group, Swindon, UK. See https://www.ofwat.gov.uk/wp-content/uploads/2015/12/rpt_com201305drainagestrategy1.pdf (accessed 21/07/2021).

IWA (International Water Association) (2003) *Performance Indicators for Wastewater Services*. IWA, London, UK.

MAFF (Ministry of Agriculture, Fisheries and Food) (2001) *Flood and Coastal Defence Project Appraisal Guidance*. MAFF, London, UK.

Moubray J (1997) *Reliability-centred Maintenance*. Butterworth-Heinemann, Oxford, UK.

Ofwat (The Water Services Regulation Authority) (2008) *Setting Price Limits for 2010–15: Framework and Approach*. Ofwat, Birmingham, UK.

Ofwat (2021) *Asset Management Maturity Assessment*. Ofwat, Birmingham, UK. See https://www.ofwat.gov.uk/wp-content/uploads/2021/03/Asset-management-maturity-assessment-March-2021.pdf (accessed 09/08/2023).

OPSI (Office of Public Sector Information) (2008) The Town and Country Planning (General Permitted Development) (Amendment) (No. 2) (England) Order 2008 SI 2008/2362.

Plihal H, Kretschmer F, Bin Ali MT *et al.* (2016) A novel method for rapid inspection of sewer networks: combining acoustic and optical means. *Urban Water Journal* **13(1)**: 3–14.

Post J, Pothof I, ten Veldhuis MC, Langeveld J and Clemens F (2016) Statistical analysis of lateral house connection failure mechanisms. *Urban Water Journal* **13(1)**: 69–80, 10.1080/1573062X.2015.1057175.

Schilperoort R, Rudelle E, Vollertsen J *et al.* (2014) CCTV versus DTS: comparison of results in the search for infiltration and inflow. In *ICUD Congress*, Kuching, Malesia.

Stützel T, Bosseler B, Bennerscheidt C and Schmiedener H (2004) *Wurzeleinwuchs in Abwasserleitungen und Kanäle*. See http://www.ikt.de/down/f0108langbericht.pdf (accessed 05/08/2023).

Syed Z and Lawryshyn U (2020) Multi-criteria decision-making considering risk and uncertainty in physical asset management. *Journal of Loss Prevention in the Process Industries* **65(2020)**: 104064, 10.1016/j.jlp.2020.104064.

Tuccilo ME, Wilmut C, Feeney C, Martelk K and Selvakumar A (2011) Field demonstration of electro-scan defect location technology for condition assessment of wastewater collection systems. *Proceedings of the Water Environment Federation* **2011(5)**: 265–281.

UKWIR (UK Water Industry Research) (2002) *Capital Maintenance Planning: A Common Framework*. UK Water Industry Research Ltd, London, UK. See https://ukwir.org/eng/reports/02-RG-05-3/66808/Capital-Maintenance-Planning-A-Common-Framework-Volume-1-Overview (accessed 09/08/2023).

UKWIR (2005) *Capital Maintenance Planning Common Framework: Review of Current Practice*. Report Ref 05/RG/05/14. UK Water Industry Research Ltd, London, UK. See https://ukwir.org/eng/reports/05-RG-05-14/66813/Capital-Maintenance-Planning-Common-Framework-Review-of-Current-Practice (accessed 09/08/2023).

WAA/WRc (Water Authorities Association/Water Research Centre) (1983) *Sewerage Rehabilitation Manual*, 1st edn. Water Research Centre, Swindon, UK.

Water UK (2018) *A Framework for the Production of Drainage and Wastewater Management Plans*. Water UK, London, UK. See https://www.water.org.uk/wp-content/uploads/2018/12/Water-UK-DWMP-Framework-Report-Main-Document.pdf (accessed 09/08/2023).

Water UK (2021) *A Framework for the Production of Drainage and Wastewater Management Plans*, revised edn. Water UK, London, UK. See https://www.water.org.uk/wp-content/uploads/2021/10/DWMP_Framework_Report_Main_Report_September_2021.pdf.

WRc (Water Research Centre) (2001) *Sewerage Rehabilitation Manual*, 4th edn. WRc Publications, Helsby, UK.

WRc (2009) *Sewerage Risk Management*. Water Research Centre, Helsby, UK. See http:/srm.wrcplc.co.uk.

Ashley R, Smith B, Shaffer P and Caffoor I
ISBN 978-0-7277-4178-3
https://doi.org/10.1680/icehudp.41783.277

Chapter 11
Urban drainage analysis and modelling

Will Shepherd, Simon Tait, Richard Kellagher and Richard Ashley

11.1. Introduction

Computational models are now a standard tool for urban drainage engineers. This chapter focuses primarily on the modelling of the time-varying hydrodynamic performance of urban drainage networks. It also introduces other types of models that are being used for urban drainage design and operation. The majority of models used for urban drainage purposes use physically based numerical equations to describe hydraulic behaviour and other phenomena, such as sedimentation, in this way representing the complex in-sewer processes. It needs to be understood that models are used to develop approximate solutions to the sets of relationships used to describe the phenomena of interest (e.g. Argyropoulos and Markatos, 2015). Many of the numerical relationships are drawn from historical and classic hydraulics, empirically derived and adapted for computational analysis. No further details of these are given here, as there are numerous texts and practice publications that can be used for further information (e.g. Mutel *et al.*, 2008).

The most commonly used urban drainage models are one-dimensional (1D) physical representations of the piped network. These first became commercially available in the early 1980s. Since the early 2000s these have developed to include 1D to two-dimensional (2D), which includes representation of the surface in two dimensions. However, there is a range of other types of models: lumped (simplified) models, data-driven models, machine learning models, each appropriate depending on the data available and modelling requirements. The analyst should recognise that a model is merely a tool used to represent reality approximately, it cannot model systems in their entirety, nor can it consider all uncertainties. In the often-quoted words of George Box: 'All models are wrong, but some are useful' (Box, 1976); and more recently applied to urban drainage models by Pedersen *et al.* (2022) when looking at the transferability of models developed for specific sites and objectives: 'Even the simplest mathematical relationship may be an approximation, even if physically based, and many models utilised for urban drainage are empirical, statistical, and based on processing large amounts of data.'

This chapter provides an overview of the development and application of urban drainage modelling approaches in order to provide guidance as to how software tools should be used and the inherent limitations.

11.2. Purpose of urban drainage models

Urban drainage models allow the user to estimate the performance of a system under a range of conditions. Physically based models are commonly used but rely on large amounts of system data. Non-physically based models often require limited or no system data, but are either data-driven or conceptual. This chapter focuses mainly on physically based models and introduces the other types of model.

There are two primary roles for drainage modelling related to performance. The first is hydraulic analysis to assess the hydraulic capacity and performance of a drainage network (or a planned network) to convey wastewater and associated solids, and the second is to assess the water quality within drainage networks and how those networks might impact the natural environment. There are models used for other aspects of designing, operating and managing drainage systems. Examples include economic models for capital and operational costs, and asset management models to plan and control the operation of systems. More recently, models for social aspects of interactions with natural drainage systems have been developed (e.g. Lin *et al.*, 2022). In summary, in the urban drainage domain, as in other infrastructural areas, if there is information, then models can be developed and used as tools for understanding and predicting the performance of systems.

Models are used to develop an understanding of the capability of a drainage system to meet certain criteria; the most common investigations are the impacts on the environment through untreated overflows, flows into wastewater treatment plants and assessment of flooding problems. The main models used for urban drainage therefore have two fundamental outputs: hydraulics and quality; the latter being dependent on the former and potentially comprising many different parameters and not just flows, flow depths and velocities.

Prior to commissioning the construction of a model, the analyst should ensure that the objectives for the model are clear and that the right type of model is used to achieve those objectives. Constructing a model can often be a costly, time-consuming and data-hungry exercise. For existing systems, it is possible to gain significant insight into performance through monitoring of flows and loads alone. However, as noted by Osborne (2022), monitoring data are unlikely to include infrequent rainfall events, nor can they help with an understanding of the implications of future changes, such as significant new developments, network modifications, or change in loading conditions such as climate change. Physically based models are needed to represent these changes in order to predict the change in performance that will result.

The required accuracy and resolution of model predictions will depend on the objective of the analysis. The following are examples of the various situations in which the results from a numerical model of a drainage network may be used.

- Ensuring building drainage systems effectively drain storm and surface water.
- Ensuring the conveyance effectiveness for solids transported in building drainage systems and in connecting pipework.
- Sizing of pipes or storage tanks when designing new parts of a network or to improve hydraulic capacity of an existing network.
- Evaluating the flooding risk for a development site or a catchment area to develop an asset management strategy to reduce flooding.
- Assessing the opportunities and benefits for disconnecting contributing runoff surfaces to reduce flood risk.
- Determining the spill frequency and discharge volumes from combined sewer overflows (CSOs) to demonstrate compliance with environmental regulations.
- Developing real-time control rules for an actively managed network.
- Understanding real-time performance of the network, either to inform real-time control, or to generate system alerts.
- Integrated catchment planning, where several models are used in conjunction to assess the overall flows, discharges, interactions, flooding and pollution consequences.

Historical models of large drainage systems were often simplified due to the then computational constraints (e.g. the earliest versions of the Wallingford Procedure models; DoE/HR Wallingford, 1983), and significant simplifying assumptions were made for particular types of flows; for example, modelled above-ground flows were based on significant spatial simplifications, including lumping large subcatchment areas. Current software tools can use a much more detailed set of data.

11.2.1 Ensuring the model works as required

Computational models of urban drainage systems are built and calibrated for a specific purpose (e.g. local flood risk reduction, or CSO spill reduction) for a given location, at the time the work was done, so great care should be taken when using a previously calibrated model for anything other than the original state of the sewerage system and the objectives of the work (e.g. Pederson *et al.*, 2022). The reasons for this are as follows.

- Data collection for model calibration is optimised for a specific purpose and at locations relevant to that purpose.
- Model calibration focuses on the major output parameters for that study, and may have accepted poorer predictive performance for parameters not required for that study.
- Drainage networks can deteriorate, and even the configuration of networks can change with time.
- Paved contributing areas tend to increase.
- Best practice modelling techniques and software tools improve with time.

All models require some form of calibration and verification to confirm that they replicate the behaviour of the system. In the simplest sense, calibration is the process of setting up or modifying the model using data compared with model results. Verification should simply be the comparison of model results with observation to confirm the veracity of the model, and the data used should be different from that used for the calibration. When verification has failed, then the model may require further calibration by adjustment of one or more parameters, although the failure may indicate erroneous input data, such as incorrect connections or areas drained. There are standardised methods for calibration, depending on the type of model being used. When using commercial models, the calibration parameters are limited to the user input data. Open-source models, however, may allow the user to modify the relationships used in the model. The process of verification and calibration for urban drainage models is set out in CIWEM (2017) and Figure 11.1 illustrates this. Figure 7.3 in Chapter 7 shows an example of a hydrograph verification plot.

Fig. 11.1 Model verification and calibration process

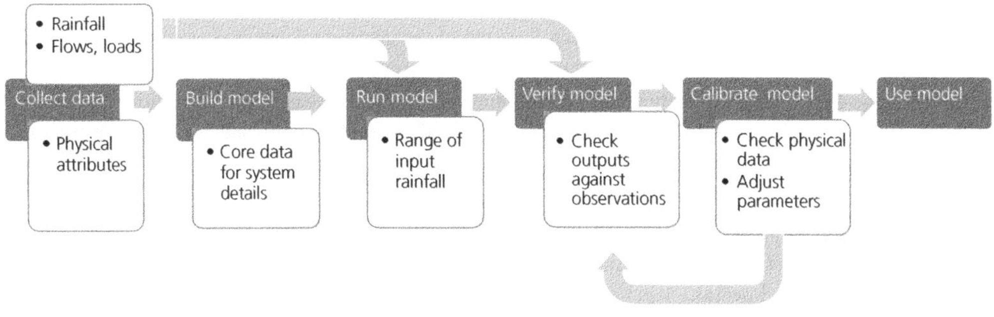

An illustration of the complexities of calibration of an urban drainage model is given by Radinja *et al.* (2021). This study utilised an 'equation discovery tool' to compare analytical models for pipe flow that included estimations of infiltration. Various models were automatically calibrated, including the runoff models and ground infiltration model (GIM) used in UK practice (Chapter 7). These were assessed for their veracity in flow predictions. The objective was to decide which of the models, based on the same hydrological processes, was most suitable for the study. The authors pointed out that the potential for human error increased with the complexity of the models. This was compounded by the difficulties of deciding which models were the most suitable for a particular application. Even the selection of monitoring locations and the period of data collection is fraught with difficulties and introduces further uncertainties when collecting calibration data (e.g. Vonach, 2017). For this reason, procedures are set out in industry standards that are aimed at minimising uncertainties and, when followed, can be 'deemed to comply with best practice'. In the UK, for example, CIWEM (2017) provides guidance on modelling best practice.

11.3. Development of physically based urban drainage models

There is a wide range of analytical models used to design, manage and operate urban drainage systems. These can range from explicit mathematical representations of the physical processes at high levels of detail (from roof runoff, over catchment surfaces, through piped networks to the treatment plant and the receiving waters), through to very simple water balance models, or models that incorporate complex statistical and optimisation tools to help identify and deal with model uncertainties and automate the design processes.

The methods used for designing drainage systems have evolved in response to new knowledge and technological capability, often due to advances in domains other than drainage (Chapters 2 and 3). This means that as drainage systems have developed in response to new knowledge or changing regulations, the different parts of a system might have been designed with very different analytical tools. One type of model or design might have been used for the surface processes, another for the inlets and property drainage, and yet another for the main drainage.

The performance of building drainage systems, comprised mainly of small-sized pipes, is determined by essentially the same hydraulic phenomena as for 'main drainage' systems, but the computational models available for building or main drainage can differ. The larger commercial network models are usually suitable only for the larger-sized conduits, above 150 mm in diameter. Building drainage models are usually bespoke and deal with the pressurised transient or pulsing flows from appliance inputs. Specialist analytical tools are required for the transport of solids in plumbing systems and small-bore pipes (e.g. McDougall and Swaffield, 2000).

In the UK, USA and Denmark, the major step forward in drainage performance assessment occurred consequentially from the availability of large mainframe computers in municipalities in the 1990s. This provided for the first time the ability to carry out simulation of the interacting hydraulic behaviour of pipes in large drainage systems. The following describes the technological development of drainage design and analysis.

11.3.1 The development of pipe-based hydraulic modelling tools

Sanitary inputs to drainage systems are described in Chapter 2 together with the appropriate codes and standards for analysis and design. Much of the analysis and modelling of sanitary flows has been based on empirical defined flow rates and solids transport rules. Flows in small-bore and plumbing systems are unsteady and pulsatile, with varying free surface and pressurised pipe-full flow. There are dynamic models available that are being used for complex or multistorey buildings, where pipe junctions and air locks are significant (e.g. Campbell, 2012). Models for flows in larger main drainage systems include both sanitary flows and stormwater. These can be operated using a dry weather flow component only. There has been renewed interest in the modelling of dry weather flows recently with the realisation that reduced domestic and other water uses can lead to problems in transporting solids in plumbing and small piped systems as well as in main drainage (e.g. McDougall and Swaffield, 2000; Saunders, 2019).

11.3.1.1 The rational method

Historically, drainage design focused on the hydraulic capacity of a system. The issues related to water quality in drainage networks have been of relatively recent interest, caused by the better understanding of pollution arising from emissions of wastewater, together with advances in wastewater treatment and better understanding of the potential for better management of wastewater. Environmental protection regulations have helped to encourage the clean-up of wastewater and minimisation of pollution (see Chapter 1 for an outline of the evolution of wastewater management).

The rational method was perhaps the first objective methodology for sizing of pipes by taking into account the intensity of rainfall and the contributing area draining to a pipe (Chapter 3). This was the primary prediction tool for surface runoff available for much of the twentieth century. Even in the 1980s, when simulation modelling of drainage systems first became possible, the rational method was used in the majority of drainage design and analysis across the world, albeit in various forms. Subsequently, a modified form has been developed as explained in DoE/HR Wallingford (1983). This introduced two 'coefficients' into the rational formula, replacing the single coefficient, in order to account for runoff volume and flow routing better.

As with all hydraulic models of drainage systems, the principal driver is the rainfall input into the model. In 1936 the concept of intensity–duration–frequency (IDF) curves for rainfall was developed by Bilham (1936) based on rain gauge information (Chapter 6). This not only enabled an objective selection of rainfall intensity to be made based on the likelihood of a rainfall event, but it also introduced the concept of modifying the intensity used in analysis based on the concept of time of concentration, so incorporating the characteristics of the contributing catchment. The time of concentration is the average time period from the rain falling on the catchment surface until the runoff passes the point of interest in the drainage system, such as a CSO or the outlet of the catchment. However, this is complicated by the variability in the times of concentration for different parts of the sewer network.

The design of drainage systems has relied on this methodology for many years and design guidance documents, such as those produced in the UK (see Chapter 2) – including *Sewers for Adoption*, first introduced in 1981 (WRc, 2018) – have explicitly specified the return periods and runoff coefficients to be used. Traditionally, the return period criterion had been of the order of 1 to 5-year return periods for estimating the pipe capacity, implicitly accepting that piped systems operated in surcharge for larger events before flooding on the surface catchment starts to occur (see also Chapters 2 and 3). The rational method is still in use and can provide a very rapid albeit approximate estimate of the runoff rates from rainfall on any surface (Chapter 7). More sophisticated analyses can produce more accurate and time-varying predictions of runoff and flows in systems, but these still have a degree of inaccuracy and should be used with care (Section 11.6).

11.3.1.2 Early simulation methods

The Wallingford Procedure was produced in the UK in 1983 to enable the simulation of surcharged drainage systems, this was used for the development of the Wallingford Storm Sewer Procedure (WaSSP) and subsequently WALRUS computer simulation models (DoE/HR Wallingford, 1983). Similar tools, such as MOUSE, produced by Danish Hydraulic Institute (DHI) (Mark, 2019), and the Storm Water Management Model (SWMM) (USEPA, 2023a) in the USA, appeared around the same time, reflecting the growth in computational analytical power. Rainfall inputs to these models were used in more sophisticated rainfall–runoff models applied to sub-catchments, which provided inputs to the pipe network. Rather than the constant intensity block rainfall used in the rational method, these early simulation tools were based on a pipe routing methodology, which involved calculating flows in surcharged groups of pipes separately from pipes with free flow. A limitation of this was that analysing looped networks, rather than dendritic systems, was not practically possible and the effects of downstream flow constraints could not be accounted for. Thus, models of looped networks, although attempted, caused analysts considerable difficulty. In addition, the available computational power in the early 1980s meant that only the smallest drainage networks could be handled in any simulation. Nevertheless, there were many effective models set up at this time (e.g. Dring, 1985).

11.3.1.3 Full solution hydrodynamic network models

As computing power increased in the late 1980s and early 1990s it was possible to develop matrix solvers to calculate the hydraulic state of sewer systems at many points in a network simultaneously. The method overcame the limitations of the previously used pipe flow routing methodology. Drainage networks were represented by a series of nodes (manholes and other assets) and links (pipes, weirs, orifices). Flows entered the networks at nodes, the flow being generated in sub-catchments, which included calculation of 'dry weather' wastewater flows, rainfall runoff, and infiltration. Full solution hydrodynamic models solved the Saint-Venant equations, which are still the foundations of the shallow wave theory (Li *et al.*, 2021) in a single dimension (a rigorous mathematical representation of flow velocity, pressure and energy losses) at all nodes and along the links simultaneously. However, certain approximations had to be made to ensure numerical stability, particularly when flows approach supercritical conditions. Other approximations included having to provide a free surface by putting a virtual narrow, vertical slot (known as a Preissman slot, e.g. Sanders and Bradford, 2011) in the tops of pipes to approximate to an unpressurised, unsurcharged state, and the practical need always to have a small 'virtual' base flow in all pipes to ensure numerical stability.

11.3.1.4 Model size and simplification

Current computational modelling power is now such that there is effectively no practical limit to the size of networks that can be modelled using full solution hydrodynamic network models to predict runoff flows and to assess the flows and depths in pipes throughout either individual rain events or long rainfall time series. For example, in the 2010s, the entirety of the drainage network in the Birmingham area (UK) had been simulated – a model with over 100 000 nodes and links (Hall *et al.*, 2016). The model comprised some 135 000 nodes with a total sewer length of 6670 km, including the surface water system for the catchment covering a total area of 44 750 hectares serving a population of *c.* 1.5 million people.

The availability of a number of software packages with high-resolution geographical information system (GIS) based user interfaces for data input and interpretation, such as InfoWorks ICM (Innovyze, 2023), MIKE URBAN (now MIKE+; DHI, 2023) and SOBEK (Deltares, 2023), has resulted in an almost complete move away from the use of simplified models to full solution hydrodynamic models for drainage design. This has the advantage of ensuring that the actual system is modelled as realistically as possible and the storage implicit within a network is represented in any hydraulic capacity assessment (e.g. Yang *et al.*, 2022).

11.3.2 Modelling overland flow

One of the main limitations of urban drainage modelling until the late 2000s was that, although flooding locations could be predicted, once water was discharging onto the surface from manholes, models could not properly take account of the topography and hence the flood flow processes taking place over the surface. At that time, the approach for dealing with flooding from manholes in a 1D model was either to define the water as 'lost' – that is, it flowed away to an unspecified location – or for it to pond above the manhole. Clearly, neither of these was a good representation of reality, nor did this allow for any interpretation of the extent offlooding. Furthermore, the reasonable assumption of contributing sub-catchments all draining to a point became more problematic as more extreme events needed to be considered. Analysts did try to get around overland flooding limitations by representing streets as 1D open channels – essentially as high-level sewers – but not only were there computational stability issues with this approach, it was also clear that where the flows were not constrained within the road, this approach was inadequate (see for example various past papers from the CIWEM Urban Drainage Group; CIWEM, 2023).

The continuing advancements in computational power allowed solution of 2D shallow water equations and thus for the modelling of flows over the catchment surface to be feasible. This, combined with the increasing availability of light detection and ranging (LiDAR) digital elevation models, enabled the next major breakthrough in drainage modelling. In the 2000s, numerous modelling tools became available, as reviewed by the UK Environment Agency (Néelz and Pender, 2009). LiDAR data are collected from an aircraft and provide cost-effective and sufficiently accurate ground-level information for urban drainage modelling purposes. While these models were initially used for river floodplain modelling, 1D–2D models were soon developed, allowing interaction between urban drainage networks and the urban catchment surface, as illustrated in Figure 11.2.

Figure 11.2 Illustration of surface flows from a one-dimensional (1D) to two-dimensional (2D) model in the city of Bradford, UK (William Shepherd)

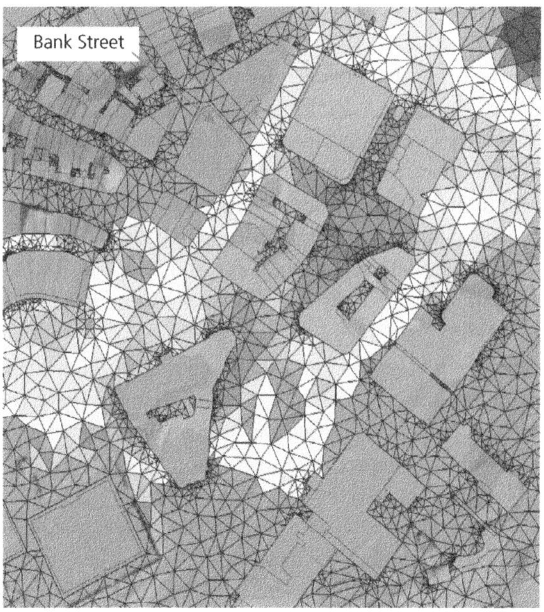

The development of integrated 1D–2D hydraulic models was the first major change in drainage modelling since simulation of individual pipes in drainage networks became possible in the 1980s. The previous practice of defining sub-catchments to model surface flooding is now becoming obsolete as models can provide a direct representation of overland flow with direct interconnectivity with the underground drainage system. However, this requires runoff models to evolve significantly from those currently used (Chapter 7), along with more detailed information on catchment surfaces and soils, and better hydraulic models to represent the interconnectivity between the 1D and the 2D networks. Current developments are providing the hydraulic relationships that represent more complex linkages between the above-ground surface and the underground system (representing gully behaviour), as these are affected by road gradient, camber, type of gully, computational mesh size and other factors. Also, the relationship of flows entering the drainage system is very different to the flow leaving the sewer system. Modelling of above-ground processes, such as the effect of surface obstructions, including road kerbs, and surface hydraulic roughness, is also an important feature of such modelling (e.g. Yang *et al.*, 2022).

11.3.3 Water quality modelling for drainage networks

The previous sections considered flow hydraulics specifically; however, as urban drainage systems convey sanitary and other waste products, modelling the conveyance of dissolved and particulate materials is also important. These pollutants derive from domestic, industrial and commercial wastewater and wash-off from the catchment surface (Chapters 2 and 3). The environmental conditions of receiving waters are an ever-increasing concern, which is reflected by the requirements of the design of CSOs as far back as the early 1960s (Chapter 3). Much of the pollution was known to originate from wastewater treatment works, CSOs and surface water discharges. Therefore, considerable efforts have been made to develop predictive relationships to add water quality simulation into the full solution hydrodynamic network drainage models described above. Stricter environmental legislation, such as the EU Water Framework Directive (Chapters 2 and 3), required a better understanding of pollution sources, especially untreated discharges from urban drainage networks. This has enabled models to be used to target changes to assets to improve the water quality of receiving surface waters to comply with such regulation (e.g. Rauch *et al.*, 2002).

Pollution from combined and stormwater systems has mainly occurred from sanitary sewer overflow (SSO) and CSO spills into rivers. These discharges have three principal characteristics that cause impacts: high levels of suspended solids; high levels of chemical oxygen demand (COD)/biochemical oxygen demand (BOD), which often results in depressed levels of dissolved oxygen for a significant stretch of a watercourse; and high levels of ammonia and phosphorus, such that fish and invertebrates could be seriously affected, especially by nutrient impacts on oxygen concentrations. However, they also contain a wide range of other persistent, bioaccumulative and toxic substances, such as polyaromatic hydrocarbons (PAHs), which are conveyed into water bodies. More information is given in Chapter 3.

There was a concerted effort to develop the knowledge required to understand and model the water quality behaviour of sewerage systems in industrialised countries in the 1990s (Ashley *et al.*, 2004; Gent *et al.*, 1996). Adding the deposition, erosion, transport and impacts of conveyed solids and pollutants into the available models required detailed investigations and simplifications due to the complex physical and biochemical processes involved (e.g. Jack *et al.*, 1996; Chapter 3). The investment in knowledge and understanding of these aspects of the models has never matched the requirements or the precision of the hydraulic models, although considerable progress has been made in biochemical process modelling (e.g. Ashley *et al.*, 1999; Hvitved-Jacobsen *et al.*, 2013). Thus, even now, these models are only approximations when used to consider the behaviour of pollutants in sewerage networks.

Despite this, the models are often used in inappropriate ways. Spurious precision in, for example, setting weir overflow levels to millimetre tolerances based on sewer flow quality modelling is not only unscientific, but also unnecessarily costly for the investments needed (Ashley *et al.*, 2003; Sriwastava *et al.*, 2018, 2021). There is thus a danger of the use of models without a proper understanding of their limitations, especially where these are masked by clever model interfaces that can help the untrained user to overlook the reality of the approximations in the science and numerical modelling processes (Schellart *et al.*, 2010). This is further considered in Section 11.6.

Dynamic models for the wastewater treatment processes were already in use in the late 1980s. These included key physical and biochemical processes. From the earliest sewage treatment optimisation model (STOM), which was developed as a stand-alone mainframe package, to STOAT (WRc, 2023), the model now used in most UK applications. Worldwide there are numerous other models available for treatment plant processes that can be linked to sewer influent models (e.g. Nemcik *et al.*, 2022).

The MOUSE hydrodynamic network model from DHI first developed the software to simulate a wide range of water quality parameters (Crabtree *et al.*, 1994), while other software packages of the time modelled only BOD, ammonia and suspended solids. MOSQITO was the equivalent model being used in the UK as a component of WALRUS, the then hydraulic model (e.g. Vélez Quintero, 2012).

Accurately representing the water quality characteristics of wastewater in a large sewer network is ambitious. Organic material can be present both in dissolved form and in the form of solids, and oxygen levels are significantly depressed by the biological processes taking place. Solids can be present both in the form of light organic material and as sediments that deposit and erode, depending on the in-pipe shear stress levels. Solids concentrations are therefore dependent on the local hydraulic conditions and the characteristics of the available solids, which can be highly variable in sewers (Ashley *et al.*, 2004). There are also other potential pollutants created by the biological reactor processes, such as ammonia and hydrogen sulphide, which are produced if the in-sewer flow and chemical conditions are conducive (Hvitved-Jacobsen *et al.*, 2013).

Although various models have been developed to represent these processes, commercial software products have generally limited their tools to representing 5-day biochemical oxygen demand (BOD_5) ammonia and sediment behaviour, treating these as conservative processes and not subject to any time dependency. These tools are deficient in that they do not represent the complex, interlinked physical, chemical and biological processes very well, but they can be useful in providing an estimation of the water quality processes that are taking place, particularly the high concentrations of any first flush associated with sediment erosion at the beginning of some rainfall events, as shown in Figure 11.3.

Although such foul flushes are observed, these do not occur in every rainfall event. There is evidence that flushes occur later in some events and not at all in others (Al Mamun, 2020).

Models for the biochemical processes in separate sanitary and, in some instances, combined sewers are being used routinely for the analysis of septicity, corrosion and odours, as outlined in Chapter 2. For example, WATS is a model developed originally by the University of Aalborg in Denmark (The WATS Guys, 2023) that is being used worldwide to examine problems of sulphide generation and the consequences thereof (Vollertsen *et al.*, 2015). WATS has been used, together with an earlier version of InfoWorks ICM (HydroWorks) to investigate the problems of balancing in-sewer storage volumes with

Figure 11.3 Example of a foul flush of pollutants in a combined sewer (Richard Ashley)

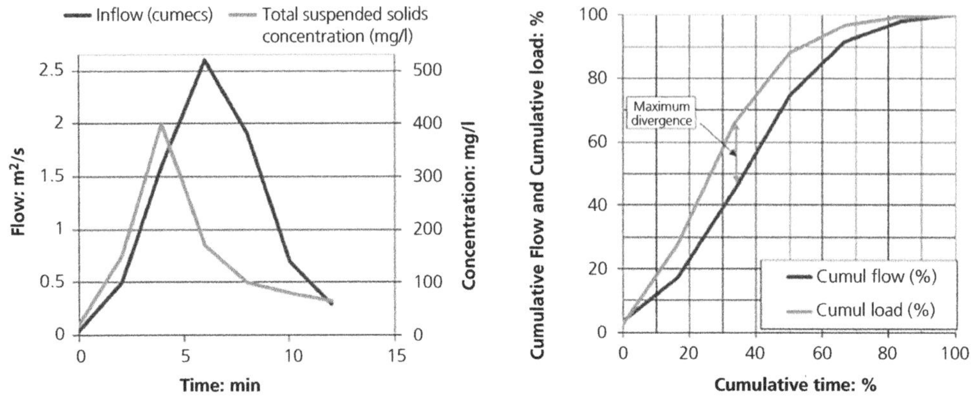

total emissions from wastewater systems. STOAT was used for the treatment plant performance assessment. Protocols were developed to avoid the various problems of CSO spills, storage and long drain-down times and dilute dry weather flows to treatment (Ashley *et al.*, 2001).

Despite the acknowledged lack of accuracy in being able to represent all the processes taking place within sewer systems, verification of models against data collected can provide some confidence in using sewer flow quality models, but limitations need to be recognised. It is unlikely that verification data will cover all the types of rainfall event the model will be used for. Nor is it likely that the duration of data collection will cover a long enough time series. Some guidance on quality modelling is provided by CIWEM (2009), but this is limited in scope and not as detailed or helpful as the equivalent for hydraulic modelling (CIWEM, 2017). The *Urban Pollution Management Manual* (FWR, 2018) provides more information on quality modelling to assess how best to upgrade assets to meet regulatory requirements (Chapter 3).

11.3.4 Modelling of nature-based systems (NbSs)

The use of non-piped urban drainage systems has grown significantly in industrialised countries in the past few decades. Many of these use forms of natural systems, comprising open bodies of water and vegetated areas, although there are systems that include below-ground measures, such as soakaways. The UK government has initiated legislation for England and Wales (Flood and Water Management Act 2010) that specifically includes sustainable drainage systems (SuDS), and this is reflected in national planning procedures. Elsewhere, various terms are in use for NbSs (Fletcher *et al.*, 2014) and a fuller description of these techniques is given in Chapter 8, together with basic design details. See also an accompanying manual that deals with blue-green infrastructure (BGI) (Washbourne and Wansbury, 2023). Here, the place and nature of the analysis and models being used for NbSs related to urban drainage are outlined.

Unlike piped drainage systems, for which flow and water quality are important, NbSs have many other aspects of performance to consider. Hydraulic performance is an important aspect of NbS performance, but in some places water quality enhancement is more important. For example, in Australia, diffuse urban water pollution was identified in the 1990s as a major urban problem, more so than flood risks (Chapter 2). Secondary to pollution is water stress; Australia has long been water stressed and the direct use of rainfall or runoff water is therefore attractive. Together, pollution and water stress led to the development of water-sensitive urban design (WSUD) concepts, which include rainwater harvesting. There are accompanying computational models for component design and analysis (e.g. MUSIC; eWater, 2023). In the USA, the Clean Water Act 1970 encouraged the use of source controls and the slowing down of urban runoff to improve discharge quality (WEF/ASCE, 1998) and to manage flooding. Accordingly, new techniques were developed for the analysis and modelling of the component processes, termed (at the time) Best Management Practices (BMPs), which have since been included in standardised computer modelling packages (e.g. SWMM; USEPA, 2023a). Despite these endeavours, the analysis of NbSs for hydraulic performance and water quality improvements is still a developing field.

NbSs provide multiple benefits beyond flooding and water pollution management (Chapter 8). The primary objective for using NbSs varies around the world. Even in unified states, such as the UK, the primary aim in England is utilisation for flood risk management (e.g. Thorne, 2020), whereas in Scotland, the main aim is to manage water pollution. There are other objectives elsewhere in the world, including promoting nature and the quality of the urban environment (e.g. Balmforth *et al.*, 2011). More recently, the realisation that NbSs are a key component in mitigating and adapting to climate change has led to a plethora of initiatives to promote use around the world (e.g. House of Lords, 2022). Several associated models for this have also emerged that include NbSs in planning for urban adaptation (e.g. McEvoy *et al.*, 2020).

Here, the discussion focuses mainly on SuDS – still the primary term for UK practice; the analysis and modelling of the processes primarily related to surface-based systems. Chapter 8 provides more detail on NbSs and SuDS.

The primary aim of SuDS use has been to manage the origin and conveyance of runoff flows, including retention and attenuation, and to provide a significant level of treatment before flows discharge to the natural environment. This section introduces analysis and modelling for the hydraulic and hydrological performance of SuDS, for water quality performance and for other aspects of SuDS. The concepts are outlined here; some design information is provided in Chapter 8 and various publications (e.g. Haested Methods, 2007; Woods-Ballard *et al.*, 2015a).

Whereas pipes are normally well defined by diameter and gradient, SuDS are much more variable in terms of their shapes and sizes, as many utilise the natural topography for blue or green features of the landscape and are often shaped to fit the contours of the landscape. The characteristics of the way in which surface water runs off, into or from SuDS are, therefore, often complex, and more difficult to analyse than where conventional urban drainage systems are used. Nevertheless, the rational formula or the modified form outlined above is frequently used to determine the runoff from contributing surfaces and simple flow balances used for the inflows into and outflows from SuDS (Woods-Ballard *et al.*, 2015b). Each individual SuDS unit typically is of small enough scale for this to be a reliable analysis.

The smaller rainfall events may result in no outflows due to the ability of some SuDS to absorb a significant amount of wetting before runoff occurs. For other SuDS, infiltration or deeper percolation allows the inflows to be absorbed into the ground, a specific design aim for infiltrating SuDS, such as soakaways. One main difference between SuDS analysis and piped drainage modelling is that even the smallest rainfall events are significant for SuDS. These are the primary wetting events that provide plant

Figure 11.4 Example of wash-off and solids loads from road surface into and through a road gully in Dundee (Richard Ashley)

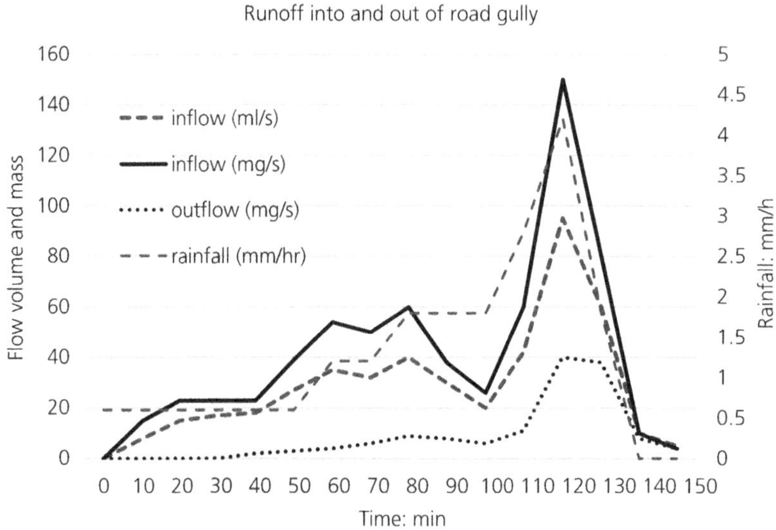

irrigation (Figure 3.2) and are also often the conveyors of the first flush of pollutants washed off from hard surfaces such as roads, as illustrated in Figure 11.4 (see also Deletic *et al.*, 2000).

Analyses of individual SuDS measures such as green roofs utilise mainly empirical, bespoke and specific relationships to estimate their hydraulic behaviour. See CIRIA's *The SuDS Manual* (Woods-Ballard *et al.*, 2015b) and Chapter 8. As regards hydrological and hydraulic behaviour, the flows in NbSs (or SuDS) are invariably free surface (although there are below-ground SuDS, Chapter 8). Modelling of free surface flows in channels and open water bodies has been a well-developed field of knowledge since the eighteenth century, with the development of empirical relationships that are still in use now (e.g. Wang and Yang, 2014). However, the majority of NbSs are small in scale and therefore modelling of flows is usually simple and localised, with the inflow and outflow rates based on water balances using, for example, reservoir routing procedures. Internal flow patterns and the effects of vegetation on flows can, however, require detailed analysis using, for example, computational fluid dynamics (CFD) where pollutant and solids movements and retention are important (e.g. Stovin *et al.*, 2022).

In general, there is a good understanding of the hydraulic and hydrological performance of SuDS measures, although where the measure is formed from natural land surface features, precise estimation of the flow behaviour is hampered by the need to account for the complex geometry of the topography unless the unit is constructed to be regular in shape. Figure 11.5 illustrates a typical SuDS unit, utilising natural landscape space. It is usually the outlet flow control that limits the flow and, for example, where a swale is designed with a flow control at the outlet the limiting factor is the performance of the flow control, which is well understood.

Considerable work has been done over the past decades on understanding the hydraulic and hydrological performance of SuDS, and development continues (e.g. Stovin *et al.*, 2022). However, in common with

Figure 11.5 A typical sustainable drainage systems (SuDS) unit utilising natural landscape (Richard Ashley)

our understanding of natural runoff processes (Chapter 7), the infiltrating behaviour of SuDS is less well understood. As for piped drainage systems, the modelling of the performance of the system for individual or discrete storm events will not provide an adequate performance assessment of the entirety of SuDS performance, which requires time series analysis. Replicating the hydrological processes that convey runoff from natural surfaces, many SuDS perform as drainage features by a complex flow balance between inflows and outflows, with more than a single inflow and outflow by way of both direct discharges and infiltration into the ground. Intermittent and recurrent rainfalls, rising and falling groundwater levels and rising and falling receiving water levels all need to be considered when modelling SuDS. As most SuDS temporarily store or slow down the inflows, complex three-dimensional (3D) internal flow processes are occurring during a rainfall event and even between events. SuDS storages may be full, partially filled or empty at the start of a rainfall event and groundwater may be high or low. Therefore, modelling needs to use rainfall events in series, mimicking nature. The time series rainfall described in Chapter 6 should therefore be used to assess the overall hydraulic performance of SuDS, and there should be consideration of individual extreme events. Details are provided in Woods-Ballard *et al.* (2015b).

There has been a growth in the availability of 3D models for routinely analysing SuDS storage systems, now linked with surface-based GIS information, visually informative and easy-to-use software packages can now help model the complex inputs to, topology of and multidimensional outflows from SuDS units. Table 11.1 lists some of the currently available software packages that are being used to design and model SuDS and NbSs. Most of these require access to external software for the geographical system requirements. Now these models are usable for both the water quantity and quality aspects of SuDS, although STORM-SEWSYS was the first such model to focus on the origin and fate of pollutants passing into and through drainage systems, including SuDS. Of those shown, only SUSTAIN includes optimised economic assessment as part of a sustainability assessment process (Chapter 12).

Table 11.1 Examples of current design and modelling software in use for SuDS and NbSs

Software	InfoDrainage (Autodesk, 2023)	MUSIC (eWater, 2023)	STORM-SEWSYS (Ahlman, 2006)	SWMM (USEPA, 2023a)	SUSTAIN (USEPA, 2023b) Uses SWMM
Country of origin	UK but now international	Australia	Sweden	USA, now used worldwide	USA
Availability	Commercial cost	Commercial cost	Research primarily	Free in basic format	Free in basic format
Capabilities	Hydrological and water quality modelling, detailed SuDS modelling (3D)	Hydrological and water quality modelling, detailed SuDS modelling (3D), rainwater harvesting and cost simulation	Water supply, stormwater, and wastewater quality simulation. Demand management	Water quantity and quality modelling for pipes and wide range of SuDS (3D)	Locates best sites, delineates drainage areas, runs existing pre and post development scenarios, optimises SuDS dimensions for costs and defined sustainability
Primary use	Drainage design and simulation for traditional and SuDS	Drainage design and simulation for WSUD	Simulation of pollutant build-up, wash-off and fate	Design and analysis of hydraulics and water quality for drainage systems	SuDS decision support tool

The various models in Table 11.1 are in use for retrofitting, new designs, problem identification and planning for using NbSs. Examples include an exemplar project led by Scottish Water developing joint SuDS/piped drainage modelling protocols, termed 'Blueprint for Sewers', and using the Innovyze suite of models (Fleming *et al.*, 2019). MUSIC is used extensively in Australasia for modelling all aspects of WSUD – for example, in the analysis of distributed stormwater harvesting systems (Di Matteo *et al.*, 2017). SWMM has been used for assessing the potential for SuDS for various housing layouts in the Cambridge to Oxford regional urbanisation project in England (Chapman and Hall, 2022), although this did not explicitly consider the water quality aspects. SUSTAIN has been used for deciding on the relative placement of NbSs in a university campus in Indonesia to improve water quality (Warganda and Sutjiningsih, 2017).

For many small SuDS schemes it is not necessary to use sophisticated modelling to predict their performance. Where these drain areas smaller than a hectare or so, the short times of concentration and limited attenuation of flows in the routing process between SuDS components may make computer model use costly and superfluous as analysis can be done manually.

One of the most important aspects of SuDS performance is that they improve the quality of the inflows, removing a range of pollutants, including particulates, heavy metals and other organic and inorganic substances, that can have impacts on downstream receiving waters. Chapter 8 provides information on SuDS and how they can improve water quality. Although many SuDS can be designed to achieve a certain pollutant removal effectiveness, their performance is variable as this depends on a wide range of factors, such as ambient vegetation and diligence of any maintenance. The modelling of this performance is therefore typically based on relationships derived from experiments on SuDS 'units', usually carried out at model scale, and field studies which have led to the development of semi-empirical relationships for the water quality performance (e.g. Charlesworth and Booth, 2017). Continual updating of information showing the latest findings for SuDS performance is provided by way of a US BMP database (Clary *et al.*, 2020).

Studies in the USA and Scandinavia in the latter half of the twentieth century and elsewhere since have shown that potentially the most polluting rainfall events are the more frequent, smaller events. Consequently, control of these smaller rainfall events is as important as the larger events in order to manage pollution risk rather than flooding (e.g. Durrans *et al.*, 2007). This is illustrated in Figure 11.6 and, in design and analysis, should be taken together with the four-points approach shown in Figure 3.2. In many countries, a water quality design storm return period is set at somewhere around 1 in 2 years, whereas the surface water flooding events used in design are set as less frequent events (e.g. Tourbier, 2002).

Figure 11.6 Illustration of the need for sustainable drainage systems (SuDS) to control smaller rainfall events (data illustrative of typical temperate country)

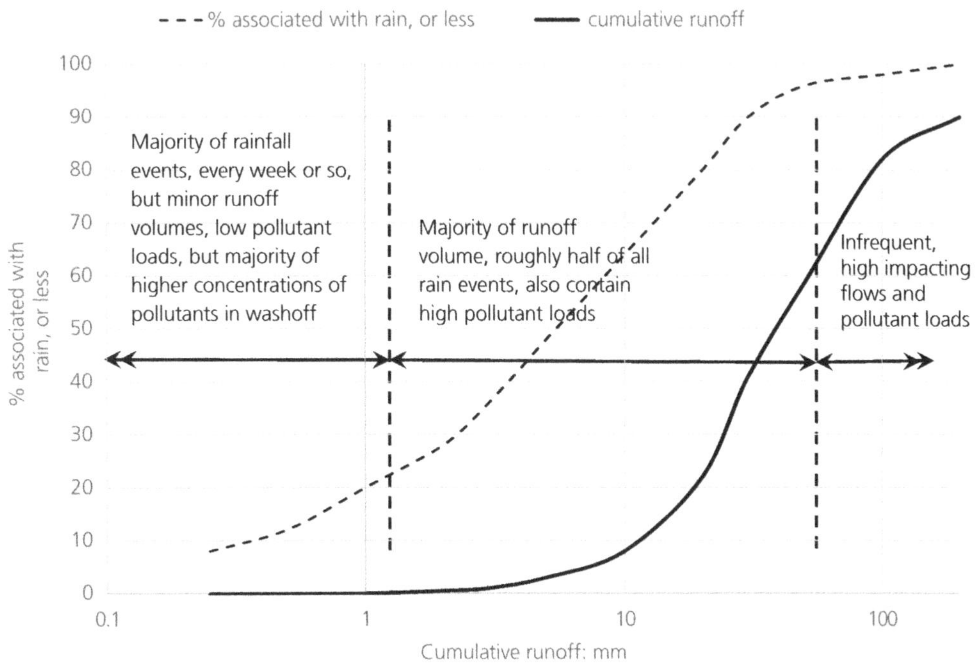

Despite the empirical nature of the understanding and prediction of SuDS and water quality effectiveness, there is no greater certainty in the predictability of the quality of flows in sewers than for SuDS. Sewer flow quality models are routinely used in integrated catchment studies (Section 11.5) with a presumption of precision (e.g. Schellart *et al.*, 2010) and SuDS can be used in the same analyses with as much confidence.

As many SuDS comprise green infrastructure the nature of the flora and vitality will depend on the time of year, temperatures, ambient rainfall and, especially, ground moisture. SuDS are therefore more susceptible than piped systems to changes due to climate and land use, with the performance of both SuDS and piped systems being affected by the changes in consequential inflows, but only SuDS being affected by any changes in ground conditions and the nature of the flora within the SuDS measure itself (e.g. Sohn *et al.*, 2019). Changes may occur to species types and diversity, affecting the runoff, infiltration, conveyance, retention capacity and treatment effectiveness of the SuDS measure. Lisenbee *et al.* (2021, 2022) review models, and use SWMM, with an agricultural drainage tool called DRAINMOD-urban, demonstrating applicability to urban bioretention systems. The model provides a detailed soil–water accounting procedure that estimates water level fluctuations and can represent the performance of bioretention systems (Chapter 8).

Climate and land use will continue to change; hence any SuDS design should consider how, over the typical lifetime of the measure, changes such as these could affect the performance. In the future, maintaining SuDS performance may be dependent on irrigation during periods of the lowest rainfall to maintain the vegetation. Even a relatively small change in mean temperature could mean that within its lifetime a surface SuDS could experience conditions more like those prevailing in southern Europe now (Washbourne and Wansbury, 2023). Despite this, the approach to the design and modelling of SuDS is often to consider only single rainfall events and to add to these a simple uplift factor for future rainfall changes, allowing for presumed planned growth in contributing areas.

The economic value provided from using main drainage systems has been defined for flood risk reduction, pollution control and facilitating development activity. As a result, there are numerous economic accounting procedures that may be used as part of economic modelling. In Section 3.2 in Chapter 3, the approach defined for the Wallingford Procedure is set out, based on utility concepts. This also utilises asset management principles, as explained in Chapter 10.

The growing understanding of the potential multifunctional performance of SuDS/NbSs has promoted the development of a range of tools to support the analysis of these benefits. It has been possible to estimate the economic value of the flood risk reduction benefits of all types of urban drainage systems, and most industrialised countries have standardised databases and methodologies for this. In the UK, for example, the 'multicoloured manual' has been the standard for economic appraisal of flood and coastal erosion risk reduction (Penning-Rowsell *et al.*, 2013). Although aimed mainly at main river and coastal flooding, it is also applicable to urban drainage. Many of the benefits assessed in economic terms are intangibles, in that these are assigned an indirect economic value, such as a benefit to human health. The Millennium Ecosystem Assessment (UK NEA, 2011) developed for the first time a worldwide set of ecosystem services (ESs); defining the economic value of ecosystems to humanity (see Chapters 5 and 8). From this, several ESs analytical models have been developed that can be used to assess the value of the many benefits arising from the use of NbSs.

Examples of valuation tools in use for urban drainage systems include: B£ST (CIRIA, 2023) in the UK; the Green Values Calculator (CNT, 2023) in the USA; TEEB (Nature Squared, 2023) in the Netherlands; the INFFEWS value tool (WSC, 2023) in Australia. Each tool can be used to assess the multifunctional benefits of either traditional buried piped infrastructure or SuDS/NbSs. The use of B£ST is described in Ashley *et al.* (2018a) and contrasted with TEEB for a Swedish case study (Hamann *et al.*, 2020). For background to the Green Values Calculator, see SNT and SB Friedman (2020), and for an application of INFFEWS, see Iftekar and Pannel (2022).

There are many other tools based on ESs that include other components of social valuation specifically to evaluate the economic benefits of green infrastructure and nature. As this is a rapidly advancing domain, it is important to ensure that where these models are required, access is obtained to the most up-to-date versions of the tools. Many now are commercial and require subscription for use, often being linked to databases that may require additional licence access.

11.3.5 Other models being used for urban drainage

Physically based models, as described in the earlier sections, are most used for urban drainage analysis as these provide information on the performance of a system throughout its extent and help to identify the effectiveness of optional interventions to improve performance. Models are also imperative for the design of new drainage systems, although these need to be proportionate in that detailed modelling can be costly and often simple models may be suitable. There are disadvantages of using physically based models, especially the need for very detailed data for the existing drainage network and, where the model is large, potentially long simulation run times. It is also necessary to collect extensive verification data (Section 11.4) which can be costly.

Non-physically based models can be conceptual (e.g. lumped) models, or data-driven models. Each type of model has value in specific applications. Conceptual models can be useful for network scale planning using long time series, or, where system knowledge is poor, they can aid understanding of where the system may require upgrading and give an indication of the efficacy of potential changes. These models take the physical layout of the system into account, but simplify it significantly by, for example, grouping together many pipe lengths, thus reducing the need for detailed network knowledge and reducing run times (e.g. Kroll *et al.*, 2017).

Data-driven models range from simple empirical models through to complex multilayered artificial neural networks. Data-driven models are especially useful for real-time control and management of networks, a key benefit here being short run times. Data-driven models require minimal knowledge of the network but need time series data to derive or train the model (e.g. Mignot and Dewals, 2022).

Generic models have been utilised for urban drainage analyses. Many of these have been used for water resources purposes or for flood and coastal engineering analysis, and can be important in specific circumstances. Examples of these include: CFD for determining flow patterns in water bodies, sediment and pollutant movements (e.g. Sonnenwald *et al.*, 2019); real options for optimising benefits and costs in relation to future uncertainty (e.g. Gersonius *et al.*, 2012); life cycle analysis (LCA) to assess the environmental impacts of constructing and using drainage assets (e.g. Sagrelius *et al.*, 2020); agent-based models to assist with decision-making (e.g. Moradikian *et al.*, 2022); and water resources simulation that can include multiobjective functionality (Tomlinson *et al.*, 2020). Standard techniques, including multicriteria decision analysis (MCDA) support tools, are also in routine usage in the urban drainage field (e.g. Thurston, 2012).

In summary, there is a core group of physically based models available to help assess the functional performance of urban drainage systems, including the capture, retention, runoff, fate and potential impacts of sanitary flows and stormwater. Associated models can be used to determine the costs and wider impacts and benefits of urban drainage systems, both traditional and nature based.

11.4. Model build, calibration, validation and use

The development of urban drainage models requires three major stages (Figure 11.1)

- model build
- model calibration
- model verification.

In the first stage, information and data are collected so that the layout and connectivity of the system can be determined. For a 1D hydrodynamic model of a sewer system this would involve collection of information on manhole locations, contributing areas (i.e. areas where wastewater is created and the surfaces from which runoff occurs), pipe invert and soffit levels, pipe diameters and materials, structural condition and any deposits. Where flood risk assessment was the aim of the model then accurate topography is required, although in industrialised countries, national databases are usually available for geographical information, including populations, paved area extent, green space and a variety of other information.

In the second stage, once the geometry of the system to be modelled has been incorporated into the modelling tool, the hydraulic and, if needed, water quality parameters will usually need to be calibrated. In the third stage, model validation or verification, an independent set of data, not used in the calibration stage, is used to assess the performance of the model, in which the predicted outputs are compared against the measured output data. A measure of how well the outputs fit the data, such as a root mean square error (RMSE) or a correlation coefficient, is often used to quantify the performance of the model. Other approaches may be used dependent on the model type and objectives for usage (e.g. Kroll *et al.*, 2017; Radinja *et al.*, 2021; Szelag *et al.*, 2022).

In the UK, the CIWEM Urban Drainage Group has produced a code of practice to encourage consistent practice when modelling urban drainage systems. This explains in detail the roles and tasks involved in carrying out a modelling study. Although not regulation, many water utilities expect that models developed should be compliant with this code of practice (CIWEM, 2017). This section further focuses on the technical aspects of model development in terms of model build, data collection calibration and verification.

The CIWEM (2017) code of practice recommends that any drainage modelling study should follow a particular workstream, which can be divided into the following stages

- definition and scoping
- model build and verification
- model use, documentation and maintenance.

In the first stage, it is necessary to identify all stakeholders and understand any links with other projects and activities within the catchment, then to obtain and review relevant data, and finally to develop project objectives for the scheme under investigation. Once the project objectives have been agreed and set, it is necessary to identify the type of model required to meet those objectives and then to create a plan to build the model.

The model can be either a new one, built specifically for the study, or based on a pre-existing model. When using a pre-existing model, careful checks should be undertaken to ascertain its suitability and where it might need improvement and updating. The first stage in the model build is to collect the plethora of data that will describe the system operation. This will include data describing the physical drainage assets, such as the pipes, manholes and ancillary structures, which is usually available from GIS-based asset databases. Data to allow calculation of flows in dry and wet weather are vital, including hydrological data describing the catchment surfaces and soil properties, for runoff calculations, data on the population and industrial flow inputs, as well as groundwater data for the dry weather flows. Operational data are needed for pump systems to understand how they operate. Finally, a flow survey (WRc, 1987) is a vital part of the data gathering to understand how the system operates.

The collected data are used to build the model. The modelled area is separated into sub-catchments where the dry and wet weather flows are generated. Dry weather flows should be relatively straightforward to model, through understanding the contributing population and the types of commercial inputs, although there can be complications such as seasonal population changes and variability in large commercial discharge inputs. Infiltration commonly varies seasonally, and in coastal areas, sewers that are below sea level can be affected by the tide. Information from wastewater treatment plant inflows is usually obtained along with intermediate monitored flow data. Wet weather runoff from urban catchments may be included by way of one of a wide range of different runoff models that are available (Chapter 7). Runoff is affected by many parameters, including the slope, roughness and infiltration rate of the ground surface, and the connections (gullies) from the surface to the sewer are also important. Most sewer system modelling has some degree of simplification; 1D sub-catchment runoff models are still most commonly used, but modelling tools now have the capability for variably distributed rainfall to be applied directly to 2D surfaces and routed to the sewer inlets.

After the model has been built, it needs to be checked and verified against measured data collected for a number of rainfall events. These data are normally collected in a short-term flow survey in which flow and depth monitors are placed at key locations in the network (Chapter 9) and data are collected for a limited number of storm events. Checks must be carried out on this information to ensure that it is suitable for use in verifying the model. The verification procedure will usually start with dry weather flow data before proceeding to wet weather data. The modelled results should be compared with the measured results to assess the general shape of the time series data, the timing of peaks and the general shape of both depth and flow hydrographs (Figure 7.3). Any differences should be noted and investigated; it is vital that any changes made to the model are justified and that parameters are not simply changed to fit the data. This is referred to as force-fitting, which might not actually reflect the cause of the poor fit and is unlikely to result in a model that fulfils the intended purpose. When any changes are made, these should be confirmed by an audit trail where possible. Historical information on flooding can be used for verification against larger events than might have been obtained in the flow survey.

Once the model is verified, if the model meets the accuracy requirements stated in the project objectives, then it is ready for use in modelling and design studies. However, even when using a verified model it is possible there will be errors in predictions, due to limitations in the dataset used to calibrate the model; for example, the limited severity of the rainfall used, measurement uncertainty in the collected data and any simplifications of the modelled network in the model. The constructed model should not be used for rainfall events that are larger than or significantly different to the events used to calibrate and verify the model unless such extrapolation can be justified.

Real-time control systems which are installed to enhance performance of the network can be incorporated into the model. Simple rule-based controls or proportional integral derivative controllers are written into most software; however, other types of control algorithm, such as fuzzy logic-based controllers, might not be directly available in the software, thus might need to be approximated in the available tools or incorporated through external application-programme interfaces (APIs).

11.5. Integrated modelling

Chapter 12 considers how there has been a growing awareness of the need to take a more holistic approach to water management in all forms, and even to take an integrated approach to all the services and infrastructure required by society. Here, only the aspects related to analysis and modelling for urban drainage systems are considered. There are numerous conceptual models for analysing integrated systems, but we consider only those aspects directly relevant to urban drainage.

There is an increasing use of integrated urban drainage models to predict, manage and optimise the performance for flooding, water quality, water resource management and treatment activities at different spatial and temporal scales across urban and rural catchments. Such an integrated modelling approach is seen as an important technique for managing the impact of agriculture, urban drainage and wastewater systems on the environment. The use of integrated models can also identify more cost- and energy-effective measures when dealing with the management of flooding and the quality of surface waters. For example, Gilboa *et al.* (2023) included the separate sanitary and surface water drainage in an integrated model study for a 'model' city in Israel. The objective of the study was to determine the most effective water system (reuse or otherwise) in terms of environmental and economic factors. Combined simulation models were used for life cycle analysis and life cycle costing, testing six variants of types of water supply and reuse. The best option was to use treated greywater or treated wastewater for non-potable use. The most impacting option was found to be the traditional supply to wastewater disposal routes.

Integrated modelling studies use a set of interdependent models that are linked together to simulate the complex linkages between urban drainage systems and the natural aquatic systems they can impact on. Integrated urban water modelling means the joint modelling of two or more of the components of the urban water system, as illustrated in Figure 11.7, from a study of the non-point pollution sources for the Jeonju catchment in southwestern Korea. Integrated modelling can be represented by a linked group of sub-models for each of the various elements of the overall system. In general, integrated urban water modelling studies can be characterised by three main features as below.

- The modelling of several processes within sub-systems (e.g. sewer network and receiving river) and the interactions of the processes across sub-system boundaries.
- The use of long time series simulations so that acute, chronic and delayed water quality impacts on the environment can be estimated.
- The ability to predict both local and catchment-wide impacts better to inform decision-making.

Figure 11.7 Illustration of linked models for simulating urban drainage catchments (Jung *et al.*, in press)

Key:
HSPF is a model used for urban hydrological simulation
SWMM is the Storm Water Management Model used for urban drainage modelling

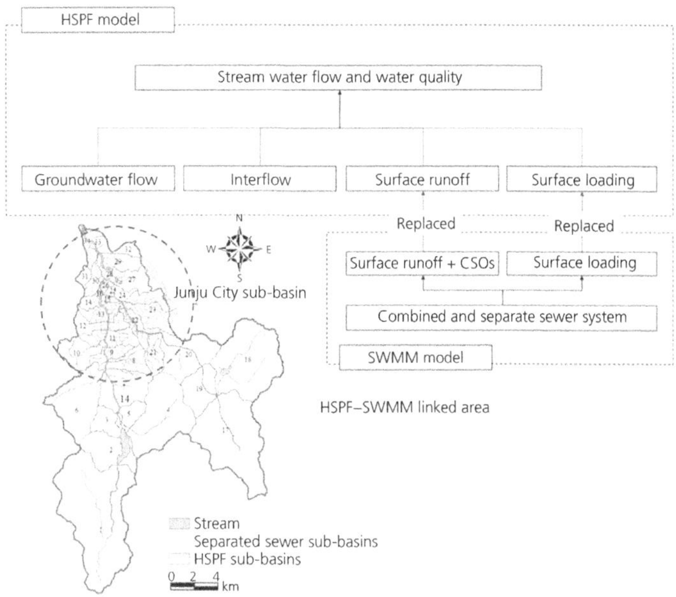

Integrated catchment modelling approaches were first developed in the late 1990s in an attempt to gain a better understanding of water quality impacts at a catchment scale. These efforts mainly comprised the use of existing models and exchanging data between them; that is, run one model, then feed the outputs into another model. Alternatively, individual component models were simplified to produce conceptual or lumped models that could be combined to model the catchment system as a whole.

The early integrated models were aimed at managing receiving water pollution (e.g. Muschalla *et al.*, 2009; Rauch *et al.*, 1998). In industrialised counties, specifications were set out on the use of models for urban pollution management, such as in the early editions of the *Urban Pollution Management Manual* (UPM Manual) (latest edition: FWR, 2018) for the UK. In Section 11.3.3, physically based models for simulating the interactions between wastewater networks and treatment plants were introduced. The UPM Manual (FWR, 2018) provides detailed guidance on modelling catchments, treatment plants and receiving watercourses in a UK context.

There are now numerous software packages, commercial or open source, that purport to provide integrated catchment modelling. These may each contain models that take different approaches and use different process relationships, often calibrated to model the pollutant sources and the transport and transformation processes. More sophisticated modelling approaches require larger amounts of data and more detailed calibration. Simple stochastically based models used in the UK, such as SIMCAT for river flow quality and SAGIS for pollution source apportioning (Comber *et al.*, 2013), have limitations due to incomplete representation of the pollution generation and transformation processes and inability to deal with the complex water quality processes during time-varying events. Commercial software companies are continually revising their high spatial resolution modelling packages to deal with integrated catchment studies, but the model water quality predictions are rarely going to be completely reliable due to the complex processes taking place in the physical world. The need to deal explicitly with uncertainty in integrated water quality models has been obvious, but quantitative studies remain rare (Section 11.6).

Notwithstanding these limitations, as computing power has grown, more model integration is being used for planning, designing and managing wastewater systems. In a Chinese example, Laio *et al.* (2019) used integrated models to help decide on which upgrade to make to a combined drainage network to achieve the most effective and cost-beneficial solution over the entire life cycle of the system. A combination of established computer models was used for this, in conjunction with bespoke software for optimisation. Mannina (2019) includes numerous papers on all aspects of urban drainage modelling, with several dealing with various aspects of integrated modelling. These papers illustrate the growth of developments in this area and the varying types of integrated models now being used in urban drainage.

11.6. Model uncertainty

The processes that impact on surface water quality are complex and occur over a wide range of temporal and spatial scales. Models and the relationships used to describe them have often been developed using data collected by individual studies for situations at specific spatial and temporal scales. For example, the understanding of sewer solids and processes outlined in Chapter 3 has been based on limited studies and often individual and localised field work. The relationships developed from such investigations may be employed at spatial and temporal scales very different from those for which they were originally developed. For example, some early commercial simulation models were found to be utilising established hydraulic relationships outside the limits of their applicability (Ashley and Jefferies, 1987). Willems (2008) concluded that sewer flow quality modelling was no more effective using the extant computational tools than a 'random number generator' due to various erroneous assumptions, with poor and inadequate data being collected for calibration.

Models for entire catchment systems may utilise a combination of tools of varying complexity to simulate the various runoff, production, transport and transformation processes. More complex models can be capable of providing more detailed results, but they often require large and expensive datasets to calibrate and validate successfully. Hence, while in some cases uncertainty may be reduced by collecting more data and rigorous model calibration, this is only possible in some instances and depends on the inherent variability of the processes modelled. In the context of integrated water quality modelling, Deletic *et al.* (2012) identified three main sources of uncertainty (Figure 11.8)

- model input uncertainties
- model calibration parameter uncertainties
- model structure uncertainties.

Model input uncertainties concern the input data (e.g. rainfall intensity) and the model parameters (e.g. pipe geometry) that are required to build and run a model. Uncertainties in input data can in part be compensated during the calibration process by adaptation of the model, but model parameters should reflect reality. 'Force fitting' of model parameters to fit measured data should be avoided as it is likely to lead to misrepresentations and errors. Calibration parameter uncertainties are related to the data used for calibration of the empirically calibrated relationships that comprise integrated models (e.g. the pipe hydraulic roughness coefficient) and to the calibration methods used to find the optimum calibration parameter values. Model structure uncertainties depend on how well the numerical model represents the systems and processes. They include uncertainties caused by the mismatch of the temporal and spatial

Figure 11.8 Diagrammatic representation of model uncertainties (modified from Deletic *et al.*, 2012)

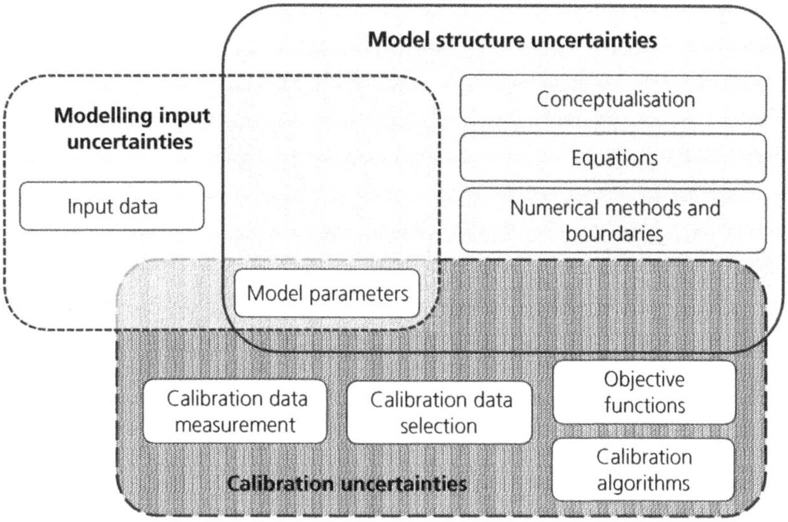

resolutions of the different sub-models, the numerical solution method used to solve equations and model conceptualisation errors, such as omitting or poorly describing key processes.

Tscheikner-Gratl *et al.* (2021) reviewed the uncertainties in integrated catchment modelling for water quality. Identifying the uncertainties due to spatial and temporal scale complexities, sub-model linkage and connectivity issues, and levels of detail, it is suggested that uncertainty assessment should be an ongoing and parallel process to any integrated modelling study.

As modelling is an expensive and time-consuming exercise, tools are being developed for analysts so that it is possible to determine what type of model to use and the requisite calibration measurements needed (i.e. in which locations and at what temporal and spatial scale), in order to balance the data collection costs and reduce uncertainty. These tools are usually based on the statistical description of the input data, the model input parameters and calibration parameters, and propagation of the uncertainties between the sub-models. The objective is to analyse how uncertainties in model input data, model input parameters and model calibration parameters propagate to model outputs. The commonest approach to uncertainty propagation analysis utilises Monte Carlo (MC) stochastic simulation (e.g. Korving *et al.*, 2002). In order to limit the computational cost of using MC methods with integrated catchment models which requires the processing of large amounts of data, the outputs of the uncertainty analysis can be linked with alternative approaches, for example, Schellart *et al.* (2010). This allows an estimate of the overall uncertainty in a water quality variable to be described for a fraction of the computational cost and so can make uncertainty analysis achievable for integrated modelling studies.

With the development of economic and social analysis and computational modelling to support the planning, design and operation of urban drainage systems, there is a need to ensure that such analyses are also subject to appropriate uncertainty assessment. As an example, the uncertainties associated with the

Figure 11.9 Uncertainties in valuing blue-green infrastructure (BGI)

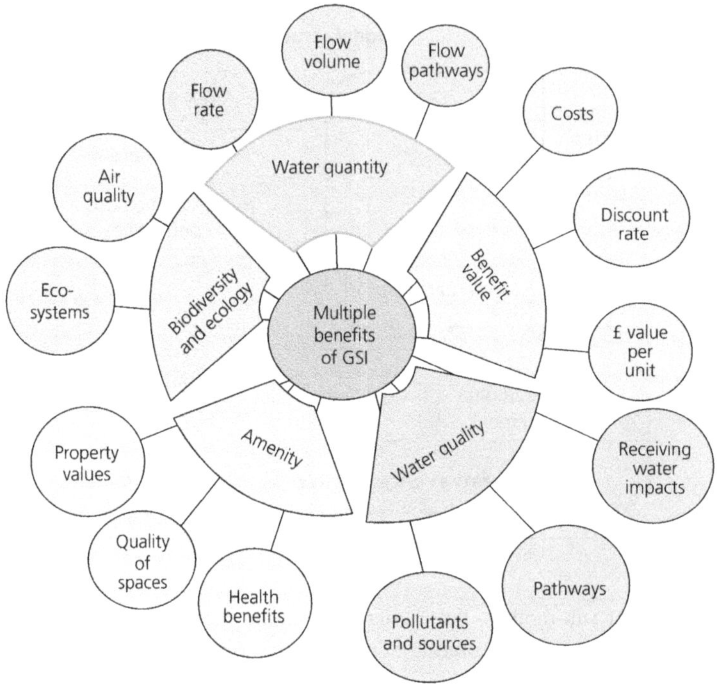

UK ecosystem services valuation tool B£ST are due to a lack of knowledge about the benefits, their extent, and how best to monetise them (Ashley *et al.*, 2018b). Figure 11.9 illustrates the multiplicity of benefits provided by BGI and, hence, where there will be uncertainties in valuation.

Box 11.1 illustrates the scale of the uncertainties in valuing BGI using B£ST. Although illustrating one sub-component of an integrated assessment of using NbSs, Box 11.1 demonstrates that as more aspects of the performance of urban drainage systems are part of an analysis, the greater need there is to be able to track and understand where and what the uncertainties might be. B£ST is often used to compare options for an NbS with the equivalents for a buried piped system. Each option requires a number of models to be used. The rainfall models may be similar, but the unit process models for runoff, infiltration, storage, conveyance, transformation processes and fate of any discharges will be different, as will the environmental and social impacts. Cost estimates may be more or less precise for the options being considered (e.g. NIC, 2022; Chapter 8).

Benefit identification will depend on the latest knowledge for the NbS, but there can be more confidence in the piped system's benefits, albeit these will be more modest. Monetisation of the benefits will also be imprecise for the BGI (Box 11.1). Social aspects such as acceptability and changes over time will be uncertain. As yet, there are no templates that can be followed to quantify well enough the uncertainties and relative reliability of comparing a piped drainage option with an NbS option, although examples of such analyses are provided in the literature (e.g. Montalto *et al.*, 2012). At the present time the focus of

Box 11.1

Ranges of uncertainties in the valuations of the multiple benefits from using BGI in B£ST

Figure 11.10 provides an illustration of the ranges of uncertainty in valuing each individual benefit in the B£ST analysis. Users of the tool set confidence scores (0–100%) on both the representation of the physical processes in the model (e.g. numbers of properties protected from flooding – Y%) and the financial unit value of those processes (e.g. £ per property protected – X%). Hence the combined confidence score is Y% × X% = Z%. The possible ranges of these combined scores are shown in Figure 11.10 based on the understanding of the contribution of BGI to the benefit category. Using the flooding risk reduction example, the figure shows that the range of possible uncertainties in the financial valuation using B£ST can be from 15% to 100%. Users set Y% and X% and this gives a confidence score overall of Z%. In practice, confidence in the physical parameters where detailed models have been used to estimate the properties benefiting from the use of BGI can be higher than the confidence in the economic value of this. The difficulty is often in predicting the numbers of beneficiaries in the future when the BGI has been constructed and is in use. More information about these uncertainties is provided in Ashley *et al.* (2018a).

Figure 11.10 Ranges of uncertainties in B£ST benefits from assigning these to physical and monetary estimates

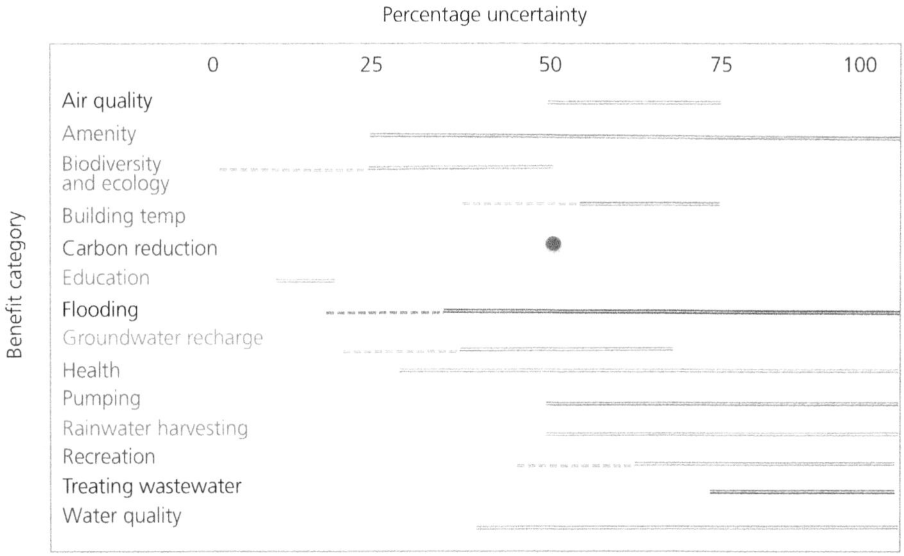

the analysis should be on the objectives of any scheme. For example, where flood risk reduction is the main aim, then the aspects around the analysis of the performance of the systems in this regard should be looked at in detail to understand the uncertainties in the analysis (e.g. Willems *et al.*, 2022).

As a final cautionary note on uncertainty, there is the ubiquitous user error problem. In addition to the potential for users to not understand the key elements of the way in which a model works, there is the danger of users taking a different perspective on the model data inputs. This is a relatively new source of uncertainty and especially applicable to the types of analysis illustrated in Box 11.1. Uncertainty about the future makes any estimates of, for example, how many people will utilise the new educational or recreational opportunities created by a new NbS unreliable, in turn making precise valuations impossible. As an example, in an analysis of the benefits of using BGI to alleviate CSO problems, compared with using traditional drainage systems (Ashley *et al.*, 2017), ten case studies were examined and benefits determined using B£ST. Each case study was undertaken by more than one expert. The results were that not one of the cases examined was estimated to have a similar benefit to cost ratio by each of the alternative experts, and in many instances the experts identified different categories of benefits as of value. This is not a criticism of the study, rather a warning of the difficulties ahead when moving from relatively straightforward physically based modelling into the realms of subjective multibeneficial modelling and analysis.

REFERENCES

Ahlman S (2006) *Modelling of substance flows in urban drainage systems*. PhD thesis, Chalmers University of Technology, Göteborg, Sweden.

Al Mamun A (2020) Review on uncertainty of the first-flush phenomenon in diffuse pollution control. *Applied Water Science* **10**: 53, 10.1007/s13201-019-1127-1.

Argyropoulos CD and Markatos NC (2015) Recent advances on the numerical modelling of turbulent flows. *Applied Mathematical Modelling* **39(2)**: 693–732, 10.1016/j.apm.2014.07.001.

Ashley RM and Jefferies CJ (1987) Sewerage simulation are standard methods appropriate? In *Proceedings of the 4th International Conference on Urban Storm Drainage*, Lausanne, August/September.

Ashley RM, Hvitved-Jacobsen T and Bertrand-Krajewski J-L (1999) *Quo vadis* sewer process modelling? *Water Science & Technology* **39(9)**: 9–22, 10.2166/wst.1999.0430.

Ashley RM, Dudley J, Vollertsen J *et al.* (2001) The effect of extended in-sewer storage on wastewater treatment plant performance. *Water Science & Technology* **45(3)**: 239–246.

Ashley RM, Dudley J, Saul AJ, Blanksby JR and Vollertsen J (2003) Trouble in store – what's going on in our wastewater systems? In *CIWEM Meeting of East Midlands Branch*, Birmingham, 11 March.

Ashley RM, Hvitved-Jacobsen T, Bertrand-Krajewski J-L and Vernbanck M (2004) *Solids in Sewers*. IWA Scientific and Technical Report No. 14. IWA Publishing, London, UK.

Ashley RM, Digman C, Futter M *et al.* (2017) *Valuing the Benefits of Storm Discharge Improvements for use in Cost-Benefit Analysis: Practitioners' Guide*. Report Ref. No. NL5946 Final Report. Water UK, London, UK. See https://www.water.org.uk/wp-content/uploads/2018/12/Storm-Overflow-Benefits-Valuation-Final-Report.pdf (accessed 06/08/2023).

Ashley RM, Digman CJ, Horton B, *et al.* (2018a) Evaluating the longer-term benefits of sustainable drainage. *Proceedings of the Institution of Civil Engineers – Water Management* **171(WM2)**: 57–66. https://doi.org/10.1680/jwama.16.00118. Paper 1600118.

Ashley RM, Gersonius B, Horton B, *et al.* (2018b) Including uncertainty in valuing blue and green infrastructure for stormwater management. *Ecosystems Services* **33(2018)**: 237–246, 10.1016/j.ecoser.2018.08.011.

Autodesk (2023) 10 things you can do in InfoDrainage that you can't do in MicroDrainage. https://blogs.autodesk.com/innovyze/2022/11/30/10-things-you-can-do-in-infodrainage-that-you-cant-do-in-microdrainage/ (accessed 06/08/2023).

Balmforth DJ (2011) *Comparing the Arrangements for the Management of Surface Water in England and Wales to Arrangements in Other Countries*. MHW UK. See https://webarchive.nationalarchives.gov.uk/ukgwa/20150624091829/https://www.ofwat.gov.uk/future/sustainable/drainage/rpt_com_201102mwhswd.pdf/ (accessed 06/08/2023).

Bilham EG (1936) Classification of heavy falls of rain in short periods. *British Rainfall* **1935**: 262–280.

Box GEP (1976) Science and statistics. *Journal of the American Statistical Association* **71(356)**: 791–799, 10.1080/01621459.1976.10480949.

Campbell D (2012) Developments in mathematical simulation of fluid flow in building drainage systems. *Building Services Engineering Research and Technology* **33(3)**: 307–317, 10.1177/0143624411413534.

Chapman C and Hall JW (2022) The potential for sustainable drainage systems (SuDS) in a regional urbanization project. *Frontiers in Sustainable Cities* **4**: 922890, 10.3389/frsc.2022.922890.

Charlesworth SM and Booth CA (eds) (2017) *Sustainable Surface Water Management*. Wiley Blackwell, Chichester, UK.

CIRIA (Construction Industry Research and Information Association) (2023) *B£ST Benefits Estimation Tool*. CIRIA, London, UK. https://www.susdrain.org/resources/best.html.

CIWEM (Chartered Institution of Water and Environmental Management) (2009) *Wastewater Planning Users Group Guide to the Quality Modelling of Sewer Systems*. CIWEM, London, UK. See https://www.ciwem.org/assets/pdf/Special%20Interest%20Groups/Urban%20Drainage%20Group/Guide-to-the-Quality-Modelling-of-Sewer-Systems.pdf (accessed 06/08/2023).

CIWEM (2017) *Code of Practice for the Hydraulic Modelling of Urban Drainage Systems*. CIWEM, London, UK. See https://www.ciwem.org/assets/pdf/Special%20Interest%20Groups/Urban%20Drainage%20Group/Code%20of%20Practice%20for%20the%20Hydraulic%20Modelling%20of%20Ur.pdf (accessed 06/08/2023).

CIWEM (2023) Urban Drainage Group. Past Papers. http://ftp2.ciwem.org/ (accessed 06/08/2023).

Clary J, Jones J, Leisenring M and Hobson P (2020) *International Stormwater BMP Database: 2020 Summary Statistics*. The Water Research Foundation, Alexandria, VA, USA. See https://www.waterrf.org/system/files/resource/2020-11/DRPT-4968_0.pdf and see the main database https://bmpdatabase.org/ (accessed 06/08/2023).

CNT (Center for Neighbourhood Technology) (2023) Green Values Calculator. https://cnt.org/tools.

CNT and SB Friedman (2020) *Green Stormwater Infrastructure. Impact on Property Values*. Center for Neighbourhood Technology, Chicago, IL, USA. See https://cnt.org/sites/default/files/publications/GSI-Impact-on-Property-Values.pdf (accessed 06/08/2023).

Comber SD, Smith R, Daldorph P, *et al.* (2013) Development of a chemical source apportionment decision support framework for catchment management. *Environmental Science & Technology* **47(17)**: 9824–9832, 10.1021/es401793e.

Crabtree R, Garsdal H, Gent R, Mark O and Dórge J (1994) MOUSETRAP – a deterministic sewer flow quality model. *Water Science and Technology* **30**: 107–115, 10.2166/wst.1994.0011.

Deletic A, Ashley RM and Rest D (2000) Modelling input of fine granular sediment into storm drainage systems. *Water Research* **34(15)**: 3836–3844, 10.1016/S0043-1354(00)00133-0.

Deletic A, Dotto CBS, McCarthy DT *et al.* (2012) Assessing uncertainties in urban drainage models. *Physics and Chemistry of the Earth* **42–44**: 3–10, 10.1016/j.pce.2011.04.007.

Deltares (2023) SOBEK and Delft3D FM Suite 1D2D. https://www.deltares.nl/en/software-and-data/products/sobek-and-delft3d-fm-suite-1d2d (accessed 09/08/2023).

DoE (Department of the Environment)/HR Wallingford (1983) *Design and Analysis of Urban Storm Drainage: The Wallingford Procedure*. Standing Technical Committee Report No. 28. Department of the Environment, National Water Council, Hydraulics Research Ltd, Wallingford, UK.

DHI (Danish Hydraulic Institute) (2023) MIKE URBAN Integrated urban water modelling. https://www.mikepoweredbydhi.com/products/mike-urban (accessed 06/08/2023).

Di Matteo M, Dandy GC and Maier HR (2017) Multiobjective optimization of distributed stormwater harvesting systems. *ASCE Journal of Water Resources and Planning Management* **143(6)**: 10.1061/(ASCE)WR.1943-5452.0000756.

Dring D (1985) The Buxton Case Study. In *Proceedings of the CIWEM Urban Drainage Group Conference*, Birmingham. http://ftp2.ciwem.org/1985/1985-11-05/01%20Paper%20Au%201985.pdf.

Durrans SR, Ahmad M, Barnard TE, Hjorth P and Pitt R (2007) *Stormwater Conveyance, Modeling and Design*, 1st edn. Haested Methods Water Resources Modeling Collection. Bentley Institute Press, Exton, PA, USA.

eWater (2023) MUSIC (Model for Urban Stormwater Improvement Conceptualisation). https://ewater.org.au/products/music/ (accessed 06/08/2023).

Fleming R, Rhodes G, Mitchell R and Digman C (2019) A blueprint for modelling sewers and implementing SuDS. In *Proceedings of the CIWEM UDG Conference*, Autumn. See http://ftp2.ciwem.org/2019/2019-Autumn/Paper%202%20A%20Blueprint%20for%20Modelling%20Sewers%20and%20Implementing%20SuDS.pdf (accessed 06/08/2023).

Fletcher TD, Shuster W, Hunt WF *et al.* (2014) SUDS, LID, BMPs, WSUD and more – the evolution and application of terminology surrounding urban drainage. *Urban Water Journal* **15(7)**: 525–542, 10.1080/1573062X.2014.916314.

FWR (Foundation for Water Research) (2018) *Urban Pollution Management Manual*, version 3.1. Foundation for Water Research, Marlow, UK. http://www.fwr.org/UPM3/ (accessed 06/08/2023).

Gent R, Crabtree R and Ashley RM (1996) A review of model development based on sewer sediments research in the UK. *Water Science Technology* **33(9)**: 1–7, 10.2166/wst.1996.0162.

Gersonius B, Ashley R, Pathirana A and Zevenbergen C (2012) Adaptation of flood risk infrastructure to climate resilience. *Proceedings of the Institution of Civil Engineers – Civil Engineering* **165(6)**: 3–4, 10.1680/cien.2012.165.6.3.

Gilboa Y, Friedler E and Schütze M (2023) Assessing water use and reuse options – a holistic analysis of a model city, coupling dynamic system modelling with life cycle assessment. *Urban Water Journal* **20(7)**: 844–858, 10.1080/1573062X.2023.2211959.

Haestad Methods (2007) *Stormwater Conveyance Modelling and Design*. Bentley Systems, Exton, PA, USA.

Hall A, Hunt C and Margetts J (2016) How to build the biggest hydraulic model in the world. In *Proceedings of the CIWEM UDG Autumn Conference*. November. See http://ftp2.ciwem.org/2016/2016-Autumn/Paper%2022%20-%20How%20to%20Build%20the%20Worlds%20Largest%20Model.....Presenters%20Alastair%20Hall%20&%20Charlotte%20Hunt.pdf (accessed 06/08/2023).

Hamann F, Blecken G-T, Ashley RM and Viklander M (2020) Valuing the multiple benefits of blue-green infrastructure for a Swedish case study: contrasting the economic assessment tools B£ST and TEEB. *ASCE Journal of Sustainable Water in the Built Environment* **6(4)**: 05020003, 10.1061/JSWBAY.0000919.

House of Lords (2022) *Nature-Based Solutions: Rhetoric or Reality?* Science and Technology Select Committee. HL Paper 147. See https://publications.parliament.uk/pa/ld5802/ldselect/ldsctech/147/14702.htm (accessed 06/08/2023).

Hvitved-Jacobsen T, Vollertsen J and Nielsen AH (2013) *Sewer Processes: Microbial and Chemical Process Engineering of Sewer Networks*. CRC Press, London, UK.

Iftekar MS and Pannel DJ (2022) Developing an integrated investment decision-support framework for water-sensitive urban design projects. *Journal of Hydrology* **607**: 127532, 10.1016/j.jhydrol.2022.127532.

Innovyze (2023) InfoWorks ICM Help Documentation. https://help.innovyze.com/space/infoworksicm (accessed 06/08/2023).

Jack A, Petrie MM and Ashley RM (1996) The diversity of sewer sediments and the consequences for sewer flow quality modelling. *Journal of Water Science and Technology* **33(9)**: 207–214, 10.2166/wst.1996.0213.

Jung KW, Jeong Y-H and Jung I-K *et al.* (in press) Application of a linked model to simulate catchment water quality in complex land-use area. *Proceedings of the Institution of Civil Engineers – Water Management*, 10.1680/jwama.21.00081.

Korving H, Van Gelder PHAJM, Van Noortwijk J and Clemens FHLR (2002) Influence of model parameter uncertainties on decision-making for sewer system management. *Fifth International Conference on Hydroinformatics* (Cluckie ID, Han D, Davis JP and Helsop S (eds)). Cardiff, UK, 1–5 July 2002, vol. **2**, pp. 1361–1366.

Kroll S, Wambecq T, Weemaes M, Van Impe J and Willems P (2017) Semi-automated build-up and calibration of conceptual sewer models. *Environmental Modelling & Software* **93**: 344–355, 10.1016/j.envsoft.2017.02.030.

Laio Z, Gu X, Xie J, Wang X and Chen J (2019) An integrated assessment of drainage system reconstruction based on a drainage network model. *Environmental Science and Pollution Research* **26**: 26563–26576, 10.1007/s11356-019-05280-1.

Li W, Kiaghadi A and Dawson C (2021) High temporal resolution rainfall–runoff modelling using long-short-term-memory (LSTM) networks. *Neural Computing and Applications* **33**: 1261–1278, 10.1007/s00521-020-05010-6.

Lin C-Y and Wi S (2022) HydroCNHS: a Python package of hydrological model for coupled natural–human systems. *Journal of Water Resources Planning and Management* **148(12)**: 10.1061/(ASCE)WR.1943-5452.0001630.

Lisenbee WA, Hathaway JM, Burns MJ and Fletcher TD (2021) Modelling bioretention stormwater systems: current models and future research needs. *Environmental Modelling and Software* **144**: 105146. https://doi.org/10.1016/j.envsoft.2021.105146.

Lisenbee WA, Hathaway JM and Winston RJ (2022) Modelling bioretention hydrology: quantifying the performance of DRAINMOD-Urban and the SWMM LID module. *Journal of Hydrology* **612(B)**: 128179, 10.1016/j.jhydrol.2022.128179.

McDougall JA and Swaffield JA (2000) Simulation of building drainage system operation under water conservation design criteria. *Building Services Engineering Research and Technology* **21**: 41–51, 10.1177/014362440002100107.

McEvoy S, Van de Ven SHM, Brolsma R and Slinger JH (2020) Evaluating a planning support system's use and effects in urban adaptation: an exploratory case study from Berlin, Germany. *Sustainability* **12**: 173, 10.3390/su12010173. www.mdpi.com/journal/sustainability

Mannina G (ed.) (2019) *New Trends in Urban Drainage Modelling. UDM 2018*. Springer, Cham, Switzerland.

Mark O (2019) *Deterministic Modelling of Urban Stormwater and Sewer Systems*. Aalborg University Press, Aalborg, Denmark. See https://vbn.aau.dk/en/publications/deterministic-modelling-of-urban-stormwater-and-sewer-systems (accessed 06/08/2023).

Mignot E and Dewals B (2022) Hydraulic modelling of inland urban flooding: recent advances. *Journal of Hydrology* **609**: 127763, 10.1016/j.jhydrol.2022.127763.

Montalto FA, Behr CT and Yu Z (2012) Accounting for uncertainty in determining green infrastructure cost-effectiveness. In *Economic Incentives for Stormwater Control* (Thurston HW (ed.)). CRC Press, Boca Raton, FL, USA, pp. 71–100.

Moradikian S, Emami-Skardi MJ and Kerachian R (2022) A distributed constraint multi-agent model for water and reclaimed wastewater allocation in urban areas: application of a modified ADOPT algorithm. *Journal of Environmental Management* **317(2022)**: 115446, 10.1016/j.jenvman.2022.115446.

Muschalla D, Schütze M, Schroeder K *et al.* (2009) The HSG procedure for modelling integrated urban wastewater systems. *Water Science and Technology* **60(8)**: 2065–2075, 10.2166/wst.2009.576.

Mutel C, Daly DL and Ettema R (2008) Hunter Rouse's historical writings and the history of hydraulics collection. *Journal of Hydraulic Engineering* **134(6)**: 683, 10.1061/(ASCE)0733-9429(2008)134:6(683).

Nature Squared (2023) TEEB for Cities. https://www.nature-squared.org/case_study/teeb-for-cities/ (accessed 06/08/2023).

Néelz S and Pender G (2009) *Desktop Review of 2D Hydraulic Modelling Packages*. No. SC080035. Environment Agency, Bristol, UK.

Nemcik A, Krupa F, Ozana S and Slanina Z (2022) Wastewater treatment modeling methods review. *IFAC-PapersOnLine* **55(4)**: 195–200, 10.1016/j.ifacol.2022.06.032.

NIC (National Infrastructure Commission) (2022) *Reducing the Risk of Surface Water Flooding*. NIC, London, UK. See https://nic.org.uk/studies-reports/reducing-the-risks-of-surface-water-flooding/surface-water-flooding-final-report/ (accessed 06/08/2023).

Osborne M (2022) Do we need models? DWMP Blog, episode 4. https://www.linkedin.com/pulse/dwmp-blog-episode-4-do-we-need-models-martin-osborne/?lipi = urn%3Ali%3Apage%3Ad_flagship3_profile_view_base_post_details%3Bby4sp%2FHnRSSrjUXO5MaBpQ%3D%3D (accessed 06/08/2023).

Pedersen AN, Brink-Kjær A and Mikkelsen P-S (2022) All models are wrong, but are they useful? Assessing reliability across multiple sites to build trust in urban drainage modelling. *Hydrology and Earth System Sciences* **26**: 5879–5898, 10.5194/hess-26-5879-2022.

Penning-Rowsell E, Priest S, Parker D *et al.* (2013) *Flood and Coastal Erosion Risk Management. A Manual for Economic Appraisal*. Routledge, Abingdon, UK.

Radinja M, Škerjanec M, Šraj M *et al.* (2021) Automated modelling of urban runoff based on domain knowledge and equation discovery. *Journal of Hydrology* **603**: 127077, 10.1016/j.jhydrol.2021.127077.

Rauch W, Aalderink H, Krebs P, Schilling W and Vanrolleghem P (1998) Requirements for integrated wastewater models – driven by receiving water objectives. *Water Science Technology* **38(11)**: 97–104, 10.1016/S0273-1223(98)00644-1.

Rauch W, Bertrand-Krajewski J-L, Krebs P *et al.* (2002) Deterministic modelling of integrated urban drainage systems. *Water Science & Technology* **45**: 81–94, 10.2166/wst.2002.0059.

Sagrelius PO, Blecken G, Hedström A, Ashley RM and Viklander M (2020) Environmental impacts of stormwater bioretention systems with various design and construction components. *Journal of Cleaner Production* **359**: 132091, 10.1016/j.jclepro.2022.132091.

Sanders BF and Bradford SF (2011) Network implementation of the two-component pressure approach for transient flow in storm sewers. *Journal of Hydraulic Engineering* **137(2)**: 10.1061/ASCEHY.1943-7900.0000293.

Saunders C (2019) *Modelling Deposition and Solid Transport in Urban Drainage Systems*. UKWIR report 19/SW/01/19. UK Water Industry Research Ltd, London, UK. See https://ukwir.org/modelling-solid-transport-and-deposition-in-urban-sewerage-systems-0 (accessed 06/08/2023).

Schellart A, Tait SJ and Ashleyk RM (2010) Towards quantification of uncertainty in predicting water quality failures in integrated catchment model studies. *Water Research* **44(2010)**: 3893–3904, 10.1016/j.watres.2010.05.001.

Sohn W, Kim J-H, Li M-H and Brown R (2019) The influence of climate on the effectiveness of low impact development: a systematic review. *Journal of Environmental Management* **236**: 365–379, 10.1016/j.jenvman.2018.11.041.

Sonnenwald F, Guymer I and Stovin V (2019) A CFD-based mixing model for vegetated flows. *Water Resources Research* **55(3)**: 2322–2347, 10.1029/2018WR023628.

Sriwastava AK, Tait SJ, Schellart A *et al.* (2018) Quantifying uncertainty in simulation of sewer overflow volume. *ASCE Journal of Environmental Engineering* **144(7)**: 04018050, 10.1061/(ASCE)EE.1943-7870.0001392.

Sriwastava AK, Torres-Matallana JA, Schellart A, Leopold U and Tait SJ (2021) Implications of model uncertainty for investment decisions to manage intermittent sewer overflows. *Water Research* **194**: 116885, 10.1016/j.watres.2021.116885.

Stovin VR, Sonnenwald F, Golzar M and Guymer I (2022) The impact of cylinder diameter distribution on longitudinal and transverse dispersion within random cylinder arrays. *Water Resources Research* **58**: e2021WR030396, 10.1029/2021WR030396.

Szelag S, Lagód G, Musz-Pomorski A *et al.* (2022) Development of rainfall-runoff models for sustainable stormwater management in urbanized catchments. *Water* **14(13)**: 1997, 10.3390/w14131997.

The WATS Guys (2023) Mega-WATS the model. https://thewatsguys.com/WATSmodel (accessed 06/08/2023).

Thorne CR (ed.) (2020) *Blue-Green Cities*. ICE Publishing, London, UK.

Thurston HW (ed.) (2012) *Economic Incentives for Stormwater Control*. CRC Press, Boca Raton, FL, USA.

Tomlinson JR, Arnott JH and Harou JJ (2020) A water resource simulator in Python. *Environmental Modelling and Software* **126**: 104635, 10.1016/j.envsoft.2020.104635.

Tourbier JT (2002) *Blue-Green Technologies: Integrated Practices to Manage Stormwater as an Asset*. The Great Swamp Watershed Association, Madison, NJ, USA. See https://www.greatswamp.org/Newsletter/newsletterfall2002.pdf.

Tscheikner-Gratl F, Bellos V, Schellart A *et al.* (2021) Recent insights on uncertainties present in integrated catchment water quality modelling. *Water Research* **150**: 368–379, 10.1016/j.watres.2018.11.079.

UK NEA (UK National Ecosystem Assessment) (2011) UK National Ecosystem Assessment. http://uknea.unep-wcmc.org/#:~:text=The%20UK%20National%20Ecosystem%20Assessment%20(UK%20NEA)%20was%20the%20first,and%20reported%20in%20June%202011 (accessed 29/07/2023).

USEPA (United States Environmental Protection Agency) (2023a) Storm Water Management Model (SWMM). https://www.epa.gov/water-research/storm-water-management-model-swmm (accessed 06/08/2023).

USEPA (2023b) System for Urban Stormwater Treatment and Analysis IntegratioN (SUSTAIN). https://www.epa.gov/water-research/system-urban-stormwater-treatment-and-analysis-integration-sustain (accessed 06/08/2023).

Vélez Quintero CA (2012) *Optimization of Urban Wastewater Systems using Model Based Design and Control*. PhD Dissertation, UNESCO-IHE, Delft, the Netherlands. See https://repository.tudelft.nl/islandora/object/uuid%3A248022cb-4395-4178-8eb8-74e8adfff346/datastream/OBJ/download.

Vollertsen J, Revilla N, Hvitved-Jacobsen T and Nielsen AH (2015) Modelling sulfides, pH and hydrogen sulfide gas in the sewers of San Francisco. *Water Environment Research* **87(11)**: 1980–1989, 10.2175/106143015X14362865226752.

Vonach T, Kleidorfer M, Rauch W and Tscheikner-Gratl F (2017) An insight to the cornucopia of possibilities in calibration data collection. *Water Resources Management* **33**: 1629–1645, 10.1007/s11269-018-2163-6.

Wang LK and Yang CT (eds) (2014) *Modern Water Resources Engineering*. Handbook of Environmental Engineering, vol. 15. Springer Science+Business Media, New York, USA, 10.1007/978-1-62703-595-8_2.

Warganda K and Sutjiningsih D (2017) Placement of BMPs in urban catchment area using SUSTAIN model: case study at Universitas Indonesia Campus, Depok, West Java, Indonesia. In *MATEC Web of Conferences* 138, 06007 EACEF 2017, 10.1051/matecconf/201713806007.

Washbourne C-L and Wansbury C (eds.) (2023) *ICE Manual of Blue-Green Infrastructure*. ICE Publishing, London, UK.

WEF/ASCE (Water Environment Federation/American Society of Civil Engineers) (1998) *Urban Runoff Quality Management*. WEF Manual of Practice No. 23, ASCE Manual and Report on Engineering Practice 87. WEF/ASCE, Reston, VA, USA.

Willems P (2008) Quantification and relative comparison of different types of uncertainties in sewer water quality modelling. *Water Research* **42(13)**: 3539–3551, 10.1016/j.watres.2008.05.006.

Willems J, Minkman E, Veerbeek W, Ashley R and Van Buuren A (2022) Policy transfer in urban water management. In *Routledge Handbook of Urban Water Governance* (Bolognesi T, Pinto FS and Farrelly M (eds)). Routledge, Abingdon, UK, pp. 257–271.

Woods-Ballard B, Wilson S, Udale-Clark H *et al.* (2015a) *The SuDS Manual*. CIRIA C753. Construction Industry Research and Information Association, London, UK.

Woods-Ballard B, Wilson S, Udale-Clark H, *et al.* (2015b) Hydrology and hydraulics: design methods and calculations. In *The SuDS Manual*. CIRIA C753. Construction Industry Research and Information Association, London, UK, pp. 506–541.

WRc (Water Research Centre) (1987) *Guide to Short Term Flow Surveys of Sewer Systems*. WRc, Swindon, UK.

WRc (2018) *Sewers for Adoption*, 8th edn. WRc, Swindon, UK. See https://www.water.org.uk/wp-content/uploads/2018/10/SfA-8-Master-2.pdf (accessed 06/08/2023).

WRc (2023) STOAT: Wastewater Plant Modelling Tool. https://www.wrcgroup.com/services/stoat-wastewater-works-modelling-freeware/ (accessed 06/08/2023).

WSC (Water Sensitive Cities) (2023) *INFFEWS Value tool: Guideline (Version 4)*. Cooperative Research Centre for Water Sensitive, Melbourne, Australia. See https://watersensitivecities.org.au/content/inffews-value-tool-guideline-version-3/ (accessed 06/08/2023).

Yang S-Y, Chang C-H, Hsu C-T and Wu S-J (2022) Variation of uncertainty of drainage density in food hazard mapping assessment with coupled 1D–2D hydrodynamics model. *Natural Hazards* **111**: 2297–2315, 10.1007/s11069-021-05138-1.

Ashley R, Smith B, Shaffer P and Caffoor I
ISBN 978-0-7277-4178-3
https://doi.org/10.1680/icehudp.41783.309

Chapter 12
Framing urban drainage

Richard Ashley, Brian Smith and Paul Shaffer

12.1. Introduction

There are many ways of framing how the world is seen, and hence how urban systems are managed. Framing is a way of defining a perspective and for setting boundaries for how systems like water are perceived and managed. Frames are a way of ensuring that there is consistency in approaches to how systems are managed, irrespective of the viewpoints or duties of the stakeholders involved. This chapter outlines and reviews some of the frames used for water and urban drainage management.

Water is experienced in many forms in the hydrological cycle (Chapter 7). With or without urbanisation, the cycle continues to function. But, for expediency and convenience the management of the water cycle is invariably 'parcelled' into easy-to-manage parts, potentially leading to siloed thinking. There is rural water management, urban water management, droughts, flooding, storage, drainage, utilisation, all interacting with each other and creating challenges when there is too little or too much water. As explained in Chapter 1, the piecemeal management of the water cycle has led to many problems, not only for water management, but also in land management (Chapter 5). Urban drainage is a somewhat out-moded term and concept (Chapter 1) as it focuses on water as a problem rather than a resource, and is but part of the water cycle and water system. In contrast with past perspectives, urban drainage cannot be treated separately from the rest of the water cycle or the natural environment.

Water and the water cycle needs to be integrated into our built and natural environment and brought into the heart of our communities. The challenges of doing this are enormous in that it requires change in all aspects of our interactions with the water cycle and even in the systems that interact with it. For example, when reviewing the progress towards water, drainage and systems being managed sustainably in Australia, Brown (2003) found that only those municipalities that had changed their policy, regulatory, professional and public systems had made any progress towards integrated water management. This work was seminal in that it became part of the understanding of how best to manage water in Australia. The water sensitivity programmes being widely implemented Australia are world-leading, and the management approach – to consider the water cycle as a whole – is now being brought into the mainstream in many other countries (e.g. Sharma *et al.*, 2019).

The water management challenges in Australia heralded the need to take an integrated approach, with all parts of the water cycle being seen as useful (Chapter 1). The Australian search for 'sustainable water' systems began in the late 1990s when the scientific community tried to understand how sustainability could inform policymakers and those responsible for delivering water services sustainably (Section 12.2). Figure 1.1 illustrates diagrammatically the components of a water-sensitive city and Figure 1.8 some of the key elements of water sensitivity and water-sensitive urban design (WSUD) in the context of natural systems and land use management. Figure 12.1 shows some of the current frameworks that are being used to help to make water systems more sustainable. Frameworks include integrated

Figure 12.1 Frameworks being used to help ensure water systems are more sustainable

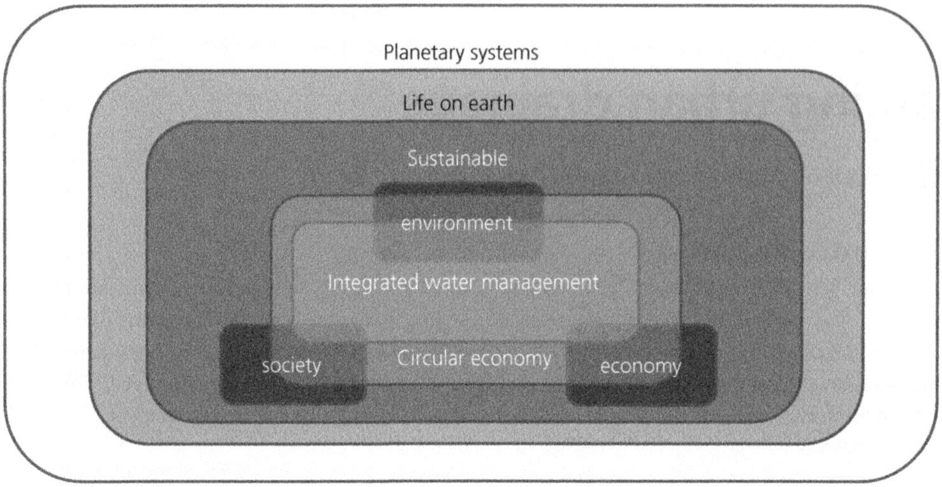

water management (IWM) and WSUD, together with ones for economies and resource management, including the circular economy and water circularity. Systems of systems are conceptualised to help understand and manage the interactions across infrastructure types.

There are many terms in use that can be related to some form of integrated approach to water systems management, as listed in Box 12.1. There are numerous other terms aside from those in Box 12.1, but each sets out to express the same general approach, which is one of integration.

Sustainability should be the over-arching framework, or paradigm, for setting infrastructure provision needs. This refers to sustainability of service provision, in terms of performance, longevity and impacts on the natural world, through the life cycle. Sadly, sustainability and the assessment of sustainability have become degraded both in many publications and in practice. Unsubstantiated use of terminology, labelling virtually everything as 'the sustainable option', has made the term almost meaningless, when it should be assessed and prioritised. Professionals and practice have fallen into the trap of claiming everything they do is 'sustainable'. This has now evolved into everything and every professional striving for resilience, arguably as this is easier to determine and achieve. However, 'resilience' and 'resilient systems' also have the risk of being misrepresented. 'Resilience' is only about recovery from shock, not about long-term functionality. Brinsmead and Hooker (2005) explain in detail how sustainability encompasses resilience, and this demonstrates why this is much easier than aiming for sustainability. Figure 12.2 illustrates the differences between sustainability, which is all encompassing, together with various other terms related to system functionality currently in use. Resilience in a system is important, but one that is resilient is not necessarily sustainable.

What then is sustainability? It is perhaps only definable in terms of components. The 'three pillars' of sustainability have been related to environment, economy and people, as defined in the latter part of the twentieth century and still pertaining today. Loucks and Gladwell (1999) were among the first to apply the principles to water systems. The four Natural Step sustainability principles (Upham, 2000) shown in

Box 12.1

Terminology relevant to the integrated management of water systems: some examples

Water sensitivity (WSUD, WSCs) see Chapter 8, Box 8.3.

Integrated catchment management (ICM) This might be the oldest of the terms currently in use. The original terminology was focused on water management within a catchment to balance needs, then it included other aspects such as water pollution. More recently ICM has become used to define a process covering the entire ecosystem within a catchment and is the foundation of the EU Water Framework Directive, which aims at IWM through river basin management planning (Chapter 5) (e.g. WISE, 2008).

Integrated water management (IWM) The Chartered Institution of Water and Environmental Management (CIWEM, 2011) reviews IWM and other terminology, concluding that IWM is as defined by the Global Water Partnership in 2000: 'the coordinated development and management of water, land and related resources in order to maximise the resultant economic and social welfare in an equitable manner without compromising the sustainability of vital ecosystems'. CIWEM in the report focuses on IWRM, adding 'resources' to the base term.

Integrated urban water management (IUWM) This is an adjunct to the above, but, being aimed at urban systems, encompasses urban planning processes and ways in which water is supplied and used and covers opportunities for decentralised systems. IUWM was the forerunner of WSUD (e.g. Mitchell, 2006).

Water cycle management (WCM) An old term that came to encompass more as water resources became stressed in countries such as Singapore. Now, the term stresses water reuse, including from wastewater sources (e.g. Wang, 2015).

Circular economy (CE) of water (CEW) This is one of the more recent terms. Morseletto *et al.* (2022) consider that water is available both as a stock and a flow in economic terms but argue that 'water is a unique element because it is a resource, a product and a service with no equivalent in the economic system'. Water is hard to set in terms of CE as it occurs in different states (liquid, solid and gas); it can variably contain different types and quantities of chemical elements and living forms. They suggest that 'the CEW is an economic framework for reducing, preserving and optimising the use of water through waste avoidance, efficient utilisation and quality retention while ensuring environmental protection and conservation'.

Figure 12.2 Illustration of what sustainability encompasses

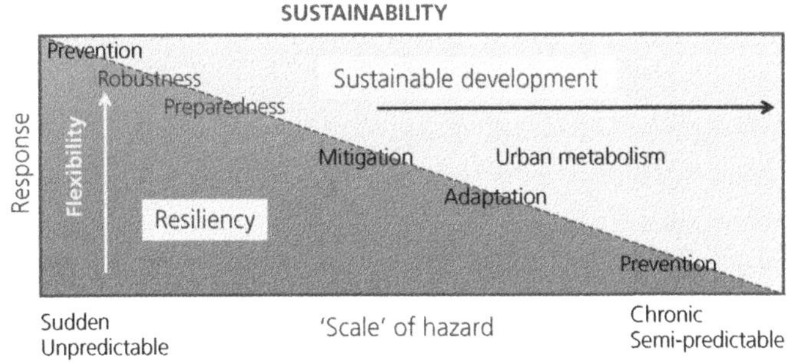

Table 12.1 are often considered for application more than the often-quoted Brundtland (1987) definition of sustainability, which relates to ensuring that what we do now does not compromise security and welfare for future generations. These are shown in comparison with the four principles defined by Fenner *et al.* (2023a), who set out a manifesto for engineers by way of a set of operational principles derived from the absolute principles.

Table 12.1 Examples of sustainability principles

The three pillars of sustainability	The Natural Step (Upham, 2000)	Fenner *et al.* (2023a) modified from 1st edition	
		Absolute principles (original numbering)	Operational principles
Environment	■ Effective protection of the environment ■ Prudent use of natural resources	A1: Respect planetary limits	O1.1 Minimise project impacts
			O1.2 Apply a sustainability hierarchy and circularity in selecting options
		A3: Regard the future (intergenerational stewardship)	O3.1 Plan long term
			O3.2 Consider all life cycle impacts
Society	■ Social progress which recognises the needs of everyone	A2: Improve human wellbeing (socio-economic sustainability)	O2.1 Seek co-generation of solutions
Economics	■ Maintenance of high and stable levels of growth and employment		O2.2 Respect people and human rights
			O2.3 Serve social needs
			O2.4 Assess project outcomes against the SDGs
		A4: Understand system interactions (complexity)	O4.1 Recognise the wider context
			O4.2 Acknowledge uncertainties
			O4.3 Ensure responsive, flexible and adaptive designs
			O4.4 Explore integrated needs

Further examples of practices for wastewater management related to the first version of the principles for sustainability are given in Johnson (2016) and for drainage systems by Balmforth (2016). For example, 'environmental sustainability – within limits' entails protecting the local water environment by limiting waste discharges from urban drainage systems. There is further discussion of the need for innovation to address the current challenges, especially regarding carbon and energy and the need to take a systems approach.

There are of course concerns about the impossibility of sustaining high and stable levels of economic growth in a finite world, as specified by The Natural Step, but, as discussed in Chapter 8, life in a neoliberal economic world demands this presumption. Fenner *et al.* (2023b) take a different approach, preferring to subordinate economics into an all-encompassing human welfare principle.

It may be that the obsession with economic 'growth' that began in the nineteenth century will be overtaken by the understanding that this obsession is responsible for most of the unsustainable impacts and practices, and a new economic paradigm will emerge that is subordinate to the environment and to human welfare, as indicated by Fenner *et al.* (2023b), rather than seeing people and other planetary systems as simply components of the current economic hegemony – that is, in terms of 'labour' and 'capital' as explained by Erickson (2022).

The challenge is to turn the various principles of sustainability into action. This is usually considered in terms of defining objectives, setting measurable outcomes and, with these, the criteria and attributes that can be used to assess achievement of the defined outcomes and objectives.

12.2. Sustainability

Sustainability is something of a 'holy grail' and often more maligned in action than adhered to. Sustainability is about ensuring the future is secure for humanity and, in order to support this, for ecology and environmental systems. Unfortunately, defining what this means in a way that can be used by practitioners has proved elusive, despite numerous studies, reports and recommendations.

There are many examples of sustainability. There are also claims for sustainable systems and products; all naive and often misleading, as sustainability is about context, and the way in which the terminology is being used is unhelpful, especially in influencing policymakers to embed sustainability (e.g. Giampietro, 2023). Any attempt to assess how more or less sustainable one option is compared with another should encompass all the aspects of life and the planet. Professionals must be more precise about terminology, especially when communicating with clients and the wider stakeholder community, not simply jumping on the latest 'buzz-word train' (Kirchherr, 2022). For example, there are signs that the term 'urban transitions' is becoming the next 'fashion'.

For water services, sustainability needs to be considered and assessed in terms of nested scales, as illustrated in Figure 12.3. Each of these needs to be considered in terms of the interactions with the wider scales. In infrastructure provision there is also a need to consider the sustainability of the processes being used to plan, deliver and manage systems. This includes the governance arrangements, institutions, regulatory aspects and users' needs, behaviours and demands.

Contemporary ideas on how to make moral and just judgements about problems with no clear 'answer' are presented by Sen (2009) and are increasingly being referred to as attempts to tackle the wicked and

Figure 12.3 Scales of sustainability

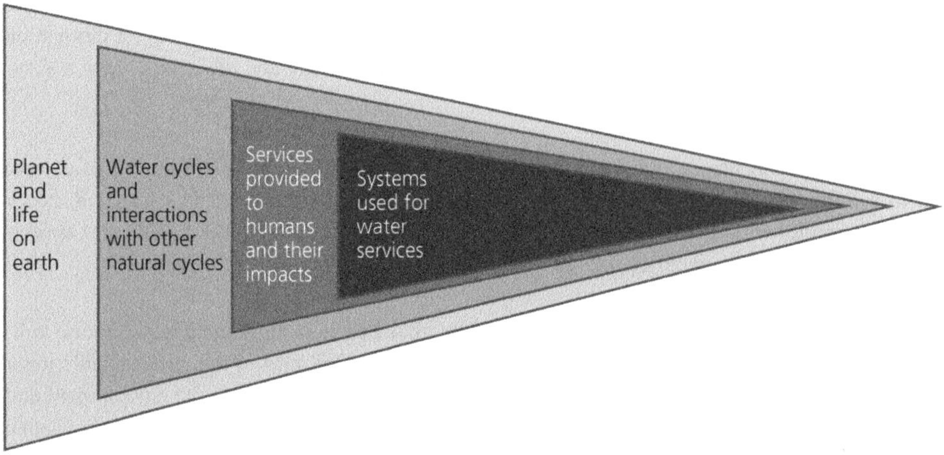

complex problems of providing services, systems and infrastructure for a fairer contemporary society (e.g. Groves *et al.*, 2015). The challenge of making decisions that are fair in society is particularly poignant when deciding on where to allocate scarce resources to manage flood risk (see also Chapter 3). For example, Sayers (2023) advocates the need disproportionally to support the most vulnerable in society, based on Rawlsian principles of social justice in preference to utilitarian (as used in the UK) or egalitarian (as used in France) approaches.

Typically, the focus of defining sustainability, albeit not always explicitly, is anthropocentric – that is, sustainability is about humanity. Decisions are now made almost exclusively on the 'bottom line', based on benefit to cost assessments, in which the costs component dominates (Erickson, 2022). Few decisions pertaining to water service options are made on moral or equity grounds, as discussed in Chapter 8 on valuing nature-based systems (NbSs).

Progressing to systems that are more sustainable is a challenge (e.g. Scheffer, 2009). Even when it is an aspiration, in developed countries the asset base of existing infrastructure cannot simply be abandoned, although it may in some instances be repurposed. What is clear is that any new or modified assets need to be planned and built to be as sustainable as possible, and this means these need to be flexible and readily adaptable. Without this, systems may be 'locked in', in that their use, despite being impacting, may be prolonged unwisely into an unsustainable in use path dependency (e.g. Palmer, 2000; Thompson and Beck, 2015). Examples of this include desalination plants in temperate climates and the use of large resource-intensive infrastructure, such as the Thames Tideway Tunnel, both in use in London.

The Sustainability Transitions Research Network (STRN, 2023) is dedicated to understanding transitions to greater sustainability. Debates around private versus public sector effectiveness for moving to systems that are more sustainable suggest resources and innovation need to come from enterprise and that the business case is clear (e.g. Ginn, 2020). However, this is problematic. For example, Lieberherr and Truffer (2015) compared the opportunities for transitioning to more sustainable water systems in Berlin, Zurich and Leeds. While the private sector appeared to have more opportunity for innovation, the public sector providers were found to be more engaged in the essential dynamic sustainability processes. Similarly, van der Jagt *et al.* (2020) found that private sector organisations had more opportunities for technological innovation, but regarding NbSs innovation, the roles of place-based dynamics, agency and governance structure were more central to nature-based innovation. However, over-reliance on private enterprise can have unforeseen consequences. For example, experience with wastewater private finance initiative (PFI) projects in Scotland shows how such funding arrangements can lock in unsustainable technologies for decades, incurring high economic and environmental costs (Dunion, 2011).

Given the understood complexity of the present world and society, the 'sustainability' challenge is about moving beyond simply sustaining what much of the world already has – safe, secure and healthy water, wastewater and drainage systems – into worldwide provision in a form that can be sustained without literally 'costing the earth' in terms of resource needs and impacts.

This is the vision for the Sustainable Development Goals (SDGs) as outlined in Chapter 5. However, consideration is given only to specific aspects of water supply and sanitation, for example, the main water asset SDG (SDG 6) focuses on drinking water, sanitation and hygiene, ignoring urban drainage other than as a co-benefit (e.g. UN, 2021). Hence pluvial flooding is not considered explicitly in the SDGs despite influencing many of the other goals (Ashley and Horton, 2022). Water is the common thread running through all the SDGs. Thus, although about sustainability, the SDGs fail at integration, focusing instead on component elements of human life, probably as being simpler to attain (e.g. Grison *et al.*, 2023).

One of the frameworks being used globally, which is of particular use for developing countries, but also in use elsewhere, is the 'water cycle safety plan' (WCSP). This is a framework for risk assessment and management (e.g. Almeida *et al.*, 2013), and is included in the World Health Organization guidelines for drinking water quality (WHO, 2017). The WCSP is based on standardised hazard analysis and critical control point analysis as used in the food industry but adapted for water systems.

All-encompassing sustainability assessments (planet to site, Figure 12.3) would require exceptional effort and, in the context of urban drainage systems, be disproportionate other than for the largest schemes. Any sustainability assessments are consequently carried out within set boundaries to limit the effort required, often focused on the development site with certain, but not all, system interactions (Figure 12.4) included. For urban drainage systems, this could encompass an analysis of the interacting systems of systems that affect or are affected by urban drainage systems and the way in which these are provided, as in Figure 12.5. At the very least, the entire water cycle should be considered even at the development scale (Marlow *et al.*, 2013).

Sustainability for water comprises

- security of water: enough water, in the right place at the right time
- security from water: not to be unduly inconvenienced by there being too much water (flooding) and of the wrong quality
- enough water and the right quality to support ecosystems (and other services)
- recognising the value of water and not abusing it.

Sustainability of the systems used to provide water services relates to the asset systems. These include nature as an asset. This is considered further in Section 12.2.1.

Figure 12.4 The boundary considerations of sustainability assessment

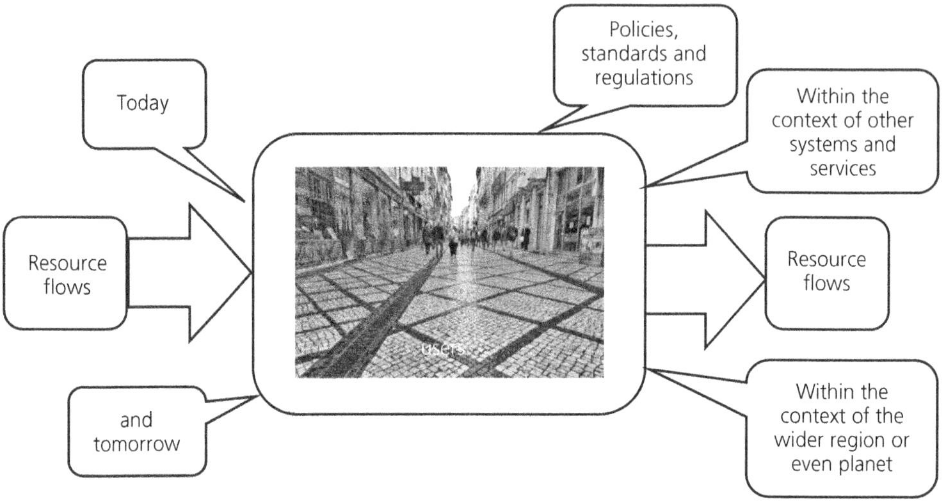

Figure 12.5 Examples of the interacting system of systems relevant to water systems (from Ashley, 2016)

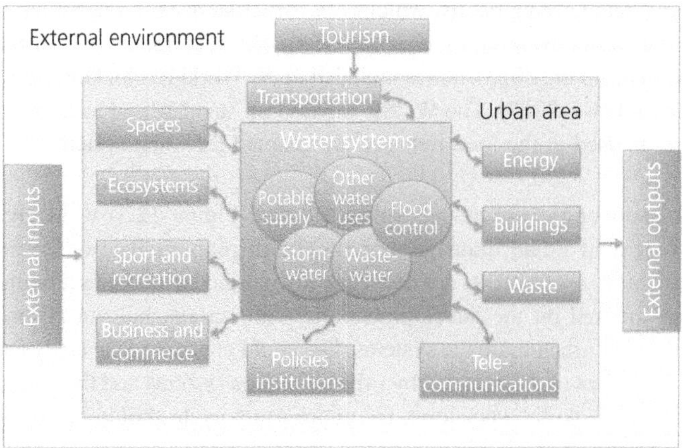

Fenner *et al.* (2023a) considered what sustainability means in civil engineering practice, as illustrated in Figure 12.6, and how this is only slowly being seen as an essential objective (Fenner and Ainger, 2020). Many professionals are unable or unwilling to challenge practices that are unsustainable in terms of the impacts from construction or usage on the principles listed above. Such behaviour is unprofessional and needs to be challenged. The client always is society at large and not the immediate client with funding.

Deriving action from principles is operationalised by Fenner *et al.* (2023b), and some examples of the way in which the principles in Table 12.1 may apply to the delivery of urban water projects are given in Table 12.2, modified from Ashley (2016).

Figure 12.6 A sustainability framework for engineers (Fenner and Ainger, 2020)

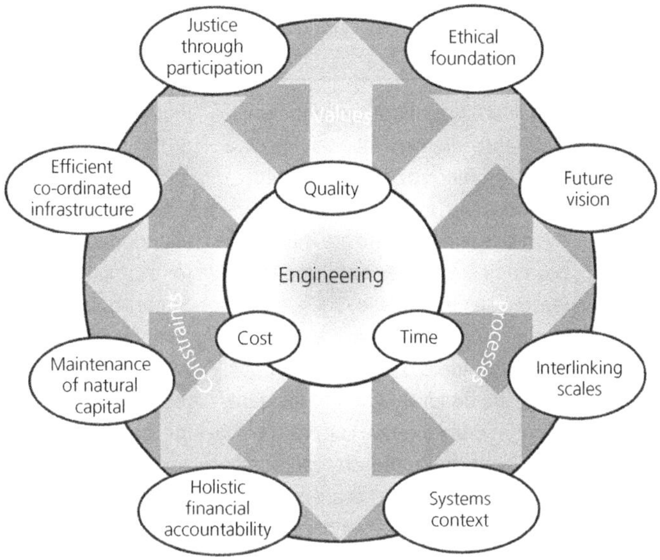

Table 12.2 Examples of application to urban water systems of some of the operational principles in Table 12.1 (Fenner *et al.*, 2023b)

Examples of operational principles (numbering from Fenner *et al.*, 2023b)	What this means for how urban water should be planned, designed and managed	Practical example
O1.2 Apply a sustainability hierarchy and circularity in selecting proposal options	This is about defining project scope and type such that it is as sustainable as possible. Scope can range from modest to all-encompassing, and each option may have more or fewer impacts. Is it possible, for example, to repurpose existing infrastructure? Or construct many small, even decentralised projects, rather than a large one? Whatever options are selected, their construction and operation should be embedded in closed-loop systems based on the principles of circularity (Section 12.5). This perspective should also be applied to the business models used for delivery and operation.	As outlined in Chapter 2, there are initiatives to use more decentralised water and wastewater systems in many places. Evidence shows these are typically more sustainable than large-scale systems. However, it is important to verify this using LCA and multicriteria analysis. Materials, services such as energy, etc. need to be considered in terms of use, reuse and minimum waste, and innovatory ways of providing resources for construction and operation need to be considered. This could involve local people in construction, operation and even the business model, e.g. for flood prevention.

(*continued on next page*)

Table 12.2 (*Continued*)

Examples of operational principles (numbering from Fenner *et al.*, 2023b)	What this means for how urban water should be planned, designed and managed	Practical example
O3.1 Plan long term	Sustainability is about the long term and hence intergenerational equity. However, much of what is done in countries such as the UK now is ignoring this. For example, there are many large private finance schemes for managing wastewater in Scotland. These are hugely expensive and lack any of the innovatory ideas outlined in this book. These are disadvantaging future generations in order to serve current communities. Here, the approach should be one in which no investments are made that cannot be altered or abandoned – what is known as a 'no-regret' approach.	Probably the most important example of this in the water sector is to avoid 'lock-in' to large and costly infrastructure that cannot be modified easily or abandoned in the future. In London, for example, the new large Thames 'super-sewer' is being built to manage combined sewer overflows. This sewer is a major asset and will have to be used for decades in order to pay for the costs. It is difficult to modify and requires expensive and carbon impacting pumping every time it is used.
O2.2 Respect people and human rights	This is embedded in professional codes of conduct and should be a given. Ideally, as wide a range of people as possible need to be effectively engaged in change, and this requires the right professionals for this task; engineers are not social scientists and need to recognise the need to work with other professions/disciplines where and when necessary.	Engagement is not always easy, for example, when planning flood schemes or reservoirs that may disadvantage or displace certain residents or use precious land space. In some parts of the world, indigenous peoples will have specific views and knowledge about what is acceptable. Here the main points are 'whose sustainability?' and 'for what purpose?' In some places, such as South African townships, any other WSS provision than individual flush toilets would not be acceptable.
O3.2 Consider all life cycle impacts	This is the temporal boundary. All infrastructure schemes need to be considered fully throughout their entire life cycle unless they are too limited in scale for this to be important. Within a sustainability perspective there is a need to consider the entire supply chain, by way of, e.g., LCA. This is increasingly complex due to the rapidity in change of the drivers impacting infrastructure systems.	The application of LCA, EIA, etc. has become almost standard, with the latter being mandatory for large projects. It is now possible to assess the impacts, for example for SuDS, using LCA.

EIA, environmental impact assessment; LCA, life cycle analysis; SuDS, sustainable drainage systems; WSS, water supply and sanitation.

12.2.1 Sustainability of water and urban drainage systems

Ensuring the sustainability of urban water, which includes urban drainage, simply means that it is provided for the long term in a manner that is reliable. This also means making sure that potential environmental impacts are reduced or mitigated. Traditionally, only the local impacts have been considered, despite environmental impact assessment (EIA) being used routinely; the entirety of impacts is often not considered, only selective impacts such as carbon footprints.

For the urban drainage practitioner, taking such a global view to the design, planning and operation of new and existing systems is a difficult task. Clients require straightforward drainage provision that is known to work and therefore has few risks. This is why the tried and tested traditional systems (Chapters 2 and 3) that have been used for millennia are usually considered the best and often 'the common-sense option', frequently mislabelled as 'sustainable'. A recent example of this is the Thames Tideway Tunnel (TTT), for which many erroneous 'sustainability' claims have been made without evidence (e.g. Stride, 2014). In London, there are severe water resource deficits (and recently constructed desalination plants), yet the new sewer continues to collect overflowing rainwater, ultimately discharging rather than using it, as is being done in many other parts of the world (Ashley, 2020). Loftus and March (2019) explain how the financial model used in England to incentivise private investment in water services has defined how the combined sewer overflow (CSO) problem in London has been dealt with using traditional approaches and infrastructure. This case is an example of the economic pillar of sustainability overriding the environmental and social pillars. In this case the economics is narrow and only relates to the profitability of the scheme. A major problem with this new sewer is the irreversibility. It will need to remain in use for at least a century, locking in high energy use. It is not adaptable and will require alternative measures to adapt to climate and other changes into the future.

In summary, traditional drainage schemes are resource hungry and environmentally impacting in both construction and use over long periods of time. The piped drainage perspective also considers the conveyed flows as 'wastes' rather than as resources to be utilised. However, there may be instances where piped systems are the more sustainable option. Therefore, it is important to try to understand how, where and when a particular drainage option may or may not be more or less sustainable than another.

What is sustainable drainage? The Construction Industry Research and Information Association (CIRIA, 2010) defined sustainable drainage systems (SuDS) as a sequence of management practices and control structures designed to manage (drain, treat and/or contain) surface water at or near its generation point in a more sustainable fashion than other more conventional techniques. This has evolved to include objectives and to be more defined as an approach (Chapter 8).

A SuDS study undertaken as part of the TTT (Thames Water, 2013) considered the feasibility of using SuDS as an alternative or supplement to the tunnel. The study concluded that SuDS would be feasible and would potentially bring added benefits (Chapter 8). SuDS retrofits were found to be useful (Stovin *et al.*, 2013), but more costly and difficult to implement due to the need to engage with the myriad of landowners across London for access. Despite EIAs having been carried out for the tunnel option, no sustainability assessments have been reported.

There have been numerous sustainability assessments reported for urban drainage systems, each of varying scope and with differences in the frames used. For example, Lundin *et al.* (2000) used life cycle analysis (LCA) as part of sustainability assessment to show how sewer separation was more sustainable than using traditional piped systems, through lower pollutant emissions to receiving waters and a more efficient recycling of nutrients to agriculture, especially of nitrogen and phosphorus. An LCA has been used to assess the environmental impacts of part of the Thames Tideway sewer tunnel (Arena, 2019). However, this considered construction only and the more significant long-term impacts of operation and of resources, but much of the material use was not included. The study did provide information on where the biggest impacts from the construction might be, including from the material used in the lining, potentially allowing the selection of alternative materials and methods. In a global study, Poustie (2014) showed how centralised wastewater systems will not sufficiently meet the needs of rapidly growing centres in the developing world, concluding that hybrid systems are the best option – some decentralised and some centralised (Chapter 2). Sagrelius *et al.* (2023) compared the relative sustainability of various constructed Swedish rain gardens, including by use of an LCA. Of 11 different units, it was concluded that not one of the designs could be concluded as being more sustainable than another.

The various studies so far published cover different aspects of urban drainage, and all have in common that the traditional approaches are found to be typically less sustainable than the SuDS, blue-green infrastructure (BGI) or NbS alternatives. This is often because the boundaries in the analyses are much wider than for the traditional systems, which are only required to drain areas effectually and safely, with the NbS schemes including many more impacts and opportunities than simply piping the water away. The growing use of real-time control (RTC) (Chapter 3) or artificial intelligence (AI) systems (Chapter 13) may also help to provide systems that are more sustainable. For example, Ashagre *et al.* (2020) in reviewing publications on automation in wastewater management conclude that scant interest has been given to the relative contribution that this can make to providing systems that are more sustainable. Most studies consider only very limited components of sustainability (see Section 12.6.2). Given the now increasing use of automated systems in wastewater management (Chapter 13), there is an urgent need for a better understanding of how to assess the relative sustainability of the approach.

Although attractive, the more sustainable options are often difficult or impossible to implement due to inflexible regulations, problems of landowners and rights, institutional responsibility boundaries and requirements to engage effectively with communities. For example, as in London for the TTT, the English water companies may not be able to implement SuDS because this will necessitate construction on the surface, bringing land ownership and other issues that a buried pipe system may not have. See also Chapter 5, where developmental and planning issues are considered.

Innovative drainage technologies (e.g. Chapter 3 and Chapter 8) applied in either residential or commercial developments, besides being technically and economically viable, as well as acceptable to the public, should be designed to be as sustainable as possible. The need for sustainable drainage is driving change within all areas of development and construction, despite the strong reluctance to change from ways in which 'we have always done it'. Nonetheless, with increasing pressure on companies and others to evaluate the environmental and social impacts of their businesses and services, there is a growing demand in both water and wastewater sectors to understand how NbSs, and hence systems that are more sustainable, can best be implemented. There are numerous examples of the strength of business cases for such investment (e.g. Ginn, 2020, and Chapter 8). In this context, sustainable drainage systems should include the design, installation and operation of pipelines in addition to the more common interpretation of the term 'SuDS' as usually presumed. The second edition of CIRIA's *The SuDS Manual* (Woods-Ballard *et al.*, 2015) sets out how SuDS can be used to help make developments 'more sustainable'. It also states that 'sustainable drainage systems (SuDS) are designed to maximise the opportunities and benefits we can secure from surface water management'. This is done by using natural drainage to manage rainwater close to where it falls, on the surface and incorporating vegetation as much as practicable (Chapter 8).

Despite the above, what are known as SuDS in UK practice and legislation may or may not be the more sustainable option in a given situation. Often, the more sustainable option will consist of an appropriate combination of NbSs and piped drainage, and even a piped drainage option alone might be the more sustainable in certain circumstances (e.g. Alves Beloqui, 2020).

12.2.2 Sustainability assessment

Any analysis of the relative sustainability of options needs to be proportionate to the scale of the development, the drivers, and the context in which it is set. Minor changes to existing drainage areas, for example, should not be seen as needing detailed sustainability assessment, but rather the adoption of sensible arrangements and best practice. However, where there are widespread issues or where the development is on a large scale, then some sort of sustainability assessment may help in choosing the best way forward, as set out in the water cycle management for new developments (WaND) guidance (CIRIA, 2010). This helps to address the somewhat vague aspirations of policy instruments, such as the English national planning guidance, National Planning Policy Framework (DLUHC, 2023; Chapter 5), where it is stated that: 'The purpose of the planning system is to contribute to the achievement of sustainable development' and there is 'a presumption in favour of sustainable development'. Rather rather ambitious statements given that there is no consensus on what this sustainable development might be. Such policies are found worldwide and illustrate the catch-all that the word 'sustainable' has come to mean. CIRIA (2010) presents a different view, which is that sustainability or sustainable development is something aspirational and requires a journey in the right direction based on selecting options that are as sustainable as feasible and avoiding those options that are clearly not sustainable.

Frameworks and methodologies for sustainability assessment of use for urban drainage systems are further considered in Section 12.6.

12.3. Water security

In terms of the overall water cycle, urban drainage can be formed in ways to maximise the resource opportunities and alleviate water stresses. These stresses are increasingly everywhere due to climate, population growth, urbanisation, the demand for irrigation water and unremitting pollution of freshwaters.

> Water security is one of the most tangible and fastest-growing social, political and economic challenges faced today. It is also a fast-unfolding environmental crisis. In every sector, the demand for water is expected to increase and analysis suggests that the world will face a 40% global shortfall between forecast demand and available supply by 2030.
>
> (Gulabchand, 2013)

The UN 2023 Water Conference in New York concluded that more focus was needed on water supplies in climate planning, with half the world's population facing acute water stress by 2030 (UN, 2023). The UN Secretary General, António Guterres, advocated for integrating the approach on water, ecosystems and climate.

The Organisation for Economic Co-operation and Development (OECD, 2013) report *Water Security for Better Lives* proposes a fundamental shift in our approach to tackling water security (see Box 12.2).

Box 12.2

Achieving water security objectives means maintaining acceptable levels for four water risks (OECD, 2013)

Risk of shortage (including droughts): lack of sufficient water to meet demand (in both the short- and long-run) for beneficial uses by all water users (households, businesses and the environment).

Risk of inadequate quality: lack of water of suitable quality for a particular purpose or use.

Risk of excess (including floods): overflow of the normal confines of a water system (natural or built), or the destructive accumulation of water over areas that are not normally submerged.

Risk of undermining the resilience of freshwater systems: exceeding the coping capacity of the surface and groundwater bodies and their interactions (the 'system'); possibly crossing tipping points and causing irreversible damage to the system's hydraulic and biological functions.

All four risks must be assessed in an integrated way as interventions to reduce one risk can increase other risks.

For example, floods and droughts both affect water quality and the provision of safe drinking water, and contribute to degradation of freshwater ecosystems. Polluted water resources, without treatment, are effectively excluded from human consumption and utilisation by industry and agricultural sectors, thereby increasing the risk of water scarcity. Climate change is exacerbating existing water risks, due to altered precipitation and river flow regimes, more frequent and severe extreme weather events, altered thermal regimes and sea level rise. Moreover, the inherent uncertainty in climate change projections makes it more challenging to assess how these risks will evolve in the future (OECD, 2013).

12.3.1 Urbanisation

Urbanisation has dramatically changed the built environment, with 82% of the UK's population now living in urban environments and growth predicted to continue. Increased development all too often impacts on many sources of water supply. Lakes and reservoirs can be strategic sources for water supply, and often also serve as energy production, flood risk management and recreational areas, but can also be polluted from runoff from rural and urban areas. Their ecosystems, particularly in tropical regions, are often fragile and vulnerable to changes in land and water use due to various anthropogenic pressures. An example is illustrated in Box 12.3.

The SDGs highlight the importance of water and, as such, the increasing demand for innovative and effective approaches to water management. Decision-makers will therefore need to invest in water security, innovate and adapt, without being limited to the solutions adopted in the past. It is possible to make a positive impact and create functional, energy-efficient, low-carbon and climate-resilient cities, with integrated water resource management as an effective way to make the transition. As part of this, individual assets, including buildings and other components, will need to adopt low-carbon principles (e.g. Construction Leadership Council, 2023) within the integrated approaches. Unfortunately, the major cities of the world are failing to adopt integrated approaches. For example, Grison *et al.* (2023) used a city water blueprint approach for the largest 200 cities in the world, which together serve some 95% of the world's urban population. Only Amsterdam and Singapore had effective integrated water resources management, with the majority of cities not 'managing their water wisely', especially cities in Africa,

Box 12.3

Water resource degradation

The Pampulha Reservoir, located in the city of Belo Horizonte, Brazil, was built in the 1930s to supply drinking water to the city. However, since the 1970s the frequency of algal blooms, particularly cyanobacteria, has increased, mainly due to eutrophication caused by contamination by the input of nutrients from ongoing urbanisation, resulting in degradation of the water quality (e.g. Friese *et al.*, 2010). The reservoir is now dead. This is a clear demonstration of a need to increase the resilience of strategic sources to anthropogenic pressures and climate change, through improved land use regulation and soil and water conservation.

Of course, there is a broader set of issues; and a need to consider the role that urban drainage infrastructure must play in the delivery of water security. The use of blue-green infrastructure techniques can help mitigate the impacts of agriculture, mining and urban sprawl on the hydrological cycle and water pollution, contribute to reducing risks of water scarcity and flooding, and promote biodiversity. While Europe has made significant efforts to improve its institutions to protect water and implement NbSs, developing countries are not quite as advanced with their decision-making processes on water management.

Asia and Latin America. This failure is also the reason why the goals for SDG 6 (cleanwater and sanitation) and SDG 11 (sustainable cities and communities) are not being met.

12.3.2 Migration

Bangladesh has been a vulnerable state for much of its short existence (Park, 2019). Floods occur every year in Bangladesh and are therefore not unexpected, but people adapt their lives to live with them. They may not be able to manage the floods, but their ability to adapt their homes and lives to live with floods is impressive, but not something that would be acceptable to western societies. Flooding and saltwater intrusion affect their environment, making it increasingly difficult to live and grow crops on land saturated with saline water. With the predicted sea level rise, 50 million people will leave the Bangladesh basin by 2050. Migration will occur not only from Bangladesh but worldwide. It is likely that there will be mass migration on a global scale, with hundreds of millions of people looking for a new country. The World Bank's *Groundswell* report (Clement *et al.*, 2021) projects that climate change-related disasters, such as floods and droughts, could cause up to 216 million people to move within or out of their countries by 2050. This then poses the question of the fitness of purpose of aging water and wastewater infrastructure in the places that they move to. A breakdown in wastewater infrastructure could lead to the re-emergence of water-borne disease, such as cholera, and to water scarcity and inadequate resources.

12.3.3 Future supply demand

The growing demands placed on water supplies are not merely the result of a growing population, but also of the way in which economies develop. Many of the problems now faced in the use of water are economic, social, environmental and political. To prepare more intelligently for current and future needs requires changes in structures governing water use. Looking beyond the sewer systems at water balance and future supply demand will require increasing complexity in thinking for how the management of stormwater as a potential resource could be used to address demand for water for a variety of non-potable purposes, such as livestock, agriculture and irrigation. Not only should stormwater be managed in a manner that is natural and sympathetic to the environment, but it also has the potential to provide sustainable water resources for now and the future.

Looking at future options for managing demand for water will require an examination of the evidence for the effectiveness of alternative interventions that could help to reduce demand in times of drought. There is also a greater need to raise awareness on water efficiency, promote and develop more innovative ways to optimise water use and reduce the demand for potable water supplies. Globally there are many initiatives to try to reframe water supply demand to be more effective, by way of new ideas and techniques. Unfortunately, in most of the global south, populations expect the same types of service as in the first industrialised countries of the north, and despite appropriately aimed technologies their use is often not acceptable.

In the UK there is a perception by most of the population that there is no shortage of water. This is perhaps compounded by the fact there are seemingly high levels of rainfall, a barrier to persuading the populace to utilise rainwater harvesting or recycling of wastewater. Most of the public seem unaware of the need for water conservation. Complacency and taking water for granted are all too commonplace, with many people having little understanding of the difference between the perceived and actual availability of water. Much of the problem stems from political and educational complacency about water and many critical services, in which water, along with many other systems, is seen as of secondary importance as 'it can always be managed'. There needs to be more co-design of infrastructure, particularly at the most local level. It is clear how to do this, but rarely are funds made available for such effective engagement (e.g. Bell *et al.*, 2021).

A better understanding of the place and value of water, its potential circularity and opportunities for recycling, is vital to secure future supplies (Chapter 4). Many places in the world, including in countries where it has not been necessary before, will have to accept the use of recycled water. Stormwater harvesting, at both property and community levels (as used in parts of Australia), will not only help create capacity in wastewater networks, with the added benefits of reducing flooding, stormwater overflows and other pollution, but will also have a positive benefit on demand management and pressure management for mains water supply systems; and help to reduce leakage.

Water management strategies do not necessarily have to be complex or employ expensive capital solutions. It is often about simplicity. Decentralisation and new technology make onsite treatment to potable standards feasible. Advances in filtration, recycling and treatment technologies and processes have the potential to reduce water discharges to drainage systems, thereby increasing the availability of non-potable water.

Water is arguably more important than any other natural resource, sustaining as it does all human life and the ecosystems on which human life depends. 'Countries can live without an aerospace industry, but they cannot live without water' (Oded Distel, Israeli water official; Anon, 2016). However, we are seeing some very encouraging and innovative approaches, from Scotland's aspirations for building a Hydro Nation (Scottish Government, 2023) to Israel's hydro-diplomacy: 'In the middle of a desert and in the world's driest region, Israel has managed and innovated its way to a water surplus' (former US Secretary of State, George P Shultz).

Many countries are now experiencing water shortages and drought. Desalination is one option now being considered and more talked about. Israel has demonstrated that this is possible (see Box 12.4), albeit desalination is costly and polluting (Chapter 1) and should only be used as a last resort when there are no other options.

Box 12.4

Israeli desalination plants (Siegel, 2017)

Israel constructed five coastal desalination plants in the time it took California to get planning permission for one due to overcoming legal issues and being tied up in planning, regulation and legislation. The mentality of waiting until everything is perfect creates long delays (Abraham Tenne, the head of the desalination division for Israeli governmental water and sewage authority). 'Worse' he says, 'frequently nothing gets started. The need for water is growing and natural water resources are being depleted with potentially terrible environmental consequences. Not taking action is also a kind of taking action. It is opting for the status quo.'

12.3.4 Governance

'If you don't know where you are going, any road will take you there', wrote Lewis Carroll in *Alice in Wonderland*. This is not an approach to be advocated! So, have we failed in terms of long-term planning? It is important to ensure that we have the correct policy, legal and regulatory structures to incentivise and maximise the use of integrated water management, including education, water conservation and harvesting or recycling. In many countries, metering and social tariffs may also incentivise responsible usage and reduce unnecessary waste (Chapter 5). Globally, it is the failure of governance systems that is primarily responsible for failure to ensure that water is managed effectively and in an integrated way. The governance effectiveness indicator of the World Bank is considered the most important parameter in assessing water management performance: 'if you want to fix the water pipes, start with the institutions' (Grison *et al.*, 2023). In assessing the Blue City Index performance, the latter also highlights the importance of effective wastewater treatment systems. The time to act is now, but this requires leadership and a will to make it happen.

As well as being vital to human health and life, water has a fundamental role in supporting the value to society of ecosystems and biodiversity. Re-amplifying the systemic shift in physical environments, population redistribution and changing demographics means that a change in our approach to water management is needed. 'An integrated approach to water management across media and sectors is needed, as there are potential trade-offs between land and water use, ecosystems, GHG emissions, soil degradation, etc.' (UNEP, 2012). Box 12.5 gives an example of a manifesto from the Netherlands on the urgent need to move to IWM. There are many others available globally.

Through a better understanding of what water service users require and what influences their desires, together with an appreciation of the choices, behavioural science can influence behaviour. Numerous pilot projects are being used empirically to measure the effectiveness of proposed interventions or alternative solutions to provide an evidence-based case for future decisions (e.g. Bell *et al.*, 2021).

12.4. Integrated water management and water sensitivity

Integrated urban water management (IUWM) has been defined in the context of the UK water industry as: 'a comprehensive approach to urban water services, viewing water supply, stormwater and wastewater as components of an integrated physical system which sits within an organisational framework and a broader natural landscape' (Caffoor, 2009). According to the Global Water Partnership (GWP, 2012):

> The goals of urban water management are to ensure access to water and sanitation infrastructure and services; manage rainwater, wastewater, stormwater drainage, and runoff pollution; control waterborne diseases and epidemics; and reduce the risk of water-related hazards, including floods, droughts, and landslides. All the while, water management practices must prevent resource degradation.

Box 12.5

Manifesto for moving to integrated water management in the Netherlands (PBL, 2023)

The interrelated global water, climate change and biodiversity crises show that current value systems, policies and economic practices are not suitable for bending the trend. It is clear, now more than ever, that water needs to be at the top of everyone's agenda, from local actors to global organisations and from public to private enterprises around the world. . . . A shared water agenda needs to address the short-term urgencies and organise a process to optimise long-term High ambition pathways, reducing water and climate-related risks. Not as a blueprint but as a mission statement and a promising perspective across landscapes, regions, countries and cities.

The following are recommendations for the short- and long-term water agenda for cities, complementing those for river basin, deltas, coasts and islands, and drylands.

1st Tier	2nd Tier
■ Always combine sanitation systems with wastewater treatment systems	■ Implement regulations to reduce nutrient emissions and other pollutants as well as water use by households and industries
■ Integrate available green and blue infrastructure in urban development	■ Reduce sand mining and innovate in building materials
■ Stimulate water- and heat-proof construction	■ Improve urban–rural interactions
■ Increase flood protection levels, apply zoning and building regulations	

Box 12.6 illustrates some of the measures used as part of IUWM and Figure 12.7 how the urban water system is interconnected (Ashley *et al.*, 2013).

As outlined in Section 12.1, there are numerous frames that are relevant to IUWM. Several of these are reviewed by Rentachintala *et al.* (2022). The definitions used above and elsewhere focus on the water system, which is understandable; however, many definitions fail to recognise interaction with urban systems (urban design, infrastructure, etc.) as this is likely to define the limitations and boundaries of IUWM (Ashley, 2020). This is why WSUD, which is defined contextually within urban spaces, is a more useful term.

Caffoor (2009), in developing an argument for the business case for IUWM for the UK, points out some of the barriers to adopting a full IUWM approach. These include self-interested protectionism, in which institutions fear that other organisations may encroach on their area of activities and a fear by businesses of being compromised by the approach. In common with sustainability, dealing with the barriers to IUWM requires cultural, policy and institutional change and reform.

Integrated urban drainage (IUD) sits within IUWM, and aims to manage urban drainage systems in a coordinated way. This has been the way forward in England since the early 2000s (Digman *et al.*, 2006; EA, 2009). IUD has been reflected in the modelling approaches developed to support the planning and design of IUD, as described in Chapter 11.

Box 12.6

Examples of measures used in IUWM (Caffoor, 2009)

- Water conservation and efficiency
- Water-sensitive planning and design, including urban layout and landscaping
- Utilisation of unconventional water sources, including roof runoff, stormwater, greywater and wastewater
- Stormwater and wastewater source control and pollution prevention
- Stormwater flow and quality management
- Use of mixed soft (e.g. behaviours) and hard technologies (assets)
- Use of non-structural tools, including education, pricing incentives, regulations and restrictive regimes and stakeholder capacity building.

Figure 12.7 How urban water systems are interconnected (Ashley *et al.*, 2013)

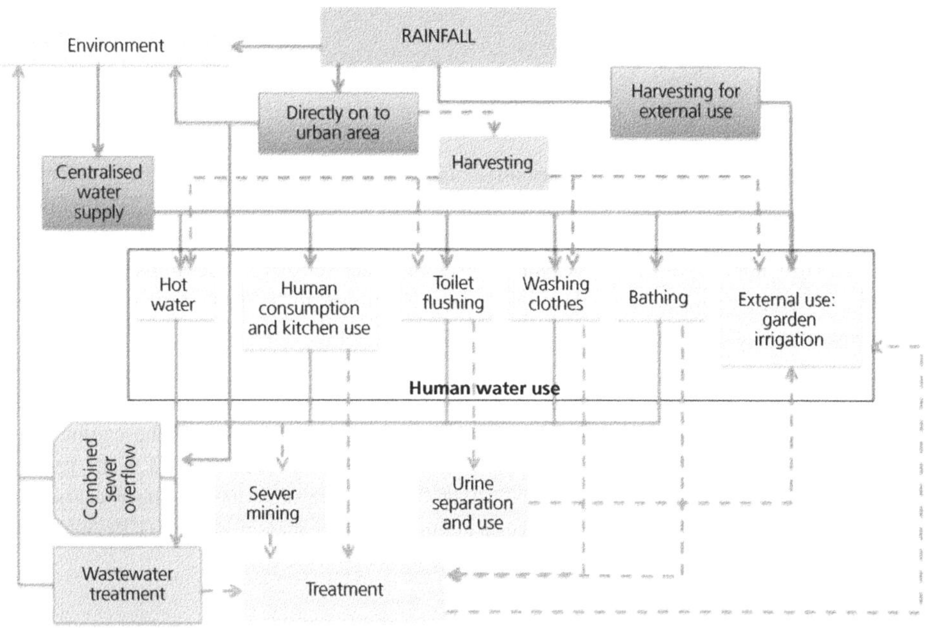

Initially, the primary focus of IUD was flood risk management, but it has since become the main framework for use in England and many other places, albeit aimed at stormwater and not necessarily including sanitary flows, although there are frameworks for planning stormwater management that sit within the 'one water approach'.

Water sensitivity is now an established concept and the subject of major developments in this century in order to operationalise the concept, as outlined in Chapter 1. The concept of 'water sensitivity' provides a vision to identify the most viable options for managing water availability (too little or too much), together with water quality. In urban areas it is vital to link this with urban design, place-making, and liveability (i.e. urbanism) (e.g. Howe and Mitchell, 2012), potentially moving towards sustainable urban water

Figure 12.8 The interactions between water-sensitive urban design (WSUD), the built environment and the urban water cycle (Ashley *et al.*, 2013)

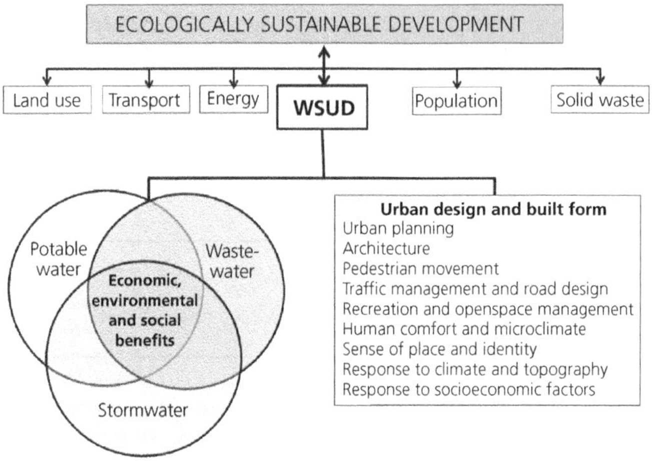

management within a sustainable 'water-sensitive' city (Sharma *et al.*, 2019). This is illustrated in Figure 12.8, which show the interactions between WSUD and the built environment. The rationale for water sensitivity is that water needs to be at the heart of urbanism, by way of planning and operation of urban systems. The Cooperative Reseach Centre (CRC) for Water Sensitive Cities in Australia provides a wide range of resources to support the realisation of water sensitivity, for example, the overall frameworks for sustainable urban water systems and their relationship to circular economies (CRCWSC, 2023a).

Delivery of WSUD in practice covers a wide range of functions within a process that includes the natural environment as well as urban design. The term 'water-sensitive' integrates both the engineering and ecological professions associated with protection of urban water resources. WSUD is the process for facilitating interaction between the urban built form, the water cycle and water management needed for water sensitivity (Wong, 2006). A water-sensitive city comprises three main principles (CRCWSC, 2023a)

- the city as a water supply catchment
- provision of support to ecosystem services
- water-sensitive communities.

The principles above understate the need also to manage too much water and to maximise the opportunities from other aspects of ecosystem services, such as heat mitigation.

Interpretations of water sensitivity and WSUD within a European context have been made in EU projects, for example, SWITCH (Hoyer *et al.*, 2011) and SKINT (Pötz and Bleuzé, 2012), and for the UK (Abbott *et al.*, 2013; CIRIA, 2013). However, in England the responsibilities for the implementation of water sensitivity approaches and WSUD processes are dispersed between mainly private water services, with necessary encouragement of and interaction with the planning function of local authorities (Chapter 5). Bide and Coleman (2019) provide guidance on how water management may be better managed in the planning system in the UK, and Krivtsov *et al.* (2019) provide perspectives on the use and effectiveness of BGI as part of WSUD. Table 12.3 gives examples of the roles that various stakeholders can play in delivering WSUD and in moving towards urban water sensitivity. Although aimed at England, these apply more generally worldwide.

There are signs that IWM approaches are beginning to gain ground in England, with a new understanding that value is not necessarily related to being the cheapest, or even as defined by ES valuation. For example, the Enabling Water Smart Communities project (EWSC, 2023) is framed around shared concepts of 'value' working with communities and major stakeholders. The project is re-evaluating the entire process and concepts essential if IWM is to be established and sustained.

Table 12.3 Examples of opportunities for major stakeholders in taking a WSUD approach in the UK (adapted from Ashley *et al.*, 2013)

Organisation/key stakeholders	Opportunities afforded by a WSUD approach
Government	Develop national policy to ensure delivery of WSUD is embodied across all activities and relevant policies. Linking water with green infrastructure, planning, sustainable growth, place-making, cities of the future, energy, carbon, waste management, ecosystem health and community health and welfare can save money and demonstrate commitment, vision and caring. It can also create green and community-related jobs, contributing to security of supply, climate change adaptation and safety.
Municipalities, local authorities	Implement local planning policies and processes that comprehensively include water in all occurrences and seek to maximise the opportunities from this to enhance places and people's lives, i.e. properly embed WSUD in urban design and planning processes and bring together local flood risk management and urban runoff pollution control and the delivery of local environmental outcomes. Manage highway drainage better, in such a way as to mitigate and help adapt to climate change and support the health and welfare of communities.
Regulators	Enforce compliance with regulations and standards and help to encourage business plans of relevant utilities to take a wider view than just cost-effectiveness. Identify and manage unintended consequences of policy tensions/conflict, support and enable the integration of planning activities across water resources and water cycle planning. Use ES valuations to give credibility to policy decisions.
Water service providers (may be the same as municipalities in some places)	Use business planning as a vehicle to engage potential partners and provide clear estimates of the value of taking a WSUD approach. This can help address some supply security and energy use issues and may provide new business opportunities. WSUD can be utilised for a range of multiple benefits reducing the need for water resource exploitation, wastewater transport and treatment and energy and carbon impacts.

(*continued on next page*)

Table 12.3 (*Continued*)

Organisation/key stakeholders	Opportunities afforded by a WSUD approach
Developers	Aim to support and enhance society and places. Embrace and drive better quality standards for sustainable construction, to comply with codes and assessments, such as BREEAM. Use NbSs in preference to grey infrastructure.
Communities and individuals	There needs to be a much greater awareness of the importance of water in urban areas. From largely ignoring water in the past, individuals need to be informed and engaged in order to make better choices and understand how their interaction with water is fundamental to urban living. WSUD can help make communities and individuals more aware of and able to manage their own interactions with the water cycle better and take advantage of opportunities, e.g. WSUD techniques could provide alternative water and nutrient sources for non-potable use.

Each stakeholder needs to engage effectively with communities and people involved in local decision-making about water in their environment, contributing local knowledge to urban planning and design (Chapter 4).

After more than a decade of development and promulgation of the concepts of water sensitivity, Rashetnia *et al.* (2022) reviewed the successes and limitations of WSUD applications worldwide and some of the frames and models used in application. The main limitations found were related to the (mainly NbSs) approaches in use (Chapter 8), including the performance of ancillary assets and failure of infiltration systems. Each was related to poor maintenance or inadequate design. The limitations of being able to identify life cycle implications and costs were also identified as impediments to use. The challenges of interdisciplinary connections needed to implement WSUD concepts across urban planning, engineering, science, environmental economics and landscape architecture require collaboration between stakeholders, which for many is difficult, a problem that is occurring widely (e.g. Cettner *et al.*, 2014).

However, it is concluded by Rashetnia *et al.* (2022) that overall WSUD approaches have been shown to achieve long-term sustainability in urban water cycle management, although nowhere in the review is long-term sustainability defined. Other benefits from the WSUD approach include the following.

- Minimisation of the hydrological impacts of urban developments on local/regional environment.
- Stormwater quality improvements, and quantity management, flood risk management and stormwater harvesting.
- More sustainable and liveable cities, by way of
  - management of the whole water cycle
  - mitigation of heat island impacts
  - protection of stream ecosystems
  - urban area sustainability
  - social benefits, including landscape amenity, enhancement in economic value, mitigating climate change impacts
  - management of sewer overflows, mainly in combined sewer systems.

In another review, Rentachintala *et al.* (2022) conclude that there is a need for standardisation of sensitivity-relevant measures and practices that should consist of innovative technologies, climate change impact assessment and uncertainty analysis. Treatment and reuse should adopt sustainable and resilient management adaptations, although again, what is meant by sustainable in this context is not explained. It is recommended that real-time monitoring and use of the Internet of Things (IoT), artificial intelligence (AI) and machine learning (ML) need to be applied for more efficient urban stormwater management.

Getting from where water service paradigms are now to what are understood as necessary changes towards water sensitivity requires significant changes to governance and institutions in most jurisdictions, moving away from the problem-centred concepts to opportunity-focused paradigms. 'Leapfrogging' ideas suggest that 'jumps' can be made from one condition in the water systems continuum (Figure 1.1) to a better one, avoiding the gradual change between. However, this can only be achieved through effective governance and policies and effective behaviours of key institutional players. As an illustration, Flores *et al.* (2023) applied a framework to study leapfrogging – the Urban Water Management Transitions Framework (UWMTF) developed by Brown and Keath (2008), together with the Governance Assessment Tool (GAT) – to the water services in the City of Zwolle, the Netherlands. GAT was used to assess how the governance context supported or restricted the implementation of BGI in terms of five dimensions: (*a*) levels and scales; (*b*) actors and networks; (*c*) problem perspectives and goal ambitions; (*d*) strategies and instruments; and (*e*) responsibilities and resources. Each of these were assessed based on the qualities of governance: (*a*) extent; (*b*) coherence; (*c*) flexibility; and (*d*) intensity. In Zwolle, the overall assessment indicated that the governance processes were moderately to highly supportive of transitioning to greater use of BGI (or NbSs). This illustration has been used here to demonstrate the crucial importance of the governance processes in moving to water sensitivity.

12.5. Water circularity

Over the past decade the concept of a circular economy (CE) has developed (e.g. BSI, 2017; Giampietro and Funtowicz, 2020). As another closed-loop system, like the water cycle, there are opportunities from developments of the CE concepts into specific applications. CRCWSC (2023a) relates the CE to water sensitivity, specifically for the circular economy for water (CEW), citing the following principles

- design out waste and externalities
- keep resources in use
- regenerate natural capital.

The CE concepts are also considered by Morseletto *et al.* (2022) in terms of the CEW. It is contended that water, as a unique element in the CE, is a resource, a product and a service which has no equivalent in economics and should be considered and valued differently.

The original concept of the CE originated from the Ellen MacArthur Foundation and is now stated to be

> A systems solution framework that tackles global challenges like climate change, biodiversity loss, waste, and pollution. It is based on three principles, driven by design: eliminate waste and pollution, circulate products and materials (at their highest value), and regenerate nature. It is underpinned by a transition to renewable energy and materials. Transitioning to a circular economy entails decoupling economic activity from the consumption of finite resources.
>
> (EMF, 2023)

In other words, human activity can be decoupled from the natural world if the technology is right. This is an illusion, and while the CE concepts are useful in the appropriate context, these will never produce a resource bank that provides infinite economic growth opportunities. Resources will always be expended in the CE cycle, including energy and additional materials, needed to cope with the entropy in the processes.

The CE is now entering more aspects of how infrastructure systems are managed, including asset management, for which a closed-loop system is aspired to (e.g. ISO/TC 251; ISO, 2023) by effective asset management in terms of BS ISO 55000 (ISO, 2014) (Chapter 10).

Despite the challenges in decoupling economic activity from resource consumption, there are some valuable insights from the CE frame. The term CEW is being used in various forms. Giampietro and Funtowicz (2020) identify the singular importance of water in the CE, and Morseletto *et al.* (2022) emphasise the uniqueness of water systems, suggesting a working definition of CEW as

> an economic framework for reducing, preserving, and optimising the use of water through waste avoidance, efficient utilisation and quality retention while ensuring environmental protection and conservation. In this model, water is seen as a precious resource, product, and service to be managed sustainably within the natural water cycle. These aims should be met in all processes in which water is used, consumed, or treated.

Water process management is often considered using frames that are established in other domains. The primary CE strategy is aimed at manufacturing and industrial processes and is based on the 'R-strategies': refuse, rethink, reduce, reuse, repair, refurbish, remanufacture, repurpose, recycle and recover, which may not be readily useful for water. Table 12.4 shows a reinterpretation of what CEW comprises. As with other framings, the CEW should not be considered in isolation from other 'economies', natural or infrastructure systems.

Many sources related to the CEW have simply rebadged the information using the term CE. These use the frames IWM, sustainability, etc., as discussed above. But, as 'CE' is the latest 'on trend' concept, many must be on it. More correctly, there are information sources on water in the CE (e.g. Delgado *et al.*, 2021; Smol *et al.*, 2023; World Bank, 2021) that consider the place and contribution of water within the CE. For UK applications the UK Water Partnership has produced a white paper (Arup *et al.*, 2019) on the contribution of water within a CE and how this promotes IWM.

12.6. Assessing sustainability and water sensitivity

Whichever of the frameworks outlined in the previous sections are being used to plan urban drainage systems, 'the devil is in the detail'. Ultimately, the scheme on the ground must be designed, constructed and operated, irrespective of the bigger context. This section introduces some practical information which will link to the other chapters in this book where details of urban drainage, planning, design and operation are set out. The various frames so far introduced provide the contextual settings for defining project principles, see Table 12.1, and the key components for defining project objectives and outcomes – for example, Table 12.4 for strategies related to water in the CE. Delivery requires various types of assessment and a clear definition of boundaries (Figure 12.4). There are numerous assessment methodologies and here there is space only for examples.

Table 12.4 CEW strategies (adapted from Morseletto *et al.*, 2022)

Strategies		Approach	Examples
Actions	Category		
	Rethink	Reconfigure and reconceptualise how water is used to ensure a more circular utilisation	Multiple types of water supplied for different uses
Decreasing	Avoid	Use no water at all	Vacuum toilets
	Reduce	Minimise usage, reduce abstractions, change technology, but some use is needed to maintain life	Low flush toilets
	Replace	Use something other than water or change to a lower quality water	Compost toilets
Optimising	Reuse	Use water again directly for a further purpose (without treatment)	Domestic greywater used to irrigate gardens
	Recycle	Use water again after additional treatment	NeWater supplies in Singapore
	Cascade	Combinations of reuse, recycling and recovering in sequence – cyclical use	NeWater arises from a cascade of supply, wastewater generation, treatment and back to supply
Retaining	Store	After use water is conveyed to storage for subsequent usage (with or without treatment)	Irrigation water collection and retention to be reused
	Recover	Recovering water and materials in solution or conveyed, together with energy	Thermal energy from showers and urban sewer networks is becoming viable

CEW, circular economy for water.

Assessment of the relative sustainability of options as part of option appraisal may be important, especially for larger projects, or to assess compliance with objectives, for example, to minimise environmental impacts. While sustainability assessments are relevant to all infrastructure systems, IWM focuses on the water cycle, and assessments of the place of urban drainage within this are best illustrated through the ongoing applications of WSUD.

Any assessment approach should be one that aligns with the decision-maker's perspective (as well as the designer and supporting professionals and the community). The approach may challenge the perspective of any of the stakeholders involved but should be clear enough to take the decision-maker and others along with the analytical and decision-making process, ensuring that all aspects of the problem are considered, even if these are later deemed to be not relevant or too difficult to include.

12.6.1 The challenge of change

The world is seemingly changing at an ever-increasing rate. How can this change be considered in ensuring safe, secure, and sustainable water and drainage?

> In reality the . . . actual service life for a scheme alters constantly, depending on . . . the maintenance activities that are undertaken and any changes in the function of the site. . . . the required performance of the scheme may change over its lifetime. Hence, the flexibility to allow change is also an important consideration in scheme selection.
>
> (Woods Ballard *et al.*, 2015)

There are many types of computational model available that can be used to simulate the performance of systems under different conditions that could occur during the lifetime of the infrastructure. However, the challenges are in specifying what conditions are to be used in the models. Traditionally, these would be to ensure robustness (Figure 12.2) of performance, both under current conditions and in a most likely future when the asset will still need to function. In a review of historical developments of wastewater systems, Neumann *et al.* (2015) found that when using this traditional 'predict and provide' approach to infrastructure planning and design, significant adaptation was needed over the lifetime of many systems but was applied only by way of a 'muddling-through' process, rather than through a planned and logical robust design and operational strategy. A more structured approach needs to be taken to ensure that such assets are sufficiently flexible that they can be modified or adapted in a planned way.

Figure 12.9 illustrates the need to adapt infrastructure and assets to cope with change, despite routine and major maintenance activity. With greater uncertainty about the future, there are now imperatives to try to ensure performance is sustained irrespective of the future.

There are two main components of this.

- Ensure that the scheme and component assets are robust and can be modified with acceptable costs and impacts; this also entails not closing off future change options.
- Develop an adaptation plan, usually comprising several options that can be used to establish adaptation pathways.

These need to be set within a planning process that envisages future coherent scenarios.

12.6.1.1 Flexibility and adaptability

Asset or infrastructure flexibility is not straightforward to assess or to plan (e.g. de Neufville and Scholtes, 2011); however, flexibility can be equated to adaptability and how easy it is to modify an asset. In general, NbSs are easier to modify than buried piped systems, and hence are more flexible and adaptable (e.g. Eckart *et al.*, 2012). The system flexibility needs to be in the adaptability of both the individual assets and the collective scheme – for example, by ensuring that the embedded flexibility is in the appropriate geographical places for best return for least cost.

A way of understanding and promoting the flexibility of infrastructure, and hence resilience, to be built-in from the inception of a scheme is to use real options analysis (ROA), as suggested by the Environment Agency in England (e.g. Brisley *et al.*, 2016). A real option is defined as 'the right – but not the

Figure 12.9 Asset performance changes over time and adaptation (modified from FAIR, 2020)

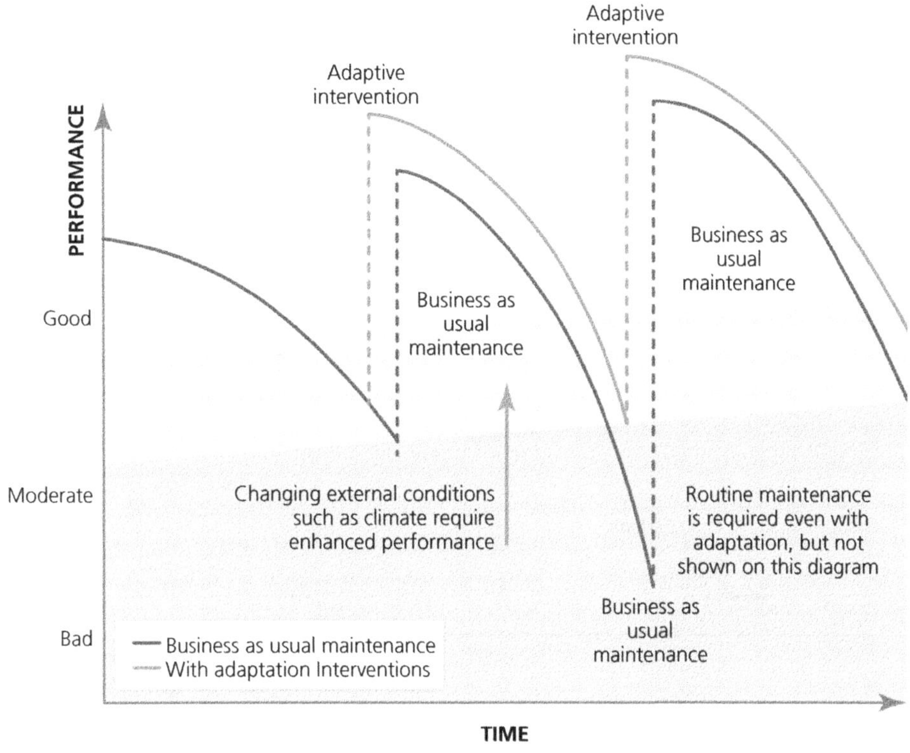

obligation' to adjust the infrastructure system in ways likely to be more resilient, as needed to continue to function as expected. These options represent physical choices about the system that ensure the flexibility to deal with uncertainty. Formal ROA uses the models from financial options analysis to inform how best to manage infrastructure systems in the face of uncertainty. This provides a way to quantify the value of flexibility in infrastructure systems; that is, of finding out which flexibilities are worth their cost. Optimisation determines the flexibility value based on a probability distribution of the uncertainties in future time periods. ROA also assumes that in the future, knowledge advances will to some extent offset the uncertainties being managed.

There are several examples of ROA being applied to schemes planned to deal with flooding. Gersonius *et al.* (2015) describe the application to an urban drainage catchment in West Garforth, Leeds, UK, as part of the Integrated Urban Drainage pilot programme for the Department for Environment, Food and Rural Affairs (Defra), illustrated in Box 12.7, where ROA was used to determine the optimum adaptation strategy.

Box 12.7

Ensuring flexibility of flood risk management assets in West Garforth (adapted from Gersonius *et al.*, 2012, 2013)

The drainage area considered for West Garforth is shown in Figure 12.10.

Figure 12.10 West Garforth drainage area (from Gersonius *et al.*, 2012)

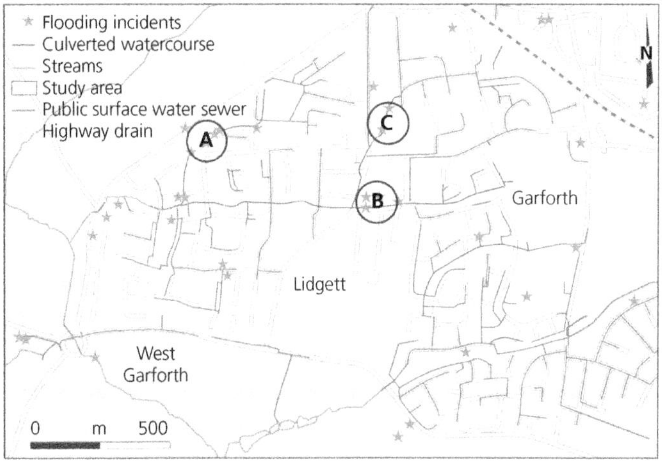

The area has a long history of flooding problems. Many properties were flooded internally in June 2007 during the project. The foci of the most at-risk areas are shown as A, B and C in Figure 12.10. The backbone of the drainage infrastructure is a system of inadequate culverted watercourses, passing through the gardens of more than 200 private properties, with numerous households having individual responsibility on their own property.

Flood risks were assessed in terms of expected annual damage (EAD) using modelling for storms of 1 in 30-year return period, with uplifts of 1.05, 1.1 and 1.2 to represent future rainfall at the short (2025), medium (2050) and long terms (2085). The responses considered in the pilot study comprised: new underground storage; pumping of excess flows to alternative locations; retrofit SuDS, including stormwater disconnection; enlargement or addition of pipes to transfer excess flows to alternative locations.

Discussions with the community in public meetings and subsequent modelling were undertaken together with ROA, to identify the feasible options over some 100 years. ROA allowed modifications to the selected assets and overall scheme to be modelled during three sequential periods, each of 30 years, with three optional scenario pathways for the second and third time periods. The objective was to optimise the expected life cycle cost for the managed adaptive strategy. The expected costs of the set of adaptive strategies were averaged over all climate change paths based on the probabilities derived from a stochastic process for the managed adaptive strategy.

The analysis resulted in three groups of feasible sequential options, A1, A2 and A3, that maximised the flexibility and adaptability over the lifetime of the scheme:

- A1: in the first 30-year time period enlarge sewer conduits, remove adverse pipe gradients and build six new (SuDS) storage areas.
- A2: built only if the intensity of the design storm has increased by some 13% in either the second or third 30-year time period, with two new storage areas, expansion of four of the stage A1 storage areas and disconnection of 213 back roofs.
- A3: built only if the rainfall intensity has increased by 28% in the third 30-year time period, including a new storage area, expanding the existing storage areas and disconnection of a further 137 back roofs.

Options A2 and A3 would only be required when the rainfall (and possibly other factors) had reached a specified level. The 'adaptation tipping point' is the time when the weakest link fails, that is, one of the assets no longer copes. This is an 'if-then-do-it' process, similar to that being followed for the Thames Barrier in London (Thames Estuary 2100 plan; EA, 2012). Such approaches require continuous performance monitoring in order to be prepared to adapt as needed and dynamic decision-making as decisions need to be continuously reviewed.

The expected net present cost of implementing the managed adaptive strategy was some £1.7 million. The traditional static, robust strategy, which is associated with a fixed (re)design that accounts for the worst-case path of climate change, had a net present cost of £2.13 million. The managed adaptive strategy avoided potentially irreversible investments by building a more modest configuration initially, with the option to adapt in the future. In this case the managed adaptation strategy saves some 20% of the costs of the traditional 'static' solution and avoids maladaptation.

12.6.1.2 Adaptation planning

The section above illustrated the process of ensuring systems have flexibility and can be readily adapted as the future unfolds. It is also expected that knowledge and technology will advance into the future, as will expectations and behaviours. However, this will be unpredictable and can only be seen as a contingency that might come about. Adaptive management is a process developed for systems that are complex, uncertain and unpredictable, including in the water sector. For example, in an EU project PREPARED, an adaptive management process was developed for use by Welsh Water that included institutional as well as technical processes (Westling *et al.*, 2019), Box 12.8.

To plan for the future, it is necessary to develop some narratives of how this may develop. One way of doing this is using scenario planning, a well-established process, but seldom used routinely in urban drainage system planning and design (Ashley *et al.*, 2023; Hunt *et al.*, 2012). The process of scenario planning is subjective, although informed by analysis, and sometimes seen as a set of discrete storylines about plausible futures. However, these allow for only a few stories and an alternative approach is to use a continuous scenario space. In this the consequences of various policy options are explored within the main dimensions of uncertainty that are set by stakeholders (e.g. Lempert *et al.*, 2003). In this way, each option can be assessed to see how it performs by way of quantified analysis.

Box 12.8

Developing adaptive capacity for Welsh Water (Westling *et al.*, 2012)

As part of the project, a series of workshops were undertaken with staff at all levels of the organisation. These were in sequential stages: aspirations; scenarios; roadmapping. As part of this process, four frames emerged to express different ideas about how water should ideally be managed in a balanced way in Wales (Table 12.5). These represented the aspirations for the future service of the organisation and set the foundation for contextualising plans to become more adaptive.

Table 12.5 Four frames used in developing an adaptive water management strategy

Frame	Sustainable	Water and drainage	Public	Nature	Responsible	Primary adaptation mechanisms
Market	Economic driven	Commodity	Customer	Nature for people to exploit	Market through the individual	Economic incentives for behaviour change, economic valuation used
Environment	Environment driven	Natural capital	Exploiter	People are stewards of nature	Everyone using water	Regulation to ensure no adverse effect on the environment
People of Wales	Social driven	Human right	Citizen	People are part of nature	Everyone collectively	Collective inclusive action for the benefit of people in Wales
Technocracy	System driven	Source of demand	Demander	Nature as an asset to fulfil demand	Experts	Prediction and technical innovation to address anticipated impacts and opportunities

The above enabled Welsh Water to understand where their future aspirations lay in balancing the frames for service delivery. Following this, the challenges of the future were considered by way of scenario workshops where four socio-economic scenarios were defined, with DPSIR chains used to understand potential impacts and then responses. An example is shown below.

An example of a specific response to a driver is shown below. The effectiveness is across scenarios and scores are from workshop consensus.

Driver, pressure, impact	Response	Effectiveness (+,++,+++)		Robustness score	Rank
		2020s	2040s		
Climate change – changing pattern of weather – changes in land use	Legislation to control use	++	+++	8	1
	Education and capacity building	++	+		

Figure 12.11 Example of a response to the climate change driver.

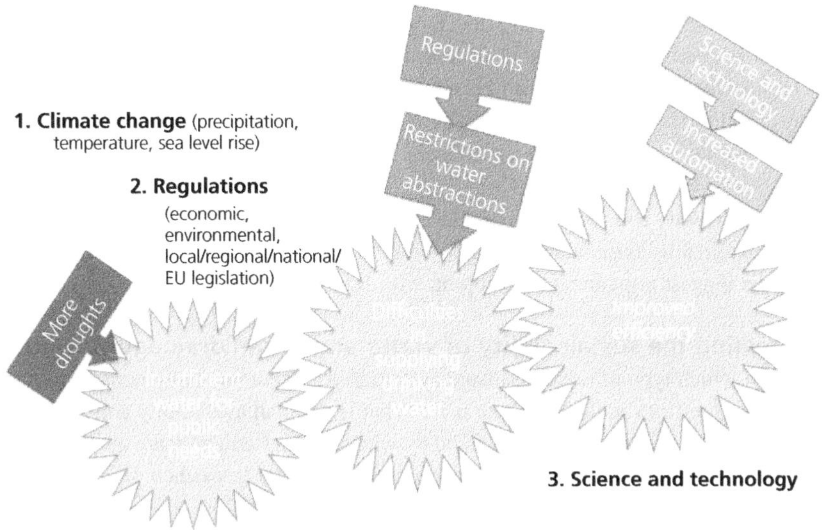

Figure 12.11 shows a response to a climate change driver. Where a response is found to be effective, then it is considered further, with other selected responses in a planned roadmapping process for implementation, with short-, medium- and longer-term actions. The entire process needed to be shared with other stakeholder groups, especially as many responses required partnering and even legislative change.

Scenario planning is usually linked with the driver–pressure–state–impact–response (DPSIR) framework as a way to consider logically the effects of and responses to change drivers, such as climate, urbanisation and expectations.

- Drivers are the reasons, needs and motives for change. They may be 'external' in that they are creating conditions that now or in the future require changes to infrastructure to preserve desired functioning of systems and services. 'Internal' drivers are all those that relate to the delivery of systems, infrastructure or services, including business and institutional practice, regulations, financing, partnerships, people and communities.
- Pressures are exerted by the drivers on the systems and may alter the state.
- State is the condition that the system is in.
- Impacts (or consequences) result from the effects of the drivers and pressures on the state of the system.
- Responses can be effected at any stage of the D–P–S–I chain.

DPSIR is used as part of the adaptation planning process through asking: 'If this, then what will be the consequences?' From this, the follow-on is: 'If we respond this way, will it work?' Futures can be considered using logical, coherent and plausible stories about future conditions. New ideas are now linking these various frames into a 'whole-of-systems' framework for developing dynamic adaptation approaches (termed pathways), as illustrated in Figure 12.12 from Sayers (2023).

Box 12.8 shows how several 'frames' were developed with Welsh Water as scenarios against which to ask questions for a number of water and wastewater infrastructure options, to try to ensure future adaptability. Ultimately, the outcomes should result in the development of a roadmap for moving into the future in a way that embeds and facilitates the planning and implementation of infrastructure systems and individual assets that are as adaptable as feasible.

This also requires flexibility and adaptability in the organisation itself as it will need to be able to respond in an agile way as circumstances change. Inflexibility within organisations, regulations and policies remains one of the biggest impediments to coping with future changes.

12.6.2 Assessing the sustainability of water and urban drainage systems

When deciding on which type of urban drainage system to use there are many considerations. Each of these is contextual. Questions include: Where is it? What is the land availability and type? What is the water cycle locally? How will it be maintained and by whom? Less frequently are questions asked about impacts on the planetary ecosystems, although environmental impact assessment (EIA) processes can be used to assess this (e.g. Glasson and Therival, 2012) and may be mandatory for certain types of projects. However, for most urban drainage projects, EIA will not be mandatory, unless the project is part of a much broader development scheme. There is also the question of whose sustainability is being considered? For example, the US Environmental Protection Agency (USEPA, 2023a) provides resources to assist in planning sustainable water infrastructure specifically for water and wastewater utilities. The emphasis is on the assets being planned and used in a way that ensures the sustainability of the services for the communities served. Guidance helps to ensure cost-effective delivery. This is supported in the USA by tools such as SUSTAIN (Chapter 11) to model the performance of urban drainage systems and help select the options that are as sustainable as possible (USEPA, 2023b). This is an example of the importance of understanding the objectives, scope and boundaries to such assessments and tools, and of selecting the appropriate methods for a particular application.

Strategic environmental assessment and sustainability appraisal is required in England in relation to local plans and development strategies, with the objective of promoting sustainable development (DLUHC, 2020). In this case, the sustainability assessment is at a high level for strategic planning, with subsequent EIA procedures considering details of schemes.

In the rest of this section the use of sustainability assessment (SA) for scheme options is considered. Developed in the early stages of defining sustainability, assessments use variations of the principles–objectives–criteria–indicators–attributes (POCIA) cascade, an example of which is shown in Figure 12.13 from the UK SWARD project (Ashley *et al.*, 2008). POCIA framing accounts for the increasing levels of detail in an assessment. The MATISSE EU project reviewed and developed a standardised methodology for integrated sustainability assessment (ISA) (Weaver and Rottmans, 2006), which is based on the POCIA approach. The methodology is defined as 'a cyclical, participatory process of scoping, envisioning, experimenting and learning through which a shared interpretation of sustainability for a specific context is developed and applied in an integrated manner in order to explore solutions to persistent problems'. This can be interpreted as illustrated in Figure 12.14. ISA is an ongoing continuous stakeholder-focused process, potentially driven by a learning alliance (Chapter 4).

Figure 12.12 A whole-system framework (adapted from Sayers, 2023)

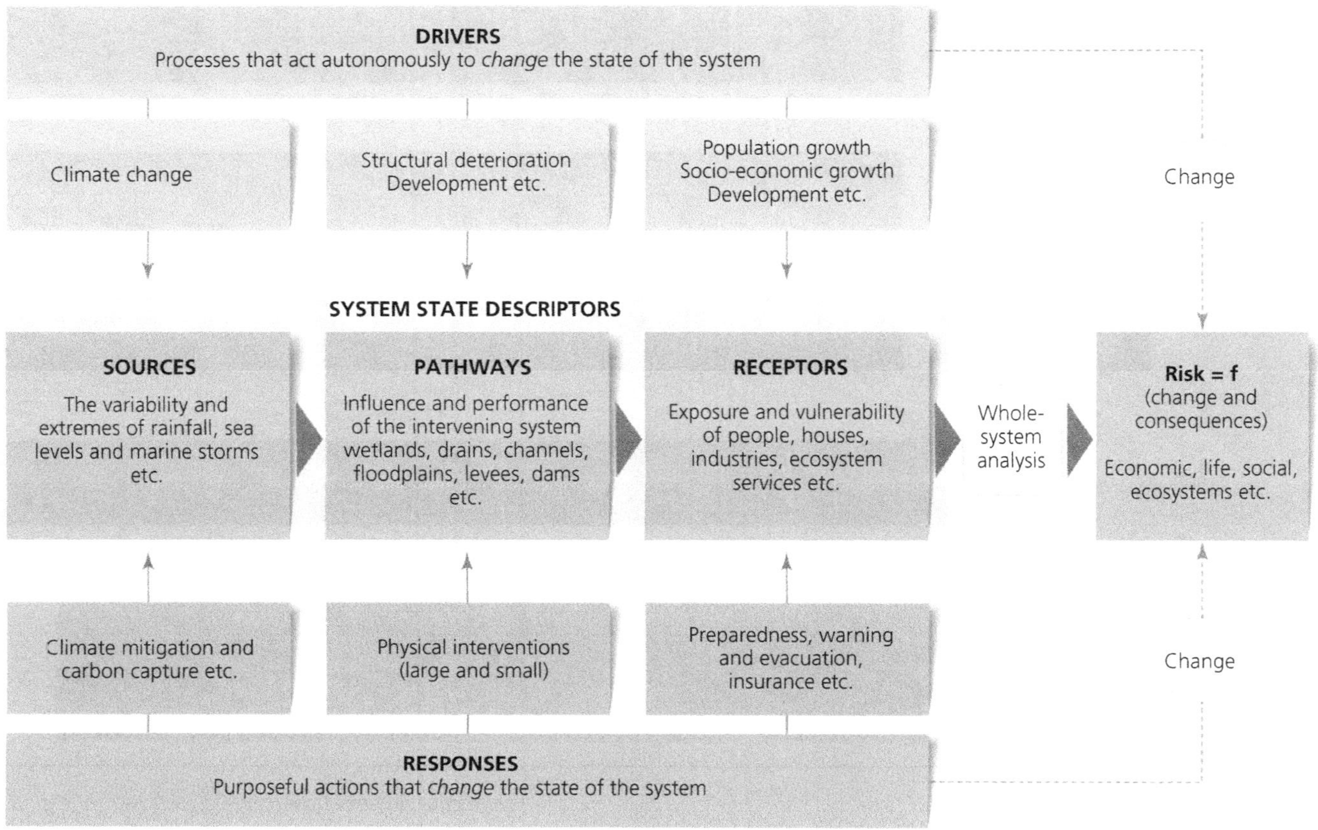

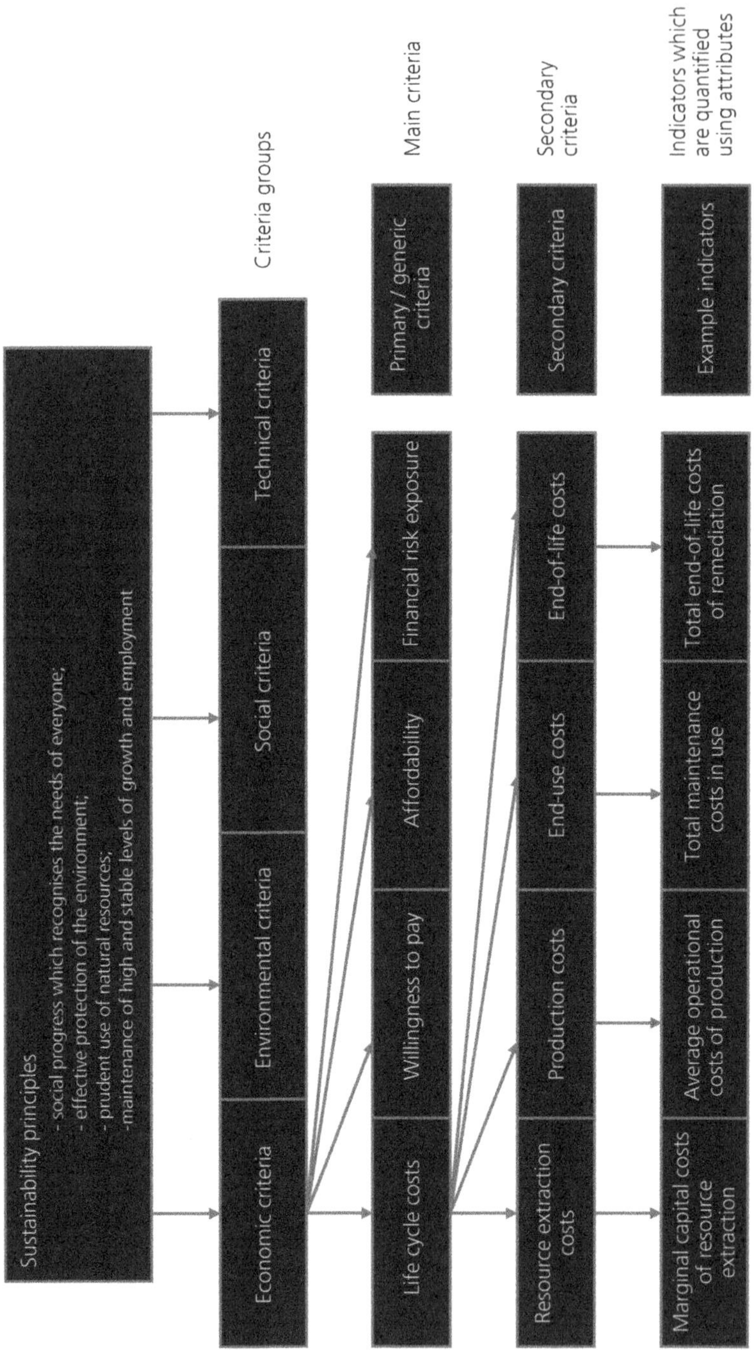

Figure 12.13 POCIA cascade (SWARD) with example indicators

Figure 12.14 Integrated sustainability assessment

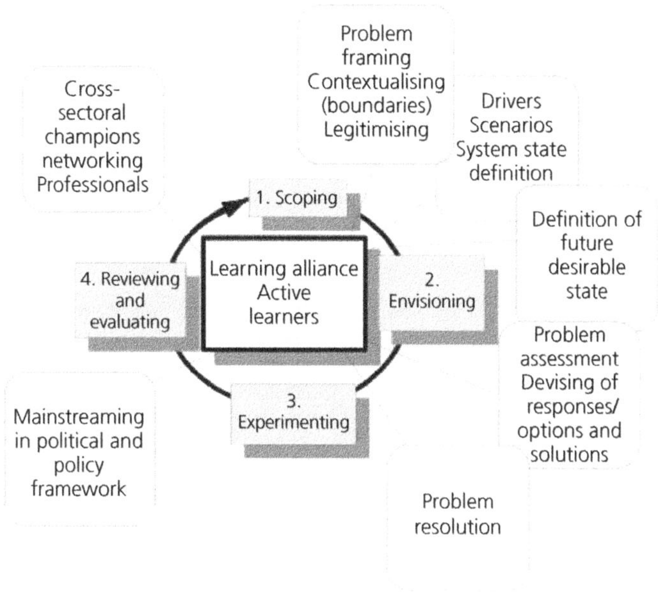

There are numerous publications dealing with SA. Many deal with only one of the three pillars of sustainability, such as 'environmental sustainability'. Although well intentioned, such analyses miss the importance of considering all three of the pillars in any assessment, which is essential for a full understanding when making decisions. For example, when considering the alignment between SA and CE in assessment processes being utilised within businesses to help deliver the SDGs, Walker *et al.* (2021) found that while sustainability indicators were being used, very few were related to the societal aspects of sustainability. There were also concerns about the lack of standardised frameworks to assist the businesses in these assessments; many frameworks were available, from academic and other development work, but they were perceived to be too complex to use. The concerns about the complexity in the use of SA frameworks were also found in municipalities undertaking urban drainage projects in the UK WaND project (CIRIA, 2010) and EU project MARE (Hurley *et al.*, 2008), for which simplified SA frameworks had to be developed.

Whichever approach is used, all methods utilise some form of indicator assessment (Figure 12.13). Some use all three of the criteria groups: environmental, economic and social, whereas others focus more on one or other of these or include additional criteria groups. In Figure 12.13, which has been developed for assessing the relative sustainability of water industry assets, a fourth group has been added to ensure that the technical aspects are included.

The SWARD framework has been used in the UK and other countries for SA of water industry assets (Ashley *et al.*, 2003, 2004, 2005). Table 12.6 illustrates some of the criteria and indicators used in SWARD and other assessment frameworks and provides information on how to obtain the relevant data.

Table 12.6 Examples of criteria and indicators used in sustainability assessments and sources of data

Criteria group	Primary criteria	Secondary criteria	Indicators	Attributes	Potential data sources
Economic	Life cycle costs	Resource extraction costs	Marginal cost of resource extraction	Unit cost of extraction	Minerals and mining sources (specifics), e.g. concrete manufacturers' data (e.g. Sagrelius *et al.*, 2023)
Environmental	Impact	Greenhouse gas emissions	CO_2 caused by scheme	Tonnes of CO_2	Life cycle analysis (e.g. Sagrelius *et al.*, 2020). EIA or footprints, (e.g. Vanham *et al.*, 2019)
Social	Stakeholder views	Acceptability to stakeholders	Public acceptability	Numbers supporting scheme	Public meetings, surveys, questionnaires, etc. (Chapter 4)
Technical	Performance	Expected outcome	Reduced CSO spills	Numbers of spills	Modelling (Chapter 11)

CO_2, carbon dioxide; CSO, combined sewer overflow; EIA, environmental impact assessment.

When assessing sustainability, it is important to ensure that all of the criteria groups are used for every option considered and that consistent primary criteria are used. Every scheme will require specific secondary criteria, but when undertaking SA for water industry assets the primary criteria defined for the SWARD project, as shown in Table 12.7, or their equivalent, should be used as these have been tested with various water service providers around the world. These have been found to be inclusive and usable in industrialised and industrialising countries.

An example of the use of the SWARD framework for the SA of the disposal of sanitary waste is provided in Chapter 14.

Assessing the sustainability of drainage options is a complex and multifaceted issue. Sustainable development is the laudable goal, but not at any cost. There is a need always to make a business case for doing things more sustainably, not least for water in the urban system. This of course skews the triple bottom line towards the economic leg. Many innovations, however, lack track record of performance in application and have originated from niche applications where the business case has often not been important in demonstrating the effectiveness of the new processes, hence business cases may be difficult to formulate (Chapter 8).

SA is a relative process in that it can be used only to determine the relative sustainability between options under consideration. Moving from the determination of the attributes to the primary criteria in Table 12.6 is complex. The quantified information will be non-commensurate in that some of the criteria, such as costs and benefits, will be in monetary units, some will be in basic form, such as numbers of sanitary

Table 12.7 Primary criteria used in sustainability assessment for water industry assets (Ashley *et al.*, 2004)

Criteria group	Primary criteria
Economic	■ Life cycle costs ■ Willingness to pay ■ Affordability ■ Financial risk exposure
Environmental	■ Resource utilisation ■ Service provision ■ Environmental impact
Social	■ Impact on risks to human health ■ Acceptability to stakeholders ■ Participation and responsibility ■ Public awareness and understanding ■ Social inclusion
Technical	■ Performance of the system ■ Reliability ■ Durability ■ Flexibility and adaptability

items prevented from overflowing, and other attributes will be qualitative, perhaps expressed in terms of preferences. Aggregating the data, even into the required information for even individual primary criteria, may also not be straightforward. From this, closing on a decision as to which option may be the more sustainable will not be easy. For some decision-makers, there is a preference for taking a subjective view by considering the entirety of the criteria and indicator results, to then decide which is their preference. This approach has the value that the information used and decision process are transparent. An alternative that is being used more often is a formalised multicriteria decision analysis (MCDA) process (e.g. Brans and De Smet, 2016; Kheybari *et al.*, 2020). The scope of the MCDA may extend beyond the SA and include other systems and sectors. For example, Zhang *et al.* (2019) consider the food–energy–water nexus for urban sustainability, for which a framework for conducting assessments is presented.

An alternative to SA is the 'reflexive' process (e.g. Dijk *et al.*, 2017), seen as a way of moving away from the need to define sustainability as some end-state, but rather to view it as the capacity of society to learn about the conditions of its future existence and wants. For this, the analytical framework considers the interlinkage of problems and scales, as well as the long-term and indirect effects (often unexpected) of possible actions. This may require new types of institution, policies and approaches to uncertainty, with ambivalence about the multiplicity of goals and the distribution of power. There are many similarities here with the 'reflection in practice' of the professional as described by Schon (2009), in which design and development processes evolve through a process of iterative reflection as the design progresses.

12.6.3 Delivering and assessing water sensitivity

As the world-leading framework for moving to more sustainable urban water systems, WSUD is being implemented in many countries worldwide. As both process and practice, WSUD includes many of the elements discussed in this chapter. WSUD is now seen within the context of the water-sensitive city

(WSC), one which contributes to an urban area's development towards more sustainable urban water (and other system) futures by the application of WSUD measures to address ongoing challenges, appropriately integrated into urban planning processes, while simultaneously providing many co-benefits, as illustrated in Figure 12.15 (e.g. Alves *et al.*, 2019; Sharma *et al.*, 2019; Wong *et al.*, 2020).

The primary source of supporting information for WSCs is the CRC for Water Sensitive Cities, in Australia (CRCWSC, 2023b). The SWITCH project utilised water sensitivity for applications around the globe (Howe and Mitchell, 2012) showing the widespread applicability. There are now numerous sources of guidance for WSUD, especially in Australian conurbations (e.g. Melbourne Water, 2023). Application in the UK (CIRIA, 2013) has also been reviewed and recommendations made. However, UK practice prefers to focus on only the components of WSUD, especially SuDS, rather than the entirety of the opportunities available from taking a WSUD approach, which could support the delivery of IWM.

Australian lessons from delivering WSUD and WSCs clarify the challenges, opportunities and processes involved. Starting with existing assets for water and wastewater systems, there was a need to move towards more sustainable practice. This transition, required in all aspects of WSC functions, has required targeted programmes of education and training as well as representations at the highest levels of government to influence policymaking and focus resources (Brown and Clarke, 2007; Wong, 2006). Now established as the 'common-sense' approach, WSUD and WSCs are deemed de rigeur in Australian urban planning (Wong *et al.*, 2020). Tools are available for assessing the state of implementation and requirement to move to greater water sensitivity for individual urban areas, such as the Water Sensitive Cities Index (Rogers *et al.*, 2020). Table 12.8 illustrates some of the criteria used in the assessment.

The full set of indices (examples in Table 12.8) has been developed after application to 50 urban areas in Australia. The framework can be used to determine where there are gaps in application and to help to target changes to the way in which the WSC ideas are being implemented locally.

In advocating a strategic approach to flood risk management, Sayers (2023) recommends a 'blending' of natural and built infrastructures as a means of addressing the practical, policy and implementation challenges in order to progress to a systems-based approach. In line with many other recommendations, there are four key characteristics of this strategic approach.

- The adoption of a dynamic whole-system long-term perspective.
- The promotion of built and natural infrastructure.
- The acceptance of a trade-off driven decision process.
- The use of a continuous, collaborative and inclusive development process.

This is thus a continuous and dynamic process, rather than the traditional 'build and walk away'. In defining strategic flood risk management, Sayers (2023) provides the means to formulate a new definition for strategic urban drainage that could be seen as a manifesto for this book.

Strategic urban drainage delivers multiple outcomes for people and nature over the short and long terms. The approach is inherently collaborative, inclusive and systems-based, with choices based on an understanding of risks, benefits and uncertainties. A continuous process of review and adaptation actively manages change, incorporates new knowledge and modifies plans, assets and actions.

Figure 12.15 The relationship between water-sensitive urban design (WSUD), water-sensitive city (WSC) and stormwater management in the urban water cycle (adapted from Avazpour, 2022)

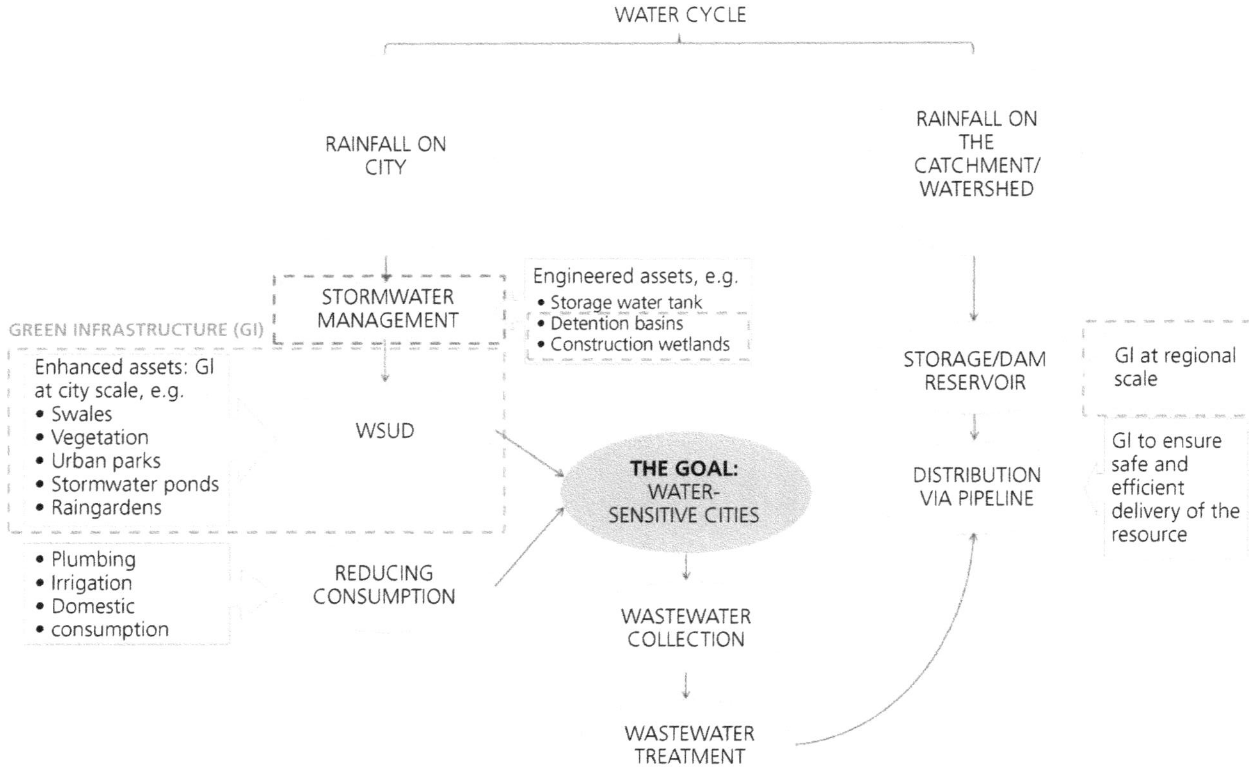

Table 12.8 Water Sensitive Cities (WSC) Index goals, indicators and strategic objectives (adapted from Rogers *et al.*, 2020)

Key goals	Examples include	Need
1: Ensure good water-sensitive governance	1.1 Knowledge, skills and organisational capacity	Strengthen the capabilities of individuals and organisations to adopt water-sensitive practices
	1.2 Water is key element in city planning and design	Improve urban planning and design frameworks and processes to drive the implementation of water-sensitive solutions through urban development
2: Increase community capital	2.3 Shared ownership, management and responsibility	Empower community to be an active participant in creating, operating and maintaining parts of the water system
	2.4 Community preparedness and response to extreme events	Support citizens to cope with and recover from impacts associated with storms, floods, drought and heatwaves
3: Achieve equity of essential services	3.2 Equitable access to safe and reliable sanitation	Provide safe, reliable and affordable sanitation services that meet defined standards for sanitation
	3.3 Equitable access to flood protection	Manage flood risk in a way that is affordable, including reducing nuisance flooding and protecting citizens and infrastructure
4: Improve productivity and resource efficiency	4.2 Low GHG emission in water sector	Maximise the use of alternatives to high carbon emitting energy sources in water system infrastructure
	4.5 Broad community benefits from water services	Stimulate beneficial outcomes of water-related services for other sectors beyond water
5: Improve ecological health	5.1 Healthy and biodiverse habitat	Design water systems to help protect, restore and create well-functioning ecosystems that contribute to ecological resilience
	5.2 Surface water quality and flows	Improve and protect the quality of surface waters and marine environments
6: Ensure quality urban space	6.1 Activating connected green-blue space	Plan and design the urban form to create many distributed, connected and well-maintained green spaces and waterways
	6.2 Urban elements functioning as part of the urban water system	Plan and design the urban form to function as an integral part of the water system
7: Promote adaptive infrastructure	7.2 Multifunctional water infrastructure	Provide multifunctional water infrastructure that seamlessly integrates into the urban landscape
	7.5 Infrastructure and ownership at multiple scales	Optimise water system performance through the integration of centralised and decentralised infrastructure

GHG, greenhouse gas.

12.7. Summary

This chapter has presented several frameworks that are being used to plan and implement better urban water management and hence urban drainage systems. Frames are specific and developed for particular perspectives.

Having been the primary driver around the turn of the millennium, sustainability is a term and a concept that is now more frequently abused than used or applied meaningfully. This chapter has shown how, from ambitious beginnings, there has been a gradual corruption in the meaning of the term, so that now at best schemes need only to be resilient and justified simply by their costs and benefits. Sustainability has become consigned to being tokenistic in many policies, regulations, reports, guidance and practice documents. Nevertheless, there have been considerable efforts to understand how best to provide infrastructure and, within this, urban drainage systems, that are as sustainable as possible. There is no consensus about what this means, although there is an understanding that the journey to sustainability is the most important consideration, even if the destination is rather unclear. In this journey it is important to define the boundaries. Spatial boundaries often include 'the development', despite supply and waste chains being important, whereas temporal boundaries may be limited to, for example, an economic horizon of, say, 25 years to make a discounted present value assessment. With rapidly changing drivers, it is essential that organisations, systems and assets are agile so that these can be adapted as necessary and in a timely way. Some form of structured futures planning is needed for this, including scenario planning for testing the robustness and flexibility of selected options.

Because of the many difficulties of even a simple sustainability assessment, intermediate concepts are being used. These include assessing the resilience of aspects of a development and, in some cases, the assessment of the contribution to liveability this makes for the here and now quality of urban environments. Importantly for decision-makers, the focus has switched almost exclusively to the financial value of the benefits of a scheme. But engineers and others need to question what value means in this context. Whose value? There are hopeful signs that alternative frames for 'value' are being used that are breaking away from strict monetisation (e.g. EWSC, 2023).

Acting on information about sustainability in whatever format, including the added benefits a scheme may provide, means that prevailing urban drainage practices may need to change. This usually happens only in response to a crisis or catastrophe or when there are clear indicators that the current way of doing things is no longer able to deliver what is required. For example, the supplanting of the demand management and reuse movement for water supplies in Australia by desalination plants illustrates how political expediency and 'magic bullet' technology (desalination) can override sustainability considerations and the need for a shift in behaviour and expectations by all concerned. This is despite more than a decade of development of the water-sensitive alternatives that are less costly and more sustainable. Changing how we manage urban water is difficult. The prevailing practices, inertia of having existing assets that operate in a way that we may wish to change, political inertia, protective institutions and even professionals, together with vested interests all act as a huge resistance to change.

Water and drainage are one part of the vital systems and services essential to human existence in urban areas and are inseparable from the many systems and services that make up the overall urban system. The realisation that water cannot be managed in isolation from all the other infrastructure systems and services provided to support urban living has also been a breakthrough in how we are now thinking about the best ways to manage urban water. There is now general understanding of the need to take a cross-disciplinary and cross-sectoral approach, to engage all the actors in the process of devising the most sustainable systems, and demonstrate what can be done and how it might be done.

REFERENCES

Abbott J, Davies P, Simkins P *et al.* (2013) *Creating Water Sensitive Places – Scoping the Potential for Water Sensitive Urban Design in the UK.* CIRIA C724F. Construction Industry Research and Information Association, London, UK.

Almeida M, Vieira P and Smeets P (2013) *Water cycle safety plan framework.* EU PREPARED project. See https://www.researchgate.net/publication/260450498_Water_cycle_safety_plan_framework_Final (accessed 08/08/2023).

Alves A, Gersonius B, Kapelan Z, Vojinovic Z and Sanchez A (2019) Assessing the co-benefits of green-blue-grey infrastructure for sustainable urban flood risk management. *Journal of Environmental Management* **239**: 244–254, 10.1016/J.JENVMAN.2019.03.036.

Alves Beloqui AI (2020) *Combining Green-Blue-Grey Infrastructure for Flood Mitigation and Enhancement of Co-benefits.* PhD thesis, Delft University of Technology, the Netherlands. See https://pure.tudelft.nl/ws/portalfiles/portal/68662526/2020_IHE_PHD_THESIS_ALVES_BELOQUI_i.pdf.

Anon (2016) Oded Distel, Head of New Tech Program, Ministry of Economy, and Director of Foreign Investments and Eco Systems. *The Source.* 18 July. https://www.thesourcemagazine.org/oded-distel/ (accessed 08/08/2023).

Arena N (2019) Life-cycle assessment applied to construction of Thames Tideway east tunnel, London, UK. *Proceedings of the Institution of Civil Engineers – Engineering Sustainability* **172(8)**: 416–423, 10.1680/jensu.18.00012.

Arup, Antea Group and Ellen MacArthur Foundation (2019) *Water and Circular Economy: A White Paper.* The UK Water Partnership. See https://www.arup.com/-/media/arup/files/publications/w/water_and_circular_economy_whitepaper.pdf (accessed 10/08/2023).

Ashagre BB, Fu G and Butler D (2020) Automation and real time control of urban wastewater systems: a review of the move towards sustainability. *Journal of Water Supply: Research and Technology-AQUA* **69(8)**: 751–768, 10.2166/aqua.2020.074.

Ashley RM (2016) Water in the urban system. In *Sustainable Water* (Ainger C and Fenner R (eds)). ICE Publishing, London, UK. pp. 51–76.

Ashley RM (2020) Changes in the way we live and value urban spaces. *Journal of Delta Urbanism* **1(1)**: 1–22, 10.7480/jdu.1.2020.5456.

Ashley RM and Horton B (2022) Characterizing financing needs and financing capacities in different regions: a global perspective on water-related financing flows and drivers for investment needs. In *Financing Investment in Water Security – Recent Developments and Perspectives* (Leflaive X, Dominique K and Alaerts GJ (eds)). Elsevier, Amsterdam, the Netherlands.

Ashley RM, Blackwood DJ, Butler D *et al.* (2003) Sustainable decision making for the UK water industry. *Proceedings of the Institution of Civil Engineers – Engineering Sustainability* **156(ES1)**: 41–49, 10.1680/ENSU.2003.156.1.41.

Ashley RM, Blackwood DJ, Butler D and Jowitt P (eds) (2004) *Sustainable Water Services – A Procedural Guide.* International Water Association, London, UK.

Ashley RM, Blackwood DJ, Souter N *et al.* (2005) Sustainable disposal of domestic sanitary waste. *ASCE Journal of Environmental Engineering* **131(2)**: 206–215.

Ashley RM, Butler D, Hurley L and Memon F (2008) Criteria and indicators in delivering sustainable water systems: from Sustainable Water Asset Resource Decisions (SWARD) to Water cycle management for New Developments (WaND). In *DayWater: An Adaptive Decision Support System for Urban Stormwater Management* (Thevenot D (ed.)). International Water Association, London, UK. pp. 275–288.

Ashley RM, Lundy L, Ward S *et al.* (2013) Water-sensitive urban design: opportunities for the UK. *Proceedings of the Institution of Civil Engineers – Municipal Engineer* **166(ME2)**: 65–76, 10.1680/muen.12.00046.

Ashley RM, Horton B, Walker L *et al.* (2023) The benefits of nature-based systems in a changing and uncertain world. *Proceedings of the Institution of Civil Engineers – Engineering Sustainability* **176(2)**: 61–71, 10.1680/jensu.21.00102.

Avazpour B (2022) *From Water Sensitive Urban Design to Water Sensitive Cities: Challenges and Enablers*. PhD thesis, University of New South Wales, Sydney, Australia. See https://unsworks.unsw.edu.au/entities/publication/ce08e450-f9e0-4778-9ce3-f79d261cf049/full.

Balmforth D (2016) Drainage and flood resilience. In *Sustainable Water* (Ainger C and Fenner R (eds)). ICE Publishing, London, UK. pp. 147–172.

Bell S, Johnson C, Austen K, Moore G and Teh T-S (2021) *Co-designing Infrastructures – Community Collaboration for Liveable Cities*. UCL Press, London, UK, 10.14324/111.9781800082229.

Bide P and Coleman A (2019) *Delivering Better Water Management through the Planning System*. CIRIA C787F. Construction Industry Research and Information Association, London, UK.

Brans J-P and De Smet Y (2016) PROMETHEE methods. In *Multiple Criteria Decision Analysis* (Greco S, Ehrgott M and Figueira JR (eds)). Springer, New York, NY, USA, pp. 187–219. See https://link.springer.com/book/10.1007/978-1-4939-3094-4.

Brinsmead T and Hooker C (2005) *Sustainabilities: A Systematic Framework and Comparative Analysis*. Cooperative Research Centre for Coal in Sustainable Development. Research Report 53. University of Newcastle, Queensland, Australia. See https://mafiadoc.com/sustainabilities-a-systematic-framework-and-_5b2ee8db097c47f0488b45e4.html (accessed 08/08/2023).

Brisley R, Wylde R, Lamb R *et al.* (2016) Techniques for valuing adaptive capacity in flood risk management. *Proceedings of the Institution of Civil Engineers – Water Management* **169(2)**: 75–84, 10.1680/itfr.63938.063.

Brown R (2003) *Institutionalisation of Integrated Urban Stormwater Management: Multiple-Case Analysis of Local Management Reforms Across Metropolitan Sydney*. PhD Thesis, University of New South Wales, Sydney, Australia. See https://unsworks.unsw.edu.au/entities/publication/04be1adb-04ad-44f4-8275-3d9e45836485/full (accessed 08/12/2023).

Brown R and Clarke J (2007) *Transition to Water Sensitive Urban Design*. Report 07/1. Faculty for Advancing Water Biofiltration, Monash University, Melbourne, Australia.

Brown R and Keath N (2008) Transitioning to water sensitive cities: historical, current and future transition states. In *11th International Conference on Urban Drainage*, Edinburgh, Scotland, UK, pp. 1–10.

Brundtland G (1987) *Report of the World Commission on Environment and Development: Our Common Future*. United Nations, New York, USA. See http://www.un-documents.net/our-common-future.pdf (accessed 10/08/2023).

BSI (British Standards Institution) (2014) BS ISO 55000:2014. Asset management – Overview, principles and terminology. BSI, London, UK.

BSI (2017) BS 8001:2017. Framework for implementing the principles of the circular economy in organizations. Guide. BSI, London, UK.

Caffoor I (2009) *Integrated Urban Water Management*. Priority Technology Area 6, Part 1. Knowledge Transfer Network, UK. N8 Water Group and UKWIR, UK.

Cettner A, Ashley R, Hedstrom A and Viklander M (2014) Sustainable development and urban stormwater practice. *Urban Water Journal* **11(3)**: 185–197, 10.1080/1573062X.2013.768683.

CIRIA (Construction Industry Research and Information Association) (2010) *WaND – Guidance on Water Cycle Management for New Developments*. CIRIA C690. CIRIA, London, UK.

CIRIA (2013) *Water Sensitive Urban Design in the UK – Ideas for Built Environment Practitioners*. CIRIA C723F. CIRIA, London, UK.

CIWEM (Chartered Institution of Water and Environmental Management) (2011) *Integrated Water Management – A CIWEM Briefing*. CIWEM, London, UK. See https://www.ciwem.org/assets/pdf/Policy/Reports/Integrated-Water-Managment-Report.pdf (accessed 08/12/2023).

Clement V, Kumari RK, De Sherbinin A *et al.* (2021) *Groundswell Part 2: Acting on Internal Climate Migration*. World Bank, Washington, DC, USA. http://hdl.handle.net/10986/36248.

Construction Leadership Council (2023) PAS 2080:2023. Carbon management in buildings and infrastructure. BSI, London, UK.

CRCWSC (Cooperative Research Centre for Water Sensitive Cities) (2023a) *Frameworks for sustainable urban water management*. CRCWSC. See https://watersensitivecities.org.au/wp-content/uploads/2022/08/Frameworks-for-sustainable-UWM.pdf (accessed 10/08/2023).

CRCWSC (2023b) Realising the vision of a water sensitive city. https://watersensitivecities.org.au/ (accessed 08/08/2023).

Delgado A, Rodriguez DJ, Amadei CA and Makino M (2021) *Water in Circular Economy and Resilience (WICER)*. World Bank, Washington, DC, USA. https://www.worldbank.org/en/topic/water/publication/wicer.

de Neufville R and Scholtes S (2011) *Flexibility in Engineering Design*. MIT Press, Cambridge, MA, USA.

Digman CJ, Balmforth DJ, Shaffer P and Butler D (2006) The challenge of delivering integrated urban drainage. In *Proceedings of the WaPUG Autumn Conference*, Blackpool, UK. See http://ftp2.ciwem.org/2006/2006-11%204_Autumn%20Blackpool%20November%202006/Paper%202%20-%20DEFRA%20Integrated%20Urban%20Drainage%20Pilots%20Scoping%20Study%20Chris.pdf (accessed 9/12/23).

Dijk M, De Kraker J, Van Zeijl-Rozema A *et al.* (2017) Sustainability assessment as problem structuring: three typical ways. *Sustainability Science* **12**: 305–317, 10.1007/s11625-016-0417-x.

DLUHC (Department for Levelling Up, Housing and Communities) (2020) Strategic Environmental Assessment and Sustainability Appraisal. https://www.gov.uk/guidance/strategic-environmental-assessment-and-sustainability-appraisal (accessed 10/08/2023).

DLUHC (2023) *National Planning Policy Framework*. DLUHC, London, UK.

Dunion K (2011) Decision 166/2011 Unison and Scottish Water. PFI/PPP Wastewater contracts. Reference No. 200701447. Decision Date: 15 August 2011. See https://www.itspublicknowledge.info/sites/default/files/Decision166-2011.pdf (accessed 10/08/2023).

EA (Environment Agency) (2009) *Research Framework – Integrated Urban Drainage*. Project: SC070064/R1. Environment Agency, Bristol, UK. https://assets.publishing.service.gov.uk/government/uploads/system/uploads/attachment_data/file/291184/scho1009brce-e-e.pdf.

EA (2012) Thames Estuary 2100 (TE2100). https://www.gov.uk/government/publications/thames-estuary-2100-te2100/thames-estuary-2100-te2100 (accessed 08/08/2023).

Eckart J, Sieker H, Valravamoorthy K and Alsharif K (2012) Flexible design of urban drainage systems: demand led research for Hamburg-Wilhelmsburg. *Reviews in Environmental Science and Bio/Technology* **11**: 5–10, 10.1007/s11157-011-9256-5.

EMF (Ellen MacArthur Foundation) (2023) *Towards a Circular Economy: Business Rationale for an Accelerated Transition*. Ellen MacArthur Foundation. See https://www.ellenmacarthurfoundation.org/towards-a-circular-economy-business-rationale-for-an-accelerated-transition (accessed 07/12/2023).

Erickson JD (2022) *The Progress Illusion – Reclaiming our Future from the Fairy Tale of Economics*. Island Press, Washington, DC, USA.

EWSC (Enabling Water Smart Communities) (2023) *Enabling Water Smart Communities*. https://ewsc.org.uk/ (accessed 08/08/2023).

FAIR (2020) *Adaptive Asset Management for Flood Protection*. FAIR project end report. EU Interreg North Sea Region. See https://northsearegion.eu/media/13662/fair_end_report-03-06-2020.pdf (accessed 08/08/2023).

Fenner R and Ainger C (2020) A review of sustainability in civil engineering: why much more commitment is needed. *Proceedings of the Institution of Civil Engineers – Civil Engineering* **173(2)**: 69–77, 10.1680/jcien.19.00036.

Fenner R, Sykes J and Ainger C (2023a) *Sustainable Infrastructure: Principles into Practice*, 2nd edn. ICE Publishing, London, UK.

Fenner R, Sykes J and Ainger C (2023b) Key principles for sustainable infrastructure. In *Sustainable Infrastructure: Principles into Practice*, 2nd edn. ICE Publishing, London, UK. pp. 45–73.

Flores CC, Müller APR, Dolman N and Özerol G (2023) Assessing the leapfrogging potential to water sensitive: the Dutch case of Zwolle. *Journal of Water and Climate Change* **14(5)**: 1638, 10.2166/wcc.2023.493.

Friese K, Schmidt G, De Lena JC, Nalini HA and Zachmann DW (2010) Anthropogenic influence on the degradation of an urban lake – the Pampulha reservoir in Belo Horizonte, Minas Gerais, Brazil. *Limnologica* **40**: 114–125, 10.1016/j.limno.2009.12.001.

Gersonius B, Ashley RM, Pathirana A and Zevenbergen C (2012) Adaptation of flood risk infrastructure to climate resilience. *Proceedings of the Institution of Civil Engineers – Civil Engineering* **165(CE6)**: 3–4.

Gersonius B, Ashley RM, Pathirana A and Zevenbergen C (2013) Climate change uncertainty: building flexibility into water and flood risk infrastructure. *Climatic Change* **116**: 411–423, 10.1007/s10584-012-0494-5.

Gersonius B, Ashley R, Jeuken, Pathirana A and Zevenbergen C (2015) Accounting for uncertainty and flexibility in flood risk management: comparing Real-In-Options optimisation and Adaptation Tipping Points. *Journal of Flood Risk Management* **8**: 135–144, 10.1111/jfr3.12083.

Giampietro M (2023) Reflections on the popularity of the circular bioeconomy concept: the ontological crisis of sustainability science. *Sustainability Science* **18**: 749–754, 10.1007/s11625-022-01267-z.

Giampietro M and Funtowicz SO (2020) From elite folk science to the policy legend of the circular economy. *Environmental Science and Policy* **109**: 64–72, 10.1016/j.envsci.2020.04.012.

Ginn WJ (2020) *Valuing Nature*. Island Press, Washington, DC, USA.

Glasson J and Therivel R (2012) *Introduction to Environmental Impact Assessment*, 4th edn. Routledge, London, UK.

Grison C, Koop S, Eisenreich S *et al.* (2023) Integrated water resources management in cities in the world: global challenges. *Water Resources Management* **37**: 2787–2803, 10.1007/s11269-023-03475-3.

Groves DG, Bloom E, Lempert RJ *et al.* (2015) Developing key indicators for adaptive water planning. *ASCE Journal of Water Resources Planning and Management* **141(7)**: 10.1061/(ASCE)WR.1943-5452.0000471.

Gulabchand A (2013) *Make Every Drop Count*. Water Resources Group. World Bank, Washington, DC, USA. See https://2030wrg.org/make-every-drop-count/ (accessed 08/08/2023).

GWP (Global Water Partnership) (2012) *Integrated Urban Water Management*. TEC Background Papers No. 16. GWP, Stockholm, Sweden. See https://www.gwp.org/globalassets/global/toolbox/publications/background-papers/16-integrated-urban-water-management-2012.pdf (accessed 10/08/2023).

Howe C and Mitchell C (eds) (2012) *Water Sensitive Cities*. International Water Association, London, UK.

Hoyer J (ed.) (2011) *Water Sensitive Urban Design: Principles and Inspiration for Sustainable Stormwater Management in the City of the Future*. Jovis, Hamburg, Germany.

Hunt DVL, Lombardi DR, Atkinson S *et al.* (2012) Scenario archetypes: converging rather than diverging themes. *Sustainability* **4**: 740–772, 10.3390/su4040740.

Hurley AL, Ashley R and Mounce S (2008) Addressing practical problems in sustainability assessment frameworks. *Proceedings of the Institution of Civil Engineers – Engineering Sustainability* **161(ES1)**: 23–30, 10.1680/ensu.2008.161.1.23.

ISO (International Organization for Standardization) (2023) ISO/TC 251. Asset Management and the Circular Economy. Committee. ISO, Geneva, Switzerland. See https://committee.iso.org/home/tc251.

Johnson A (2016) Wastewater management. In *Sustainable Water* (Ainger C and Fenner R (eds)). ICE Publishing, London, UK. pp. 123–146.

Kheybari H, Rezale FM and Farazmand H (2020) Analytic network process: an overview of applications. *Applied Mathematics and Computation* **367(15)**: 124780, 10.1016/j.amc.2019.124780.

Kirchherr J (2022) Bullshit in the sustainability and transitions literature: a provocation. *Circular Economy and Sustainability* **3**: 167–172, 10.1007/s43615-022-00175-9.

Krivtsov V, Arthur S, Allen D and O'Donnell E (eds) (2019) *Perspectives on Blue-Green Infrastructure*. CIRIA C780F. Construction Industry Research and Information Association, London, UK.

Lempert RJ, Popper SW and Bankes SC (2003) *Shaping the Next One Hundred Years: New Methods for Quantitative, Long-Term Policy Analysis*. RAND Corporation, Santa Monica, CA, USA. See https://www.rand.org/pubs/monograph_reports/MR1626.html (accessed 08/12/2023).

Lieberherr E and Truffer B (2015) The impact of privatization on sustainability transitions: a comparative analysis of dynamic capabilities in three water utilities. *Environmental Innovation and Societal Transitions* **15**: 101–122, 10.1016/j.eist.2013.12.002.

Loftus A and March H (2019) Integration for whom? Financialisation and the Thames tideway tunnel. *Urban Studies* **56**: 2280–2296, 10.1177/0042098017736713.

Loucks DP and Gladwell JS (eds) (1999) *Sustainability Criteria for Water Resource Systems*. Cambridge University Press, Cambridge, UK.

Lundin M, Bengtson M and Molander S (2000) Life cycle assessment of wastewater systems: influence of system boundaries and scale on calculated environmental loads. *Environmental Science and Technology* **34**: 180–186, 10.1021/es990003f.

Marlow DR, Moglia M, Cook S and Beale DJ (2013) Towards sustainable urban water management: a critical reassessment. *Water Research* **47(20)**: 7150–7161, 10.1016/j.watres.2013.07.046.

Melbourne Water (2023) Introduction to WSUD. https://www.melbournewater.com.au/building-and-works/stormwater-management/introduction-wsud (accessed 10/08/2023).

Mitchell VG (2006) Applying integrated urban water management concepts: a review of Australian experience. *Environmental Management* **37**: 589–605, 10.1007/s00267-004-0252-1.

Morseletto P, Mooren CE and Munaretto S (2022) Circular economy of water: definition, strategies and challenges. *Circular Economy and Sustainability* **2**: 1463–1477, 10.1007/s43615-022-00165-x.

Neumann MB, Rieckermann J, Hug T and Gujer W (2015) Adaptation in hindsight: dynamics and drivers shaping urban wastewater systems. *Journal of Environmental Management* **151**: 404–415, 10.1016/j.jenvman.2014.12.047.

OECD (Organisation for Economic Co-operation and Development) (2013) *Water Security for Better Lives*. OECD Studies on Water. OECD Publishing, Paris, France. See https://www.oecd.org/env/resources/Water%20Security%20for%20Better%20Lives-%20brochure.pdf (accessed 08/08/2023).

Palmer R (2000) From the inside out. *Environmental Values* **9**: 411–418, 10.3197/096327100129342128.

Park W (2019) The country disappearing under rising tides. *BBC Future*. 2 Sept. https://www.bbc.com/future/article/20190829-bangladesh-the-country-disappearing-under-rising-tides (accessed 08/08/2023).

PBL (2023) *The Geography of Future Water Challenges – Bending the Trend*. PBL Netherlands Environmental Assessment Agency, The Hague, the Netherlands. See https://www.pbl.nl/en/publications/geography-of-future-water-challenges (accessed 08/08/2023).

Pötz H and Bleuzé P (2012) *Urban Blue-Green Grids: Manual for Resilient Cities*. Atelier Groenblauw, Delft, the Netherlands. See (for 2022 edition) https://www.naibooksellers.nl/green-blue-grids-manual-for-resilient-cities-groenblauwe-netwerken-handleiding-voor-veerkrachtige-steden.html?___store=english&___from_store=default (accessed 21/12/2023).

Poustie M (2014) *Enabling Socio-Technical Transitions to Sustainable Urban Water Management in the South West Pacific Region*. PhD thesis. Monash University, Melbourne, Australia. https://bridges.monash.edu/articles/thesis/Enabling_socio-technical_transitions_to_sustainable_urban_water_management_in_the_South_West_Pacific_Region/4670059.

Rashetnia S, Sharma AK, Ladson AR, Browne D and Yaghoubi E (2022) A scoping review on water sensitive urban design aims and achievements. *Urban Water Journal* **19(5)**: 453–467, 10.1080/1573062X.2022.2044494.

Rentachintala LRNP, Gangireddy MG and Mohapatra PK (2022) Urban stormwater management for sustainable and resilient measures and practices: a review. *Water Science & Technology* **85(4)**: 1120, 10.2166/wst.2022.017.

Rogers B, Dunn G, Hammer K *et al.* (2020) Water Sensitive Cities Index: a diagnostic tool to assess water sensitivity and guide management actions. *Water Research* **186**: 116411 10.1016/j.watres.2020.116411.

Sagrelius PO, Blecken G, Hedström A, Ashley R and Viklander M (2020) Environmental impacts of stormwater bioretention systems with various design and construction components. *Journal of Cleaner Production* **359**: 132091, 10.1016/j.jclepro.2022.132091.

Sagrelius PO, Blecken G, Hedström A, Ashley R and Viklander M (2023) Sustainability performance of bioretention systems with various designs. *Journal of Environmental Management* **340**: 117949, 10.1016/j.jenvman.2023.117949.

Sayers PB (2023) *Strategic Flood Risk Management: A System-Based Approach*. PhD Thesis, Technical University of Delft, Delft, the Netherlands. https://research.tudelft.nl/en/publications/strategic-flood-risk-management-a-system-based-approach.

Siegel SM (2017) *Let There Be Water*. Thomas Dunne Books, New York, NY, USA.

Sharma AK, Gardner T and Begbie D (eds) (2019) *Approaches to Water Sensitive Urban Design*. Elsevier, Amsterdam, the Netherlands. See https://www.sciencedirect.com/book/9780128128435/approaches-to-water-sensitive-urban-design#book-description (accessed 08/12/2023).

Scheffer M (2009) *Critical Transitions in Nature and Society*. Princeton University Press, Woodstock, UK.

Schon DA (2009) *The Reflective Practitioner*. Ashgate, London, UK.

Scottish Government (2023) *Water*. https://www.gov.scot/policies/water/hydro-nation/ (accessed 08/08/2023).

Sen A (2009) *The Idea of Justice*. Belknap Press, Harvard University, Cambridge, MA, USA.

Smol M, Prasad MNV and Stefanakis AI (eds) (2023) *Water in Circular Economy*. Springer, Cham, Switzerland.

Stovin VR, Moore SL, Wall M and Ashley RM (2013) The potential to retrofit sustainable drainage systems to address combined sewer overflow discharges in the Thames Tideway catchment. *Water and Environment Journal* **27**: 216–228, 10.1111/j.1747-6593.2012.00353.x.

Stride P (2014) Building for 22nd century London: Thames Tideway Tunnel. In *Presentation at Water in the Sustainable City Conference*, City of Gothenburg, Sweden.

STRN (Sustainability Transitions Research Network) (2023) https://transitionsnetwork.org/about/ (accessed 10/08/2023).

Thames Water (2013) *Thames Tideway Tunnel*. Application for Development Consent. Application Reference Number: WWO10001 Needs Report. January. Doc Ref: 8.3. Appendix E – Potential Source Control and SuDS Application APFP Regulations 2009: Regulation 5(2)(q). Thames Water Utilities Limited, London, UK.

Thompson M and Beck B (2015) *Coping with Change: Urban Resilience, Sustainability, Adaptability and Path Dependence*. Future of Cities: working paper. Foresight, Government Office for Science, London, UK. http://www.gov.uk/government/publications/future-of-cities-coping-with-change.

UN (United Nations) (2021) *UN-Water, 2020: Summary Progress Update 2021 – SDG 6 – water and sanitation for all*. Version: 1 March 2021. UN, Geneva, Switzerland. https://www.unwater.org/publications/summary-progress-update-2021-sdg-6-water-and-sanitation-all.

UN (2023) *Water Conference*. See https://sdgs.un.org/sites/default/files/2023-03/Closing%20press%20release_waterconference_FINAL_24Mar.pdf (accessed 08/08/2023).

UNEP (UN Environment Programme) (2012) *Measuring Water Use in a Green Economy*.UNEP, Paris, France. See https://www.resourcepanel.org/reports/measuring-water-use-green-economy (accessed 08/08/2023).

USEPA (US Environmental Protection Agency) (2023a) *Planning for Sustainable Water Infrastructure*. https://www.epa.gov/sustainable-water-infrastructure/planning-sustainable-water-infrastructure (accessed 10/08/2023).

USEPA (2023b) System for Urban Stormwater Treatment and Analysis IntegratioN (SUSTAIN). https://www.epa.gov/water-research/system-urban-stormwater-treatment-and-analysis-integration-sustain (accessed 10/08/2023).

Upham P (2000) Scientific consensus on sustainability: the case of the natural step. *Sustainable Development* **8**: 180–190, 10.1002/1099-1719(200011)8:4%3C180::AID-SD143%3E3.0.CO;2-R.

van der Jagt APN, Raven R, Dorst H and Runhaar H (2020) Nature-based innovation systems. *Environmental Innovation and Societal Transitions* **35**: 202–216; 203, 10.1016/j.eist.2019.09.005.

Vanham D, Leip A, Galli A *et al.* (2019) Environmental footprint family to address local to planetary sustainability and deliver on the SDGs. *Science of the Total Environment* **693(25)**: 133642, 10.1016/j.scitotenv.2019.133642.

Walker A, Opferkuch K, Lindgreen ER *et al.* (2021) Assessing the social sustainability of circular economy practices: industry perspectives from Italy and the Netherlands. *Sustainable Production and Consumption* **27**: 831–844, 10.1016/j.spc.2021.01.030.

Wang XC (2015) *Water Cycle Management – A New Paradigm of Wastewater Reuse and Safety Control.* Springer Briefs in Water Science and Technology. Springer, Cham, Switzerland.

Weaver P and Rottmans J (2006) Integrated sustainability assessment: what is it, why do it and how? *International Journal of Innovation and Sustainable Development* **1(4)**: 10.1504/IJISD.2006.013732.

Westling EL, Sharp L, Scott D *et al.* (2019) Reflexive adaptation for resilient water services: lessons for theory and practice. *Global Environmental Change* **57**: 101937, 10.1016/j.gloenvcha.2019.101937.

WHO (World Health Organization) (2017) *Guidelines for Drinking-water Quality: Fourth Edition incorporating the First Addendum.* WHO, Geneva, Switzerland. See https://www.who.int/publications/i/item/9789241549950.

WISE (Water Information System for Europe) (2008) *Water Note 1 – Joining Forces for Europe's Shared Waters: Coordination in International River Basin Districts.* European Commission (DG Environment), Brussels, Belgium.

Wong T (2006) Water sensitive urban design – the journey thus far. *Australian Journal of Water Resources* **10(3)**: 213–222, 10.1080/13241583.2006.11465296.

Wong T, Rogers BC and Brown RR (2020) Transforming cities through water-sensitive principles and practices. *One Earth* **3(4)**: 436–447, 10.1016/j.oneear.2020.09.012.

Woods-Ballard B, Wilson S, Udale-Clark H *et al.* (2015) *The SuDS Manual*, 2nd edn. CIRIA C753. Construction Industry Research and Information Association, London, UK.

World Bank (2021) *Groundswell Part 2: Acting on Internal Climate Migration.* World Bankl, Washington, DC, USA. See https://openknowledge.worldbank.org/entities/publication/2c9150df-52c3-58ed-9075-d78ea56c3267 (accessed 21/12/2023).

Zhang P, Zhang L, Chang Y *et al.* (2019) Food-energy-water (FEW) nexus for urban sustainability: a comprehensive review. *Resources, Conservation and Recycling* **142**: 215–224, 10.1016/j.resconrec.2018.11.018.

Ashley R, Smith B, Shaffer P and Caffoor I
ISBN 978-0-7277-4178-3
https://doi.org/10.1680/icehudp.41783.357

Chapter 13
Future outlook for urban drainage

Jacqueline Diaz-Nieto, Brian Smith, Richard Ashley and Jeroen Langeveld

Throughout this book we consider concepts for the urban drainage and water services sector that would be required to support a transition to the contemporary and effective management of urban water systems. After all, water is essential to all life, and the natural water environment provides the raw materials that the sector depends on. Water supply and its return to the water cycle must be safe, secure and as low risk as possible, and ultimately work towards achieving the UN Sustainable Development Goals (SDGs) in the process.

The message here for practitioners is as follows.

- Remember that, as professionals, your client is society, and the environment that sustains society, no matter who is paying for the scheme.
- Begin by thinking as broadly as possible – even for small drainage schemes – both spatially and temporally; from the carbon impacts to people, and how the scheme will be used and sustained over time. Only then should you concentrate on the detail.
- Challenge received 'common-sense' and traditional 'this is how we have always done it' thinking. Innovate to create schemes that are as sustainable as possible.
- Wherever and whenever possible, work with (and not against) natural systems in preference to manufactured resources and the need for complex construction.

13.1. Current and future challenges

13.1.1 Climate change and carbon reduction

Climate change is the single biggest strategic issue facing the sector, followed at pace by affordability. Uncertain and changing rainfall and weather patterns are placing an increasing strain on drainage systems. Combined with population and development growth, and urban creep, this will result in a greater flood risk, increased flows to treatment and more frequent storm overflow discharges with consequent environmental impact to the point that urban drainage systems are not sustainable from a long-term, hydraulic, environmental and financial viewpoint. This will affect the ability to achieve compliance with environmental and regulatory obligations, and to provide safe, secure and healthy environments. Understanding future drainage needs is challenging, due to the diverse range of external pressures and the highly localised nature of networks.

Uncertainty, changing societal demands and needs, combined with accelerated environmental changes and their consequences define the need for a more structured, longer-term approach to ensure that urban drainage systems are sufficiently flexible and resilient to future pressures. Current systems lack flexibility in the face of a changing climate and population, compounded by the fragmented way in which water is presently managed and regulated in most parts of the world. Wastewater and drainage systems are dynamic and have an intricate interaction with other systems and so face a multifaceted set of challenges. Current and future challenges are different in scale, nature and complexity to those previously overcome; in short, the low-hanging fruit has been picked! Continued operation of and investment in traditional approaches – as a way to meet the population's requirements as determined in for example, the EU Water Framework Directive, the SDGs and, in the UK, the Environment Act 2010 (HMG, 2021) – will mean that energy demands will also continue to increase, at a time when reducing energy usage has become the key objective, as will carbon footprints associated with construction, and a recognition to reduce embedded carbon in infrastructure projects (e.g. ICE, 2022).

Sewerage utilities should understand the condition of their assets – including any deterioration – and show leadership in working with the wide range of actors with an impact on drainage systems to assess future pressures (e.g. Defra, 2016). Innovation will be required to enable future pressures to be addressed. One key way to address these challenges is to apply innovative stormwater management techniques, making use of advances in both smart systems and engineering solutions that mimic natural sustainable systems. This is intricately linked to water resources, long-term investment planning and alignment to regulatory cycles. In England, Water UK, the organisation representing the water services providers, established a development programme, 21st Century Drainage, to ensure drainage systems were fit for the future. This has and is developing tools to enable long-term planning and ensure that action is taken ahead of failures that affect services or the environment (see Water UK, 2016).

13.1.2 Ageing systems

Ageing systems refers both to the physical infrastructure that was designed in a bygone era and to historical legacy issues relating to changing frameworks for drainage responsibilities. In Europe, some drainage infrastructure pre-dates 1850. The combined sewer systems built before the 1960s are relatively inflexible to future pressures and present an increasing maintenance and repair burden to sustain a minimum level of service. Ageing infrastructure has resulted in existing legacy issues affecting the underlying performance of drainage systems (e.g. OECD, 2020). Increasing volumes of rainwater and surface runoff entering sewers due to climate change and other factors result in hydraulic incapacity and increased energy costs and therefore increased operational carbon emissions (Zhou *et al.*, 2023). Radical changes are required to address the root causes, and how stormwater is managed. In the UK, it is estimated that 30–70% of flows in combined sewers originate from highway drainage, and it will require a huge effort to reduce volumes of stormwater entering the drainage networks. To reduce the amount of flow to the network will require improved management of the system so that the default disposal route for surface water is not into a collection system for sewage treatment, but into a useful resource.

An ageing physical asset base presents additional pressures for the industrialised world, and arrival at the end of asset life is resulting in large programmes of work being required for reinforcement or replacement. Deteriorating performance poses a significant risk to flooding, pollution and river water quality. New asset management practices, that incorporate advances in digitalisation (see Section 13.6), are required to optimise existing and future investment to enhance system efficiency and serve the needs of current and future generations. Improved planning, response and recovery will be required to keep the drainage systems and services functioning effectively. This will also require new ways of doing things. Instead of replacing like with like, future options should start with the question 'What can we make on the surface to do the job of the previous buried infrastructure?' (e.g. WWAP and UN, 2018).

Unless there is a concerted and joined-up intention to reduce the volumes of surface water entering sewers, the current problems of flooding, pollution and a big operational and embedded carbon footprint will continue to grow. Achieving a sustainable and viable asset management plan as a component of a long-term drainage strategy, and an economic case that can be articulated to end users and other stakeholders, is not straightforward. Volatility in the market, with increases in energy and chemical costs in the long term and changes to carbon accounting, are additional unknown factors.

There is a unique opportunity to manage urban water systems in a holistic way that considers the diversity of society's needs and the importance of water to sustain the health of the natural environment. We need to prepare for the unexpected and respond quickly to change and external circumstances. In recognising the changing dynamics and increasing external stresses that control and surround the provision of urban drainage services, we and our infrastructure will need to become more flexible and adaptive, and capable of evolving, while maintaining predictability and stability. Having a resilient and sustainably managed urban drainage network is an imperative.

The socio-economic and natural environment in which we operate is changing. The issues are complex and challenging. Preparing more intelligently for current and future needs is critical in protecting, and fundamental to providing, affordable and resilient drainage infrastructure to mitigate against climate impacts. Understanding complex systems and having the ability to predict how water will behave under different circumstances as it moves through the land phase of the cycle will help mitigate water-related problems in society and improve environmental protection. Transboundary water management will become an increasing issue. So what can we learn from these practices? How do we get agreement between neighbouring upstream countries that it is in both of their interests to adopt a catchment-based approach and develop increased collaboration, as opposed to competition, over the use of water resources within their catchments, such as where there are tensions between agriculture and energy production?

Energy prices and the costs of resources such as fuel and chemicals will continue to rise. Energy price forecasts are highly volatile, creating uncertainty which affects business and economic growth. We will need to be even more efficient in the use of energy and think differently about how we obtain resources and how to use them more sustainably. A circular economy – that is, tapping into all the energy and resources contained within the wastewater we collect, rather than considering it a waste product – will help minimise the use of limited natural resources (Chapter 12). Changing lifestyles and service user behaviour have led to increased sewer abuse, with the disposal of inappropriate materials to sewers in many parts of the world. Coping with future challenges raises not only the question of 'what to do?' but also 'who does what?', 'why?', 'at what level?' and 'how?' A viable response to the current and future challenges requires sufficient capacity, integrity and transparency (achieved by way of coherence between properly engaged stakeholders; see Chapter 4), well-designed engineering and supply chain frameworks to be in place, and adequate and accessible information.

13.1.3 Societal needs and expectations

In the developing world in the 2020s, some 4.5 billion people defecate in the open or have unsafe sanitation, which is responsible for more than 1200 deaths per day of children aged under 5 years. Flooding is endemic and increasing, albeit most countries have not effectively addressed main river flooding and urban drainage is low on agendas. In towns and cities, collection and disposal of municipal wastes are often badly planned and managed, impacting on stormwater systems, and poorly managed sanitation compromises the ability to move to nature-based systems (NbSs) due to the risks to human health. This is the case for even middle-income countries such as India. The challenges of the developing world's needs, as expressed in the SDGs, of providing basic water and sanitation services to all, call for different ways of provision, if this is to be affordable (Leflaive *et al.*, 2022). Projections of global financing needs for water infrastructure estimated in 2015 ranged from US $6.7 trillion by 2030 up to US $22.6 trillion by 2050 (Winpenny, 2015), for all water services, not just sanitation. Innovatory initiatives such as those underway by the Bill & Melinda Gates Foundation (2023) to develop new types of toilet are essential if safe water supply and sanitation is to be provided globally.

In the industrialised world, unlike the developing world, alongside decaying infrastructure there are increasingly stringent environmental quality standards and regulatory requirements and growing societal expectations. New thinking and alternative, more innovative ways to deliver a robust, resilient and affordable operating model for water and drainage services are required. Effective asset management principles are needed to safeguard the long-term drainage security and support sustainable social, economic and environmental outcomes. Social and environmental changes are happening at a faster rate than ever before and need to be considered in decision-making when selecting the best operational approach or capital solutions. These issues are becoming of increasing concern and significant drivers for future investment. No longer can we predict with certainty what the major social and environmental drivers will be. Decision-making should be founded on sound evidence, engineering science and economic judgement, but integrated drainage planning and catchment-wide governance will help in planning for future pressures (Chapter 12). Sewers and water mains are often considered to be minor players in the world of urban infrastructures. Gas networks, electricity and telecommunication cables are prioritised in urban areas, as are roads and parking spaces. Urban green areas are also infrastructures occupying surfaces, interacting with stormwater management systems (Tscheikner-Gratl *et al.*, 2019). Van Riel *et al.* (2017) show that despite the various utility players operating independently, using traditional silo thinking, they are often willing to change practice. There is an urgent need for more effective transdisciplinary and trans-utility working (e.g. Leach and Rogers, 2020). However, multifunctional systems can also bring added risks, where they are co-dependent and may add costs (e.g. Winpenny, 2015).

In England, Drainage and Wastewater Management Plans (DWMPs) provide a framework for more collaborative and integrated long-term planning (Chapter 5). DWMPs encourage working with other organisations that have infrastructure responsibilities that may interact with drainage, flooding and environmental protection systems (Water UK, 2021) (Chapter 10).

13.2. Emerging responses and sector evolution

Uncertainty, changing societal demands and needs, combined with accelerated environmental changes and their consequences, define the need for a more structured, longer-term approach to ensure that the urban drainage networks are sufficiently flexible and resilient to cope with future pressures. It is important to consider both present uncertainties and how the future may unfold in respect of 'shocks' to the drainage systems and impacts on performance and consequences. Future shocks could have a damaging effect on financeability and investor and donor confidence. Employing a dynamic 'adaptive' incremental approach will help manage future uncertainty in predictions of potential risks. In developing an understanding to inform a planned and logical robust design strategy, it is important to think about how things may change over time, particularly climate and socio-economic changes. Due consideration should be given to the following (Chapter 12).

- System: spatial and temporal
- Drivers: social and environmental
- Response: insights and skills.

Box 13.1

Business continuity management

It is important to recognise the role that business continuity management has in the context of urban drainage practice; that is, planning for any event that could threaten, or put at risk, one's business. Risk management underpins all business continuity planning, including drainage and, perhaps more importantly, stormwater management. In simple terms

- identify business risks
- analyse them
- manage them.

It is about Mitigation → Adaptation → Optimisation

It is possible to outsource design, operations and maintenance, but not to outsource risk, a concept that many fail to recognise or understand. Loss of money may be forgivable, but loss of, or damage to, reputation is not – and that is why it is important to embrace future challenges.

Scenario planning, while not intended to predict the future, is a tool for thinking about the future based on the following assumptions.

- The future is unlike the past, and is shaped by human choice and action.
- The future cannot be foreseen, but exploring the future can inform present decisions.
- There are many possible futures, and scenarios map a 'possibility space'.
- Scenario development involves rational analysis and subjective judgement.

In the water sector, uncertainty and business continuity management often go hand in hand. There is there for the need for a flexible, adaptive approach to risk management, as outlined in Box 13.1.

There are three key areas where the water sector is already adapting and evolving

- people: skills and capacity planning
- places: planning, finance, and regulatory framework
- artefacts: technology.

Ongoing developments and observations in these areas are discussed in the following sections.

13.3. People: skills and capacity planning

Playing a key role in infrastructure provision, public health, development growth and the economy, 'having the right skills' is critical if we are to meet the long-term needs of society. There is an emerging view that the skills shortage is a very real concern; with a pressing need to address appropriate resources and skills within the sector. In the UK, like many industrialised countries, much of the industry-specific expertise is concentrated in an ageing workforce. With retirement there is a loss of knowledge, so succession planning is important, to upskill early career employees and ensure that graduates and other new entrants are suitable to work in the water industry.

'To deliver our responsibilities safely and effectively at a sustainable cost, takes an exceptionally skilled and substantial workforce, the sustainability of which is under pressure' (Energy & Utility Skills Partnership, 2017). Across the world, the water service providers face intense competition for many traditional core skills, while at the same time having to keep pace with change. This is leading to a growing increase in highly specialist expertise and growing complexity within roles. Technology is evolving, and skills need to evolve too. A rapidly evolving technological environment creates the need for more diversity of skills and the people who perform them.

In the developed world, there is an ongoing technological innovation deficit in the water sector, with a continuous need to embrace new technologies to improve operation and performance. With high levels of technical skills and experience, risk can be controlled very effectively. Failure of business planning is all too often the result of poor management, which ultimately comes back to lack of skills. Conversations within the sector repeatedly say that we can 'grow our own' skills, but we often fail to do this and there is limited management control over those 'bought in'. Strategies will only be as good as the infrastructure that supports them, and this includes skills. To target and address potential barriers to implementation and uptake properly will require transition and transformation management with careful incremental change, possibly spanning time frames of several decades.

Elsewhere in the world, many skills deficits occur due to the migration of many of the more capable members of society into developed countries in search of better education and more prosperous lifestyles. This leaves those most committed to their indigenous societies and those who may be poorly educated or are unable or unwilling to leave to undertake the delivery of the highly complex systems and services needed. Although many highly skilled migrants may ultimately return to their home countries, this is the major problem in developing nations, alongside failure to allocate sufficient importance or resources to services such as urban drainage (e.g. DWA, 2012).

Providing a suitably skilled workforce to meet growing aspirations is crucial as an important enabler for the water sector, the overall economy and society, with a need to develop a new, inclusive approach to attracting skills and talent. Many other sectors, especially IT, are seen as far more attractive to the most talented compared with the 'dirty' water industry. The water industry is faced with tensions between, on the one hand, short-term affordability and resource prioritisation and, on the other hand, long-term investment and, where businesses are involved, investor returns. Many competing priorities need to be managed. Addressing the skills deficit through a planned programme of replacement will provide continuity and reduce the risk of disruption, including, crucially, risks to financial stability. Having the right skills will enable the sector to remain cost-efficient and competitive, stimulating productivity, investment and, where needed, profitability. Furthermore, any skills strategy also needs to address retention.

A deficit in appropriate skills will restrict the ability of the sector to operate effectively and cost-efficiently. This will be further compounded in a more competitive market for people. Any significant shortage of skills is likely to have an adverse impact on future development and essential wastewater infrastructure projects. A purely sectoral approach will not work, due to international and regional labour markets and the need for many in the workforce to have cross-sectoral expertise (e.g. digital, data, mechanical, electrical, civil and hydraulic knowledge, etc.). For example, the water industry in the UK is competing against high-profile digital, energy, flood resilience, waste and transport projects for the best engineers.

In the UK, the National Infrastructure Commission (NIC) has highlighted several areas of concern, where urgent action is needed to avoid serious risks to delivering the objectives of the National Infrastructure Strategy (NIC, 2023). These relate to reduced capacity caused by significant changes to both physical infrastructure and organisational structures. The skills shortage makes it difficult to attract and develop young professionals. As well as engineers, many people from a range of technical backgrounds are employed in the sector, particularly in science. It is also the same across Europe: 'Being able to attract and retain young people with STEM [science, technology, engineering and mathematics] qualifications continues to be an issue' (Energy & Utility Skills Partnership, 2020). Consideration must be given to how long it takes to introduce such changes, including hiring and training people. With changing technology and processes there may be a need to consider what the 'operator of the future' will look like and how to provide the appropriate training and support to respond to the inevitable challenges.

Perhaps there is also a need to rethink the principles of human resource management and capacity planning for the modern era, to ensure the sector can deliver value and meet changing demands. Capacity development can be achieved through a process of learning, transferring, acquiring and using new abilities to replace the old way of doing things (DWA, 2012). To maximise available resources it is important to understand the impact these changes may have to enable adjustment to capacity in a timely manner.

> Broader development of technical capacity for the design and operation of water infrastructure and wastewater treatment systems is essential in greening the economy and transition to social sustainability. Promoting adequate technical capacity to support water and wastewater management is thus an important part of the policy package for addressing water challenges in many countries.
>
> (DWA, 2012)

Box 13.2 highlights an example of how, within the UK, the concept of marketisation and the issue of skills affect strategic planning.

13.4. Places: planning, finance and regulatory framework

13.4.1 Current situation

There are very real external pressures that we must prepare for now, to ensure intergenerational fairness (the essence of sustainability, Chapter 12). As a society, we must think very differently about the future – about how we manage too much or too little water; how we reduce, reuse and recycle water and how we begin to value wastewater as the precious resource that it is, to ensure that it is not wasted and seen as something to dispose of as quickly as possible. We need to prepare for a more sustainable future and find the balance between valuing and restoring the natural capital and supporting society's aspirations for

Box 13.2

Capacity and skills planning

The answer to the overarching key question of 'What is the market failure we are trying to address?', is that it is about capability, including user service, continuous improvement, transformation and expected long-term economic growth. There is an urgency to inculcate the importance of the development of a sectoral skills strategy. In terms of strategic positioning, any skills strategy would need to address retention so as not to impact on business or other strategic plans and levels of service.

better places to live and work (Water UK, 2016). The planning process is integral to successful achievement of more effective management and control of stormwater draining to the wastewater networks (Chapter 5), ensuring that it is managed efficiently, effectively and sustainably. With foresight and planning, we can deliver more reliable services and substantive improvements to preserve and enhance the environment. It is therefore imperative that we develop a shared understanding of new and best practices for stormwater management that deliver for public health and planning policy objectives (Chapters 3 and 8).

Capital improvements and infrastructure construction programmes must conform with town planning policies and processes. This is especially important with regard to how land is allocated and utilised. The growing use of NbSs rather than underground infrastructure necessitates changes in urbanism, from a 'hidden service' to one that is celebrated in place-making (Chapter 5). Alongside changes in how urban spaces need to be planned and developed in the future, the Fourth Industrial Revolution in the twenty-first century has led to the 'environmental economy', and the Decade of Action, where new technologies and artificial intelligence (AI) are changing options and hence decision systems across human society. Popkova *et al.* (2023) state that in this context there is a need to

> ensure environmental justice, energy security, competitiveness, environmental taxation improvement and responsible production and consumption development. Society's contribution must be coupled with optimising environmental decisions, environmental awareness, improvement of environmental behaviour to protect biodiversity and preference for ESG [environmental, social and governance] investments.

These changes are fundamental to the whole ecosystem of society. The way in which planned uses of space are decided on may need to change drastically, both to maximise the opportunities, for it to be affordable, and to provide long-lasting protection to people and the environment.

Water and other infrastructure projects currently compete for funding with numerous other programmes and initiatives. There are few easy solutions to balancing the complex and sometimes contradictory demands. Upgrading the capacity of ageing infrastructure in an existing development would involve major capital investment. Future demographic changes and the impacts on available capacity will require regular reviews of asset plans to avoid under- or over-investment in the drainage infrastructure. Without a clear programme of action, securing adequate financing for long-term drainage security is at risk (OECD, 2020).

13.4.2 Future expenditure

To build more sustainable systems, there will be a need to develop a strategy for the medium and long terms, such as the DWMPs (Water UK, 2021), which that look at a planning horizon of 25 years or more (Chapter 10). Significant further investment is vital if the drainage infrastructure is to be maintained and enhanced for future generations. Finance and accessing sources of funding are an increasing challenge. Significant investment in both operating and capital programmes will be required to meet water quality standards and service level needs and expectations.

13.4.3 Financial sustainability

Wastewater networks are essential components of urban infrastructure, and their effective functioning is critical for the everyday life of people in the urban environment. This is achieved, among other things, by means of regular, planned maintenance of these systems supported by secure and sustainable investment (Druade *et al.*, 2022). The water industry in the industrialised world acknowledges that with an ageing infrastructure, its performance is highly dependent on effective maintenance of ageing assets, such as pumps, sewers and ancillary infrastructure. Prolonging the life of an asset as much as possible or adopting a phased approach to renewal would undoubtedly defer high-cost capital investments. As discussed in Chapter 10, there are many approaches to maintenance, including incident-based, preventive, condition-based or predictive, that is, reactive, proactive or planned. Currently, many maintenance budget owners and managers are actively exploring more cost-effective approaches, such as

- just-in-time maintenance
- reliability-centred maintenance (RCM)
- condition-based monitoring (CBM).

With increasing budgetary pressures, particularly on operational expenditure (OpEx), using resources to inspect and maintain assets based on regular intervals is not necessarily the best approach and a more 'intelligent', cost-efficient means of scheduling inspection and maintenance activities is required (Chapter 10). The advent of hi-tech instrumentation and availability of more sophisticated techniques now displaces the past subjective judgement of inspectors (Chapter 9), at least in industrialised countries. As the sector increasingly moves towards increased digitalisation and smart systems, there will be more 'eyes and ears' on the assets by means of instrumentation and telemetry, providing data in real time, enabling the use of AI to identify the optimal time for which to schedule inspection and maintenance prior to asset failure. However, whatever maintenance strategy is adopted, a robust management plan is imperative to ensure that the necessary plant and infrastructure requirements can be economically achieved.

Financial and performance improvements can be delivered through an integrated approach to infrastructure investment and delivery, with a principal focus on high-priority catchments and assets. Identifying 'hot spot' catchments and a range of solution options will help to prioritise investment to maximise economic benefits and identify those which will have the greatest impact in the shortest timeframe. It is essential that existing assets are exploited or 'sweated' wherever possible to augment 'capacity' before investing in new projects.

Consideration should be given to developing financing alternatives and opportunities for more equitable charging structures; for example, in Germany, this is proportionate to the amount of generated surface water runoff based on land use or impervious surface area. Balancing differences between residential, commercial and industrial users would promote a better understanding of and personal responsibility for surface water runoff and be effective in improving stormwater-related problems, such as flooding and pollution, and reducing the impervious surface area draining to sewers.

Maintaining fiscal discipline is a necessity to ensure successful delivery of any long-term strategy for urban drainage. Important interlinks between technology, regulation and finance exist but these can be both simple and complex, as demonstrated in Figure 13.1.

13.5. Artefacts: back to basics and mimicking nature

There is a current and growing movement away from hard infrastructure (e.g. energy-intensive concrete and pumping) towards solutions inspired by nature (the so-called 'soft solutions'), nature-based systems (NbSs). In most parts of the world, there are targets and requirements to reduce both embedded and operational carbon as well as a need for carbon sequestration, which NbSs can help to achieve.

Figure 13.1 Technology, regulation and finance relationships

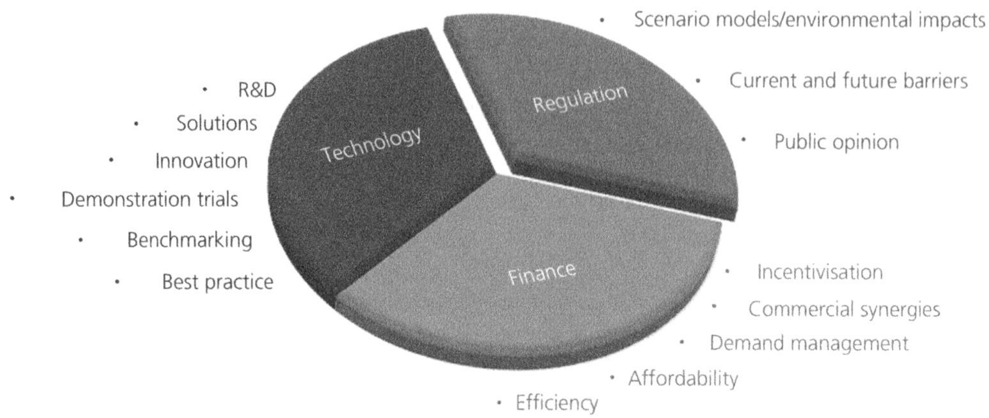

Practitioners in England are currently faced with finding a drainage option that maximises the value for multiple other criteria that also address the ecological and biodiversity crisis (e.g. Defra, 2023). When multiple drivers are addressed by a holistic solution that mimics nature, these may not necessarily be the more cost-effective and easily delivered options. Another step change for many practitioners, at least in the UK, is that urban drainage systems should now be designed to provide a required level of removal of nutrients to reduce urban diffuse pollution (Natural England, 2022). The requirement not to increase the nutrient loading in receiving watercourses from new developments, which considers the overall increase in effluent associated with new dwellings, has surprised many practitioners. Many developments in areas where nutrient neutrality in England is needed are not progressing until appropriate solutions are identified.

Reducing the volume of stormwater that enters the combined sewer network can maximise opportunities to maintain service and environmental performance, and in effect create capacity within the existing ageing network while the shift in urban drainage practice from buried to above-ground systems takes effect. More readily available capacity within the sewer network will reduce the need for future investment in piped systems and wastewater treatment works upgrades, thereby accommodating future development growth and adaptation to the effects of climate change, ensuring sustainability. However, sewer separation is not necessarily the panacea that many believe, as illustrated by the examples in Chapter 2, Box 2.1, in which separation led to greater water pollution by heavy metals and polyaromatic hydrocarbons (PAHs). The ubiquity of incorrect connections is another problem, as even a single sanitary connection can compromise an entire stormwater drainage system.

There is a growing awareness and acceptance that hybrid drainage systems that make use of blue, green and grey infrastructure are likely to be the most sustainable approach for managing stormwater in many cases

> Nature has ingenious systems for the management and delivery of water in all its phases: precipitation, evaporation, absorption by plants, wetlands, flows to rivers and lakes, and infiltration that recharges

> groundwater and aquifers. No additional infrastructure is required – the water systems are in place, naturally. Once this natural environment has been disrupted by human development, the stormwater created by this development becomes an issue that requires intervention and ongoing management.
> (Liptan and Santen, 2017)

While many of the same drivers apply around the world, the UK remains behind other developed countries in the use of such techniques to manage and prevent stormwater entering existing and ageing sewers.

The USA and Australia are more advanced in implementing sustainable stormwater management programmes. The Clean Water Act 1972 (amended in 1987) (CWA) was an important legislative driver in the USA for changing the business approach to stormwater management (Peters and Zitomer, 2021; USEPA, 2022). In response to the impacts of urbanisation on water quality, the CWA provided a regulatory mandate that prohibits the discharge of pollutants into waters of the USA unless the discharge complies with a National Pollutant Discharge Elimination System (NPDES) permit (City of Portland, 2020; *Stormwater Management Manual*). This also opened up various federal and state funding routes for remediation. While water quality was the original key driver, now a combination of both conventional and 'green' infrastructure measures are designed to manage water quality and quantity and reduce sewer rehabilitation costs. Practice in Australia has been driven by drought, water scarcity and the rising cost of potable water. Both countries have made significant progress in embedding truly separate drainage systems, in particular the retrofitting of green infrastructure and sustainable drainage systems, within mainstream practice, such that they have become a widely accepted component of the toolkit for effective drainage. Such 'sustainable' stormwater management has been successfully implemented in both large- and small-scale applications.

13.6. Artefacts: technology and smart systems

Chapters 9 and 11 outline the journey and current level of computation, instrumentation and technological applications already being used within the drainage sector. The Fourth Industrial Revolution, being referred to as Industry 4.0, is the convergence of existing digital technologies including big data and analytics (data science), Internet of Things (IoT), artificial intelligence (AI), machine learning (ML) simulation, augmented reality and other emerging technological initiatives (Schwab, 2016). In the context of urban drainage networks, all the above approaches can be applied to develop smart drainage systems. Smart systems and AI is a discipline in its own right, with applications across many sectors, with needs for digitalisation in order to achieve efficiencies and transition into the digital era and Fourth Industrial Revolution (Agrawal *et al.*, 2018; Finlay, 2018). Smart systems are not entirely new to the water industry. In the industrialised world, the water industry has been at the forefront and has been using what is now known as smart systems and AI for several decades, by means of supervisory control and data acquisition (SCADA) systems to operate assets remotely, such as treatment works and pumping stations. The textbook *Urban Drainage* by Butler *et al.* (2018) has had a chapter on smart systems, albeit previously titled 'Integrated management and control', since at least the second edition published in 2004.

Many practitioners will use or rely on smart systems as part of modern daily life; for example, many people in the UK have experienced a smart motorway and will therefore be familiar with the principles of applying this smart approach to managing infrastructure, albeit one that is now perceived to have failed.

Figure 13.2 Adaptive pathway approach towards a global optimised smart system (source Vallabhaneni and Speer, 2011)

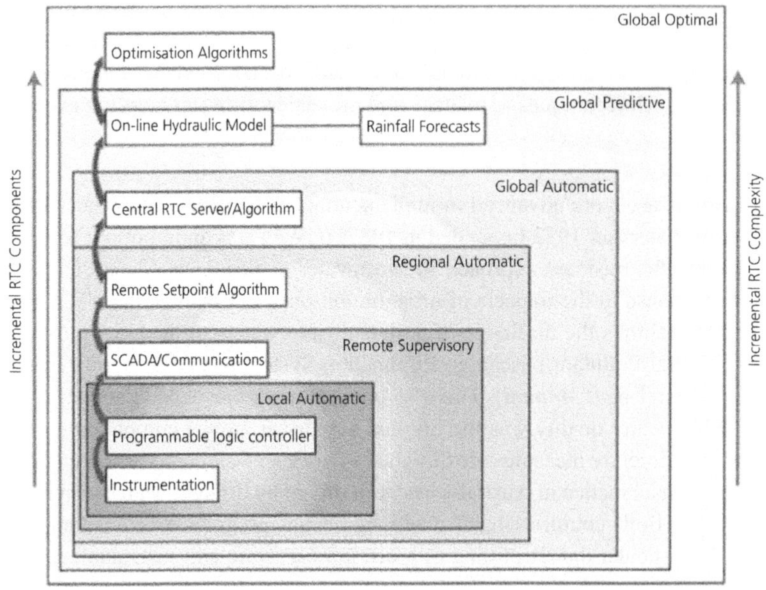

Key characteristics of a smart system are that they are cost-effective and adaptive solutions that make use of the existing infrastructure, thereby reducing costly, disruptive and inflexible major construction work, and are used to make intelligent decisions in real time using incoming data. Figure 13.2 illustrates how smart systems are an adaptive pathway approach (Chapter 12). By starting with implementation of local smart solutions at the site scale, growth and capability of smart systems can be integrated in a phased approach towards a global approach that aims towards a fully integrated digital twin (see Section 13.6.2).

A smart urban drainage system is one that makes use of hardware, communication and AI for improved decision-making, to provide any or all of the benefits given in Table 13.1, which provides examples and opportunities to provide smart systems benefits when responding to the current urban drainage challenges.

For some time now the global water industry has been data rich, with its computational records for asset data, time series monitoring, work management systems, incident and user reports and a plethora of other datasets forming an integral part of the core provider activity. However, being data rich is very different to being information rich, let alone data-driven. Emerging technologies and capabilities in data science and AI now open exciting and seemingly endless opportunities for the water industry to generate added value from its data and become an information rich and data-driven industry, in which the performance and condition of the assets is understood, enabling proactive and evidence-based interventions. The ultimate aim is to become more proactive (rather than reactive, which is disruptive, costly and unpredictable, often compared to 'firefighting'). Water industry practitioners can see the potential that AI creates through better, evidence-based targeted interventions at the optimal point in time.

Table 13.1 Examples of opportunities to implement a smart urban drainage system

Application	Challenges addressed	Description
Reduce storm overflows as a consequence of hydraulic incapacity	Hydraulic incapacity; changing rainfall patterns; increased runoff to sewers	Sewer water level sensors provide real-time visibility of capacity availability within the sewer network. Flow control devices within the network are actively controlled to hold back flow with the objective of preventing storm overflows. Rainfall nowcast products can provide additional predictive capability.
Reduce storm overflows due to causes other than lack of capacity	Blockages or capacity loss due to sewer abuse or sedimentation	Algorithms/ML detects abnormal sewer level response in dry weather and learns when this indicates a blockage formation that requires sewer cleansing. Rainfall nowcast products can provide additional predictive capability.
Targeted rehabilitation investment	Ageing infrastructure	AI applied for sewer CCTV image processing to automatically classify and prioritise defects for rehabilitation.
Proactive/just-in-time maintenance	Efficient use of resources (skilled workers, time, and money)	Asset monitoring of electrical and mechanical components is processed in real time to detect patterns that indicate when maintenance is required (compared with cyclical maintenance visits based on a given time interval).
Smart alarms	Failure detection (e.g. rising main bursts)	AI and ML applied to use existing monitoring data (e.g. pumped flows, pressure deviations) to detect and confirm failures and raise a targeted alarm for response.
Public environmental data notifications	Meeting public expectations (for bathing waters)	Environmental data is validated and verified using AI and made publicly available in real time.
Reduce energy consumption	Increased energy costs	Improved process control at pumping stations to pump flows within optimum pump efficiency range.

13.6.1 Smart, data-driven decision-making

Finlay (2018) provides a pragmatic definition of all AI (including ML, neural networks and real-time control (RTC)) as computer algorithms. The critical benefits of a smart network are in having an immediate and robust view over every asset, its status, condition and performance, and from this being able to process when and how to make fast and cost-effective decisions. Collection systems generally operate in a passive manner. Active management, using AI that is capable of multicriteria evaluation in real time, is improving overall system performance by making balanced decisions on flood risk reduction and pollution prevention (Sun *et al.*, 2020). Modern information and communication technologies have a vast array of innovative capabilities (Chapter 9). Harnessing this technology is helping to create more intelligent means of managing and protecting water and wastewater resources. AI and smart technologies, such as adaptive real-time control, ML and neural networks, will provide faster and more reliable data, enabling quicker decision-making and intervention (Kapelan *et al.*, 2020). Smart infrastructure will also help operators optimise resource allocation.

A data-driven approach to asset management must be cognisant and not lose sight of the end goal of physical asset management, as this provides the wastewater collection system. The physical assets and the level of service provided ultimately make up the value of a service provider. In the UK's private and regulated water sector this is known as the regulated capital value (RCV). All efforts to implement a smart system must not be at the detriment of investing in the asset health of the actual physical assets that deliver the core services (Chapter 10). Providing core services are the priority and technology choices must deliver improvements to these. Above all other digital transformation objectives, a smart wastewater network must assist in the following.

- Prolonging asset life by identifying the optimal point in time (often referred to as 'just in time') for interventions that avoid costly catastrophic failure. This is colloquially referred to as 'sweating the asset', or extending the asset life as much as possible and postponing high rehabilitation costs until necessary.
- Identifying and eliminating common root causes of failure, as this is the most effective means of ensuring service failures do not repeat themselves.

In developing an environment for analytics and communication, it is vital to invest in research and development in technologies that will help to develop the infrastructure that will best meet provider and user needs in the long term. Linking the physical and digital world and adopting an integrated approach will help enable a seamless transition, as will an acceptance of the use of AI and innovative technologies to develop smart infrastructure and optimise resource allocation.

13.6.2 Digital twins of physical assets

A digital twin (DT) is defined as a virtual representation of a physical asset enabled through data and simulators for real-time prediction, optimisation, monitoring, controlling and improved decision-making (Rasheed *et al.*, 2020). Section 13.6.1 described local and single-objective applications of smart systems, whereas a digital twin is a comprehensive digital version of the physical drainage system. A DT is a layered approach rather than a single piece of hardware or software, and incorporates a range of components such as pumps, pipelines and processes. The DT not only provides real-time material for more informed decision-making, but can also make predictions about how the asset will evolve or behave in the future. In an ideal setting, a digital twin will be indistinguishable from the physical asset both in terms of appearance and behaviour, but has the added advantage of making future predictions. It also provides the possibility for humans to interact physically with the asset using an avatar (Mollerup *et al.*, 2017). According to the Gartner Hype Cycle (Gartner, 2021), there are now increasing opportunities to create DTs and they are being developed for wastewater applications within academia and industry (e.g. Pedersen *et al.*, 2021). A DT can include RTC, which as a technology has been used in the water sector since the 1980s, but more recently has been implemented and demonstrated to reduce storm overflows (Mollerup *et al.*, 2017; Van Loenen *et al.*, 2012; Vezzaro *et al.*, 2014). A data-driven (statistical and stochastic) approach relies only on data and generally disregards the existing and well-established hydraulic and process models, and therefore does not advance the understanding of the physical assets and system processes in the way a DT can.

The current drainage infrastructure DT is the hydraulic model as explained in Chapter 11, requiring the next iteration to be able to run the model in real time using rainfall nowcasting (Chapter 6). While computing power has advanced immensely in the past decades, hydraulic models as built and run now still take a considerable amount of resources and time to compute. For this reason there are few, if any, catchment scale drainage infrastructure models that are running in real time, let alone predicting the future state of the system with sufficient lead time to deploy responses to flooding or pollution.

Given that there has been considerable investment in recent decades in building accurate and high-quality catchment scale hydraulic models, one challenge for IT and data science practitioners is how the sector continues to benefit from the hydraulic models within a digital era; what is required so that hydraulic models can be solved with the required speed to deploy actionable insights, akin to the often nationally owned meteorological prediction models? (Rasheed *et al.*, 2020).

13.6.3 Service transformation and strategy

There are numerous water and wastewater service providers and arrangements for institutions and responsibilities worldwide (Chapter 5). Many are public services, operated by or on behalf of municipalities (e.g. Juuti and Katko, 2005). There are also providers who offer a commercial service, as in England, where these are regulated businesses. Whichever model is being used, the service providers will need to address the forthcoming challenges and opportunities, especially those provided by the new Digital Age.

Business leaders within the English water services sector are responding to the changing market environments and the digital business growth environment and are embarking on the journey of transition towards smart networks. Often the transition forms part of wider strategies for digital business capabilities and digital enabling initiatives. In England, industry regulators also require the water sector to develop business plans that embrace the latest technology and smart networks (EA, 2021).

With such fast-moving technology advancement becoming available, the challenge for sector leaders is in selecting appropriate activity areas to which innovative technologies can be applied and that will provide the highest return on investment. From this, developing a clear roadmap to bring all stakeholders together for the digital transition or, as commonly known, the new ways of doing things. Digital transformation is a slow fuse. The rate of technological innovation is exponential and requires organisational and institutional change, which involves people and professions and requires sensitive change management. At the time of writing, the latest market disruptor, which professionals have generally been unprepared for, is the unprecedented update of widely and freely available web-based apps that can write code by taking in ordinary, common speech instructions. Specialist suppliers who produce engineering design and data analysis reports are at risk of being supplanted by web-based apps. Using such an app, a non-specialist can roughly describe their functional design specification and the app scouts the internet for previous examples and user input data to produce designs, analysis and reports.

Implementing a truly smart system requires multidisciplinary collaboration between several specialisms in the water sector, but they might not be accustomed to using the same technical language and working seamlessly to develop a fully integrated solution. Figure 13.3 shows a conceptual framework developed by the Smart Water Networks Forum (SWAN) that depicts a layered approach composed of the parts (disciplines) that make up the smart network. These disciplines include the following.

- Engineers who understand the mechanics of the physical assets.
- Instrumentation, control and automation (ICA) engineers who can understand the electronics and logic.
- Telemetry and telecommunications specialists who know how to get these data to the brain centre for processing.
- Data management and specialist architects who can design the software and systems to store data in retrievable and searchable formats.

Figure 13.3 The SWAN wastewater network management layered approach (SWAN, 2023)

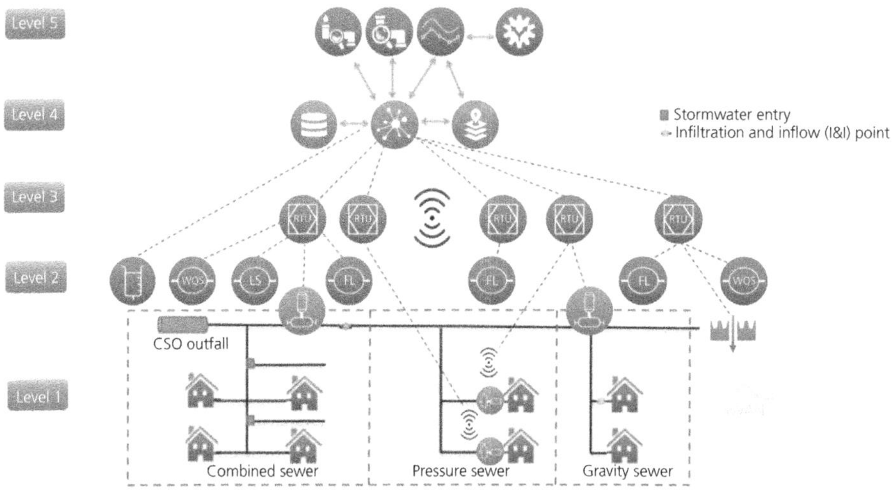

Level 1: The physical layer made up of civils, mechanical and electrical assets.
Level 2: The sensor infrastructure (rain gauges, water quality sensors (WQS), level sensors (LS), flow meters (FL), etc.).
Level 3: Communications/telemetry/outstations infrastructure (remote terminal units (RTUs), cabled, wireless, IoT, radio, etc.).
Level 4: Data management and display (SCADA, GIS, dashboard, work management systems).
Level 5: Data fusion and analysis (high-value information, algorithms, automated responses, forecasts, decision-support systems, detection, optimisation).

- Data scientists and specialist programmers who can take incoming data and build algorithms that make sensible automated decisions.

The use of smart networks in the water sector will depend on active engagement and participation of stakeholders from inception through to implementation. A carefully planned and sensitive stakeholder engagement process should be in place for achieving system acceptance by users. Inclusion of operations and maintenance staff is critical at early stages and throughout implementation. The long-term maintenance needs of all control elements are critical, and the changed approach needs to be carefully planned with affected colleagues (Perciavalle *et al.*, 2017).

Dilemmas and questions that service providers are currently faced with include the following.

- Should the sector develop in-house AI capabilities or use AI suppliers and frameworks?
- Is AI and data science a core function of the water industry and therefore will data scientists become just as essential as hydraulic modellers to the water industry?
- Can the digital transformation be delivered under current organisational structures (e.g. existing role descriptions and skills sets)?

There are different views of which type of operating model is best suited for a smart wastewater network, such as a 'federated' model, but with a centralised-expertise base or a centralised model. Federated architecture allows interoperability and information sharing between semi-autonomous decentralised organised lines of business, IT systems and applications. In this model, an integrated control centre houses experts that are capable of remotely operating workers who are 'low skilled'. An example of this would be a farmer operating a pumping station on their property, using technologies such as augmented reality or Google Glass, but being directed from a central expert command. This would result in significantly fewer people being required to operate above-ground assets. It could provide quicker and more reliable data, enabling faster decision-making, all of which helps to improve the management of risk and ensure the continuity of the critical drainage service. A federated approach is more agile and flexible, allowing much more freedom of choice. However, a centralised model provides more comprehensive end-to-end transformation and is more scalable (up or down), but relies on strong leadership.

One key challenge is to unlock the potential of AI and data-driven technologies to optimise wastewater network performance and deliver operational efficiencies compared with the current ways of working. The digital transformation to a smart wastewater network will undoubtedly require dealing with what, until now, has been unknown territory, and testing various technologies. While it is important to consider scalability at an early stage, starting small with pilots can provide invaluable learning and insight. The journey will initially require technology selection that is compatible with the existing modus operandi. See Box 13.3.

AI and ML concepts are now being used to drive wastewater process improvements, which are helping reduce the number of pollution incidents, enable targeted NbSs, provide early detection of leaks and enhance energy efficiency and will ultimately, lead to optimised processes and efficient integrated water management. The development of smart networks is an evolving process. Ideally, this is about creating a fully connected network, topologically linked, that enables digital navigability across the assets. One of the main advantages is the ability to adapt to an uncertain future through operational changes rather than construction of new infrastructure. A phased approach that builds on what has already been achieved has a higher chance of success and is easier to manage than yet unknown innovation (Figure 13.2). Furthermore, an incremental staged transition will support evaluation of the benefits at each stage and ensure increased acceptance of the technology.

13.7. Outlook for urban drainage

Urban drainage as a service and as a discipline has matured, having gradually developed from ad hoc arrangements in ancient times into a more formalised discipline. Yet, many of the ancient and traditional practices are still used successfully now. The world has changed and is changing, perhaps faster than at any time since industrialisation. This includes the form and layout of urban areas, which potentially may change significantly within the lifetime of any drainage system. This raises questions as to whether

Box 13.3

Technology and smart networks – extract from WISER (EA, 2021)

'Embracing technology and developing smart networks is essential to drive process efficiency and in achieving carbon reductions. Employing smart technology to inform drainage and wastewater planning and water resource planning will enable service providers to develop targeted solution delivery for more effective environmental outcomes, better service, and long-term efficiencies. It will also help achieve carbon reductions.'

traditional urban drainage practices are still fit for purpose. There are challenges in the immediate and longer term that need to be faced, but there are also opportunities to change what we do and how we do it.

13.7.1 Challenges

We face a multifaceted set of challenges and to meet future demands will need to consider every option to balance pressures on the drainage infrastructure. Over the past 100 years in the industrialised world the approach to managing wastewater through underground sewers, tanks and shafts has led to great benefits in public health, river and bathing water quality and flood control. However, there is a social and economic imperative to move from short-term delivery of service levels to longer-term sustainability of drainage infrastructure, or at least the services provided, for which building resilience is important to reduce vulnerability to natural hazards. In most places, significant investment in drainage capacity and storage has been provided to improve performance and meet ever-tightening regulatory obligations. Notwithstanding this, the ubiquitous use of combined sewer overflows (CSOs) for alleviating overloaded networks (and potential impacts on river and human health) has become seemingly untenable to the public. Dealing with these problems and the ageing buried infrastructure requires increasing investment in new assets and maintenance or replacement, but is adding mainly grey, buried infrastructure that is relatively inflexible to future pressures as well as costly the right course of action? In many countries the sewer networks are largely an inherited legacy of predominantly combined sewers that were not designed or envisaged to deal with the volumes of stormwater that they are now expected to convey or the range of substances being discharged. Problems from overloaded drainage systems will be worse in the future as surface water inputs increase due to climate change and urban development in most parts of the world. In addition, there is a legacy of flows entering drainage systems that may not have been designed or intended to be there, such as watercourses, highway drainage, inundation from stormwater runoff from fields and adjacent land (muddy flooding), incorrect connections and flows from informal sealing of surfaces. Continuing solely with an approach of using, maintaining and enhancing traditional drainage infrastructure is not sustainable. A point in time has been reached at which addressing future challenges requires a transformative change in approach. In the industrialised world it will be necessary to move towards one that maximises opportunities and resource efficiency, as opposed to traditional ‘problem centred’ urban drainage practices. As such, a phased approach to adaptation and mitigation is a more appropriate and sustainable response, rather than conventional solutions in which society ‘builds its way out of a problem’ by increasing conveyance or storage capacities.

However, given that it is impossible to control what is put into drainage systems by users, especially into underground piped systems, there will be a need to continue to expect a wide range of flows and an almost infinite variety of substances being discharged. Significant amounts of flows in sewers are due to groundwater ingress, in many cases most of the flow conveyed. Therefore, it is likely that there will always be a need for underground piped systems, mainly to deal with groundwater, but also to convey the most polluted flows (i.e. those with substances that cannot be managed locally, such as PAHs and microplastics) to centralised treatment plants where they can be dealt with more efficiently. However, these future sewers will, if systems are properly maintained, not be conveying surface water and the groundwater volumes will be better controlled. Much of the groundwater should be managed locally, with NbSs slowing flows and providing greater storage, and stormwater being used as a resource, either directly for non-potable use or supporting base flows in the ground.

In many parts of the world there are no or inadequate urban drainage systems. Despite the SDGs and worldwide efforts, there are still more than 5 billion people without adequate sanitation and more than

4 billion without safely managed water supplies. Although less than 50% of these live in urban areas, the main challenge in these countries, apart from financing, is to fulfil the SDG 6 aspirations by 2030. In many of these countries, such as India, drainage systems are usually malfunctioning, due to lack of prioritisation of investment and political neglect and because of abuse by other interacting infrastructure services and providers, especially highways and municipal wastes. The main challenges are in the policy sphere, and in acceptability of the many affordable technologies that could be used, other than 'western style' water and sanitation systems. For many, however, climate change is already bringing stresses to water resources and, in countries such as Bangladesh, increasing severity and durations of flooding. Dealing with urban drainage in such contexts is unlikely to be a priority for the foreseeable future.

Governance arrangements and institutions are the single most important factor in making cities water-sensitive. Worldwide there are more and more 'future of water and cities' initiatives and publications, but few actual shifts in established practice or the called-for institutional and governance reforms needed to make urban drainage fit for the future. The context for any new developments will inevitably remain within public policy and regulatory settings as these systems are public services. Neither policy nor regulations have been able to adapt to the speed of change we now experience. Despite worldwide exhortations by many respected agencies, governments worldwide are failing to manage water systems in an integrated way, let alone take whole-systems approaches. Yet, this will be essential to cope with the future pressures and impacts of water management. Policies and regulations will need to enable rather than constrain advances and innovations. However, how this can best be effected in a smart-city world is yet to be determined. Current approaches are simply 'tinkering at the edges' of the real-world systems. Yes, controlling a single CSO discharge is locally important, but it is the aggregation of the entirety of the diffuse polluting sources that needs to be controlled, aimed at specific outcomes in terms of societal benefits. Urban drainage is but one small part of the whole ecosystem of societal systems and services and prioritisation of resource allocation needs to balance the needs of all the sectors and everyone in society.

13.7.2 Opportunities

The four streams of the urban water cycle (potable water, wastewater, groundwater and stormwater) are intricately linked, with different technologies and strategies applying to each and several applying to more than one (Smith *et al.*, 2014). Improving the hydrological balance in catchments is needed for the protection of the environment and would also reduce risks of sewer, fluvial and surface water flooding. There is a need to move rapidly to some form of integrated water resource management using the optimal combination of above- and below-ground infrastructure (Chapter 12). The best approach is now more heavily influenced by the scale of intervention. NbSs are especially useful at the local and neighbourhood scale, bringing water into the public realm and providing local opportunities for direct and indirect use (Chapter 8). Management of runoff and rainfall at source like this will also provide some alleviation downstream, reducing system overload and consequent flooding and pollution. However, this will come at a price. NbSs, if not appropriately integrated into urban planning and design, can use more land. Making space to retrofit NbSs, particularly in existing developments, will become an increasing challenge into the future. Perhaps changing land uses and shared spaces can be an opportunity, in which new mobility initiatives, for example, can be utilised to implement local NbSs alongside other changes. This would also bring other benefits, including buffering the impacts from climate change.

Globally, there is evidence that wastewater 'reuse' is becoming more established, despite the 'yuck factor'. Decentralisation of water and wastewater assets for treatment and local provision is being more often considered and utilised in appropriate contexts. Onsite sanitation (Chapter 2) is therefore an option, but due to the ubiquitous problems of complex pollutants outlined above, it might not be a panacea for widespread use, at least in developed countries.

With strategic oversight, it is imperative to improve the certainty in understanding the nature of surface water threats in order to improve the robustness and resilience to responses to the risks. This may also improve preparedness for more extreme weather patterns and development pressures resulting from population movements and redistribution as well as growth. An integrated approach to stormwater management can provide opportunities to deliver innovative, intelligence-driven solutions to reduce stormwater discharges to sewers, control CSOs and help alleviate future challenges, thus contributing to economic development and growth (see Chapter 3).

Traditional urban drainage system design processes are undergoing change, perhaps more profound even than those following the advent of computational power in the twentieth century. The growing availability and ubiquity of sensors of all types, and their widespread use, and the accessible databases and real-time data streams made possible by the advancement of smart infrastructure and intelligence provide endless opportunities to track and understand the world. This, combined with agile decision-making (Chapter 9), will enable integration of stormwater networks, abstraction and discharge controls using dynamically controlled systems, which will deliver increased water supply security, river quality and environmental improvements, reduced flood risk and improved network robustness and resilience to future demands and impacts (Smith *et al.*, 2014). This requires data in real time for agile decision-making (Chapter 9). Considering the dynamic interactions of minor/major systems and bidirectional interaction – that is, water in/water out – but this would require a flexible approach to managing environmental impacts. It is therefore essential to develop quickly an integrated approach to water resource management with a diversified/sustainable water supply system, and while this is a strategic necessity, it requires an agile, long-term, regulatory and policy environment to work.

The development of smart infrastructure and intelligence is mirrored by a variety of computational models. Most users of they models for all parts of the water cycle and beyond are presuming that the core science and algorithms on which they are based are robust and fit for purpose, or at least that the 'trained' relationships from the data streams are reliable. The innovative and attractive graphical user interface (GUI) for model users can disguise inaccuracies and inappropriate use of scientific relationships in the embedded analytical engines. Often, with tight deadlines to produce outputs, users are trained on how to produce outputs using the GUI, but are not trained or given the interfaces to question and challenge the embedded assumptions. The approach is to monitor a few storm events and the flows and loads using the new generation of sensors and then to compare the model results with the monitoring data, in the verification process. The model is then used to 'predict' future events, usually of a different scale to the monitored verification events, and then to design measures to handle flows or to improve performance.

Worrying as this is, the next step in the evolution of urban drainage practice will be even more profound and introduce much greater opacity to model use. This is the fruition of the development of ML and AI, which will transform how urban drainage systems are planned, designed and operated. In smart cities there will no longer be a need for human to input in each of these aspects of urban drainage. Planning, design and (once built by humans) operation will all be automated. Although, maintenance and management of local NbSs will require human expertise and intervention, as will physical asset management and renewal. The effectiveness of these changes will only be apparent when the outcomes can be seen in the performance.

A further complication will be that every system and service will also undergo this digital transformation as smart cities develop. Fortunately, AI systems will readily operate across a multisector domain, and readily assimilate and respond to changing circumstances, be they external to systems (e.g. climate) or internal (e.g. service demands), informed by smart sensing. The challenge then is to try to understand how to establish AI systems that will be able to make the decisions that we understand to be best for urban drainage practice and with no unintended consequences within smart urban systems. The current and future generations of infrastructure specialists will no longer consider only one domain, such as urban drainage, but will need to be cross-systems experts for smarter cities. How expert knowledge and capabilities will be assimilated into AI processes is not yet clear, but work is currently being carried out on this, so it will be. In the future, the primary challenges will not be in the physical water domain, but in the cyber world, where cybersecurity will be paramount in order to ensure sustainable system and service continuity.

There is and always will be a need for the physical infrastructure and assets required to manage wastewater safely. However, changes in need and possibilities, linked with advancing technologies, provide new opportunities to move away from the 'dispose safely away from urban areas' model to one where every drop of rainwater and runoff is utilised as a resource and every sanitary flow considered for utilisation. This may be for anthropogenic purposes or to support flora and fauna.

Current legislation in England, the Water Industry Act 1991 (HMG, 1991), states that water companies have the following general duty to provide a sewerage system.

> It shall be the duty of every sewerage undertaker to provide, improve and extend such a system of public sewers (whether inside its area or elsewhere) and so to cleanse and maintain those sewers as to ensure that that area is and continues to be effectually drained.

With the increasing use of SuDS, blue-green infrastructure (BGI) and NbSs, this then poses the question as to whether water utilities have a role to play in providing natural capital for society as a whole and, if so, what and how. As we know, SuDS, BGI and NbSs can perform a drainage function, so could potentially fall within the general duties of providing a sewerage system. This then leads on to the question of the role of water utilities in the public health arena. Water is essential for life, and the water environment has clear demonstrable benefits to wellbeing and mental health. Water professionals are health service providers. Should water regulation therefore be reformed as health policy, just as when the forerunners of the Chartered Institution of Water and Environmental Management (CIWEM, Chapter 1) were established with their principal aims about public health? And just as the first drainage engineers were in fact 'public health engineers', finding ways to prevent the public coming into contact with sewage. Embracing such a mindset may accelerate the advancing use of SuDS, BGI and NbSs.

In summary, the advancement of urban drainage practice, in categories of practice, people, and policy and institutions, with a need to do the following.

(*a*) Practice for the future

- Ensure urban drainage is appropriately integrated with both the water cycle and other urban infrastructure systems and services.

- Use and make the most of all types of water, at source, stored, harvested, conveyed, used and recycled.
- Treat 'wastewater' as a resource wherever possible.
- Encourage and develop sound hydrological cycle management through capture, treatment and use of rainwater.
- Make better use of water resources in response to long-term environmental, economic and social objectives.
- Endeavour to provide urban drainage systems that are as sustainable as possible, ensuring that the future and how it may change is appropriately considered in option selection, including who will manage the assets.
- Ensure the effective identification of pollutant sources, pathways and impacts (such as microbial source tracking) in order to control these.
- Minimise pollutant impacts, especially to groundwater.
- Use systems that are part of and mimic nature wherever and whenever possible.
- Identify and manage risk by acknowledging and making explicit the uncertainties in any analyses used in planning and designing urban drainage systems.
- Select urban drainage systems that can provide as many benefits as possible to society, especially for the most vulnerable.
- Utilise appropriately scaled urban drainage systems, from local to centralised, as best to suit the purposes for use and maximise opportunities.

(*b*) People

- Work with people; not only the key professionals, but also communities.
- Facilitate multidisciplinary dialogue with stakeholders and communities to co-design solutions.
- Establish robust partnerships and alliances with other significant players relevant to urban drainage practice.
- Promote social learning, improved awareness and understanding of the challenges faced (for everyone, including society).
- Adopt a multistakeholder approach and action plan to identify and address knowledge and skills gaps.
- Identify what current urban drainage specialists must prepare for and ensure professionals become more multidisciplinary, especially in the world of smart infrastructure and intelligence.
- Give proper consideration to how, by whom and when asset management will be ensured for any new or renovated urban drainage system.

(*c*) Policy and institutions

- Work within existing policy and institutional frameworks with contracting models, but highlight any shortcomings that are inhibiting integration, collaboration, innovation (particularly at scale) and the use of best practice.
- Highlight and seek to find ways of dealing with changes and constraints from policy, funding and regulatory processes that are inhibiting innovation.
- Identify how institutions' governance and systems can engage fully with relevant stakeholders.

Many of the bullet points above should already be in place. However, for various reasons, these are often not observed in current urban drainage practice. It is possible that 'sewers' as known now may no longer be required in the future. Local and source control measures will better utilise 'wastes' directly and homes and properties may become more self-contained. There will always be a need for exceedance drainage measures, but these would be expected, as far as practicable, to utilise NbSs. Otherwise, conveyance of excess flows will be to watercourses and storage systems above and below ground for utilisation. However, for the foreseeable future, existing assets need to remain in use in developed countries, due to the unaffordability of their abandonment, thus incurring the need for huge investments for refurbishment in the current widespread neglected and decaying systems. This will provide opportunities to use them differently for additional purposes, including for heat or energy recovery. Ideally, the developing world should not follow the practices of the developed world, but should learn from the resource intensity of traditional drainage and adopt appropriate systems and technologies. Many of the technologies used in the developing world will herald the future for how the developed world will need to change if it is to become more sustainable.

13.8. To conclude: *ça donne à réfléchir*!

The idiom '*la nourriture pour la pensée*' or 'food for thought' literally translates as '*ça donne à réfléchir*' – 'it makes one think'. It is hoped that this book has, or will, make urban drainage professionals and others that interact with urban drainage consider water and the role it plays within each respective discipline and how water can be brought into designs and landscapes to reduce impacts to the natural environment from urban drainage systems and create social and natural capital.

The following quote from Victor Hugo, recognised as the most influential French romantic writer of the nineteenth century, who wrote one of the most famous books to immortalise sewers: *Les Misérables*, and is often identified as the greatest French poet, is perhaps quite apt to finish on: '*On résiste à l'invasion des armées; on ne résiste pas à l'invasion des idées*' (Dicocitations, 2023). It translates as 'An invasion of armies can be resisted, but not an idea whose time has come.' Perhaps the time of integrated water management is here. As engineers, scientists, academics, architects, landscape architects and professionals, we have a duty to our communities and society to 'integrate' our disciplines and work together. It is incumbent on us to make it happen.

REFERENCES

Agrawal A, Joshua G and Goldfarb A (2018) *Prediction Machines: The Simple Economics of Artificial Intelligence*. Harvard Business Review Press, Boston, MA, USA.

Bill & Melinda Gates Foundation (2023) Water, Sanitation & Hygiene. https://www.gatesfoundation.org/our-work/programs/global-growth-and-opportunity/water-sanitation-and-hygiene (accessed 09/08/2023).

Butler D, Digman CJ, Makropoulos C and Davies JW (2018) *Urban Drainage*, 4th edn. CRC Press, Boca Raton, FL, USA.

City of Portland (2020) *Stormwater Management Manual*. Bureau of Environmental Services, Portland, OR, USA. See https://www.portland.gov/bes/stormwater/swmm (accessed 09/08/2023).

Defra (Department for Environment, Food and Rural Affairs) (2016) *Creating a Great Place for Living. Enabling Resilience in the Water Sector*. Defra, London, UK. See https://assets.publishing.service.gov.uk/government/uploads/system/uploads/attachment_data/file/504681/resilience-water-sector.pdf (accessed 09/08/2023).

Defra (2023) Biodiversity Net Gain. https://www.gov.uk/government/collections/biodiversity-net-gain#:~:text=Biodiversity%20net%20gain%20(%20BNG%20)%20is,than%20it%20was%20before%20development (accessed 09/08/2023).

Druade S, Keedwell E, Kapelan Z and Hiscock R (2022) Multi-objective optimisation of sewer maintenance scheduling, *Journal of Hydroinformatics* **24(3)**: 574–589, 10.2166/hydro.2022.149.

DWA (German Association for Water, Wastewater and Waste) (2012) *Skills Challenges in the Water and Wastewater Industry.* UNESCO-UNEVOC International Centre, Bonn, Germany. See https://unevoc.unesco.org/fileadmin/user_upload/docs/WaterBooklet.pdf (accessed 09/08/2023).

EA (Environment Agency) (2021) *Water Industry Strategic Environmental Requirements (WISER)*. Environment Agency, London, UK. See https://www.gov.uk/government/publications/developing-the-environmental-resilience-and-flood-risk-actions-for-the-price-review-2024/water-industry-strategic-environmental-requirements-wiser (accessed 09/08/2023).

Energy & Utility Skills Partnership (2017) *Many Skills One Vision. Energy and Utilities Workforce Renewal and Skills Strategy: 2020.* Energy & Utility Skills Ltd, Solihull, UK. See https://www.euskills.co.uk/wp-content/uploads/2017/11/Workforce-Renewal-and-Skills-Strategy-2020.pdf (accessed 09/08/2023).

Energy & Utility Skills Partnership (2020) *Many Skills One Vision. Workforce Renewal and Skills Strategy 2020–2025.* Energy & Utility Skills Ltd, Solihull, UK. See https://www.euskills.co.uk/wp-content/uploads/2020/06/EUSP-Skills-Strategy-2020-2025_compressed-v0.3.pdf (accessed 09/08/2023).

Finlay S (2018) *Artificial Intelligence and Machine Learning for Business: A No-Nonsense Guide to Data Driven Technologies*, 3rd edn. Relativstic, Preston, UK.

Gartner (2021) 5 Trends Drive the Gartner Hype Cycle for Emerging Technologies. https://www.gartner.com/smarterwithgartner/5-trends-drive-the-gartner-hype-cycle-for-emerging-technologies-2020/#:~:text=Learn%20more%3A%20About%20the%20Gartner%20Hype%20Cycle%20Methodology&text=Algorithmic%20trust,Digital%20me (accessed 09/04/2021).

HMG (Her Majesty's Government) (1991) *Water Industry Act 1991.* The Stationery Office, London, UK. See https://www.legislation.gov.uk/ukpga/1991/56/section/94 (accessed 09/08/2023).

HMG (2021) *Environment Act 2021.* The Stationery Office, London, UK. See https://www.legislation.gov.uk/ukpga/2021/30/contents/enacted (accessed 09/08/2023).

Dicocitations (2023) On résiste à l'invasion des armées, on ne résiste pas à l'invasion des idées - Victor Hugo. https://www.dicocitations.com/citations/citation-7505.php (accessed 09/08/2023).

ICE (Institution of Civil Engineers) (2022) COP27: The key takeaways for civil engineers. https://www.ice.org.uk/engineering-resources/information-sheets/cop27-key-takeaways-for-civil-engineers/ (accessed 09/08/2023).

Juuti PS and Katko TS (eds) (2005) *Water, Time and European Cities – History matters for the Futures.* WaterTime project. EU Contract No. EVK4–2002–0095. See https://trepo.tuni.fi//bitstream/handle/10024/128378/juuti_katko_water_time_and_european_cities.pdf?sequence=1 (accessed 09/08/2023).

Kapelan Z, Weisbord W and Babovic V (2020) *Digital Water: Artificial Intelligence Solutions for the Water Sector.* International Water Association, London, UK. See https://iwa-network.org/wp-content/uploads/2020/08/IWA_2020_Artificial_Intelligence_SCREEN.pdf (accessed 09/08/2023).

Leach JM and Rogers CDF (2020) Briefing: embedding transdisciplinarity in engineering approaches to infrastructure and cities. *Proceedings of the Institution of Civil Engineers – Smart Infrastructure and Construction* **173(2)**: 19–23, 10.1680/jsmic.19.00021.

Leflaive X, Dominique K and Alaerts G (eds) (2022) *Financing Investment in Water Security: Recent Developments and Perspectives.* Elsevier, Amsterdam, the Netherlands.

Liptan TW and Santen JD (2017) *Sustainable Stormwater Management.* Timber Press, Portland, OR, USA.

Mollerup AL, Mikkelsen PS, Thornberg D and Sin G (2017) Controlling sewer systems – a critical review based on systems in three EU cities. *Urban Water Journal* **14(4)**: 435–442, 10.1080/1573062X.2016.1148183.

Natural England (2022) *Nutrient Neutrality and Mitigation: A Summary Guide and Frequently Asked Questions* (NE776). Natural England, UK. See https://publications.naturalengland.org.uk/publication/6248597523005440 (accessed 21/05/2023).

NIC (National Infrastructure Commission) (2023) *Infrastructure Progress Review 2023*. NIC, London, UK. See https://nic.org.uk/studies-reports/infrastructure-progress-review-2023/ (accessed 09/08/2023).

OECD (Organisation for Economic Co-operation and Development) (2020) *Financing Water Supply, Sanitation and Flood Protection: Challenges in EU Member States and Policy Options*. OECD Studies on Water. OECD Publishing, Paris, France.

Pedersen AN, Borup M, Brink-Kjær A, Christiansen LE and Mikkelsen PS (2021) Living and prototyping digital twins for urban water systems: towards multi-purpose value creation using models and sensors. *Water* **13**: 592, 10.3390/w13050592.

Perciavalle P, Woodall P, Abrera J, Vallabhaneni S and Johnson K (2017) The digital water/wastewater utility of the future: case studies in leveraging smart utility technology and best management practices. In *Proceedings of the 90th Annual Water Federation Technical Exposition and Conference*. Chicago, IL, USA. Water Environment Federation.

Peters PE and Zitomer DH (2021) Current and future approaches to wet weather flow management: a review. *Water* **93(8)**: 1179–1193, 10.1002/wer.1506.

Popkova EG, Segi BS and Bogoviz A (2023) Editorial: evolution of environmental economics and management in the age of artificial intelligence for sustainable development. *Frontiers in Environmental Science* **11**: 1176612, 10.3389/fenvs.2023.1176612

Rasheed A, San O and Kvamsdal T (2020) Digital twin: values, challenges and enablers from a modeling perspective. *IEEE Access* **8**: 21980–22012, 10.1109/ACCESS.2020.2970143.

Schwab K (2016) The Fourth Industrial Revolution: what it means and how to respond. World Economic Forum. https://www.weforum.org/agenda/2016/01/the-fourth-industrial-revolution-what-it-means-and-how-to-respond/ (accessed 09/04/2021).

Smith B, McKay G, Ashley R and Digman C (2014) Evaluating the benefits and risks of water sensitive urban design in the Yorkshire region. In *Proceedings of the 13th International Conference on Urban Drainage*. Sarawak, Borneo.

Sun C, Romero L, Joseph-Duran B *et al.* (2020) Integrated pollution-based real-time control of sanitation systems. *Journal of Environmental Management* **269**: 110798, 10.1016/j.jenvman.2020.110798.

SWAN (Smart Water Networks Forum) (2023) Wastewater Network Management. https://www.swan-tool.com/wastewater-network-management (accessed 09/08/2023).

Tscheikner-Gratl F, Caradot N, Cherqui F *et al.* (2019) Sewer asset management – state of the art and research needs. *Urban Water Journal* **16(9)**: 662–675, 10.1080/1573062X.2020.1713382.

USEPA (US Environmental Protection Agency) (2022) *Proposed 2022 Clean Water Act Financial Capability Assessment. Guidance*. USEPA, Washington, DC, USA. See https://www.epa.gov/system/files/documents/2022-02/2022-proposed-fca_feb-2022.pdf (accessed 09/08/2023).

Vallabhaneni S and Speer E (2011) Real-time control to reduce combined sewer overflows. *WaterWorld*, 1 Feb. https://www.waterworld.com/home/article/16192327/realtime-control-to-reduce-combined-sewer-overflows (accessed 11/2021).

Van Loenen A, Heeringen KJ, Nooyen RV and Velzen E (2012) Leveraging existing infrastructure for central automatic control of multiple sewer systems. In *Proceedings of the 10th International Conference on Hydroinformatics - HIC 2012 "Understanding Changing Climate and Environment and Finding Solutions"*. 14–18 July 2012, Hamburg, Germany, pp. 1–9.

Van Riel WJ, Post J, Langeveld J, Herder P and Clemens F (2017) A gaming approach to networked infrastructure management. *Maintenance, Management, Life-Cycle Design and Performance* **13(7)**: 855–868, 10.1080/15732479.2016.1212902.

Vezzaro L, Christensen ML, Thirsing C, Grum M and Mikkelsen PS (2014) Water quality-based real time control of integrated urban drainage: a preliminary study from Copenhagen, Denmark. *Procedia Engineering* **70**: 1707–1716, 10.1016/j.proeng.2014.02.188.

Water UK (2016) *21st Century Drainage Programme – the Context*. Water UK, London, UK. See https://www.water.org.uk/wp-content/uploads/2018/12/21CD-Context-doc.pdf (accessed 08/08/2023).

Water UK (2021) *A Framework for the Production of Drainage and Wastewater Management Plans*, revised edn. Water UK, London, UK. See https://www.water.org.uk/wp-content/uploads/2021/10/DWMP_Framework_Report_Main_Report_September_2021.pdf (accessed 09/08/2023).

Winpenny J (2015) *Water: Fit to Finance?* World Water Council and OECD, Marseille, France. See https://www.worldwatercouncil.org/sites/default/files/Thematics/WWC_OECD_Water_fit_to_finance_Report.pdf (accessed 09/08/2023).

WWAP (United Nations World Water Assessment Programme) and UN-Water (2018) *Nature-Based Solutions for Water. The United Nations World Water Development Report 2018*. UNESCO, Paris, France. See https://unesdoc.unesco.org/ark:/48223/pf0000261424 (accessed 09/08/2023).

Zhou S, Yu B, Lintner BR, Findell KL and Zhang Y (2023) Projected increase in global runoff dominated by land surface changes. *Nature Climate Change* **13**: 442–449, 10.1038/s41558-023-01659-8.

Ashley R, Smith B, Shaffer P and Caffoor I
ISBN 978-0-7277-4178-3
https://doi.org/10.1680/icehudp.41783.383

Chapter 14
Case studies

Having set the scene – the background, current perspectives and future outlook for urban drainage practice – this chapter presents a number of case studies, mainly from the UK. These offer a wide range of approaches, practical applications and solutions to the many, and sometimes complex, issues faced. The case studies are an extensive variety, covering some of the traditional and long-established topics, such as flooding, stormwater, blue-green infrastructure, pollution, water quality, asset management and community engagement, collaboration and partnerships; through to intelligent surface water management, digital innovation and the application of technologies new to drainage practitioners; and unconventional thinking, such as artificial intelligence, machine learning, reinforcement learning and adaptive learning. These provide a clear demonstration of the art of the possible and how urban drainage, and its practitioners, has and continues to evolve to provide a broader range of societal, environmental and economic benefits. This book also signposts the reader to PD ISO/TR 24539:2021 Examples of good practices for stormwater management, which contains 19 relevant case studies from seven countries.

14.1. Madison Valley Stormwater Improvement Project

This case study outlines an innovative solution implemented by Seattle Public Utilities (SPU) in the north west of the USA to address flood risk in the area. The project was constructed in two phases and has two locations in Seattle's Madison Valley.

- Phase 1: Madison Valley stormwater storage facility
- Phase 2: Washington Park stormwater storage facility

Together, the two sites and underground infrastructure can contain the volume of stormwater that caused flooding and sewer backups in the two largest storms in 157 years of records. This significantly reduces the potential for stormwater flooding and sewer backups in the area while at the same time creating usable open space for the community.

The city of Seattle's sewerage infrastructure is a complex mix of separate, partially separate and combined systems – about one-third of each. There had been a history of flooding at a frequency of four times per annum in the area over a period of about 30 years. The area's drainage system was built in the early 1970s and was designed for a 25-year storm event, but storms of increasing intensity were occurring more frequently. The Forward Thrust initiative approved in 1968 provided resources to separate sewage from stormwater; however, the effort was stopped in the late 1970s under the Federal Clean Water Act. In 2004, stormwater overwhelmed the combined sewer, flooding local residents, resulting in US $2.5 million in damages that had to be paid to 12 homeowners (D'Amore Law Group, 2011; *Seattle Times*, 2011).

Analysis of rainfall and modelling of the existing system was undertaken to characterise the problem and develop both an interim and long-term solution. The interim solution consisted of the creation of a detention basin to store 3785 m^3. In 2006, the interim solution was nearing completion when a 'series of extraordinary events' led to the flooding of more than 30 homes and the collapse of a privately owned retaining wall. The 2006 event heightened public awareness and concentrated political minds and resolve.

14.1.1 Phase 1

Following the catastrophic event, SPU increased the scope of phase 1 of the project to expand and improve the detention basin from 3785 m^3 to 6435 m^3 of storage. The stormwater storage facility expansion was designed in 2008–2009 and built between 2009 and 2010.

Six properties were purchased for the interim stormwater storage area. The property owners who agreed to sell were willing to do so, safe in the knowledge that their neighbours would be protected from flooding and a community amenity space would be provided. The plan was designed with the community. Many public meetings were held through the design stages to gather community and stakeholder input on the project's public involvement plan, site design and landscaping plan. The people who lived in and around the park supported and provided input to SPU to help decide what features to include.

Site enhancement of phase 1 improved the detention basin space to make it a neighbourhood natural amenity. It stores stormwater during very large storms, but most of the time it serves as an attractive open space for the community. The site includes a creative design providing a green and inviting environment with native planting, an intermittent creek flow, walking paths, open green space, a sandy play area for children, art sculptures and broad stone steps providing amphitheatre seating (see Figure 14.1).

14.1.2 Phase 2

To provide as sustainable as possible a long-term solution for stormwater storage in the Madison Valley and further protect the area against flooding from storm events, a stormwater storage facility was built at Washington Park to augment the Madison Valley Stormwater Park. The location was based on engineering factors as well as impacts to the community, cost and schedule constraints. The two sites were connected by approximately 800 m of 1200 mm internal diameter underground pipes. The storage facility was designed to hold an additional 8328 m^3 of stormwater – 4921 m^3 in a below-ground

Figure 14.1 Detention basin with amphitheatre seating

Figure 14.2 Artist's impression of storage tank and rain garden

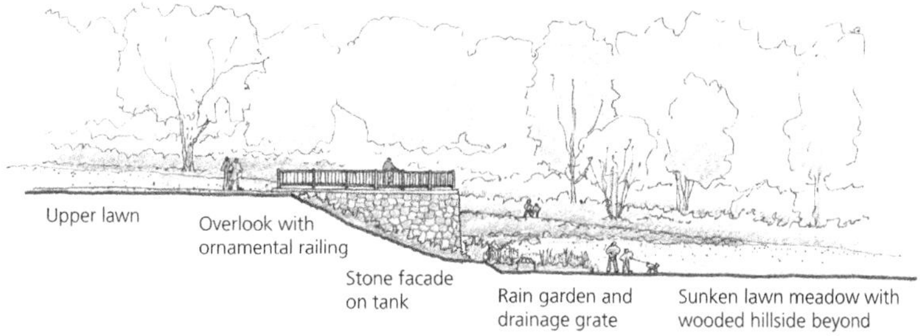

stormwater tank and an above-ground landscaped stormwater storage area to hold 3407 m^3 of excess stormwater during periods of heavy rain. Together, the infrastructure is enough to protect against flooding from storm events as big as the worst two on record in Seattle.

The above-ground storage tank is a major feature of this project, doubling its function as public infrastructure. Not only does it hold a huge amount of water, but it was designed with a plaza and scenic overlook, affording a view of the rain garden below and Washington Park's play fields (see Figure 14.2).

Artwork is fully integrated within the site's natural stormwater management system. Rainwater drains down the top of the storage tank where it is slowed and directed to trickle down the chevron-patterned rock wall. Cavities in the wall for birds enhance the integration with the natural environment.

14.1.3 Conclusion

It has become a regulatory requirement in Washington state to use green stormwater infrastructure (GSI). SPU is a strong supporter of GSI, and its stormwater code requires the use of GSI to the maximum extent feasible for all public and private projects when a landscaped area is developed above a defined threshold. However, there are other important drivers as to why SPU wants to do more with GSI and use it as a preferred method where feasible. GSI provides additional benefits and serves as a reminder to citizens that their actions matter. Like all public utilities, SPU has ongoing challenges, whether regulatory driven or to achieve service levels for its residents.

This project demonstrates what can be achieved. The stormwater park is a great example of a nature-based solution and the multibenefit approach being used by SPU. A measure of its success is that since the project was completed there have been no flooding claims. The solution is inspirational, safe, economic, operable, performing, deliverable, socially acceptable and transferrable. Storing 6500 m^3 of stormwater surface water in the middle of a residential area is no mean feat.

Community engagement was critical to the success, to gather community input into the project plan, site design and landscaping plan. The skills and resources required to achieve successful customer engagement should not be underestimated.

Madison Valley phase 1 and phase 2 provide approximately 15 000 m^3 of stormwater storage. The total cost reduced from US $80 million to US $35 million in comparison with a traditional grey infrastructure solution.

REFERENCES

D'Amore Law Group (2011) https://damorelaw.com/city-of-seattle-settles-with-homeowners-over-decades-long-flooding-and-sewage-problem/ (accessed 11/08/2023).

Seattle Times (2011) https://www.seattletimes.com/seattle-news/seattle-pays-25-million-to-flooded-madison-valley-homeowners/ (accessed 11/08/2023).

Credits: Seattle Public Utilities
Author: Brian Smith

14.2. Gosforth Valley constructed wetland

Scientific estimates show that 64% of the world's wetlands have disappeared since 1900. Inland wetlands are disappearing at a faster pace than coastal ones, but the overall trend is clear. Given the challenges we are facing, it is incumbent on us, morally and professionally, to reverse the declining loss of the world's wetlands. So, what is driving this loss? Unfortunately, wetlands are often viewed as wasteland; something to be drained, filled and converted to other purposes. However, the Lea Brook Valley Volunteers have a different view. The Gosforth Valley wetland in Derbyshire, England, is owned by Yorkshire Water but managed and maintained through the support of Dronfield Town Council and the Lea Brook Valley Volunteers. Known locally as 'the swamp', it has undergone a dramatic remodelling to improve flood attenuation and attract more visitors to the urban wildlife haven. After years of neglect, the Lea Brook Valley secured a biodiversity fund of £48 500 from Yorkshire Water to transform the wetland into an oasis for wildlife, with the help of volunteers.

The site forms part of the larger Lea Brook Valley, which is a green corridor into Dronfield town centre. An innovative creation, it serves an important dual function: to control and store stormwater to protect the town of Dronfield from flooding, and to provide habitat. Designed to store excess water during times of heavy rainfall, it collects water from surrounding built-up areas that may otherwise flood. Water levels are controlled by a balancing dam that gradually releases water into the Lea Brook (see Figure 14.3).

Figure 14.3 Arial view of the Gosforth Valley wetland (map data: Google, ©2022)

Further development and continuous improvement of the site has enhanced the beauty, amenities and wildlife habitat for the benefit of the residents and those who have interest in conservation and biodiversity. Expert wetland design advice helped to ensure the sustainability of any landscape changes, which are carried out in coherence with the operational role of the site as a retention pond and the hydrological regime.

Volunteers work with local conservation groups, the local parish council, Yorkshire Water and others, carrying out many activities, such as conserving the local ancient hedgerow and creating a nature trail. They ensure the wetland is free from rubbish and is well maintained by undertaking weekly rotas for litter picking and general maintenance. Information boards (see Figures 14.4 and 14.5) have been provided to support community, educational and environmental enhancement. Active management programmes are in place to record progress through formal monitoring of the site, consisting of plant surveys, hydrological monitoring and amphibian surveys.

The habitat has been developed to support wetland bird species such as shoveller (*Anas clypeata*), grey heron (*Ardea cinerea*) and snipe (*Gallinago gallinago*); invertebrates such as the common darter (dragonfly) (*Sympetrum striolatum*), water boatman (*Notonecta glauca*) and water flea (*Bosmina longirostris*); amphibians such as the common frog (*Rana temporaria*) and common toad (*Bufo bufo*); and mammals such as the European water vole (*Arvicola amphibius*).

Ground and conservation work, which took over a year and a half to complete, has resulted in the excavation and deepening of the open water area with the planting of common reed around it to improve the wildlife habitat. A decking area has also been created overlooking the wet pond for visitors to enjoy, with picnic tables installed near to the dry basin.

This site is a great example of how working together through partnering and trust supports community, educational and environmental enhancement. Community engagement has brought about friendship. Strangers have become friends. Looking after the environment is rewarding and local volunteers receive training on how best to manage the site. A programme to inspire and encourage schools' and young people's engagement with the natural world has been developed, including interaction with the local college/high school, which uses the site for its biology practical assessment.

Rowan Williams, a former Archbishop of Canterbury, during a speech to the UN, referred to Gosforth Valley as an exemplar site, which has also been recognised as a Construction Industry Research and Information Association (CIRIA) national best practice case study for biodiversity.

14.2.1 Next steps: treating urban pollution

Constructed wetlands have been identified as a technique for removing phosphate and nitrate from surface runoff, thus reducing the problem of eutrophication caused by excessive phosphate and nitrate content in aquatic ecosystems. Plant species, substrate composition, structure and microbial population of constructed wetlands can influence phosphorus removal effectiveness. The wetland was monitored during the summer of 2022, with water, soil and plant samples collected to understand the potential for phosphorous removal. Phosphatase activity was used as an indicator to assess the ability of roots, stems, leaves, soil and water samples to remove phosphate. The results indicated that the constructed wetland could remove phosphate, and that plants had a significant effect on the removal of phosphate.

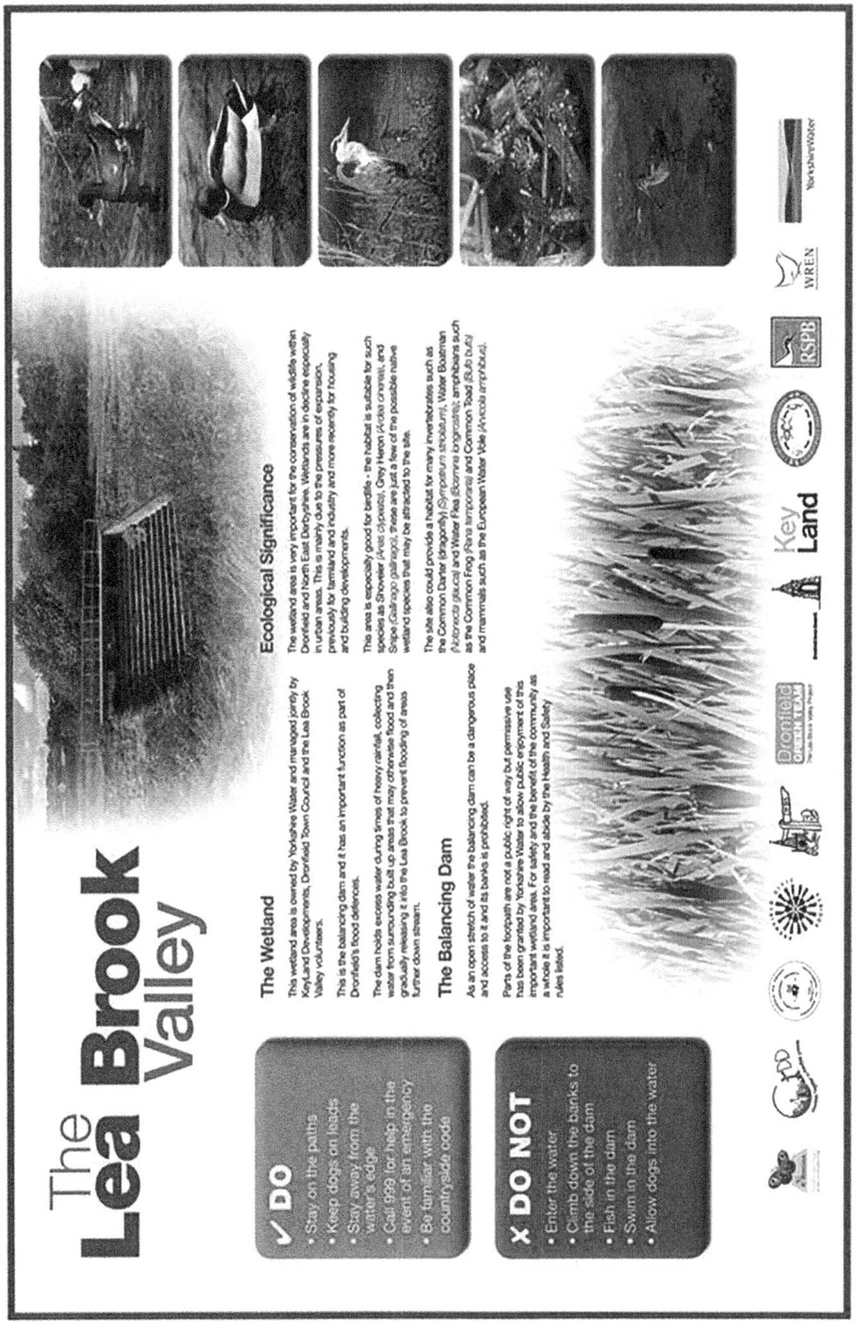

Figure 14.4 Information board

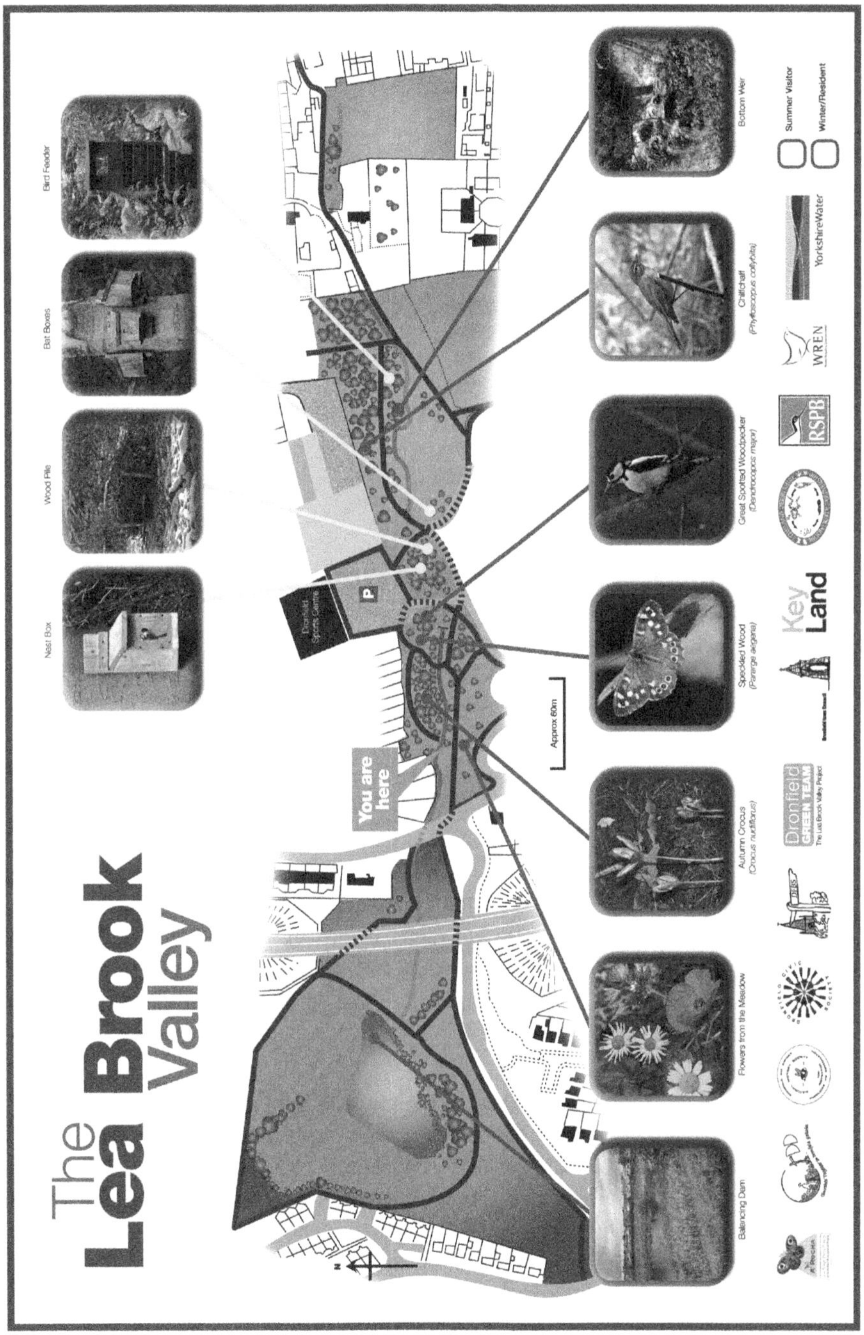

Figure 14.5 Species in the valley

By comparison, *Typha* spp. were more effective in removing phosphate than *Phragmites* spp. Moreover, roots were more effective in removing phosphorus than stems and leaves. Soil samples showed low phosphatase activity, thus a low potential for phosphorus removal when compared with plants. It is therefore feasible to use plants to remove phosphate from pollution originated by urban surface runoff. *Phragmites* spp. and *Typha* spp. both have the potential to remove phosphate, yet *Typha* spp. were the most effective plants in removing phosphate in this study.

14.2.2 To conclude

The management and maintenance of this site requires skill sets not traditionally employed by sewerage undertakers – such as horticulture, and landscape and vegetation management. The success of this joint venture is really about the aspirations and vision of individuals. This would not have been possible in isolation and demonstrates the need for, and benefits from, people, communities and partnerships. The study on phosphate and nutrient removal also demonstrates the important role that constructed wetlands have in treating urban pollution.

The key take home message is 'Nature is a gift that keeps on giving'.

Credits: Revd David Walker, Dr Isabel Douterelo Soler

Author: Brian Smith

14.3. Northallerton Flood Forum: a strategic approach to flood risk management

The Northallerton catchment in North Yorkshire, England, with the villages of Brompton and Romanby, is in a low-lying river valley, surrounded by steep, well-drained agricultural land, which makes it susceptible to flash flooding. The town is located on the confluence of the Willow Beck, Sun Beck and Turker Beck. These watercourses and their associated flood zone 2 and 3 areas, in combination with drainage and sewerage infrastructure inadequacies and surface water runoff, are the main influence on flooding within the town and impose the main restriction on development.

Largely culverted, the Sun and Turker Beck watercourses run through suburbs and the centre of the town, causing flood threats to amenities, including Grade II listed buildings, conservation areas and listed monuments as well, as the Friarage Hospital. The rivers in the catchment are constricted by bridges, culverts and their channel capacities. At times of peak flow, excess floodwaters have overtopped the watercourse in the agricultural land on their approach to the culverts, causing flooding in the east of the town.

Extensive flooding during November 2000 and the summer months of 2002 amplified the level of flood risk in the community and, along with heightened public awareness and political strength of feeling, prompted the establishment of a multiagency Flood Forum. The purpose of the Flood Forum was to engage with external organisations and develop inter-agency partnership working, identify key stakeholders and promote a shared approach to problem understanding, solution development and the implementation of flood alleviation measures to prevent disruption and damage caused by such events.

The Flood Forum included representatives from North Yorkshire County Council (NYCC), Hambleton District Council, Yorkshire Water, Northumbrian Water, the Environment Agency, the River Wiske

Internal Drainage Board and parish and town councils. The Flood Forum worked closely with consultants Mouchel Parkman and MMB to review the latest floodplain risk mapping report, and prepared a forward strategy for short- and long-term measures to alleviate flooding in Northallerton and the adjoining villages of Brompton and Romanby. Key objectives were set to

- assess the nature, scale and risk of flooding
- clarify roles, responsibilities, constraints and limitations of stakeholders and the challenges they faced
- raise awareness, stimulate dialogue, share knowledge and expertise
- identify and understand the inter-relationship of drainage systems and mechanisms for flooding
- gain a better understanding of the dynamics of flooding
- develop an effective flood risk management strategy within an integrated framework.

The review reported on the interactions of the public sewer networks and watercourses during extreme rainfall events. A two-stage programme of works was identified, aimed at individual perimeter schemes and town centre schemes. The key stakeholders agreed that the problems associated with flooding from watercourses within Northallerton and associated drainage networks should be addressed in an innovative, integrated, catchment management approach that encompassed all the known flooding drivers and concerns. Critical areas included were watercourse and sewer network hydraulic modelling and the development of a drainage management study.

Through the development of the inter-agency partnership, numerous flooding problems were identified and resolved using both operational and capital solutions, with clear benefits delivered to customers and very real and tangible benefits to all organisations and stakeholders. This was a clear demonstration of what can be achieved through willing cooperation between agencies. Improved communication and relationships led to a clear understanding of local issues, problems, risks, roles and responsibilities and proved an excellent basis for working together towards the common goal of improved customer service.

Sun Beck and Turker Beck, both of which suffer from exceedance during peak flows and urban runoff, had at some point in the past been diverted and culverted to facilitate development and erroneously added to the statutory sewer records as public surface water sewers. With no evidence to suggest that they were either constructed as public sewers or vested as public sewers, they were reclassified as watercourses to reflect their correct legal status and were enmained by the Environment Agency as critical ordinary watercourses (COWs).

The primary design ethos was to provide upstream attenuation, to slow down excess rural flows entering the urban drainage systems until the excess urban flows had dissipated, resulting in improved capacity for urban runoff thus preventing town centre flooding. Major works comprised several rural and one urban surface water storage areas to control watercourse flow rates through the urban area and provide a 1 in 100-year flood event protection. It should be noted that this flood protection scheme could not have proceeded without reclassification of the watercourses. Yorkshire Water undertook significant investment to resolve problems associated with the public foul and surface water sewerage systems.

Figures 14.6 and 14.7 illustrate the extensive number of stakeholders and consultees involved in the process.

Figure 14.6 Key stakeholders

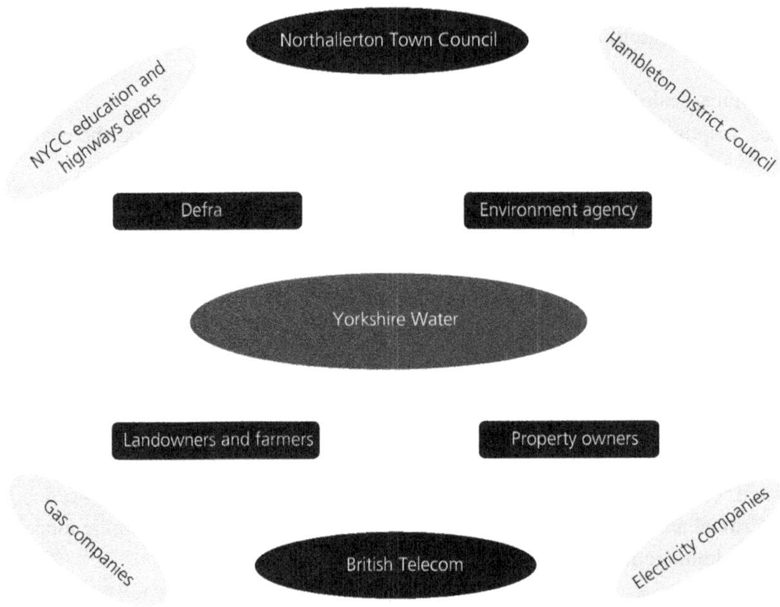

The initiative resulted in an excellent, trusted relationship with local authorities and customers. It improved operational, regulatory and financial performance and directed limited finances in the best possible way. Successful scheme delivery and problem resolution were testament to the dedication and commitment of the Flood Forum. The approach was widely acknowledged as visionary and innovative. Along with promoting a shared approach to problem understanding and solution development, this strategic approach engendered positive attitudes and cohesive teamwork. As a tried and tested, responsible, professional approach, this is a business practice that can be developed and implemented at local or regional level. The approach aligned with government proposals to develop new strategies for flood risk management and was used by the Department for Environment, Food and Rural Affairs (Defra) as the basis of the Integrated Urban Drainage (IUD) pilot studies in 2008.

To conclude, the Flood Forum was unique and visionary. The secret to its success was strong leadership and unstinting support. The project was about committing to working with others, listening to stakeholders' mutual customers, understanding the issues and developing an integrated strategy for flood risk management. It was very much about restoring the faith of customers and the community. It was about honesty and trust.

Author: Brian Smith

Figure 14.7 Consultees

Archaeologists	Contractors	English Nature	Rail companies
Association of British insurers	Geophysical surveyors	Hydraulic modellers	Countryside Heritage
County and district councillors	Church Commissioners	Geotechnical engineers	Topographical surveyors
Civil engineers	Ecologists	English Heritage	Structural engineers
Land surveyors/negotiators	Planning supervisor (H&S)	Site supervisors	Planners
Local schools	Mechanical engineers	Project co-ordinators	Visual 3D modellers
Natural history society	Quantity surveyors	Sports clubs	Panel engineers (dam specialists)
	Brompton and Romanby Parish Councils	River Wiske Internal Drainage Board	

14.4. Killingworth and Longbenton

This case study outlines the collaborative and sustainable approach carried out by the Northumbria Integrated Drainage Partnership (NIDP). This was an innovative partnership-working initiative for surface water management with Northumbrian Water, the Environment Agency and North Tyneside Council in England to reduce the risks of highway, external and internal property flooding for residents of Killingworth and Longbenton. This work has increased the capacity of the combined sewer, thereby reducing the risk.

The Longbenton area has a history of flooding, exacerbated by a main river (the Longbenton Letch) flowing into a combined sewer adjacent to the worst-affected properties. A second water body, Killingworth Lake, overflowed into the combined sewer upstream of the letch, the combination of which resulted in reduced sewer capacity. During stormwater events this resulted in

- increased flood risk to 3594 properties
- disruption to strategic transport links in the area, affecting residents and businesses and impacting the local economy
- increased levels of stormwater entering the combined sewer system and passing forward to the wastewater treatment works
- increased operation of combined sewer overflows (CSOs) discharging to the Ouseburn river, thus contributing to its status as moderate in the EU Water Framework Directive (WFD) classification
- inability of the existing drainage system to accommodate future growth and climate change.

It was not practical to assign responsibility for managing the flood risk to a single Risk Management Authority due to the integrated nature of the drainage system. A collaborative approach was required. The overall project objective was to design and implement innovative, sustainable solutions for stormwater management that reduced flood risk while increasing the potential for development in the area and increased local biodiversity.

The NIDP entered a formal partnership to develop and deliver the scheme, overseen by a project board including representatives from the three key stakeholders. This new approach required the partners to address specific challenges of stakeholder management, funding/programming and new ways of working. A detailed customer engagement plan was jointly developed by the partners to ensure effective communications with customers and other third-party stakeholders. A considerable amount of engagement was carried out with local residents and specialist groups to explain the project and seek public opinion, which informed various design changes prior to submission of a formal planning application. A special mobile customer engagement bus was designed to help facilitate this (Figure 14.8).

The philosophy adopted was that of stormwater management good practice in accordance with the sustainable drainage systems (SuDS) management train, essentially looking to deal with surface water as close to the source as possible. At an early stage, the partners chose to maximise the social, environmental and financial benefits of the scheme by promoting SuDS options for surface water management.

A number of physical constraints presented challenges for the watercourse diversion routes, including a new school build, a village green, a cemetery, a public park, the East Coast Main Line railway and farmland.

Figure 14.8 Customer engagement bus

The scheme was designed to accommodate rainfall events with a return period of once in 40 years for the sewerage system and once in 100 years for surface water. It comprised three principal components.

- Removal of Longbenton Letch (main river) from the combined sewer by diverting the flows northwards into the adjacent Forest Hall Letch (main river).
- Stormwater attenuation features at two locations
  - a managed flood plain at the upper reaches of Forest Hall Letch
  - an amphitheatre detention basin, retention ponds and wetland areas
  - providing 5000 m^3 of storage designed to facilitate future housing growth on adjacent land.
- Diversion of Killingworth Lake overflow from the combined sewer to an existing watercourse and the natural environment. This attenuation area has a capacity of 12 000 m^3. Three floating island ecosystems (Figure 14.9) were installed to improve biodiversity and provide natural habitats for wildlife, such as fish and nesting birds in the area.

Different funding and contract procurement processes were aligned to achieve the required outcomes for all parties. These were brought together in a collaborative agreement, which included the financial and non-financial contributions of each party, definition of roles and responsibilities and review processes to ensure the project objectives were met. The overall scheme benefits were assessed using the CIRIA B£ST tool, which helped in calculating the sum that the Environment Agency was able to contribute. This collaborative partnership approach with shared resources enabled the scheme to happen, otherwise it would have been unaffordable. B£ST identified total benefits in excess of £51 million, with a benefit to cost ratio of 10:1.

Figure 14.9 A floating island ecosystem on Killingworth Lake

The project initiated innovations in ways of working, most notably the integrated use of stakeholder powers for the diversion of the river and drainage assets, as well as the SuDS implementation itself. Other important aspects included the joint communications strategy, risk apportionment and management of compensation claims.

Early communication, consultation and trust were key to the success of the project. Stakeholder engagement was more protracted than for conventional schemes. However, taking the necessary time ensured that great outcomes were achieved for project partners and the public.

Public acceptance of SuDS projects cannot be taken for granted and the case for them must be explained to all stakeholders. Therefore, it is essential to factor in sufficient time for presentations and negotiations on SuDS proposals.

Table 14.1 identifies the key benefits accrued from the project.

Credits: Northumbrian Water Limited; Mike Madine, Northumbrian Water Limited
Author: Brian Smith

14.5. Improving the quality of runoff from the M56 motorway near Lymm in Cheshire

Mag Brook is a small brook that runs through a pleasant nature reserve in the town of Lymm in Cheshire, England. The runoff from a busy 1.3 km section of the M56 motorway drains into the stream, and discharges to the nature reserve when it is raining.

Table 14.1 Key benefits accrued

Social	Reduced probability of flooding to all 3594 residential properties
Environmental	■ Creation of 3 ha of water-dependent habitat ■ Improvements in water quality, sufficient to raise the Ouseburn's EU WFD status from moderate to good, achieved by – reducing peak flows and sediment transportation – reducing CSO spill volumes by approximately 82 000 m^3 per annum – reducing energy use and associated carbon footprint through a reduction in flow of 477 330 m^3 of stormwater per annum to the wastewater treatment works ■ Alignment with, and supporting delivery of, the biodiversity action plan for the area
Economic	Total scheme cost of £6.249 million, comprising ■ Northumbrian Water £3.058 million ■ Environment Agency £2.757 million ■ North Tyneside Council £0.433 million

National Highways operates the motorway network in England and uses a computer model (formerly HEWRAT) to determine if the outfalls that discharge to the water environment pose a risk of pollution. The model indicated that this outfall off the M56 creates a risk of pollution from copper and zinc in the stream. Copper and zinc levels in road runoff can be quite high, as the metals come from brake pads and tyre degradation from vehicles.

14.5.1 The existing drainage infrastructure

Before the risk of pollution was identified and quantified by the modelling, the runoff from the motorway was discharged by way of a filter drain at the side of the carriageway. A filter drain is a stone-filled trench with a drainage pipe at the bottom. As the runoff passes into the stone layer, contaminated sediment is captured in the void spaces between the stones. This system can remove up to 40% of the suspended solids and associated pollutants in road runoff. However, the risk assessment had identified a risk that was posed by the dissolved metals in the runoff. The design team sought a solution that could be incorporated into the filter drain to capture the dissolved metals.

14.5.2 Assessing the level of risk

National Highways has an ongoing programme whereby outfalls on the strategic road network are evaluated with respect to the pollution risk that they represent to receiving watercourses. The priority outfalls programme classifies the risk from highway outfalls as: A (very high); B (high); C (moderate); D (low); X (risk addressed).

These risk categories are based on a two-step process. In the first step, for a national dataset of outfalls, the initial risk category is established using a GIS-based prioritisation tool and scoring system (the 'baseline assessment'). This baseline assessment tool was developed in line with pollution risk criteria described in the National Highways *Design Manual for Roads and Bridges* (DMRB) (LA113: Road Drainage and the Water Environment) (Standards for Highways, 2023) and is based on (road) catchment area, traffic volume, river size and proximity to designated protected habitats and species.

In the second step, this risk category is verified using a process of desk-based and field-based checks and application of the National Highways Water Risk Assessment Tool (LA113: Road Drainage and the Water Environment). Once the risk type is verified, the priority outfalls programme establishes further pre-scribed courses of action, which may include, for example, identifying and implementing mitigation actions.

For this particular outfall to Mag Brook, the risk assessment tool identified a requirement to capture and remove 52% of the dissolved copper and 7% of the dissolved zinc.

14.5.3 A new solution

Working with British manufacturers, National Highways and its contractors selected a reactive treatment media that is known to capture dissolved metals. The product is a granular reactive treatment media that relies on ionic exchange to capture the dissolved metal ions. It is delivered to site loose, in 1 m^3 bags, so that it can be deployed in a number of treatment options. This material had been tested by the University of Chester in accordance with the published British Water *Code of Practice for the Assessment of Manufactured Treatment Devices Designed to Treat Surface Water Runoff* (British Water, 2016). The test measures the efficiency of the material to capture and remove dissolved copper and zinc from artificially created urban runoff. The university undertook the testing and established that the reactive treatment media captures 99.5% of copper and 99.8% of zinc in urban runoff.

14.5.4 Designing the solution

The team worked together to calculate the necessary volume of the reactive treatment media that must be included in the filter drain to deliver the required level of treatment. They had to ensure that there was adequate 'contact time' between the runoff and the reactive media. They also had to ensure that the layer of reactive media would continue to function effectively for the same maintenance interval as the stone layer above it. This was done to align the maintenance operations for the stone and for the reactive media to keep the cost of maintenance to a minimum. When the stone needs to be replaced, the reactive treatment media can be removed and replaced at the same time, using the same vacuum extractor. The stone is cleaned and returned to the filter drain, but a new layer of reactive treatment media is installed. It is expected that this operation will need to be carried out approximately once every 15 years, although 'low' sections of the motorway, which gather more runoff and more suspended solids, may need more frequent maintenance operations. Once the reactive media is 'spent', it will have to be disposed of by a licensed waste disposal contractor, although it is hoped that techniques can be developed in the future to 'recharge' the media so that it can be reused.

14.5.5 Installation of the treatment system

In March 2019, the old filter drain stone was removed using a vacuum extractor during overnight closures of the motorway lanes so that the operation could be carried out safely. Once the stone was removed, a layer of geotextile was laid in the filter drain and then overlaid by a 250 mm layer of reactive treatment media. The geotextile was then 'folded' over the top of the reactive media layer before a layer of clean stone was replaced over the top. The contractors found the installation of the reactive treatment media very straightforward and commented that it had been no different to their normal operation for installing a standard filter drain.

14.5.6 The benefits to the environment

The north west of England is criss-crossed by many motorways, all of which contribute to water pol-lution. It is important to identify those that pose a particular risk of pollution and to introduce treatment schemes like this one, to reduce the level of pollution and prevent the bioaccumulation of toxic metals. It is also essential that once treatment schemes have been designed and built, they must be maintained in accordance with the guidance from the design engineer and manufacturer's instructions. This guidance should be included in the hand-over documents when the scheme is complete. Reducing the level of toxic pollution in rivers will allow aquatic species to thrive and their populations to recover.

REFERENCES

British Water (2016) Code of Practice for the Assessment of Manufactured Treatment Devices Designed to Treat Surface Water Runoff. See https://www.britishwater.co.uk/page/Publications (accessed 10/08/2023).

Standards for Highways (2023) *Design Manual for Roads and Bridges*. LA113: Road Drainage and the Water Environment. See https://www.standardsforhighways.co.uk/search/d6388f5f-2694-4986-ac46-b17b62c21727 (accessed 10/08/2023).

Credits: Jo Bradley, Stormwater Shepherds UK; SDS Limited, National Highways
Author: Jo Bradley

14.6. Sidmouth Amphitheatre surface water management scheme

Located on the south coast of Devon, England, Sidmouth has suffered regular stormwater flooding from overland flow, with up to 64 residential properties and businesses at risk. One of the key flow paths is along Station Road leading into the town centre, resulting in flood water affecting properties along the route. Several flood events have affected the town in the past decade.

How to intercept surface water flows on a busy highway and provide a flood storage solution to attenuate flood flows, while also reducing the flow arriving in the town centre, posed a significant challenge. As the town centre consists of narrow streets, and historic buildings with shallow foundations, traditional approaches to mitigate the flood risk were not possible. A surface water storage solution therefore needed to be located further upstream, in an area of open parkland, in a way that was acceptable to the local community and other stakeholders and sensitive to the impact on the mature trees protected under a town-wide tree preservation order. The design standard for the scheme was that it should intercept the 100-year storm (1% annual exceedance probability; AEP), thereby significantly reducing the flooding impacts in the town centre.

A landscape vision for the storage area was developed, comprising an amphitheatre as the focal point of the scheme, as shown in Figure 14.10, creating a link to the site's historical use as a venue for community and festival events dating back to the 1950s.

The broader system design comprises two 'crossdrains' in Station Road. A crossdrain is a large linear drainage channel running perpendicular to the alignment of the highway, intercepting flows across the full width of the carriageway. The crossdrains intercept approximately 300 l/s of overland flow and divert it into the parkland area, outfalling nearly 50 m to the south into a 5 m wide swale. Runoff is slowed as it leaves the pipe by rip-rap and a check dam, trapping sediment in the process. The 90 m long swale then conveys flows forwards, filtering the water as it goes by way of the vegetation before the runoff reaches the final check dam upstream of the amphitheatre. Water overtopping the lower check dam spreads out across the base of the amphitheatre and infiltrates through the surface of the amphitheatre by way of a spiral shaped filter drain into a 400 m^3 subsurface geo-cellular crate tank. Once the subsurface storage is full, water will pond on the surface of the amphitheatre to a retained depth of approximately

Figure 14.10 Historical festival location and newly created amphitheatre

500 mm (providing a further 300 m^3 of storage). During dry weather or wet weather up to and including the 30-year storm (3.33% AEP) water will be stored below ground ensuring that the amphitheatre can continue to function for its dual use, as community and performance space.

Water leaves the amphitheatre through either infiltration, if groundwater is low, or by way of a piped outflow to the nearby surface water sewer network at a greatly reduced rate (40 l/s maximum). In the event of a blockage of the outlet, or an event more than a 200-year storm (0.5% AEP), a spillway has been designed as an exceedance flow path to direct flows safely overland into the adjacent area of the parkland.

Some of the key benefits of using a green infrastructure solution to solve this particular flooding problem include the water quality improvements to the runoff as a result of the water passing through check dams (to trap silt), passing over grass and through stone filter media (to remove metals, oils and other hydrocarbons), and the creation of a usable amenity space that would not have been possible if a surface solution had been used on its own.

Several innovative features and designs have been incorporated into the scheme, two examples (Figures 14.11, 14.12 and 14.13) of which are integral to the overall landscape vision.

The amphitheatre 'seating' (Figure 14.11) required the construction of almost vertical steps out of earthworks and grass, that could withstand saturation during the design flood event, as well as members of the public sitting and standing on them on a regular basis. This required a carefully developed step design that employed reinforced earth techniques, with a state-of-the-art grass product 'Lindum Grassfelt' to resist erosion. All the grassfelt used is plastic free, which was an important consideration in eliminating the future possibility of generating microplastics.

Figure 14.11 Amphitheatre seating with near vertical seating steps

Figure 14.12 Central drain cover feature in place

Figure 14.13 Central drain cover feature removed through a self-raising mechanism

The central flow control chamber (Figures 14.12 and 14.13) posed a risk by its very nature and location in the middle of a public park, away from lighting sources. There was a risk that during a flood it could be dislodged, leaving the 2.5 m deep chamber beneath as a hazard to the public. To overcome this and enable ease of access for maintenance purposes, a lockable, self-raising cover and frame was developed. The additional benefit provided was that the ornate engraved limestone surrounding the cover would not be damaged as and when the cover is removed for maintenance.

The modular geo-cellular storage crate located beneath the amphitheatre is unlined, allowing for very low ingress of groundwater. If the tank had been lined, the uplift pressure would have required a heavy counterweight/mass concrete slab to prevent buoyancy. By allowing the groundwater in at the calculated rate of less than 2 l/s, no liner and no counterweight were required. As groundwater only reaches the tank invert for a short period of time each year, for the rest of the year infiltration is possible, with a discharge to the sewer network of less than the 40 l/s permitted, with the discharge reducing to zero during some drier periods. This approach allows recharge of the underlying aquifer, is more sustainable (by reducing the use of concrete) and benefits the downstream sewer network by providing more capacity.

The project team developed a culturally relevant blue-green infrastructure solution, integrating the stormwater management needs into a creative, environmentally sensitive, climate-resilient design, which also functions as a performance venue or other community amenity space. The outcome is a scheme that enhances the local landscape, provides a cultural link to its previous use as a performance location, enhances biodiversity and recreational space, and has widespread support from the local community and local government (see Figure 14.14).

Credits: Jacobs

Authors: Paul Hargreaves, Vanessa Mourant, Gabi Steimer

Figure 14.14 Scheme information board

14.7. Use of sensors in hydrological-hydraulic model calibration

This case study investigates the performance of the railway drainage system in a historically frequently flooded site in Knockenjig, Scotland. Sensors played a vital role in helping to calibrate and develop a hydrological-hydraulic model of the catchment and drainage system (Wu, 2022). The collected data were used to evaluate the accuracy of the model and adjust parameters to improve model performance.

A hydrological-hydraulic model was built to simulate the performance of the railway drainage system. Constructed using the UN Environmental Protection Agency's Storm Water Management Model (SWMM), the model simulated movement of water through the surrounding catchment area and the railway drainage components. SWMM is a dynamic rainfall-runoff simulation model, consisting of a rainfall-runoff component and a routing module. The rainfall-runoff component represents the transformation of rainfall into runoff through the user defined catchment areas. Then the routing module transports the runoff through a drainage system that is composed of pipes, channels, storage/treatment units, pumps and regulators, into an outfall.

Water level sensors and flow rate monitors were installed for a 6-month period in locations identified in the system that have a higher risk of flooding. The collected sensor data were used to calibrate the hydraulic performance model and assist further model development.

The main objective of the model was to find the duration and scale of drainage system failures under heavy storms. Hence the calibration process focused on the peak of the water depth in catchpits (access chambers), as these values are important indicators of whether the water level would rise above ground level.

The calibration process was designed in two stages.

- Hydrology: The runoff volume was calibrated by adjusting the infiltration parameters and comparing the simulated and sensor-measured volumes of discharge.
- Hydraulics: The condition of the drainage assets was calibrated by adjusting the pipe roughness and comparing simulated and sensor-measured catchpit depths and flow rates.

For each stage, one or more objectives were considered. For the hydrology component, the two objectives were

- minimising the high flow rate error $E_1(X)$
- minimising the total flow volume error $E_2(X)$.

For the hydraulic aspects, the objective was to minimise the catchpit water depth error $E_3(Y)$. The errors are measured with mean absolute percentage error and percentage root mean square error, which are widely used measures of prediction accuracy and are presented in percentage format which is easy to understand and compare. They both provide measurement of relative difference between measured and modelled data as a percentage of the measured data.

The hydrological-hydraulic model comprised a digital replica of the drainage system and several small contributing catchments. High accuracy laser levels were used to provide detailed elevation information of the existing drainage system and surrounding topography. A digital replica of the drainage system was then built in SWMM using the digital elevation data and Network Rail's drainage asset database records. A thorough evaluation of the surrounding catchment area geology was performed based on the digital terrain model downloaded from the Environment Agency Lidar Digimap. The following sub-catchment characteristics were computed for each catchment: area, average slope, flow width and impermeability.

After preliminary tests, several hydraulic deficiencies in the drainage system were identified and decisions on sensor placement were made based on the results. The nodes circled in Figure 14.15 indicate the locations where flooding occurred during the simulation. The circles in Figure 14.16 indicate the locations of the water depth sensors and the triangles indicate the locations of the flow rate sensors.

A set of simulations was run with a range of hydrological parameters from 1 October 2020 to 31 October 2020. The parameters tested were arithmetic sequences with range and interval listed in Table 14.2.

The optimum set of parameters under the two minimising objectives are shown in Table 14.3.

When multiobjective calibration is performed, its trade-off needs to be examined. This can be done using a Pareto frontier. Without additional criteria, all Pareto optimal solutions can be considered candidates for the 'best fit' parameter set. In Figure 14.17, the Pareto frontier is marked as a line, which consists of a set of non-dominated solutions such that no one criteria can be improved without making the other poorer. In this case any point on the Pareto frontier can be chosen as the best fitted parameter values. The selection of the solution can be made based on the asset manager's objective. By choosing to prioritise minimising $E_1(X)$, the asset manager is able to reduce the risk of flash flooding due to high flow rate. However, this could potentially overestimate the total volume of runoff during a long period of less intense rainfall, and hence overestimate the potential risks.

After the hydrological calibration process was complete, the chosen parameter values were put back into the SWMM model to be used for the hydraulic parameter calibration. As described, the hydraulic parameter to be calibrated was the roughness of the pipes, and the water depth in each of the monitored catchpits was analysed to find the roughness value that minimised the water depth error $E_3(Y)$ at the monitored catchpits. The roughness coefficient used in SWMM is Manning's n, and the range of roughness values tested was from 0.01 to 0.06, with a step increase of 0.005. As shown in Table 14.4, the water depth error $E_3(Y)$ was minimal when Manning's n was equal to 0.01.

With both the optimum hydrological and hydraulic parameter values identified from the calibration process, the model was then validated with sensor data collected from 1 November 2020 to 30 November 2020. The simulated and observed flow rate at nodes 4 and D1 are shown in Figure 14.18.

As shown in the figures, both the flow rate and the water depth fit reasonably well. Most of the simulated peaks occur at the same time as in the sensor data and many have a reasonable correspondence in value. However, a lot of the flows in between the rainfall events are not captured well. As the focus was on the flood risk, the data fitting around the peak flow rate and highest water depth was more valued and the effect of the lower readings between rainfall events is less considered, which was already reflected by choices made when selecting the combination of parameter values on the Pareto frontier as the ones that minimise the high flow error were preferred.

Figure 14.15 Long section view of the flooded nodes

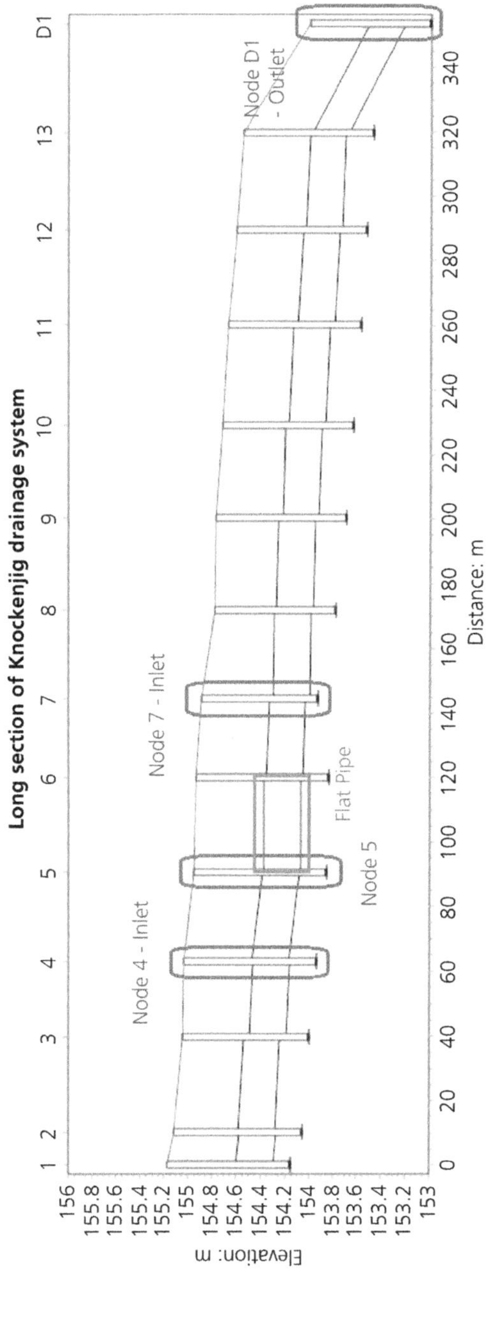

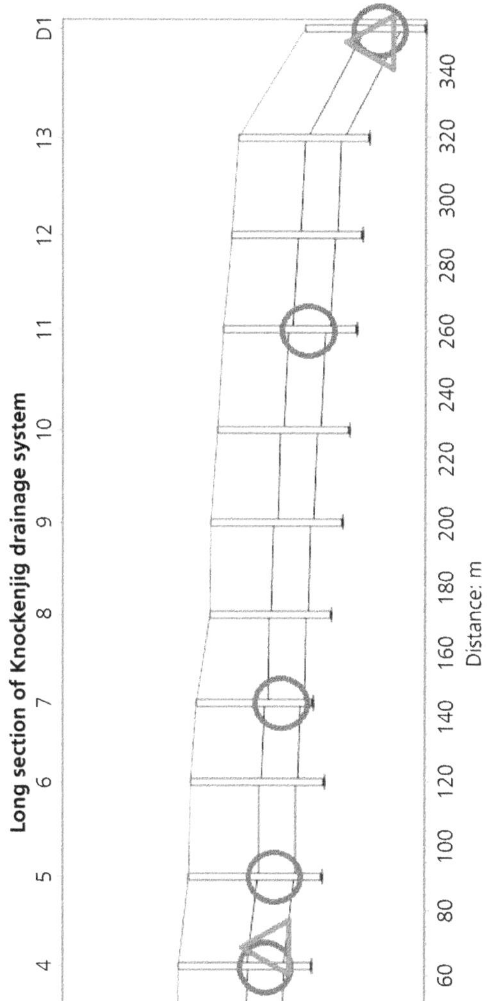

Figure 14.16 Locations of the sensors

Table 14.2 Arithmetic sequences

Parameter	Range	Interval
Maximum infiltration rate (mm/h)	(8, 76)	1
Minimum infiltration rate (mm/h)	(0.1, 11.5)	0.1
Decay constant	(4, 7)	0.5

Table 14.3 Optimum hydrological parameters for objectives $E_1(X)$ and $E_2(X)$

	Maximum infiltration rate (mm/h)	Minimum infiltration rate (mm/h)	Decay constant
Minimising high flow rate error $E_1(X)$	39	0.1	4
Minimising total flow volume error $E_2(X)$	76	0.1	4

Figure 14.17 Pareto frontier of the two objectives

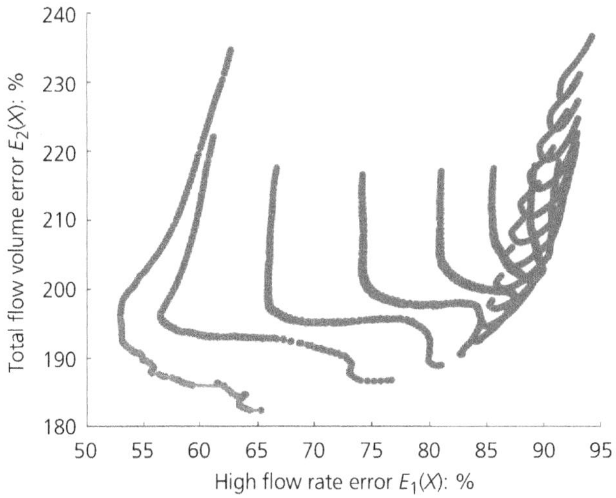

Table 14.4 Water depth error $E_3(Y)$ with various values of Manning's *n*

Manning's *n*	0.01	0.02	0.02	0.03	0.03	0.04	0.04	0.05	0.05	0.06	0.06
$E_3(Y)$	0.4	0.46	0.59	0.75	0.92	1.12	1.32	1.5	1.66	1.81	1.95

Figure 14.18 Flow rate at nodes 4 and D1 using calibrated parameters

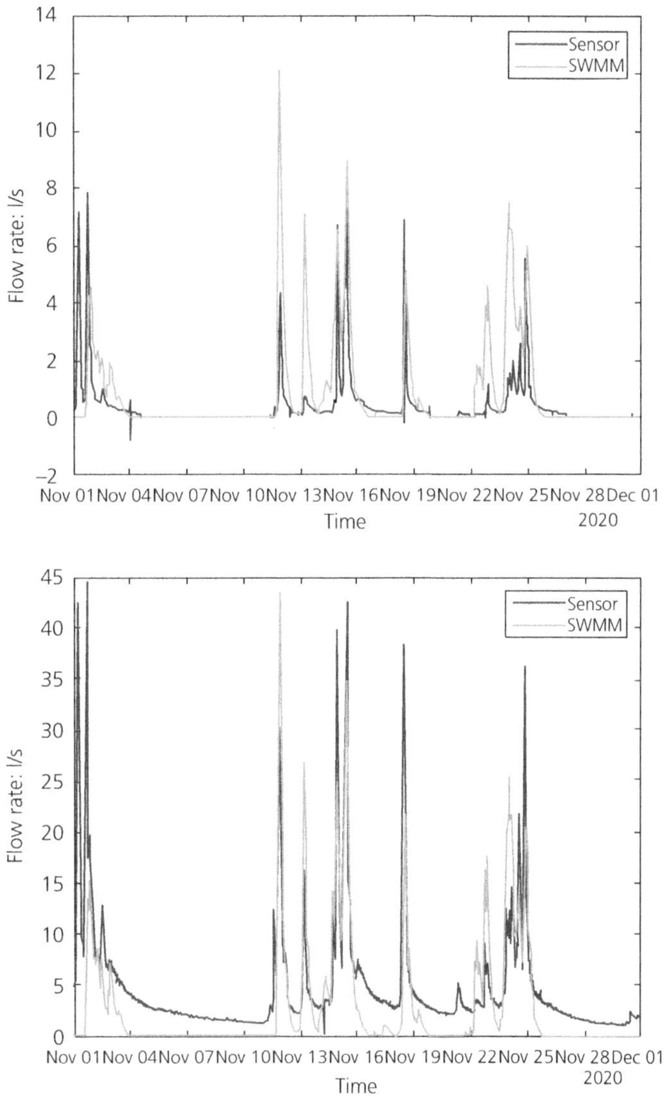

The values of each type of error for the validation set are given in Table 14.5, which shows the water depth error $E_3(Y)$ to be the same, whereas total flow volume error $E_2(X)$ is lower than that of the calibration set. However, the peak flow error is higher for the validation set. This may be due to the fact that the case study hydraulic system serves a relatively small catchment area, and the radar rainfall data were processed over a 1 km square, hence local variations that are expected to exist may have led to modest inaccuracy in forecasting runoff. Despite recent advances in radar technologies, it is still very challenging to measure rainfall accurately due to its highly variable nature over time and space. This type of error can potentially be reduced by installing numerous local rain gauges and comparing their measurements with the radar data to eliminate the uncertainties in the precipitation data.

Table 14.5 Estimation error for the calibration set and validation set

	$E_1(X)$	$E_2(X)$	$E_3(Y)$
Calibration	50%	180%	40%
Validation	110%	150%	40%

REFERENCES

Wu Y (2022) *Whole Life Cost Modelling for Railway Drainage Systems Including Uncertainty.* Doctoral thesis, University of Sheffield, Sheffield, UK.

Credits: The University of Sheffield

Authors: Dr Yiqi Wu, University of Sheffield; Dr Andy Nichols, University of Sheffield; Professor Simon Tait, University of Sheffield

14.8. Smart wastewater network monitoring

Yorkshire Water's wastewater network monitoring was the end-product of several teams, including operations, control room, operational technology (OT), information technology (IT), data science, asset management, innovation and operational maintenance, and input from framework consultants. Yorkshire Water's journey into a smart wastewater network started as early as 2003 in AMP3, when it implemented one of the industry's first predictive operational tools, known as the DG5 predictor, which evolved into the pollution predictor. In effect this was programmed in MS Excel to take incoming combined sewer overflow (CSO) telemetry data overnight and by the morning had automatically prioritised CSOs with a high likelihood of blockage formation. On a daily basis, colleagues would use the DG5/pollution predictor outputs to raise sewer cleaning work. Use of the DG5/pollution predictor was very successful and embedded into daily operational processes. By AMP6 the Yorkshire Water modus operandi, based on the predictive power of an MS Excel spreadsheet, was being evolved and formed the basis for Yorkshire Water's smart wastewater network strategy. Yorkshire Water acknowledged early on that the backbone of a smart wastewater network is the monitoring – because, as the well-known saying goes, you can't manage what you don't measure. Since the early days of the DG5/pollution predictor, instrumentation and monitoring have moved on and by AMP6, when water companies were being asked to make efficient use of digitalisation, Yorkshire Water established a baseline in terms of what was already being monitored on the wastewater network. The following statement by Donald Rumsfeld was of particular relevance.

- There are the 'known knowns'; the things we know that are known.
- The are the 'known unknowns'; the things that we know we don't know.
- And there are also the 'unknown unknowns'; the things that we don't know we don't know.

But by this point in time, following decades of installing instrumentation and telemetry on assets, a fourth category was of relevance to the water industry.

- The 'unknown knowns'; the things that we know we have, but rarely look at.

The water industry as a sector in the UK was soon to realise that in this era of regulatory requirements for all storm overflows to have event duration monitoring (EDM), and with network pumping stations and some network storage assets being designed with monitoring and telemetry as standard, a wealth of unexploited data was accumulating. While the existing monitoring and telemetry built into the wastewater network was being used with analogue alarm points, it was soon realised that as an industry more could be done to convert the incoming telemetry data into information for real-time decision-making based on a more holistic and systems-based approach. Innovation in technology and artificial intelligence (AI) have made real-time decision-making for operational activity a reality.

This case study provides an overview of the more recent monitoring initiatives implemented on Yorkshire Waters' wastewater network in line with its long-term digitalisation and smart strategy. In AMP6, Yorkshire Water deployed a large monitoring programme to address pollution and sewer flooding drivers.

- 1000 level monitors on public sewers for pollution prediction due to other causes.
- Over 1000 monitors on properties (gullies and lateral drains) at risk of internal sewer flooding due to other causes.
- Pressure monitoring installed at 60 high-risk rising mains for burst detection.

Following the success and benefits observed with increased monitoring, in AMP7 a further 500 monitors were installed at properties at risk of internal sewer flooding due to other causes. The sewer level monitoring has been taken to the next level by feeding sewer level and rainfall data into AI and machine learning (ML) for blockage detection, and operational practices were introduced to respond to predictions and prevent pollution incidents. Similarly, by AMP7 huge strides were made in improved interpretation of rising main pressure data to detect failures, and sophisticated logic was incorporated into existing alarm management processes, resulting in much improved rising main burst prevention and detection.

14.8.1 Property-level monitoring

A particular challenge for Yorkshire Water is internal sewer flooding, especially given that it has a high proportion of cellared properties. The objective of increased monitoring was to provide greater visibility and early warning of blockage formation on small-diameter lateral sewers, while getting the right balance and managing incoming data from large numbers of monitors. Initially, learning was taken from Yorkshire Waters' success in smart leakage detection on a highly instrumented clean water network. However, compared with the clean water distribution network, the wastewater network has a unique set of challenges.

- In terms of a starting point, the wastewater network has less instrumentation compared with distribution networks, which benefit from existing and fully operational water meters fitted at property level.
- Service failures tend to be triggered at the same time in wastewater networks, that is, during wet weather. Having a greater visibility of the issues during wet weather creates new challenges in prioritising, managing and resourcing work baskets that are very concentrated in a discreet point in time.
- The nature of the fluid and the aggressive environments that monitoring equipment has to endure.

14.8.2 CSOs and EDM

EDM is generally well understood and industry best practice guidance is available, such as the revised CIWEM Urban Drainage Group good practice guide (CIWEM, 2021). Yorkshire Water's objectives as part of developing a smart wastewater network are to generate additional benefits from the regulatory requirements for EDM and apply AI and ML to EDM data for improved blockage detection and optimised maintenance programmes. The pollution predictor evolved and, using rainfall data, was brought into the analysis for improved blockage detection at over 50 CSOs throughout the Yorkshire Water region. The revised approach produced significant benefits and better insights, compared with the pollution predictor that had been in use for several years and just looked at level data. The most significant benefit was that scheduling colleagues in the control room had much greater confidence in raising work orders for field colleagues to investigate anomalies in CSO behaviour identified with AI using rainfall and sewer level data.

In addition to CSOs, the revised pollution predictor based on an AI algorithm was also extended to blockage detection using the same EDM instrumentation in 1000 sewer locations identified as benefitting from similar proactive sewer cleansing activity based on AI prediction of blockage formation.

14.8.3 Property gulley monitoring

In many ways the principle for monitoring the wastewater network and using AI to detect blockage formation on lateral sewers is no different to that for main sewers. The objective is the same, in that AI needs to learn what normal behaviour looks like and then be able to detect any deviations from normal behaviour, both in wet weather and dry weather. However, there were additional challenges associated with property-level monitoring.

- Size of the devices (the size of conventional sewer level monitoring equipment not being appropriate for gulley locations).
- Programme challenges associated with the large numbers of sensor deployments.
- Gaining access to private property introduced programme timescale risks.
- Customer behaviour and curiosity/interference with devices.
- Large volumes of data generated, and time required to develop reliable algorithms for each site with the required confidence levels for colleagues to trust the data and raise non-abortive work.

Yorkshire Water learnt a lot from the first deployment of property-scale monitoring, and in AMP7 deployed a further 500 property gulley monitors. Some of the key aspects handled differently were as follows.

- A different approach was used to gain access to properties to ensure that programme timescales were met.
- Improved data visualisation and analytics to help control room colleagues act on incoming data.
- A lot was learnt about typical blockage formation lead times at the property scale, which influenced the timescales and priorities on which field jobs were raised.
- Updated specifications for the monitoring equipment
  - method of installation
  - parameters monitored
  - type and size of device
  - telecommunications.

14.8.4 Pressure monitoring of rising mains for burst detection

Pollution incidents from rising mains are a particular challenge on the wastewater network. The very nature of rising mains means that they often traverse rural locations, pumping across natural catchments and traversing watercourses, and therefore have high potential to cause high-profile pollution incidents. The fact that rising mains often pump across natural drainage catchments through sparsely populated areas also means that bursts can go unreported and undetected for long periods of time, thereby increasing the impact of the pollution incident. In AMP6, Yorkshire Water identified its critical rising mains and prioritised these based on risk of failure, which was generally associated with the age of main and type of pipeline material. Pressure monitoring was retrofitted onto 60 high priority rising mains and an operation process was put in place for colleagues to monitor the incoming data and raise work for field colleagues where anomalies were identified. Lessons learnt were then incorporated into algorithms and used to set dedicated rising main burst alarms based on sophisticated logic applied to the rising main pressure data.

14.8.5 To conclude

Yorkshire Water now has many 'eyes' on the wastewater network and is generating and collating a huge amount of data. Although it is still somewhat compartmentalised according to specific drivers, such as pollution or internal sewer flooding, Yorkshire Water has developed a sturdy backbone for an integrated smart wastewater network. There is work in progress to break down the data silos and derive additional benefits and drive greater value through a holistic approach by bringing all the eyes on the network into one place for improved decision-making at a catchment systems level for both operational and strategic activity.

REFERENCES

CIWEM (Chartered Institution of Water and Environmental Management) (2021) *Event Duration Modelling Good Practice Guide*. Urban Drainage Group, CIWEM, London, UK. See https://www.ciwem.org/assets/uploads/CIWEM_UDG_EDM_Good_Practice_Guide_2021_final.pdf (accessed 18/12/2023).

Credits: Yorkshire Water
Author: Jacqueline Diaz-Nieto

14.9. Yorkshire Water Tunnel sewer inspection

14.9.1 Project overview

This case study outlines an innovative approach by Yorkshire Water in 2004 to undertaking a structural and service condition survey of the Yorkshire Water Tunnel. This tunnel is a strategically important, high-risk asset and probably the most critical length of sewer in the catchment area, being the final link in the chain of the catchment sewer system before reaching the wastewater treatment works.

Due to the size and flow conditions, it was not possible to undertake a physical 'walk through' inspection. The 3 m diameter tunnel is critical infrastructure with the only access for introducing surveying equipment being at each end. The huge difficulty to overcome was that the two access points were more than 4 km apart.

In 2004, 'off the shelf' surveying equipment was incapable of surveying such large diameters over such long lengths, which left little alternative but to design a system and method from scratch.

14.9.2 Equipment

Due to the large diameter, it was necessary to design and manufacture a pan and rotate head camera built around a very light sensitive CCD camera module, controlled by a radio control device operated at the survey vehicle sending and receiving signals by way of a specially commissioned 2 km long, Kevlar reinforced, co-axial cable, imported from Sweden.

Because of the length involved, it was not possible to use a standard closed-circuit television (CCTV) cable through which normally travels both the picture and the power for the camera and lights. The co-axial cable allowed the transmission of high-quality images only over such a length, which meant that it was also necessary to design sealed battery packs containing NiMh high-capacity batteries to provide power to the low-voltage LED light head and camera, which was mounted on a purpose-built catamaran or floating platform to stabilise the camera.

To help manage and control the 2 km of cable, powered drums were also designed and manufactured to assist in the paying out of the cable in a controlled manner thus allowing speedy and neat rewind to the drum.

14.9.3 Method of survey

Because it would have been too difficult and expensive to survey the tunnel in a single pass with 4 km of cable, the inspection was carried out in the following four stages.

1. Survey downstream from access shaft for 2 km using CCTV only. Retrieve equipment back to the shaft on completion.
2. Survey downstream from access shaft for 2 km using sonar only. Retrieve equipment back to the shaft on completion.
3. Survey upstream from outlet structure for 2 km using CCTV only. Retrieve equipment back to the outlet on completion.
4. Survey upstream from outlet structure for 2 km using sonar only. Retrieve equipment back to the outlet on completion.

The weight of 2 km of cable laid out over such a long distance, combined with the effect of the sewage flows, was too much for manual retrieval. Capstan-type cable-assist winches therefore had to be utilised to control the pay out of the cable and tow lines as well as to pull them back on completion of each stage. Also, due to the stresses applied to both the cable and the 10 mm polypropylene tow lines, it was essential to install pulleys securely, ensuring that neither came into contact with the sewer structure itself. This ensured not only that the rope or cable was protected from damage, but also that the listed tunnel portal structure at the inlet works would also be protected from the cutting actions of the tow line.

14.9.4 The survey

The CCTV element of the survey provided video footage of the tunnel above the water line, with the sonar, mounted under the floats, providing a profile of the tunnel wall beneath the water line therefore providing 360° of critical condition information.

14.9.5 Continuous scanning sonar

Although in-sewer sonar technology had been around for almost 15 years, its obvious benefits were not being utilised enough in the UK water industry. For many years, sewerage engineers believed that the structural condition of a sewer beneath the water line was likely to be similar to that observed by a standard CCTV inspection above the water line. On many occasions, this broad assumption has been proved to be very wide of the mark, especially where the sewer structure is of brick or masonry construction. Where high flow velocities occur in this type of sewer, over time the mortar between the brick courses is gradually eroded, often leading to sections of the sewer (which would otherwise have remained hidden from view) 'dropping' as the support beneath the sewer invert disappears.

Unfortunately, sonar is unable to 'see' through silt or debris, but even then it can prove its worth in accurately determining the quantity of silt present in the sewer. This information can then be utilised in planning heavy cleaning, as very often a sewer that has been running 'over capacity' for many years probably simply requires cleaning to free up valuable volume capacity within it. Sonar can then be used not only to assess the structural condition of the now visible invert, but also to prove the quality of the cleaning by serving a quality control purpose.

One final element of the inspection was to survey vertically down a vent shaft, which, standing around 10 m above ground level, required the construction of a working platform. A CCTV camera was then lowered down the shaft, rotating through 360° in segments of around 5 m over a total vertical distance of over 80 m to provide a general condition survey of its structural state. A plumb line, placed at a point due north, was used to allow the surveyor to code the defects in the shaft as if it were a sewer 'on end', using the plumb line as the 12 o'clock reference position.

The inspection was successfully completed in around 10 days by a team of up to eight personnel.

Credits: Yorkshire Water
Author: Brian Smith

14.10. Valuing the BGI in Bradford, Yorkshire

The City of Bradford Metropolitan District Council (CBMDC), in England, is undertaking a major highway realignment scheme along the Shipley and Canal Road Corridor. The scheme, which is essentially a road widening scheme, but incorporating other transport and infrastructure improvements, runs alongside the major green corridor situated to the east of the city, known as the Canal Road Corridor. This is a large open space near the centre of the city and provides a linear access corridor running north–south. The Bradford Beck runs along or near the western boundary of the green corridor. The Shipley and Canal Road Corridor Area Action Plan sets out the importance of this corridor as a significant driver of growth for the city in terms of meeting its housing requirements, and also for promoting economic growth and development. It is proposed that the corridor will be developed as an 'urban eco-settlement'. The Bradford Beck, where it runs through the corridor, is seen as an important component of the eco-settlement. Although having problems in regard to existing flood risks, and polluting inputs into the beck (including from CSOs) that are not being adequately controlled, the beck has considerable potential to contribute to the quality of the local environment. The highway scheme has been identified as helping with leveraging to improve the local environment along the corridor; that is, the works associated with the highway have possibilities for wider enhancement of the corridor, especially for the quality of the beck and linear park; thus adding wider amenity and other benefits to be enjoyed by the growing local residential population and others. Significant environmental net gains (as defined in the National Planning Policy Framework) are possible by judicious and synergistic development of the highway where it runs alongside the beck and park. In addition, there are possibilities for the addition of green infrastructure within the highway boundary, mostly retrofit into existing spaces, that can provide environmental benefits through, for example, reduced air pollution and noise attenuation.

The beck currently is of limited water quality status and does not follow a natural hydromorphological course, having been much modified. There are current flood risks from the beck to existing properties and commercial buildings, and to highway inundation and traffic disruption. A consultants' report provides a case for considering the beck together with the highway improvement scheme as an opportunity to alleviate some of the current flood risks and also to ensure that a proposed housing development is not constrained by flood risk. A report on the ecology related to the corridor provides an ecological assessment covering the land within both the corridor and Bradford City Centre Area Action Plan areas. This suggests that appropriate measures could be used to enhance the aquatic and fish population in the parts of the beck that have been 'canalised'. This includes the removal (or infill and movement), together with renaturalisation, of a culverted section of Bradford Beck over some 75–80 m that currently runs alongside the existing highway, providing roadside structural support. This culvert built in the early part of the twentieth century is now in poor condition and if retained would need considerable renovation.

There are considerable opportunities to improve the ecology of the beck over a 120 m length of culvert and realigned watercourse in a newly renaturalised section, were the existing culvert to be abandoned. Initial ideas are illustrated in Figure 14.19. A longer length, up to 600 m, is also considered for renaturalisation in the flooding risks report.

The city cross-department team, in conjunction with key community groups, aims to utilise nature-based systems (NbSs) as part of the scheme and to consider the added benefits of so doing in addition to those from the highway realignment.

Figure 14.19 Means for improvement of ecological linkage, habitat creation and public engagement opportunities (courtesy of Friends of Bradford's Becks, given in Skinner *et al.*, 2014)

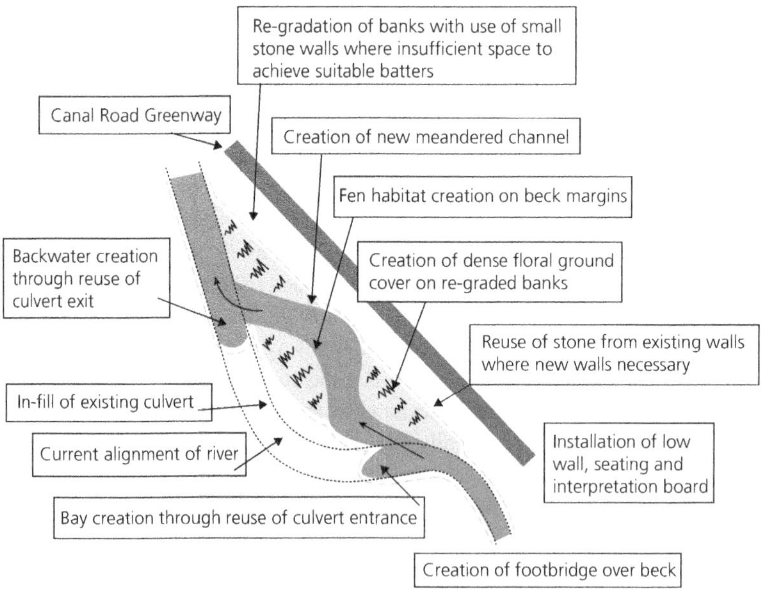

14.10.1 Assessing value

The scheme was one of the case studies in the BEGIN EU project on social innovation for blue-green infrastructure (BGI) (BEGIN, 2023). The benefits assessment framework used was based on the CIRIA B£ST (benefits of BGI) tool (CIRIA, 2023). There were initially four options appraised in successive addition of measures, as illustrated in Figure 14.20.

The results from the B£ST analysis are shown in Figure 14.21. The highway realignment (option 1) was assumed as the baseline situation and the other options are shown in comparison with this in Figure 14.21. Not surprisingly, each addition of BGI measures increases both the costs and the benefits, with option 4 showing the greatest net present value (NPV), based on an estimate of the benefits, of more than £10 million. The value of the individual benefits determined for option 4 are shown in Figure 14.22. B£ST provides the means to express uncertainty in the results by using a user-defined confidence in the data input, for both physical parameters and their monetisation. The present value estimates are shown before and after the application of confidence scores.

Figure 14.20 Options appraised

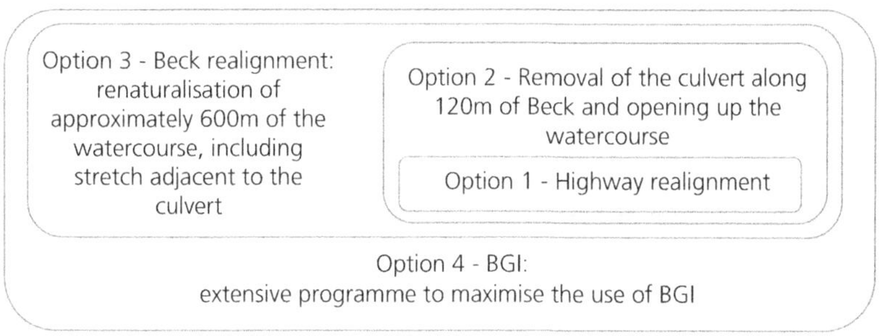

Figure 14.21 Costs versus benefits analysis

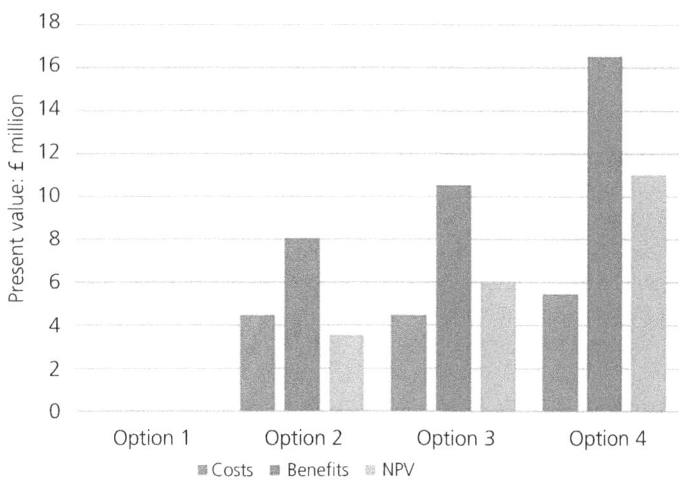

Figure 14.22 Value of benefits analysis for option 4

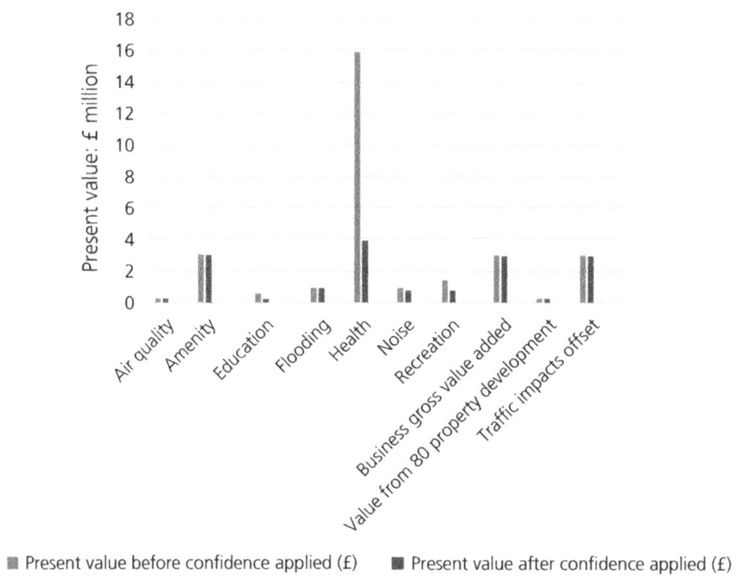

The greatest benefit is to human health, due to the various recreational and other human activities promoted by using BGI. Amenity relates to property value and street scene benefits. As this scheme contributes to business development, this is a significant value, as is traffic impact reduction.

14.10.2 Conclusion

This case study shows an example of how a non-water-related scheme may also be utilised to provide benefits related to the associated use of NbSs. It should be incumbent on all stakeholders, especially professionals, to seek out such synergies and add to the scope and value of infrastructure projects wherever feasible. This requires a willingness to cooperate across traditional boundaries, to remove 'silo' thinking and protectionism and to aim for the common good of society, both within organisations and across organisations.

REFERENCES

Ashley RM, Horton B, Walker L *et al.* (2023) The benefits of nature-based systems in a changing and uncertain world. *Proceedings of the Institution of Civil Engineers – Engineering Sustainability* **176(2)**: 61–71, 10.1680/jensu.21.00102.

BEGIN (2023) Bradford - Shipley Canal Road Corridor and Harrogate Road/New Line improvement. See https://northsearegion.eu/begin/bgi-pilot-projects/bradford/ (accessed 10/08/2023).

CIRIA (Construction Industry Research and Information Association) (2023) B£ST. See https://www.ciria.org/CIRIA/CIRIA/Item_Detail.aspx?iProductCode=W047AF&Category=FREEPUBS (accessed 10/08/2023).

Lerner D (2017) *Cleaning up the Beck for a better Bradford.* Aire Rivers Trust, Bradford, UK. See https://bradfordbeckdotorg.files.wordpress.com/2017/08/bradford-investigation-proposal-lowres.pdf (accessed 18/08/2023).

Skinner E, Price L and Masheder R (2014) *Ecological Assessment for the Shipley – Canal Road Corridor and Bradford City Centre Area Action Plans*. West Yorkshire Ecology. See https://www.bradford.gov.uk/documents/citycentreactionplan/2%20publication%20draft/evidence%20base/ecological%20assessment.pdf (accessed 18/08/2023).

WSP (2018) *Shipley Canal Road Corridor Flood Risk and Benefits Study Strategic Outline Case*. Draft Report. WSP Ltd, London, UK.

Credits: The City of Bradford Metropolitan District Council
Author: Richard Ashley

14.11. Sustainability of sanitary waste disposal

Sanitary waste comprises a variety of domestic items, including female sanitary goods, condoms, wet wipes and baby nappies, together with a range of other bathroom used personal hygiene products, such as cotton buds. Once used, these need to be disposed of if they are not reusable, and most domestic users choose to put these either into the household solid waste route or into the WC (see Figure 14.23). A small number burn this waste.

There have been various studies that have looked at the relative sustainability of using disposable or reusable baby nappies. These studies have used life cycle analysis (LCA) to show that reusable nappies are more sustainable provided they are washed collectively in a community facility. However, the majority of baby nappies in use are marketed as disposable. In an extensive study in Scotland, carried out in the 1990–2000s, it was found that disposing of these and other hygiene items is more sustainable by way of the solid waste route into household waste bins. The project, in collaboration with the forerunners of Scottish Water (East, West and North Scotland Water), took more than 5 years and received funding from the UK's

Figure 14.23 A screen choked with cotton buds

Engineering and Physical Sciences Research Council (EPSRC) and others. Since the original study, there has been much more use of wet wipes, many marketed as flushable, and a growing public interest in pollution of water bodies caused by sewer overflows. This case study shows how the original study carried out the sustainability assessment, the outcomes from this and how decisions were made as a result.

14.11.1 Sustainability assessment

Chapter 12 set out the principles and criteria required for sustainability assessment. The study was carried out in two stages. Stage 1 reviewed the key elements of sustainability appraisal in terms of using the WC or the bin for sanitary waste disposal and identified that binning was more sustainable. During this stage, the study worked with four communities (populations less than 2000) in Dundee, Scotland, each connected to a combined sewer system. Stage 2 applied the findings from stage 1 to a community situated on the west coast of Scotland, determining which of six options was likely to be the more sustainable. The three commonly used criteria categories were used to define the primary criteria in stage 1, and these were extended to the four categories of criteria defined in the SWARD project, together with the primary criteria (Table 12.6) were used to define the initial indicators in stage 2.

14.11.2 Stage 1: Think before you flush (TBYF)

A single integrative indicator was defined as representing the individual indicators in aggregate: the number (or mass) of sanitary waste items being flushed by way of the WC. The expected outcome from selecting the more sustainable option for management was that the numbers (or mass) would decline from the WC route. The individual indicators used were

- economic
  - cost-benefit expressed as net present value (NPV) per head in the catchment for (i) the wastewater route and (ii) the solid waste route
- environmental
  - numbers of sanitary items found on beaches or watercourse banks
  - public willingness to pay for cleaner beaches and watercourses
- social
  - public willingness to change from WC route to solid waste route
  - actual change in disposal habits.

Each individual indicator was assessed quantitively where possible, by data collection, surveys of local water bodies and sewers, and engagement in public meetings and door to door. It was found that a sustained reduction of some 50% in disposal by way of the WC could be achieved by targeted public engagement campaigns, where these set out the demonstrable scientific evidence that the solid waste route was more sustainable. Public opinion supported a levy of some 1–1.5% on local taxes to pay for cleaner water bodies. The overall financial value of switching from flushing to binning was found to be beneficial, although the NPV varied between the catchments studied. A major conclusion was that the national slogan being used at the time to encourage the switch: 'Bag it and bin it, don't flush it' was not liked by the public. The project therefore created a more acceptable slogan: 'Think before you flush', which is in use currently, mainly in Ireland (Figure 14.24).

Figure 14.24 Public information poster

14.11.3 Stage 2: Sustainability of wastewater assets

The stage 1 project developed confidence that a switch from disposal by way of the WC to the solid waste route would deliver a more sustainable outcome. Stage 2 considered other possible options in more detail, focused on a combined wastewater network on the coast in the west of Scotland.

A. End of pipe solution

A1. Install 6 mm screens (or equivalent) at the storm overflow at the downstream treatment plant.

B. Habit change

B1. Education to encourage habit change.

B2. Retrofit/fit low-flush small-bore outlets to existing/new WCs to constrain what can be flushed.

C. Spill reduction

C1. In-sewer storage to retain volumes.

C2. Source control of storm drainage to reduce flow volumes.

C3. Sewer rehabilitation to reduce exfiltration.

The overall objcctive was to reduce the number of sanitary waste items escaping to the environment by way of storm overflow discharges in the catchment in the way that is likely to be the most sustainable. The SWARD primary criteria were used (Table 12.6) and secondary criteria and indicators defined in conjunction with West of Scotland Water personnel. There were some 15 categories of attributes used, some quantitative and some qualitative. This generated 15 × 6 options = 90 individual items of data.

Table 14.6 Options to manage sanitary waste, their relative sustainability and decision-makers' preferences

Option	Sustainability ranking	Decision-makers' preference
B1. Education to encourage habit change	1	5
C3. Sewer rehabilitation to reduce exfiltration	2	3
C2. Source control of storm drainage to reduce flow volumes	3	4
B2. Retrofit/fit low-flush small-bore outlets to WCs	4	6
A1. Install 6 mm screens (or equivalent)	5	1
C1. In-sewer storage to retain volumes	6	2

For each category, specific analytical tools were used: sewer flow modelling; LCA for environmental impacts; economic analysis for NPV; consultation with sewerage undertaker and public engagement using a variety of means. Three different multi-criteria decision analysis (MCDA) tools were used for the final collated assessment of the options; each gave a slightly different ranking in terms of the more sustainable option. The overall rankings are shown in Table 14.6.

Despite the public behaviour change being determined as the most sustainable option, the decision-makers from West of Scotland Water preferred to install screens at the outlet of the treatment plant overflow. The reasons for this were to do with regulatory constraints – installing such screens had become mandatory, and also robustness – concern that the change in behaviour of the community would not be sustained into the future. It is notable that the decision-makers preferred the options that were within their aegis, entailing retrofits to their own assets, rather than to individual household assets.

14.11.4 Public campaigns

Notwithstanding the decision-makers' preference in this study, the sewerage undertaker subsequently commissioned widespread and very focused public engagement programmes alongside the asset retrofits in various parts of Scotland, aiming to change public behaviour. These were accompanied by surveys of sewers and beaches to assess whether these campaigns were successful. Unfortunately, due to organisational changes, these campaigns were stopped before any conclusions could be drawn.

14.11.5 The aftermath of the study

A project under the Public Understanding of Science theme of EPSRC, entitled Wastewater and Sustainable Systems: Public Understanding of the Credibility and Necessity for Research was started in 1999. This aimed to engage communities in understanding how wastewater systems function. This was by way of public meetings and events, and specific activities aimed at cognitive groups. Concerted campaigns linked to a major issue, such as the flushing of sanitary waste, were found to be a useful way of raising awareness. The success of such initiatives is dependent on community involvement and 'ownership' of the issues. Communities were found to respond much more to being presented with information that had some clear personal relevance to them.

'A Bogs Life' was developed at Abertay University based on the IMSI Multimedia Fusion games engine (Figure 14.25). This arcade game was based on the educational presentation 'The Toilet Trail', used in schools for some 5 years. The game was designed to raise awareness of the problems caused by flushed sanitary waste, acting as a learning aid to reinforce knowledge gained from 'The Toilet Trail' presentation.

Figure 14.25 Arcade computer game

Despite the public dislike, the bag it and bin it slogan is still in use, now primarily by the sewerage undertakers in England. The study reported here was based in Scotland and has largely been ignored elsewhere in the UK, where there have been further attempts to promote behaviour change by public relations consultancies; not having learnt the lessons from the detailed project work outlined here.

14.11.6 Conclusion

The sustainability assessments outlined here, together with the public engagement activities, required prodigious efforts by up to 100 researchers, practitioners and public and private bodies, together with numerous communities, schools and other organisations. This was to consider one 'simple' issue, the best way to deal with sanitary wastes. Addressing and attempting to change practices to promote greater sustainability can require major commitments, necessitating the scientific justification and appropriate and effective stakeholder engagement. Such activities are not the province of engineers alone, requiring cross-, intra- and multidisciplinary working.

Decisions about adopting the more sustainable of several options are set within the boundaries of functioning of organisations and societal acceptability. Even were a decision-maker to select the more sustainable option based on the sort of scientific analysis outlined here, the successful and sustained functioning of the scheme will depend on many factors, many of which are beyond the control of the decision-maker or lead organisation. In theory, these difficulties should be included in any MCDA that is used to inform the decision. However, where the responsibilities for ensuring the most sustainable option is used lie with community and individual behaviour, there is understandable nervousness that people will do what is best for the society and planet. This is why many behaviour-related changes are encouraged by legal constraints (do not do it or face a penalty) and by inducements, such as taxation breaks or payments. For example, in various countries, including Germany, the on-property control of rainwater drainage is rewarded by reduced local taxes. The overall conclusion is that often the scientifically demonstrable sustainable approach may be difficult to implement because of stakeholder inertia or reluctance to change what they do.

BIBLIOGRAPHY

Ashley RM, Souter N, Butler D *et al.* (1999) Assessment of the sustainability of alternatives for the disposal of domestic sanitary waste. *Water Science Technology* **39(5)**: 251–258.

Ashley RM, Hendry S, Souter N *et al.* (1999) Domestic sanitary waste disposal via the WC – unsustainable. In *Proceedings of the CIB W62 Conference in Water supply and Drainage for Buildings*. September, Edinburgh, UK. https://www.irbnet.de/daten/iconda/CIB19920.pdf.

Ashley RM, Blackwood D, Butler D and Jowitt P (eds) (2004) *Sustainable Water Services: A Procedural Guide*. International Water Association, London, UK.

Ashley RM, Blackwood DJ, Souter N *et al.* (2005) Sustainable disposal of domestic sanitary waste. *ASCE Journal of Environmental Engineering* **131(2)**: 206–215, 10.1061/(ASCE)0733-9372(2005)131:2(206).

Butler D, Jowitt P, Ashley R *et al.* (2003) SWARD: decision support processes for the UK water industry. *Journal of Management of Environmental Quality* **14(4)**: 444–459.

Clean Coasts and Irish Water (2023) Think Before You Flush. https://thinkbeforeyouflush.org/ (accessed 10/08/2023).

Foxon TJ, McIlkenny G, Gilmour D *et al.* (2002) Sustainability criteria for decision support in the UK water industry. *Journal of Environmental Planning and Management* **45(2)**: 285–301.

Southern Water (2023) Bag It and Bin It. https://www.southernwater.co.uk/help-advice/keep-it-clear/bag-it-and-bin-it (accessed 10/08/2023).

United Utilities (2023) Word search fact sheets. https://www.unitedutilities.com/globalassets/documents/pdf/8660-h2o-wordsearch-sheets_v2.pdf (accessed 10/08/2023).

Credits: The studies reported here were carried out over more than a decade by researchers, practitioners and others.

Author: Richard Ashley

14.12. Ultra-low-flush toilet demonstrator

Innovation in toilet design has gained increasing popularity in recent years. Organisations such as the Bill & Melinda Gates Foundation have driven innovation through their Reinvent the Toilet Challenge (Bill & Melinda Gates Foundation, 2023). However, despite growing awareness of, and willingness to address, global water scarcity, the scale-up of novel technologies such as ultra-low-flush toilets (ULFTs) requires 'technology demonstrators' to enable the products to pass through technology readiness levels and reach market-ready status. It is not sufficient to design a novel water-saving technology and trial it at a bench scale. A human-centred approach must be adopted to enable innovators within estates management, facilities and commercial property management to 'take the risk' and trial such technologies at scale. This case study reports on such a technology trial (TRL8) at the University of Exeter's Streatham Campus. The project was fully reported in a paper in the journal *Water* (Melville-Shreeve *et al.*, 2021a) and water meter data from the monitoring programme is freely available by way of an annex to published works.

The demonstrator site is set within a university campus with an old, combined sewer network, developed over more than 100 years around what was originally a working farm on the fringe of the city of Exeter, England. The hillside setting is sloped, and the eastern quadrant of the campus was selected to deploy

ULFTs in order to maximise their combined impact on the local drainage network; that is, to minimise flows and thus maximise risk of downstream network blockages. The project was funded as part of a Horizon 2020 project (grant agreement ID: 778742) (CORDIS, 2023) and the research questions for the demonstrator revolved around quantifying the real-world water savings, while understanding the potential for increased risk of in-sewer blockages associated with low volume toilet flushes. Propelair (2019) provides details of the installation, and campus trials are reported on the University of Exeter website (University of Exeter, 2020).

Typical washrooms on campus have 6/4 litre dual-flush toilets. In this study, 119 ULFTs were installed to replace these industry-standard units. The low water consumption for these ULFTs combined with the highest unit water costs in the UK (£5.52/m^3 delivered and treated) were sufficiently attractive to the campus estates managers to accept potential downsides, such as increased sewer blockage risks and the associated cost to clear these.

14.12.1 Starting with a pilot

With all technology roll-outs, the deployment of agile methods is a tried and tested route to reveal real-world learning and ultimately increase the chances of success by starting small and reviewing feedback from initial activities before upscaling a concept. Elements of the ULFTs deployed are different to a traditional toilet and thus the users visiting such washrooms need to find them intuitive. There are two key differences from a user perspective. First, the air-powered flush will only activate when the toilet lid is closed, which is necessary to achieve an airtight seal around the toilet pan (Figure 14.26). Second, the unit creates a short burst of noise as the cistern-mounted pumps mobilise air through the pan. In both instances, closing the toilet (i.e. touching the lid) and hearing the flush is a new experience for any novice ULFT user.

Figure 14.26 Activation instructions

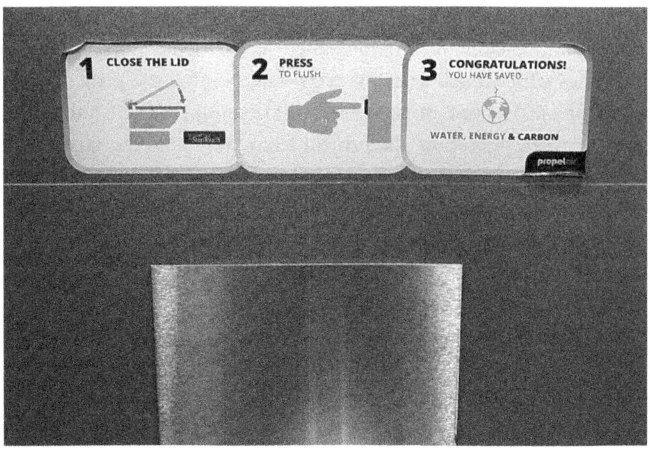

Figure 14.27 Ultra-low-flush toilet

A series of three ULFTs were installed within the maintenance operatives' washrooms (Figure 14.27) in order to build familiarity and enable the maintenance team to develop 'ownership' for the wider installation roll-out. A series of feedback sessions were run, helping the project team to learn from the installation and operation phase of these units. With 30 000 customers (i.e. students) on campus needing reliable toilet and sewer network operation, the project team facilitated the larger roll-out through early engagement with facilities managers, estates managers and campus directors. The pilot evidence was shared formally in a university Board paper in autumn 2018, whereupon the university approved installation across approximately one quarter of the central campus.

14.12.2 Designing for data acquisition

Key to the roll-out was the need to capture high-resolution water demand data. It was established that water usage at each building was available using 15-minute meters; however, the size and scale of each building meant that this was too coarse a dataset to help define incoming sewer flows. The water demand monitoring budget was fixed within the project so a relatively low cost (less than £1000/washroom) system was designed and tested at a benchtop level. The project team installed this bespoke monitoring platform, which included a flush counter in the back of every ULFT.

Furthermore, volumetric flow meters were used to enable future analysis of all toilet, urinal and bathroom tap fittings within one highly trafficked building (Harrison Building). With this detailed dataset, the project team have been able to explore additional novel research pathways using the demonstrator site, including those which could not be foreseen at the start of the trials. One such example is reported in work conducted on the campus during 2020 wherein the COVID-19 concentrations in sewer flows was monitored, and researchers explored linkages to flush data captured in real-time from the ULFT demonstrator monitoring platform (Sweetapple *et al.*, 2022).

14.12.3 Water-saving impact

The water-saving impact of ULFT installation at the University of Exeter demonstrator site was significant. A study by Melville-Shreeve *et al.* (2021b) found that using the specific ULFTs reduced mean water consumption by an average of 6 m^3/day. In turn, this equates to a reduced sewer contribution and thus the reduction in volume of flow to full treatment at the downstream wastewater treatment works. The data indicate an annual cost saving of £12 580 in water charges, which equates to approximately £100/toilet in this high-occupancy setting.

14.12.4 Hygiene and maintenance

Before the study started, one of the facilities team's concerns with the use of ULFTs was the risk that they may require more frequent cleaning or maintenance. However, the toilets were not found to be more difficult to maintain or clean than traditional toilets, and the original units remain in situ over 4 years after their initial installation. In terms of blockages, the project was able to explore this in detail over a number of months. This enabled existing faults, such as root ingress, cracks in pipes and areas where blockages were likely to develop (e.g. where pipes had shallow/negative gradient), to be identified. The hypothesis that the ULFTs would increase blockages was not borne out in the studies (Melville-Shreeve *et al.*, 2021a).

14.12.5 User acceptance and feedback

The long-term feedback from ULFT by users was an important factor in their adoption at the University of Exeter site. The installation team provided an A4 flyer on the back of each cubicle door (Figure 14.28) with a central feedback email address to ensure continued issues, compliments and complaints could be logged.

14.12.6 Conclusion

In conclusion, the ULFTs installed at the University of Exeter demonstrator site provided significant environmental benefits. The water saved and the life cycle reduction in water treatment, water pumping and wastewater flows all provide environmental benefits. The ULFT roll-out saves in the region of £10 000 per annum in water and sewerage charges. The technology has the potential to be a cost-effective and more sustainable solution for reducing water usage in commercial buildings, and should be considered by engineers and building managers as a viable alternative to traditional toilets. Project teams seeking to set up such pilot studies and demonstrator sites need to engage stakeholders as openly and early as possible. Furthermore, the value of the studies can be increased by providing open access to the emerging datasets. In this instance, the high-resolution micro-component water metering data from the ULFT roll-out has been made available for wider research initiatives and further collaborations and propositions have emerged as a direct result.

Figure 14.28 Information flyer

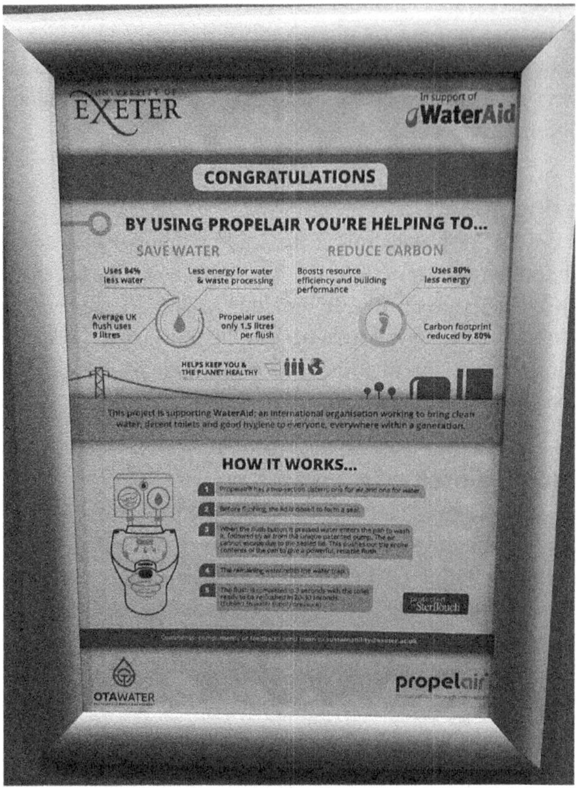

REFERENCES

Bill & Melinda Gates Foundation (2023) Reinvent the Toilet Challenge: A Brief History. https://www.gatesfoundation.org/our-work/programs/global-growth-and-opportunity/water-sanitation-and-hygiene/reinvent-the-toilet-challenge-and-expo (accessed 10/08/2023).

CORDIS (2023) The refinement, miniaturisation and demonstration of an ultra-low flush toilet capable of saving 2.8 billion litres of clean, potable water being unnecessarily wasted in Europe every day. https://cordis.europa.eu/project/id/778742/reporting (accessed 10/08/2023).

Sweetapple C, Melville-Smhreeve P, Chen AS *et al.* (2022) Building knowledge of university campus population dynamics to enhance near-to-source sewage surveillance for SARS-CoV-2 detection. *Science of the Total Environment* **806(1)**: 150406, 10.1016/j.scitotenv.2021.150406.

Melville-Shreeve P, Cotterill S, Newman A and Butler D (2021a) Campus study of the impact of ultra-low flush toilets on sewerage networks and water usage. *Water* **13(4)**: 419–419, 10.3390/w13040419.

Melville-Shreeve P, Cotterill S and Butler D (2021b) Capturing high-resolution water demand data in commercial buildings. *Journal of Hydroinformatics* **23(3)**: 402–416, 10.2166/hydro.2021.103.

Propelair (2019) The Propelair Toilet – Sustainability and Savings. See www.propelair.com (accessed 10/08/2023).

University of Exeter (2020) Propelair Toilet Case Study. https://www.exeter.ac.uk/about/sustainability/newsandevents/archive/title_688094_en.html (accessed 10/08/2023).

Credits: Centre for Water Systems, University of Exeter
Author: Dr Peter Melville-Shreeve

14.13. Reinforcement Learning: the cognitive treatment plant

In considering the outlook for urban drainage, this case study illustrates the potential application of digital innovation in the water industry. Undertaken by the STFC Hartree Centre and IBM Research, this was a government-funded project whose aim was to generate technology that is beneficial to industry by harnessing the potential of high-performance computing, data analytics and cognitive technologies (see Figure 14.29). This work was supported by the STFC Hartree Centre's Innovation Return on Research programme, funded by the Department for Business, Energy and Industrial Strategy.

The water cycle provides opportunities for innovation. The main output from the study is transferrable technology and machine learning applicable to both waste and clean water processes across all stages of the water cycle (see Figure 14.30).

Theoretical research for a cognitive treatment plant project aimed to increase the efficiency of the wastewater treatment process using an artificial intelligence (AI) technique known as Deep Reinforcement Learning. The AI is taught to minimise energy usage and disposal costs, while maintaining high quality of effluent and minimising the number of regulatory violations. Exposing the AI to a wide variety of conditions through a simulator ensures the robustness and stability of the control system. In short, this is about improving the flexibility of a controlled wastewater plant under different weather conditions.

Figure 14.29 Integrating high-performance computing, data analytics and cognitive technologies

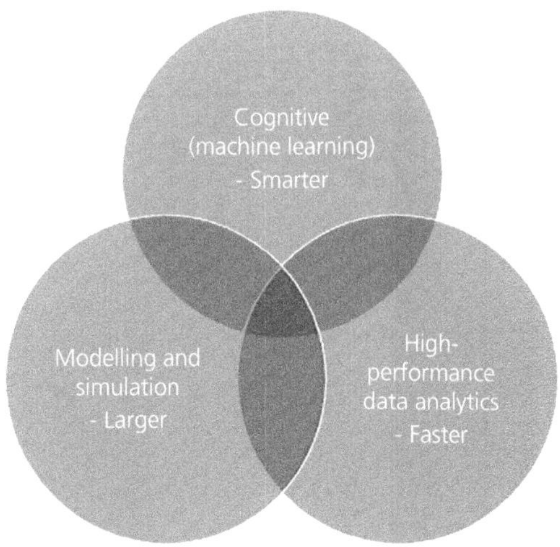

Figure 14.30 The water cycle from water resource management to wastewater treatment

Common problems for operating a wastewater plant efficiently are that sensors do not always accurately measure the process; there is uncertainty in loads and volume of influents, which are dependent on weather and domestic/industrial usage. As a result, wastewater plants are often run in a risk-averse way; for example, the plant is operated as if a storm were coming, and more oxygen is pumped into the aerated tanks to make sure that the wastewater is treated properly even under adverse conditions. This means that there are opportunities for smart pumping scheduling or reduced pumping to lower energy/electricity/operational costs based on using weather forecasts and a thorough analysis of which amount of pumping should be sufficient.

In this study, IBM implemented AI on a controller that controls the pumping rate into an aerated tank. While in operation, the controller learns how different setpoints can lead to different influent quality and energy costs. It then continuously learns and figures out how to optimise the setpoints under different states of the tank, so that it can minimise the operational cost while still ensuring that the effluents satisfy the treatment standard.

The primary purpose was to apply the latest capabilities in computing technologies to an industrial challenge and help drive economic development of the water industry using leading-edge scientific knowledge and applications. Reinforcement Learning control significantly reduced the number of issues slowing down processes and increased throughput. Sludge, disposal and treatment costs were reduced. Sludge, disposal and violation costs were improved in exchange for a slightly higher aeration cost. The 'smart' controller showed improvements over the static benchmark and significantly reduced the number of large violations (more than two standards breached) and improved the amount of runtime when no violations occurred (see Figure 14.31).

Reinforcement Learning can improve water and wastewater asset operations and customer experience through dynamic and more flexible responses to operation changes. This has a high potential for the water sector. Wastewater plants are already automated but not optimised. Sludge is a significant cost in the wastewater process. There are multiple basic control systems and sensors that could use the algorithm to improve the treatment efficiency while reducing the energy cost. Hence, the building blocks are in place within water companies. The algorithm can also further be developed to improve other processes, such as nutrient recovery.

For a digital future, machine learning, the Internet of Things and use of big data will also require different skills. There are not many people trained for these advances in computer science. Infrastructure needs computing power, but not all the time.

Overall, the approach shows that adaptive learning can be customised for individual wastewater treatment sites and systems and for different operating conditions (weather and inputs).

Credits: Dr Lan Hoang, IBM Research UKI (UK and Ireland); STFC Hartree Centre; IBM Research UKI

Authors: Dr Lan Hoang, Brian Smith

Figure 14.31 Results from the cognitive treatment plants compared with baselines

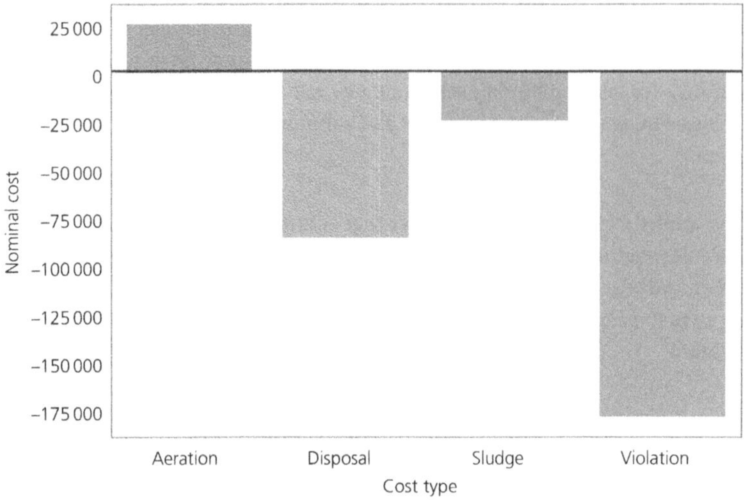

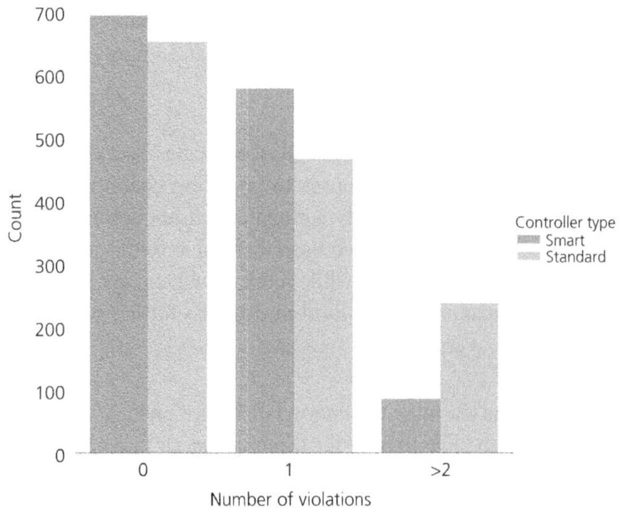

14.14. Smart blue-green roof to address plot scale drainage and water efficiency

The smart blue-green roof addresses the challenge of too much water and too little water in existing towns and cities, where ground space is at a premium and yet there are requirements to address the ecological and climate change crisis. The smart blue-green roof is a data-driven approach to optimising opportunities arising from the two main water issues related to climate change: water scarcity (by reducing water demand) and flooding (by storing rainfall close to where it falls). The smart blue-green roof turns a space into a place for communities by creating beautifully biodiverse places where people can go and enjoy the health and wellbeing benefits that blue-green infrastructure provides.

This smart sustainable drainage systems (SuDS) solution unlocks opportunities to build resilience to sewer and surface water flooding into existing crowded urban landscapes. This innovation makes use of combining meteorological and asset performance data with remotely controlled technology within place-making and community-led planning for rainfall events, flooding and preparedness for climate change. The smart blue-green roof is the result of multidisciplinary collaboration between experts from various fields. SME EPG led the innovative design that brought together industry partners in meteorology and forecasting (Kisters), technology architecture and data scientists (MashUp Analytics), agronomy (STRI), urban redevelopment (Bruntwood) and drainage products (Polypipe) (see Figure 14.32).

The smart blue-green roof was made reality as part of a pioneer urban redevelopment investment programme that aims to develop the most innovative city workspaces by putting sustainability and biodiversity at the heart. It is common knowledge now that climate change, urban creep and connections from new development are putting a huge strain on the urban drainage systems designed to protect us from flooding. SuDS and source control (holding rainfall close to where it falls) are promoted as solutions for creating capacity within wastewater infrastructure, but it is the very same towns and cities suffering from flooding where space is typically at a premium, and achieving low discharge rates from source control is a challenge. At the same time, retrofitting green infrastructure introduces additional water supply demands to keep the green infrastructure alive and thriving. This is where the smart blue-green roof is a multibenefit solution that builds resilience to flooding and increased water demands into existing towns and cities; a multipronged adaptation to climate change. Hydrological (rainfall) and hydraulic (flow attenuation) calculations are computed in real time using artificial intelligence (AI) optimisation so that the roof discharges ahead of the incoming storms. This is different to traditional drainage solutions, which discharge into sewers or watercourses during the storm or straight after the storm, when the receiving drainage system allows. With the smart blue-green roof the attenuation system will be empty when the storm arrives, and the maximum amount of rainfall is held back at source without any discharge to the receiving system. The attenuation storage is provided by a geocellular system with passive irrigation cones. The stored water is reused by the green roof component directly from the geocellular unit by way of sub-surface irrigation (Young *et al.*, 2022). By combining geocellular units with passive irrigation cones, the captured water is available for the green roof vegetation. Permavoid was integrated into the smart blue-green roof because it is a lightweight and shallow attenuation system. Load on any roof is critical and with a retrofit roof the parameters are pre-set; a solution was therefore designed in line with the roof's ability to handle the load of the substrate and water.

Construction benefits of the smart blue-green roof are that the number of outlets from the roof are reduced, because the idea is to capture and store the water. On retrofit solutions any existing outlets are blocked or converted into overflows, making it immediately visible, and the flood risk reduction benefits are understandable. Biodiversity and community benefits are provided by locally grown and carefully selected wild flowers. These are cultivated for about 12 months and arrive on a pallet to lay on the blue-green roof as quickly as possible, thereby creating a self-sustaining meadow, with minimal demands on additional water supply, that helps pollinators, bees, butterflies and even birds. What makes the smart blue-green roof smart is that that level of data from the attenuation component is combined with meteorological forecasts to be processed and optimised by artificial intelligence, which makes a decision on when to start emptying the roof at a safe discharge rate to ensure that the storage capacity is available for when a storm arrives. Put simply, the smart blue-green roof captures rainfall events and holds that water in the attenuation for reuse during dry periods to keep the vegetation alive and save having to water the vegetation. A unique intelligent part of the smart blue-green roof is its fail-safe mechanism to protect against loss of power or loss of incoming weather data for decision-making. The flood benefits are protected by ensuring the capacity is available as a priority.

Figure 14.32 Schematic diagram of the smart blue-green roof

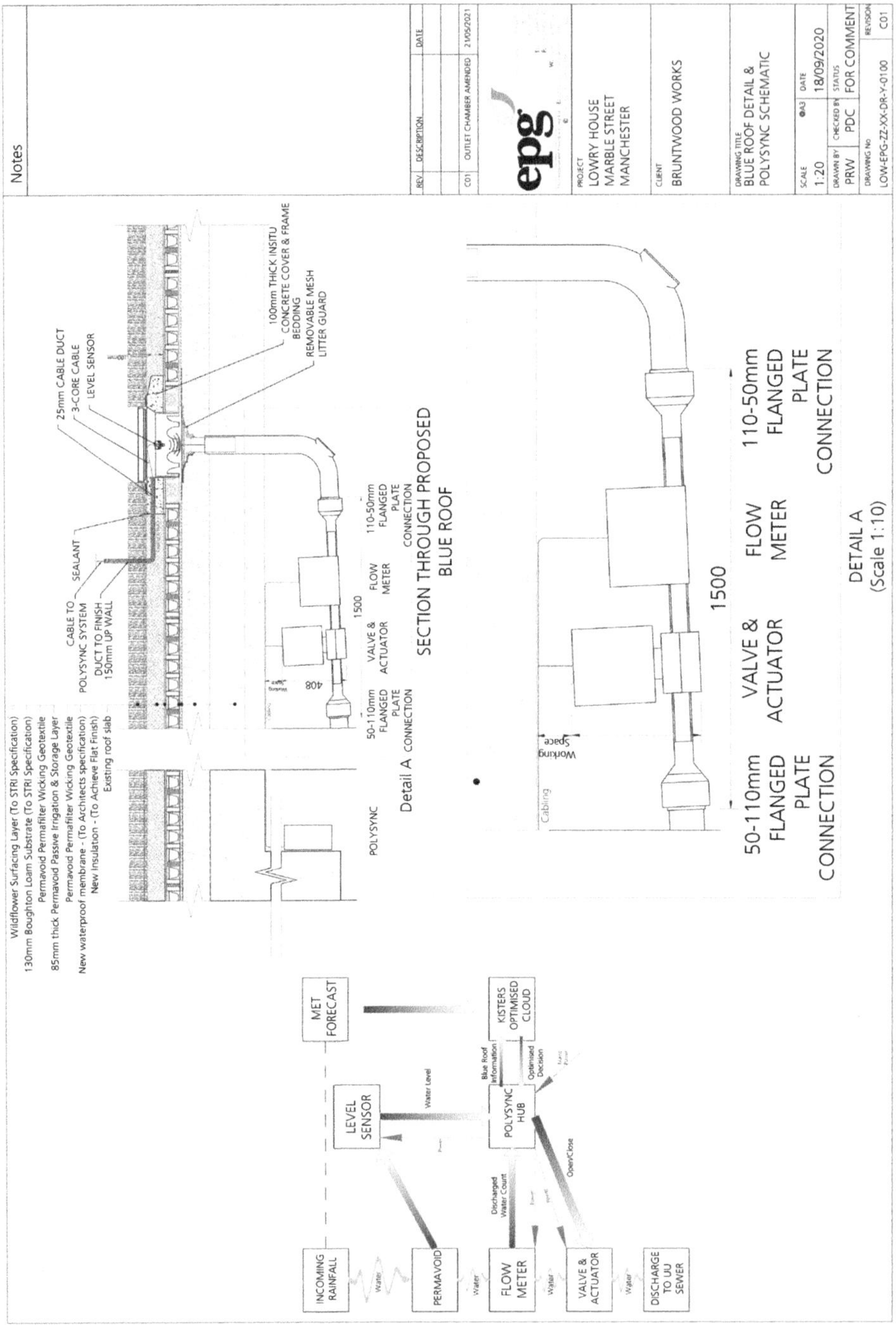

This type of solution addresses storage requirements in spaces where ground space is limited and at a premium. This solution delivers value by unlocking the potential to use multiple retrofit solutions in crowded existing urban areas to reduce the impacts of sewer and surface water flooding. Implementation of smart blue-green roofs in towns and cities across the UK and internationally can significantly impact and reduce surface water flooding and also unlock capacity in the wastewater infrastructure.

The smart blue-green roof has now been installed for over a year, the vegetation is thriving, having experienced both summer and winter climates (Figure 14.33). Post-project appraisal also now provides evidence to demonstrate successful water capture during two rainfall events in 2022, providing essential water to the vegetation during a drought period and zero discharges to wastewater infrastructure.

Figure 14.33 Pre- and post-project photographs showing the biodiversity and amenity improvements

REFERENCES

Young T, Mann R and Spring C (2022) Use of plastic geocellular modules as a replacement gravel layer for natural turf root zones. *International Turfgrass Society Research Journal* **14**: 339–351.

Credits: Phil Williams, EPG; John Griffiths, Kisters; Leo Phillips, MashUp
Author: Jacqueline Diaz-Nieto

Ashley R, Smith B, Shaffer P and Caffoor I
ISBN 978-0-7277-4178-3
https://doi.org/10.1680/icehudp.41783.437

Index

Page numbers in the form 340*t* indicate tables. Page numbers in the form 401*f* indicate figures.

Printed and bound by CPI Group (UK) Ltd, Croydon, CR0 4YY
26/01/2025
14632121-0001